Energy Metabolism of Farm Animals

Energy Metabolism of Farm Animals

Proceedings of the 14th Symposium on Energy Metabolism
Newcastle, Co. Down, Northern Ireland
14–20 September 1997

Edited by

K.J. McCracken, E.F. Unsworth and A.R.G. Wylie

Department of Agriculture for Northern Ireland
and The Queen's University of Belfast, Belfast, UK

CAB INTERNATIONAL

CAB INTERNATIONAL
Wallingford
Oxon OX10 8DE
UK

Tel: +44 (0)1491 832111
Fax: +44 (0)1491 833508
E-mail: cabi@cabi.org

CAB INTERNATIONAL
198 Madison Avenue
New York, NY 10016-4314
USA

Tel: +1 212 726 6490
Fax: +1 212 686 7993
E-mail: cabi-nao@cabi.org

A catalogue record for this book is available from the British Library, London, UK.
A catalogue record for this book is available from the Library of Congress, Washington DC, USA.

ISBN 0 85199 276 5

First published 1998
Transferred to print on demand 2002

Printed and bound by Antony Rowe Ltd, Eastbourne

CONTENTS

PREFACE

This book contains a compilation of papers presented at the 14th Symposium on Energy Metabolism of Farm Animals, held in the Slieve Donard Hotel, Newcastle, Co Down, Northern Ireland from 14-20 September 1997. The meeting was the latest in the highly successful series of Symposia dating back to 1958. The scientific programme was organised in association with the Nutrition Commission of the European Association of Animal Production by an international committee of which the membership was:

Dr K J McCracken (Northern Ireland) - Convenor
Dr Y van der Honing (The Netherlands) - General Secretary
Dr J Aguilera (Spain)
Dr C Wenk (Switzerland)
Dr D J Farrell (Australia)
Dr D E Johnson (USA)

The Local Organising Committee was responsible for all the detailed technical and social arrangements necessary for a stimulating and successful conference. The membership was:

Dr K J McCracken (Chairman)
Dr R Agnew
Dr A Carson
Dr E F Unsworth
Dr A R G Wylie

During the Symposium the help of Mrs E McCracken (Associates programme), Miss F Lewis, Mr J Birnie and Mr P McGrattan (audio and visual support) and many other members of staff from The Agricultural Research Institute of Northern Ireland and the Agriculture Group, Department of Agriculture, Newforge Lane, Belfast was invaluable.

The scientific programme was based around eight main topics concerned with various aspects of animal energetics, namely:

Session 1	Organ and Tissue Metabolism
Session 2	Lactation and Reproduction
Session 3	Regulation of Metabolism
Session 4	Methodology
Session 5	Growth
Session 6	Environmental Aspects and Maintenance
Session 7	Modelling
Session 8	Feed Evaluation and Dietary Factors.

Following the well-established format of previous symposia, each of the sessions was in the hands of its own chairman, discussion leader and two scientific secretaries as detailed below.

Session	Chairman	Discussion Leader	Scientific Secretaries
1	H.F. Tyrrell	D.S. Parker	M. Klein A.R.G. Wylie
2	H.F. Tyrrell	C.K. Reynolds	C.P. Ferris J.F. Hocquette
3	I. Ortigues	P.J. Reeds	R.L. Baldwin J.J. Ramsey
4	D.E. Johnson	N.D. Scollan	A. Mitchell J. van Milgen
5	D.J. Farrell	C. Wenk	G.E. Carstens H. Jorgensen
6	D.J. Farrell	K.A. Johnson	A.B. Hall E.M. Larson
7	K. Jakobsen	G. C. Emmans	H.C. Freetly T. Yan
8	K. Jakobsen	J.D. Sutton	J. Hutcheson J. Schrama

The International Committee is grateful to all of those who contributed to the smooth running of the meeting and in particular to the oral presentations and the lively and enthusiastic discussion sessions.

Over 100 scientific delegates, active in the field of energy metabolism, participated in the meeting and contributed the papers, based on recent research findings, which are published herein. The key-note lecture entitled '*Opportunities for application of studies on leptins in animal production*' delivered by Professor Dale Romsos, Michigan State University, USA, was one of the highlights of the week and the full text of the paper makes a significant contribution to the scientific value of this publication.

Tributes were paid at the outset to two esteemed colleagues, Professor Maria Kotarbinska (Poland) and Dr Carlos Prieto (Spain), whose deaths had occurred since the previous Symposium, and these tributes are recorded in the following section.

It was agreed at the General Assembly that the 15th Symposium would be held in Denmark in 2000, organized jointly by the Royal Veterinary and Agricultural University, Copenhagen and the Danish Institute of Agricultural Sciences, Foulum. Details may be obtained on the Internet at http.www.husdyx.kvl.dk/htm/ac/15sympem.htm.

The financial support of the bodies listed overleaf is gratefully acknowledged. Special thanks are due to the staff of the Slieve Donard Hotel for providing a comfortable and pleasant atmosphere for a stimulating conference; to Mr Tim Hardwick and Miss Rachel Robinson, CAB International, for their invaluable assistance in preparing this book; to the three ladies who undertook the editorial corrections; to my colleagues who shared the burden of editorial work and to the many others, too numerous to name, who assisted before, during and after the Symposium.

K McCracken
October 1997

ACKNOWLEDGEMENTS

The Organising Committee wishes to express its sincere gratitude to our three major sponsors and the many public and private bodies which gave support in cash or in kind. Their contributions assisted greatly in the successful organization of the Symposium.

Belfast City Council
Department of Agriculture for Northern Ireland
Irish Agricultural Wholesale Society (IAWS)

BOCM Pauls Ltd
Church and Dwight Company
Devenish Feed Supplements Ltd
Down District Council
EAAP
Finnfeeds International Ltd
Forum Products Ltd
IRTU
Livestock and Meat Commission for Northern Ireland
Novo Nordisk
Nutec Ltd
Nutrition Services International
Nutrition Society
Nutrition Society (Irish Section)
O'Kane Poultry Ltd
Roche Products Ltd
Tesco Ltd
The Queen's University of Belfast
Trouw Nutrition
United Dairy Farmers Ltd

The Editorial Committee is particularly grateful to Mrs Irene Smith, Mrs Kathleen Dowds and Mrs Carol Clarke whose patience and expertise were severely tested in seeking to obtain a high degree of uniformity throughout this publication.

IN MEMORY OF PROFESSOR MARIA KOTARBINSKA

Professor Maria Kotarbinska, an eminent researcher at the Kielanowski Institute of Animal Physiology and Nutrition in Jablonna, Poland, passed away on 5 November 1995.

Maria Kotarbinska was born in Warsaw in 1928. She graduated from the Olsztyn University of Agriculture and Technology. She began her career at the Agricultural University of Warsaw where, in 1958, she obtained a PhD in agriculture for her work on iron metabolism in suckling pigs. In 1961, Maria Kotarbinska spent one year at the National Institute for Research in Dairying in Shinfield. There she not only continued her research on young pigs, but became acquainted with modern approaches to pig husbandry and was a very keen observer of all its phases.

She moved to the Instutute in Jablonna in 1963, where for many years she headed the Pig Nutrition Department. At the Institute, she concentrated on the research delineated by Professor Jan Kielanowski on energy metabolism of growing animals. She was one of the originators of the Jablonna school that used the comparative slaughter method to study energy and protein metabolism. Maria Kotarbinska's work provided a basis for determining the maintenance energy requirements for growing pigs and the energy cost of protein deposition in the body. This was a pioneer approach at the time.

The results of her work, reported in the paper entitled "Energy metabolism in growing pigs", and the application of her research methods drew much attention worldwide and came to be used by numerous researchers outside Poland. Maria Kotarbinska helped design and develop the laboratory, which is the only one of this type in Poland, and one of the few in Europe. For over 30 years it has been providing valuable information about body composition and energy metabolism in pigs maintained under practical conditions similar to those encountered in production.

Professor Kotarbinska co-operated with many research centres abroad, and was the national co-ordinator of the Pig Production Committee within the framework of the EAAP. She took part several times in the Symposia on Energy Metabolism of Farm Animals. She was highly respected and esteemed internationally and contributed to various expert groups including FAO.

In Poland, Maria Kotarbinska was known primarily as the author of nutritional standards for pigs. She was a respected expert on nutritional evaluation of feeds and the practical aspects of feeding pigs, for which she usually advocated simple but effective production methods. She published more than 150 scientific papers and popular articles on pig nutrition and related topics. She set an example of hard work and personal dedication to solving research problems. Her profession became the focus of her life.

With Professor Maria Kotarbinska we have lost an outstanding scientist and many of us have lost a good and reliable friend.

Henryk Fandrejewski

IN MEMORY OF

DR CARLOS PRIETO

Dr Carlos Prieto and his family - his wife and their daughter and son - sadly died, victims of the tragedy that took place at the camp site 'Virgen de las Nieves' in the Pyrenees on 7 August 1996.

Carlos Prieto was born on 2 January 1952. He studied Biological Sciences at the University of Granada between 1969 and 1974. In January 1975 he was awarded a four year postgraduate grant to carry out PhD studies on protein metabolism of poultry at the Department of Animal Physiology (later Department of Animal Nutrition) of the Estación Experimental del Zaidín (EEZ) in Granada, which belongs to the Spanish Research Council (CSIC). He gained his PhD degree in March 1979 at the University of Granada. He was awarded a Postdoctoral Fellowship by CSIC and spent two very challenging and productive years at the Rowett Research Institute. There, working under the tutorship of Dr Gerald Lobley along with Drs John MacRae and John Brockway, he developed his scientific expertise and became adept in technical aspects of indirect calorimetry and in the application of tracer techniques in metabolic research.

Back in Spain he obtained a three year postdoctoral grant to collaborate in the development of research projects in progress at the Department of Animal Physiology of the EEZ. In 1984 he obtained a first appointment as Professor at the University of Granada. However, he remained working at the EEZ where he was awarded a permanent position as a Researcher in February 1986. In March 1990 he was promoted to Senior Scientist.

During the short time he was allowed to stay with us he made important contributions in the field of both energy and protein metabolism of farm and laboratory animals. The results of his studies have been published in more than 40 scientific papers and contain valuable information concerning the energy requirements of goats, new aspects of protein utilization in mongastric animals and interactions between amino acid supply and energy metabolism. He attended and/or presented contributions to the last six Symposia on Energy Metabolism.

His solid scientific background, broad interest and great capacity of work made Carlos Prieto one of the leading scientists in the field of animal nutrition in Spain. He was also highly involved in the training of young scientists to whom he sought to transmit his untiring enthusiasm for science. His sense of humour, open mind and positive view of life made of him a very stimulating colleague. We all have lost not only a brilliant scientist but, more that that, a beloved friend.

Jose Aguilera

OPPORTUNITIES FOR APPLICATION OF STUDIES ON LEPTIN IN ANIMAL PRODUCTION

D.R. Romsos

Department of Food Science and Human Nutrition, Michigan State University, East Lansing, MI 48824-1224, USA

The ob/ob mouse and other genetic mutant rodents have been used for years as models to better understand metabolic, nutritional and neuroendocrine factors central to the control of body fat and skeletal muscle accretion. Such studies have had application to human obesity and diabetes as well as to meat animal production. Since the December 1994 presentation in *Nature* of the cloning of the ob gene (Zhang *et al.*, 1994), research in this area has markedly intensified with major new findings being announced monthly or more often. This brief review will focus on selected findings emerging from the discovery of the ob gene that are of potential interest to the livestock industry. Representative references will be cited, rather than a comprehensive listing of the hundreds of papers published in the past several years, to illustrate the current status of this rapidly evolving research area. Several reviews, commentaries and perspectives on leptin have been published that provide useful background reading (Campfield, 1996; Caro *et al.*, 1996; Hamann & Matthaei, 1996; Flier, 1997; Giacobino, 1996; Rohner-Jeanrenaud *et al.*, 1996; Saladin *et al.*, 1996; Schwartz & Seeley, 1997; Taylor *et al.*, 1996; Trayhurn, 1996). There is currently a paucity of data on the ob gene in livestock, but this will undoubtedly change in the near future.

Obesity in ob/ob mice is caused by a single gene mutation (see Bray, 1984 and Bray & York, 1979 for reviews). These mice are both hyperphagic and hypometabolic; they thus develop gross obesity. They also exhibit a stunting of skeletal muscle and bone growth. Plasma insulin and corticosterone concentrations are elevated whereas plasma growth hormone, sex steroids and thyroid hormone concentrations are depressed. These mice are infertile. Clearly, the ob gene mutation adversely influences many metabolic pathways of importance to the livestock industry. I will describe how this gene mutation is currently thought to disturb growth and development.

Identification of the ob Gene, and Leptin Synthesis/Secretion

The ob gene encodes a 4.5 kilobase adipose tissue messenger RNA with a highly conserved 167-amino acid open reading frame (Zhang *et al.*, 1994). Ob protein is secreted from adipocytes into the blood. It appears to circulate as a monomer and has homology to the class I cytokine family. White adipocytes are the major site for ob protein synthesis, although at selected phases of the lifecycle other organs may be capable of leptin synthesis. During pregnancy substantial synthesis of ob protein may occur in the placenta (Hassink *et al.*, 1997; Masuzaki *et al.*, 1997b) and during neonatal development brown adipocytes may synthesize ob protein (Dessolin *et al.*, 1997).

Ob/ob mice have a nonsense mutation in codon 105 of the ob gene (Zhang *et al.*, 1994). This mutation yields a stop codon and subsequent production of a truncated inactive ob protein. Thus, ob/ob mice develop obesity because of a lack of ob protein, a protein that prevents rather than causes gross obesity. To avoid confusion in the literature Friedman and associates named this protein leptin, a word derived from the Greek root Leptos, meaning thin (Halaas *et al.*, 1995).

A major research effort is underway to elucidate mechanisms responsible for regulation of leptin synthesis and secretion from white adipocytes. Positive modulators of leptin synthesis/secretion include insulin, glucocorticoids, food intake, and increased body fatness whereas diabetes, adrenalectomy, food deprivation, exercise and cold exposure diminish leptin synthesis/secretion (Ahren *et al.*, 1997; Cusin *et al.*, 1995; Deng *et al.*, 1997; De Vos *et al.*, 1995; Fredrich *et al.*, 1995a, 1995b; Grunfeld *et al.*, 1996; Hamilton *et al.*, 1995; Leroy *et al.*, 1995; Lonnquist *et al.*, 1995; MacDougald *et al.*, 1995; Mandrup *et al.*, 1997; Masuzaki *et al.*, 1997a; Mizuno *et al.*, 1996; Pagano *et al.*, 1997; Saladin *et al.*, 1995; Zachwieja *et al.*, 1997; Zakrzewska *et al.*, 1997). Emphasis is now on understanding how these modulators trigger the cellular and molecular mechanisms within adipose tissue that control leptin synthesis/secretion.

There are now many reports describing effects of feeding, fasting, obesity and other physiological states on plasma leptin concentrations in humans, with values often in the ng/ml blood range (Considine *et al.*, 1996; Haffner *et al.*, 1997; Lahlou *et al.*, 1997; Licinio *et al.*, 1997; Lonnqvist *et al.*, 1997; Maffei *et al.*, 1995; Niskanen *et al.*, 1997; Pombo *et al.*, 1997; Ravussin *et al.*, 1997; Turpeinen *et al.*, 1997). Leptin is cleared from the circulation with a half-life of 25 min (Klein *et al.*, 1996), with the kidneys appearing to be the route for clearance of leptin (Cumin *et al.*, 1996). As soon as antibodies are available to measure leptin in livestock, it should be possible to quickly characterize effects of various physiological states on plasma leptin in these species. Attention will undoubtedly include emphasis on various feeding states, stages of growth, and lactation/reproduction.

Leptin Receptors and Signal Transduction

For leptin, a protein secreted from adipose tissue into the blood, to mediate physiological actions, it probably needs to interact with a specific receptor. Receptors for leptin were indeed identified with amazing speed, with the initial report appearing only months after the announcement of the cloning of the ob gene (Tartaglia *et al.*, 1995). Leptin receptors have a single plasma membrane spanning region and have characteristics of the class I cytokine receptor family which include receptors for growth hormone and prolactin (Tartaglia *et al.*, 1995; see review by Tartaglia, 1997). The extracellular domain of leptin receptors is large, i.e. 816 amino acids. Variations occur in the size of the intracellular domain, with a long form containing about 303 amino acids and a family of short forms of the receptor containing about 30 amino acids. The long form of the leptin receptor is thought to participate in signal transduction whereas the shorter forms of the receptor may play roles in leptin uptake by the brain and clearance from the circulation.

The long form of the leptin receptor probably regulates gene transcription. Current evidence suggests that the long form of the leptin receptor acts through JAK (janus kinases) and STAT (signal transducers and activators of transcription) proteins (Ghilardi *et al.*, 1996; Ghilardi & Skoda, 1997; Tartaglia, 1997). JAK phosphorylates an intracellular domain of the leptin receptor subsequent to leptin binding to an extracellular domain. This leads to activation and translocation of STAT proteins to the nucleus where they affect gene transcription. There are a family of STAT proteins. This raises the possibility that leptin might mediate a diverse array of signals in the nucleus. Further, some reported actions of leptin occur within seconds or minutes after exposure to leptin (Chen *et al.*, 1997; Glaum *et al.*, 1996). These findings make it probable that leptin also stimulates signal transduction pathways distinct from those involving gene transcription where more than a few seconds or minutes are likely required to mediate physiological responses.

Assuming that the long form of the leptin receptor mediates the physiological actions of leptin, the tissue location of this receptor has been explored to identify targets of leptin action. The mRNAs for both the long and shorter forms of the leptin receptor are expressed in multiple tissues (Ghilardi *et al.*, 1996; Lollmann *et al.*, 1997). However, the transcript for the long isoform of the receptor is expressed at much lower levels than the shorter forms in most tissues, except in the hypothalamus where substantial expression of the long form occurs. These data suggest that the hypothalamus is an important target for leptin action.

Leptin Regulates Food Intake

One of the most readily apparent characteristics of ob/ob mice (i.e. mice lacking leptin because of a mutation in the ob gene) is their insatiable appetite (Bray, 1984; Bray & York, 1979). An obvious hypothesis to test was whether replacement of leptin in ob/ob mice would lower their food intake. Recombinant leptin was produced and injected intraperitoneally into ob/ob mice (Campfield *et al.*, 1995; Halaas *et al.*, 1995; Pelleymounter *et al.*, 1995; Weigle *et al.*, 1995). As predicted, food intake was lowered after treatment with leptin. To test whether leptin *per se* caused this decrease in food intake (as opposed to a generalized decrease in food intake secondary to a nonspecific component of the recombinant preparation), leptin was administered to db/db mice (Campfield *et al.*, 1995; Halaas *et al.*, 1995). The db/db mouse has a mutation in the long form of the leptin receptor that completely blocks signal transduction (Chen *et al.*, 1996; Chua *et al.*, 1996; Ghilardi *et al.*, 1996; Lee *et al.*, 1996). Consistent with a specific effect of leptin on food intake, leptin failed to influence food intake in these db/db mice with a defective leptin receptor system. In animals with diet-induced obesity, investigators have reported a resistance to the effectiveness of leptin to inhibit food intake, or a failure of plasma leptin to increase sufficiently to lower food intake (Frederick *et al.*, 1995a; Surwit *et al.*, 1997; Widdowson *et al.*, 1997). Multiple-site abnormalities in the leptin signal transduction system may participate in the induction of hyperphagia.

Leptin appears to specifically target satiety and reduce meal size without influencing meal frequency (Kahler *et al.*, 1997). The possibility that leptin induces satiety by slowing the rate of gastric emptying has been tested (Barrachina *et al.*, 1997b). Other mechanisms are probably involved because leptin reduced food intake of mice without affecting the rate of gastric emptying.

The hypothalamus is an obvious target for leptin to regulate food intake given the central role for the hypothalamus in this process (Luiten *et al.*, 1987). Consistent with this, the long form of the leptin receptor is abundantly expressed in the hypothalamus (Ghilardi *et al.*, 1996; Lollmann *et al.*, 1997), leptin activates neurons in the hypothalamus (Elmquist *et al.*, 1997), and ob/ob mice injected intracerebroventricularly with leptin consume less food than those injected intravenously with leptin (Campfield *et al.*, 1995; Stephens *et al.*, 1995). For leptin, a 167 amino acid protein synthesized in adipose tissue, to enter the brain, a transport system must be available. A saturable transport system that unidirectionally transports leptin from blood to cerebrospinal fluid has now been characterized (Banks *et al.*, 1996). It has been speculated that this transport system utilizes one of the short forms of the leptin receptor to translocate leptin.

NPY (neuropeptide Y) and CRH (corticotropin releasing hormone) are two neuropeptides within the hypothalamus intimately involved in regulation of food intake (Glowa *et al.*, 1992; Leibowitz, 1991). NPY stimulates feeding and CRH inhibits feeding. The possibility that leptin regulates feeding by influencing NPY and CRH gene expression or peptide secretion has been investigated (Raber *et al.*, 1997; Schwartz *et al.*, 1996a, 1996b; Stephens *et al.*, 1995; Wang *et al.*, 1997). Intracerebroventricular injection of leptin into fasted rats lowered NPY mRNA abundance in the arcuate nucleus of the hypothalamus and elevated CRH mRNA abundance in the paraventricular nucleus of the hypothalamus (Schwartz *et al.*, 1996b). Thus, leptin might limit NPY synthesis and enhance CRH synthesis to constrain food intake. Leptin also constrains the effectiveness of exogenous administered NPY to enhance food intake, suggesting that leptin interferes with the NPY receptor signal transduction pathway linked to food intake regulation (Mistry *et al.*, 1997; Smith *et al.*, 1996).

While data are available to support a role for NPY in the leptin-induced regulation of food intake, the complexity of possible interactions was underscored in a recent report utilizing NPY-knockout mice. Surprisingly, mice without the ability to produce NPY, a proposed key regulator of food intake (Leibowitz, 1991), still maintained normal food intake (Erickson *et al.*, 1996). This illustrates that redundant mechanisms are able to compensate for a loss of one component of the regulatory system. Intracerebroventricular administration of leptin to NPY-knockout mice causes a reduction in food intake equivalent to that observed in leptin-treated control mice, indicating that NPY is not the exclusive system by which leptin modulates food intake (Erickson *et al.*, 1996). Clearly, leptin-induced regulation of food intake involves multiple neuroendocrine systems. Cholecystokinin may, for example, also participate in the leptin regulatory system to enhance the onset of satiety (Barrachina *et al.*, 1997a). Considerable effort will be required to fully understand the cellular mechanisms whereby leptin regulates food intake.

Leptin Regulates Metabolic Rate

The ob/ob mouse is not only hyperphagic, but also hypometabolic. Long term administration of leptin to these mice lowers their food intake and reduces their percentage body fat (Pelleymounter *et al.*, 1995). These chronically leptin-treated ob/ob mice also have a higher metabolic rate per unit body weight than control ob/ob mice (Pelleymounter *et al.*, 1995). It is, however, difficult to determine based on these studies whether leptin has direct effects on metabolic rate, or whether effects are secondary to an increased percentage lean tissue. Leptin acutely stimulated metabolic rates of fed ob/ob mice within an hour after intracerebroventricular administration (Mistry *et al.*, 1997). This 40% increase in metabolic rate occurred even though leptin concurrently lowered food intake of these mice. At least one component of this rapid-onset increase in metabolic rate can be attributed to a leptin-induced activation of brown adipose tissue thermogenesis via stimulation of the sympathetic nervous system (Collins *et al.*, 1996). Longer term treatment with leptin may also increase the capacity for uncoupling of oxidative phosphorylation by increasing uncoupling of protein-2 (Zhou *et al.*, 1997). Metabolic rates were not increased when fed lean mice were injected intracerebroventricularly with leptin, but were increased when fasted lean mice received leptin (Mistry *et al.*, 1997). It appears that leptin facilitates maintenance of diet-induced thermogenesis. Ob/ob mice, even in the fed state, are well known to have a deficit in diet-induced thermogenesis that contributes to their obesity. Fasting lowers leptin secretion and diet-induced thermogenesis in lean mice. Leptin might integrate signals associated with food intake to activate diet-induced thermogenesis. This suggestion is supported by observations that restricted fed animals administered leptin retain less body fat than identically fed control animals (Halaas *et al.*, 1997; Levin *et al.*, 1996).

Ob/ob mice fail to thermoregulate appropriately when cold-exposed. Leptin may thus also play an important role in thermoregulatory thermogenesis, which is particularly important in neonatal animals. Administration of leptin to 7-16 day old rat pups prevented the normal hypometabolic, torporlike state that neonates experience during certain periods each day (Stehling *et al.*,1996). Leptin did not influence metabolic rates of these pups at other periods during the day. This alleviation of the torporlike state increased total daily metabolic rates by about 10%, without any change in milk intake of the pups. Other studies in neonatal mice have also shown that leptin increases metabolic rate in neonates without influencing food intake (Mistry *et al.*, unpublished). Survival of neonatal animals would be compromised if maintenance of thermogenesis was tightly coupled to constraint of food intake, whereas such coupling in adult animals would assist in the prevention of obesity. Leptin-induced regulation of food intake and energy expenditure might be developmentally programmed to first enhance survival of neonates by maintaining capacity for thermogenesis and later to activate systems to constrain food intake of adults to prevent obesity. Survival of neonatal livestock is enhanced if they co-exhibit a vigorous appetite and active thermoregulatory thermogenesis. It will be interesting to determine if leptin functions to promote thermogenesis without constraining food intake in these neonates, as appears to be true in rodents.

A Role for Leptin in Fetal/Neonatal Development?

Several lines of rapidly emerging evidence raise the possibility that leptin plays a role in fetal/neonatal development. The placenta has leptin mRNA and secretes leptin (Hassink *et al.*, 1997; Masuzaki *et al.*, 1997a). Leptin is present in human cord blood (Sivan *et al.*, 1997). Rat pups have detectable plasma leptin at birth, with a pronounced peak in plasma leptin occurring at about 10 days of age (Rayner *et al.*, 1997). It is tempting to speculate that leptin might play a role in modulating some of the dynamic neuroendocrine changes occurring at this stage of development. For example, one of the very early onset abnormalities in ob/ob mice is an enhanced insulin secretion response to stimulation of protein kinase C within pancreatic islets (Chen & Romsos, 1997). This abnormally high insulin secretion persists even when pancreatic islets from 2 week old ob/ob mice are cultured *in vitro*. Might the absence of leptin in early development program islets from ob/ob mice to persistently hypersecrete insulin? We have shown that addition of leptin to islets from ob/ob mice specifically inhibits protein kinase C-stimulated insulin secretion (Chen *et al.*, 1997). These findings provide insights into one specific signal transduction system leptin might regulate during neonatal development. The physiological relevance of leptin to fetal/neonatal development remains to be fully established.

Leptin as a Sensor of Nutritional Adequacy

Leptin concentrations in plasma fluctuate with nutritional status, leptin is elevated after a meal and lowered during fasting. Evidence is now emerging to suggest that these changes in leptin provide signals to the central nervous system to mediate hormonal responses to feeding and fasting (Ahima *et al.*, 1996; Carro *et al.*, 1997; Legradi *et al.*, 1997). For example, plasma thyroid hormone concentrations decline with fasting. This decline can be largely prevented by administering leptin to fasted rats (Ahima *et al.*, 1996; Legardi *et al.*, 1997). This would help explain why ob/ob mice, deficient in leptin, have low plasma concentrations of thyroid hormones even in the fed state.

Growth hormone provides another example where leptin might serve to integrate nutritional status to appropriately modulate hormone secretion. Administration of leptin antiserum to fed rats lowers growth hormone secretion, suggesting that food-induced leptin secretion provides a signal to stimulate growth hormone secretion (Carro *et al.*, 1997). Administration of additional leptin to a fed rat does not further increase growth hormone secretion, whereas leptin administration to a fasted rat (i.e. a rat with a low plasma leptin concentration) stimulates growth hormone secretion. It will be very informative to determine how nutrient intake stimulates leptin secretion and then how this increase in leptin signals various hormonal responses.

A Role for Leptin in Reproduction

The ob/ob mouse is infertile. This infertility is corrected by administration of leptin (Barash *et al.*, 1996). Because leptin administration also diminishes the gross obesity characteristic of these animals it was not clear whether leptin was exerting direct or indirect effects on the reproductive system of these animals. This issue is being aggressively investigated with evidence forthcoming that leptin directly affects the hypothalamic/ pituitary/gonad axis.

Several studies have now shown that leptin influences the onset of puberty, findings consistent with a critical role for adipose tissue mass in onset of puberty. Treatment of prepubertal female rats with leptin reduced their food intake and body weight gain, but did not delay the onset of puberty as occurred in rats pair-fed to the leptin-treated animals (Cheung *et al.*, 1997). Treatment of mice with leptin also reduced food intake and growth rate, but these mice reproduced up to 9 days earlier than control mice (Chehab *et al.*, 1997). Leptin is speculated to be the signal indicating that fat reserves are sufficient to support reproduction.

Leptin has direct effects on the hypothalamus, pituitary and ovary. Leptin increases LHRH release from the hypothalamus and LH and FSH release from the anterior pituitary of rats (Yu *et al.*, 1997). Interestingly, the higher concentrations of leptin tested actually decreased release. Thus, the hypothalamus and pituitary may sense nutritional status and metabolic fuel reserves via leptin. Leptin is needed to optimize reproduction, but excess leptin, as would be expected to occur in obesity, may diminish reproductive efficiency.

Results of two studies suggest that leptin can directly impair ovarian granulosa cell extradiol production (Spicer & Francisco, 1997; Zachow & Magoffin, 1997). Leptin appears to function by inhibiting insulin and IGF-1 mediated augmentation of FSH-stimulated extradial production. It will be important to better understand the dose response relationships of the reproductive axis to leptin. Clearly, there is a critical minimal concentration of leptin necessary to sustain reproduction. Adequate food intake and body fat reserves appear essential to attain this level of leptin secretion. Leptin then provides a signal that is integrated within the hypothalamus and pituitary. Whether leptin influences the ovary directly under these conditions remains unclear. High concentrations of leptin would appear to negatively influence reproductive signals to each of the three sites, i.e. the hypothalamus, pituitary and ovaries.

Application of Leptin Research to Livestock Production

It is too early to fully predict the ultimate uses that the livestock industry will derive from leptin research. The discovery of leptin is leading to major advances in understanding how adipose tissue provides signals to the brain to regulate a variety of metabolic processes. It is also becoming clear that leptin may directly influence metabolism in peripheral organs independent of actions of leptin within the central nervous system. Although not discussed in this review preliminary data are available to suggest a role for leptin in liver, skeletal muscle, adipose tissue, adrenal and hemopoietic cell metabolic regulation (Bornstein *et al.*, 1997; Cohen *et al.*, 1996; Gainsford *et al.*, 1996; Muoio *et al.*, 1997). A number of studies have now shown effects of leptin on pancreatic islets, however, the full physiological significance of these studies remain to be determined (Chen *et al.*, 1997; Emilsson *et al.*, 1997; Kieffer *et al.*, 1996, 1997; Leclercq-Meyer *et al.*, 1996; Pallett *et al.*, 1997). Given the critical role that leptin must play, based on the adverse outcomes in animals either lacking leptin or resistant to leptin, in normal physiology, it seems highly likely that this basic research will at least indirectly have a major potential impact on the livestock industry.

The leptin signal transduction system is beginning to be understood. As these data emerge it will be possible to clone the genes involved and to examine the components of this system in various breeds of animals. Such studies may uncover gene-based alterations that will help explain differences in production efficiencies. Some animals may be somewhat resistant to leptin and others may be supersensitive, as appears to be the case in laboratory rodents. Appropriate strategies could then be employed to capitalize on these findings.

An obvious focus of leptin research with practical linkages to livestock production is food intake regulation. Studies with leptin may lead to strategies that will enable manipulation of food intake to either enhance the rate of production or constrain food intake as appropriate. It may also be possible to utilize the leptin signal system to optimize food-induced hormone secretion.

The involvement of leptin in reproductive efficiency represents another area with significant application to the livestock industry. Leptin may provide an important key to the reproductive failure characteristic of highly productive dairy cows and sows. These animals would be expected to have low plasma leptin concentrations, possibly too low to permit reproduction. Likewise, leptin may help explain reproductive inefficiency in overly fat animals. These animals may have high plasma concentrations of leptin and leptin resistance.

These are exciting times in science. Techniques are now available to answer questions that would have been impossible to adequately address only a few years ago. Leptin represents but one of many new discoveries with great potential. The livestock industry needs to utilize each of these emerging discoveries to full advantage to remain competitive. Only time will tell whether leptin research will contribute significantly to the future productivity of this industry.

Acknowledgements

I thank Shelli Pfeifer for preparing the manuscript. Supported by the Michigan Agricultural Experiment Station and NIH DK 15847.

References

Ahima R.S., Prabakaran D., Mantzoros C., Qu D., Lowell B., Maratos-Flier E. & Flier J.S. (1996). Role of leptin in the neuroendocrine response to fasting. *Nature* 382:250-252.

Ahren B., Mansson S., Gingerich R.L. & Havel P.J. (1997). Regulation of plasma leptin in mice: influence of age, high-fat diet, and fasting. *Am. J. Physiol.* 273 (*Regulatory Integrative Comp. Physiol.* 42):R113-R120.

Banks W.A., Kastin A.J., Huang W., Jaspan J.B. & Maness L.M. (1996). Leptin enters the brain by a saturable system independent of insulin. *Peptides* 17:305-311.

Barash I.A., Cheung C.C., Weigle D.S., Ren H., Kabigting E.B., Kuijper J.L., Clifton D.K. & Steiner R.A. (1996). Leptin is a metabolic signal to the reproductive system. *Endocrinol.* 137:3144-3147.

Barrachina M.D., Martinez V., Wang L., Wei J.Y. & Tache Y. (1997a). Synergistic interaction between leptin and cholecystokinin to reduce short-term food intake in lean mice. *Proc. Natl. Acad. Sci.* USA 94:10455-10460.

Barrachina M.D., Martinez V., Wei J.Y. & Tache Y. (1997b). Leptin-induced decrease in food intake is not associated with changes in gastric emptying in lean mice. *Am. J. Physiol.* 272 (*Regulatory Integrative Comp. Physiol.* 41):R1007-R1011.

Bornstein S.R., Uhlmann K., Haidan A., Ehrhart-Bornstein M. & Scherbaum W.A. (1997). Evidence for a novel peripheral action of leptin as a metabolic signal to the adrenal gland. *Diabetes* 46:1235-1238.

Bray G.A. (1984). Hypothalamic and genetic obesity: an appraisal of autonomic hypothesis and the endocrine hypothesis. *Int. J. Obesity 8, Suppl.* 1:119-137.

Bray G.A. & York D.A. (1979). Hypothalamic and genetic obesity in experimental animals: anautonomic and endocrine hypothesis. *Physiol. Rev.* 59:719-809.

Campfield L.A., Smith F.J. & Burn P. (1996). The OB protein (leptin) pathway - a link between adipose tissue mass and central neural networks. *Horm. Metab. Res.* 28:619-632.

Campfield L.A., Smith F.J., Guisez Y., Devos R. & Burn P. (1995). Recombinant mouse ob protein: evidence fot a peripheral signal linking adiposity and central neural networks. *Science* 269:546-549.

Caro J.F., Sinha M.K., Kalaczynaki J.W., Zhang P.L. & Considine R.V. (1996). Leptin: the tale of an obesity gene. *Diabetes* 45:1455-1462.

Carro E., Senaris R., Considine R.V., Casanueva F.F. & Dieguez C. (1997). Regulation of *in vivo* growth hormone secretion by leptin. *Endocrinol.* 138:2203-2206.

Chehab F.F., Mounzih K., Lu R. & Lim M.E. (1997). Early onset of reproductive function in normal female mice treated with leptin. *Science* 275:88-90.

Chen H., Charlat O., Tartaglia L.A., Woolf E.A., Weng X., Ellis S.J., Lakey N.D., Culpepper J., Moore K.J., Breitbart R.E., Duyk G.M., Tepper R.I. & Morgenstern J.P. (1996). Evidence that the diabetes gene encodes the leptin receptor: identification of a mutation in the leptin receptor gene in db/db mice. *Cell* 84:491-495.

Chen N.-G. & Romsos D.R. (1997). Persistently enhanced sensitivity of pancreatic islets from ob/ob mice to PKC-stimulated insulin secretion. *Am. J. Physiol.* 272 (*Endocrinol. Metab.* 35):E304-E311.

Chen N.-G., Swick A.G. & Romsos D.R. (1997). Leptin constrains acetylcholine-induced insulin secretion from pancreatic islets of ob/ob mice. *J. Clin. Invest.* 100:1174-1179.

Cheung C.C., Thornton J.E., Kuijper J.L., Weigle D.S., Clifton D.K. & Steiner R.A. (1997). Leptin is a metabolic gate for the onset of puberty in the female rat. *Endocrinol.* 138:855-858.

Chua Jr. S.C., Chung W.K., Wu-Peng X.S., Zhang Y., Liu S.-M., Tartaglia L. & Leibel R.L. (1996). Phentotypes of mouse diabetes and rat fatty due to mutations in the ob (leptin) receptor. *Science* 271:994-996.

Cohen B., Novick D. & Rubinstein M. (1996). Modulation of insulin activities by leptin. *Science* 274:1185-1188.

Collins S., Kuhn C.M., Petro A.E., Swick A.G., Chrunyk B.A. & Surwit R.S. (1996). Role of leptin in fat regulation. *Nature* 380:677.

Considine R.V., Sinha M.K., Heiman M.L., Kriauciunas A., Stephens T.W., Nyce M.R., Ohannesian J.P., Marco C.C., McKee L.J., Bauer T.L. & Caro J.F. (1996). Serum immunoreactive-leptin concentrations in normal-weight and obese humans. *N. Engl. J. Med.* 334:292-295.

Cumin F., Baum H.-P. & Levens N. (1996). Leptin is cleared from the circulation primarily by the kidney. *Intl. J. Obesity* 20:1120-1126.

Cusin I., Sainsbury A., Doyle P., Rohner-Jeanrenaud F. & Jeanrenaud B. (1995). The ob gene and insulin: A relationship leading to clues to the understanding of obesity. *Diabetes* 44:1467-1470.

Deng C., Moinat M., Curtis L., Nadakal A., Preitner F., Boss O., Assimacopoulos-Jeannett F., Seydoux J. & Giacobino J.-P. (1997). Effects of α-adrenoceptor subtype stimulation on obese gene messenger ribonucleic acid and on leptin secretion in mouse brown adipocytes differentiated in culture. *Endocrinol.* 138:548-552.

Dessolin S., Schalling M., Champigny O., Lonnqvist F., Ailhaud G., Dani C. & Ricquier D. (1997). Leptin gene is expressed in rat brown adipose tissue at birth. *FASEB J.* 11:382-387.

De Vos P., Saladin R., Auwerx J. & Staels B. (1995). Induction of ob gene expression by corticosteroids is accompanied by body weight loss and reduced food intake. *J. Biol. Chem.* 270:15958-15961.

Elmquist J.K., Ahima R.S., Maratos-Flier E., Flier J.S. & Saper C.B. (1997). Leptin activates neurons in ventrobasal hypothalamus and brainstem. *Endocrinol.* 138:839-842.

Emilsson V., Liu Y.-L., Cawthorne M.A., Morton N.M. & Davenport M. (1997). Expression of the functional leptin receptor mRNA in pancreatic islets and direct inhibitory action of leptin on insulin secretion. *Diabetes* 46:313-316.

Erickson J.C., Clegg K.E. & Palmiter R.D. (1996). Sensitivity to leptin and susceptibility to seizures of mice lacking of neuropeptide Y. *Nature* 381:415-418.

Flier J.S. (1997). Leptin expression and action: New experimental paradigms. *Proc. Natl. Acad. Sci. USA* 94:4242-4245.

Frederich R.C., Hamann A., Anderson S., Lollmann B., Lowell B.B. & Flier J.S. (1995a). Leptin levels reflect body lipid content in mice: evidence for diet-induced resistance to leptin action. *Nature Medicine* 1:1311-1314.

Frederich R.C., Lollmann B., Hamann A., Napolitano-Rosen A., Kahn B.B., Lowell B.B. & Flier J.S. (1995b). Expression of ob mRNA and its encoded protein in rodents. *J. Clin. Invest.* 96:1658-1663.

Gainsford T., Willson T.A., Metcalf D., Handman E., McFarlane C., Ng A., Nicola N.A., Alexander W.S. & Hilton D.J. (1996). Leptin can induce proliferation, differentiation, and functional activation of hemopoietic cells. *Proc. Natl. Acad. Sci.* USA 93:14564-14568.

Ghilardi N. & Skoda R.C. (1997). The leptin receptor activates janus kinase 2 and signals for proliferation in a factor-dependent cell line. *Molec. Endocrinol.* 11:393-399.

Ghilardi N., Ziegler S., Wiestner A., Stoffel R., Heim M.H. & Skoda R.C. (1996). Defective STAT signaling by the leptin receptor in diabetic mice. *Proc. Natl. Acad. Sci.* USA 93:6231-6235.

Giacobino, J.-P. (1996). Role of the α-adrenoceptor in the control of leptin expression. *Horm. Metab. Res.* 28:633-637.

Glaum S.R., Hara M., Bindokas V.P., Lee C.C., Polonsky K.S., Bell G.I. & Miller R.J. (1996). Leptin, the obese gene product, rapidly modulates synaptic transmission in the hypothalamus. *Mole. Pharm.* 50:230-235.

Glowa J.R., Barrett J.E., Russell J. & Gold P.W. (1992). Effects of corticotropin releasing hormone on appetitive behaviors. *Peptides* 13:609-621.

Grunfeld C., Zhao C., Fuller J., Pollock A., Moser A., Friedman J. & Feingold K.R. (1996). Endotoxin and cytokines induce expression of leptin, the ob gene product, in hamsters. *J. Clin. Invest.* 97:2152-2157.

Haffner S.M., Miettinen H., Mykkanen L., Karhapaa P., Rainwater D.L. & Laakso M. (1997). Leptin concentrations and insulin sensitivity in normoglycemic men. *Intl. J. Obesity* 21: 393-399.

Halaas J.L., Boozer C., Blair-West J., Fidahusein N., Denton D.A. & Friedman J.M. (1997). Physiological response to long-term peripheral and central leptin infusion in lean and obese mice. *Proc. Natl. Acad. Sci.* USA 94:8878-8883.

Halaas J.L., Gajiwala K.S., Maffei M., Cohen S.L., Chait B.T., Rabinowitz D., Lallone R.L., Burley S.K. & Friedman J.M. (1995). Weight-reducing effects of the plasma protein encoded by the obese gene. *Science* 269:543-546.

Hamann A. & Matthaei S. (1996). Regulation of energy balance by leptin. Exp. Clin. Endocrinol. *Diabetes* 104:293-300.

Hamilton B.S., Paglia D., Kwan A.Y.M. & Deitel M. (1995). Increased obese mRNA expression in omental fat cells from massively obese humans. *Nature Medicine* 1:953-956.

Hassink S.G., de Lancey E., Sheslow D.V., Smith-Kirwin S.M., O'Connor D.M., Considine R.V., Opentanova I., Dostal K., Spear M.L., Leef K., Ash M., Spitzer A.R. & Funanage V.L. (1997). Placental leptin: An important new growth factor in intrauterine and neonatal development. *Pediatrics* 100:E1.

Kahler A., Eckel L.A., Geary N., Campfield L.A., Smith F.J. & Langhans W. (1997). Chronic administration of ob protein reduces food intake in male rats by selectively decreasing meal size. Society for the Study of Ingestive Behavior Abstract, p 26.

Kieffer T.J., Heller R.S. & Habener J.F. (1996). Leptin receptors expressed on pancreatic cells. *Biochem. Biophy. Res. Comm.* 224:522-527.

Kieffer T.J., Heller R.S., Leech C.A., Holz G.G. & Habener J.F. (1997). Leptin suppression of insulin secretion by the activation of ATP-sensitive K^+ channels in pancreatic cells. *Diabetes* 46:1087-1093.

Klein S., Coppack S.W., Mohamed-Ali V. & Landt M. (1996). Adipose tissue leptin production and plasma leptin kinetics in humans. *Diabetes* 45:984-987.

Lahlou N., Landais P., De Boissieu D. & Bougneres P.-F. (1997). Circulating leptin in normal children and during the dynamic phase of juvenile obesity. *Diabetes* 46:989-993.

Leclercq-Meyer V., Considine R.V., Sener A. & Malaisse W.J. (1996). Do leptin receptors play a functional role in the endocrine pancreas? *Biochem. Biophy. Res. Comm.* 229:794-798.

Lee G.-H., Proenca R., Montez J.M., Carroll K.M., Darvishzadeh J.G., Lee J.I. & Friedman J.M. (1996). Abnormal splicing of the leptin receptor in diabetic mice. *Nature* 379:632-635.

Legradi G., Emerson C.H., Ahima R.S., Flier J.S. & Lechan R.M. (1997). Leptin prevents fasting-induced suppression of prothyrotropin-releasing hormone messenger ribonucleic acid in neurons of the hypothalamic paraventricular nucleus. *Endocrinol.* 138:2569-2576.

Leibowitz S.F. (1991). Brain neuropeptide Y: an integrator of endocrine, metabolic and behavioral processes. *Brain Res. Bull.* 27:333-337.

Leroy P., Dessolin S., Villageois P., Moon B.C., Friedman J.M., Ailhaud G. & Dani C. (1995). Expression of ob gene in adipose cells. *J. Biol. Chem.* 271:2365-2368.

Levin N., Nelson C., Gurney A., Vandlen R. & De Sauvage F. (1996). Decreased food intake does not completely account for adiposity reduction after ob protein infusion. *Proc. Natl. Acad. Sci.* USA 93:1726-1730.

Licinio J., Mantzoros C., Negrao A.B., Cizza G., Wong M.-L., Bongiorno P.B., Chrousos G.P., Karp B., Allen C., Flier J.S. & Gold P.W. (1997). Human leptin levels are pulsatile and inversely related to pituitary-adrenal function. *Nature Medicine* 3:575-579.

Lollmann B., Gruninger S., Stricker-Krongrad A. & Chiesi M. (1997). Detection and quantification of the leptin receptor splice variants Ob-Ra, b, and, e in different mouse tissues. *Biochem. & Biophy. Res. Comm.* 238:648-652.

Lonnqvist F., Arner P., Nordfors L. & Schalling M. (1995). Overexpression of the obese (ob) gene in adipose tissue of human object subjects. *Nature Medicine* 1:950-952.

Lonnqvist F., Wennlund A. & Arner P. (1997). Relationship between circulating leptin and peripheral fat distribution in obese subjects. *Intl. J. Obesity* 21:255-260.

Luiten P.G.M., ter Horst G.J. & Steffens A.B. (1987). The hypothalamus intrinsic connections and outflow pathways to the endocrine system in relation to the control of feeding and metabolism. *Prog. Neurobiol.* 28:1-54.

MacDougald O.A., Hwang C.-S., Fan H. & Lane M.D. (1995). Regulated expression of the obese gene product (leptin) in white adipose tissue and 3T3-L1 adipocytes. *Proc. Natl. Acad. Sci.* USA 92:9034-9037.

Maffei M., Halass J., Ravussin E., Pratley R.E., Lee G.H., Zhang Y., Fei H., Kim S., Lallone R., Ranganathan S., Kern P.A. & Friedman J.M. (1995). Leptin levels in human and rodent: measurement of plasma leptin and ob RNA in obese and weight-reduced subjects. *Nature Medicine* 1:1155-1161.

Mandrup S., Loftus T.M., MacDougald O.A., Kuhajda F.P. & Lane M.D. (1997). Obese gene expression at in vivo levels by fat pads derived from s.c. implanted 3T3-F442A preadipocytes. *Proc. Natl. Acad. Sci.* USA 94:4300-4305.

Masuzaki H., Ogawa Y., Hosoda K., Miyawaki T., Hanaoka I., Hiraoka J., Yasuno A., Nishimura H., Yoshimasa Y., Nishi S. & Nakao K. (1997a). Glucocorticoid regulation of leptin synthesis and secretion in humans: elevated plasma leptin levels in cushing's syndrome. *J. Clin. Endocrinol. Metab.* 82:2542-2547.

Masuzaki H., Ogawa Y., Hosoda K., Sagawa N., Matsumoto T., Mise H., Miyawaki T., Yasuno A., Nishimura H., Mori T. & Nakao K. (1997b). Non-adipose tissue production of leptin: leptin as a novel placenta-derived hormone in humans. Endocrine Society 79th Annual Meeting. Abstract, p 118.

Mistry A.M., Swick A.G. & Romsos D.R. (1997). Leptin rapidly lowers food intake and elevates metabolic rates in lean and ob/ob mice. *J. Nutr.* 127:2065-2072.

Mizuno T.M., Bergen H., Funabashi T., Kleopoulos S.P., Zhong Y.-G., Bauman W.A & Mobbs C.V. (1996). Obese gene expression: reduction by fasting and stimulation by insulin and glucose in lean mice, and persistent elevation in acquired (diet-induced) and genetic (yellow agouti) obesity. *Proc. Natl. Acad. Sci.* USA 93:3434-3438.

Muoio D.M., Dohn G.L., Fiedorek Jr. F.T., Tapscott E.B. & Coleman R.A. (1997). Leptin directly alters lipid partitioning in skeletal muscle. *Diabetes* 46:1360-1363.

Niskanen L., Haffner S., Karhunen L.J., Turpeinen A.K., Miettinen H. & Uusitupa M.I.J. (1997). Serum leptin in relation to resting energy expenditure and fuel metabolism in obese subjects. *Intl. J. Obesity* 21:309-313.

Pagano C., Englaro P., Granzotto M., Blum W.F., Sagrillo E., Ferretti E., Federspil G. & Vettor R. (1997). Insulin induces rapid changes of plasma leptin in lean but not in genetically obese (fa/fa) rats. *Intl. J. Obesity* 21:614-618.

Pallett A.L., Morton N.M., Cawthorne M.A. & Emilsson V. (1997). Leptin inhibits insulin secretion and reduces insulin mRNA levels in rat isolated pancreatic islets. *Biochem. Biophy. Res. Comm.* 238:267-270.

Pelleymounter M.A., Cullen M.J., Baker M.B., Hecht R., Winters D., Boone T. & Collins F. (1995). Effects of the obese gene product on body weight regulation in ob/ob mice. *Science* 269:540-543.

Pombo M., Herrera-Justiniano E., Considine R.V., Hermida R.C., Galvez M.J., Martin T., Barreiro J., Casanueva F.F. & Dieguez C. (1997). Nocturnal rise of leptin in normal prepubertal and pubertal children and in patients with perinatal stalk-transection syndrome. *J. Clin. Endocrinol. Med.* 82:2751-2754.

Raber J., Chen S., Mucke L. & Feng L. (1997). Corticotropin-releasing factor and adrenocorticotrophic hormone as potential central mediators of ob effects. *J. Biol. Chem.* 272:15057-15060.

Ravussin E., Pratley R.E., Maffei M., Wang H., Friedman J.M., Bennett P.H. & Bogardus C. (1997). Relatively low plasma leptin concentrations precede weight gain in pima indians. *Nature Medicine* 3:238-240.

Rayner D.V., Dalgliesh G.D., Duncan J.S., Hardie L.J., Hoggard N. & Trayhurn P. (1997). Postnatal development of the ob gene system: elevated leptin levels in suckling fa/fa rats. *Am. J. Physiol.* 273 (*Regulatory Integrative Comp. Physiol.* 42):R446-R450.

Rohner-Jeanrenaud F., Cusin I., Sainsbury A., Zakrzewska K.E. & Jeanrenaud B. (1996). The loop system between neuropeptide Y and leptin in normal and obese rodents. *Horm. Metab. Res.* 28:642-648.

Saladin R., Staels B., Auwerx J. & Briggs M. (1996). Regulation of ob gene expression in rodents and humans. *Horm. Metab. Res.* 28:638-641.

Saladin R., De Vos P., Guerre-Millo M., Leturque A., Girard J., Staels B. & Auwerx J. (1995). Transient increase in obese gene expression after food intake or insulin administration. *Nature* 377:527-529.

Schwartz M.W. & Seeley R.J. (1997). The new biology of body weight regulation. *J. Am. Diet. Assoc.* 97:54-58.

Schwartz M.W., Baskin D.G., Bukowski T.R., Kuijper J.L., Foster D., Lasser G., Prunkard D.E., Porte Jr. D., Woods S.C., Seeley R.J. & Weigle D.S. (1996a). Specificity of leptin action on elevated blood glucose levels and hypothalamic neuropeptide Y gene expression in ob/ob mice. *Diabetes* 45:531-535.

Schwartz M.W., Seeley R.J., Campfield L.A., Burn P. & Baskin D.G. (1996b). Identification of targets of leptin action in rat hypothalamus. *J. Clin. Invest.* 98:1101-1106.

Sivan E., Lin W.M., Homko C.J., Reece E.A. & Boden G. (1997). Leptin is present in human cord blood. *Diabetes* 46:917-919.

Smith F.J., Campfield L.A., Moschera J.A., Bailon P.S. & Burn P. (1996). Feeding inhibition by neuropeptide Y. *Nature* 382:307.

Spicer L.J. & Francisco C.C. (1997). The adipose obese gene product, leptin: evidence of a direct inhibitory role in ovarian function. *Endocrinol.* 138:3374-3379.

Stehling O., Doring H., Ertl J., Preibisch G. & Schmidt I. (1996). Leptin reduces juvenile fat stores by altering the circadian cycle of energy expenditure. *Am. J. Physiol.* 271 (*Regulatory Integrative Comp. Physiol.* 40):R1770-R1774.

Stephens T.W., Basinski M., Bristow P.K., Bue-Valleskey J.M., Burgett S.G., Craft L., Hale J., Hoffmann J., Hsiung H.M., Kriauciunas A., MacKellar W., Rosteck Jr. P.R., Schoner B., Smith D., Tinsley F.C., Zhang X.-Y. & Heiman M. (1995). The role of neuropeptide Y in the antiobesity action of the obese gene product. *Nature* 377:530-532.

Surwit R.S., Petro A.E., Parekh P. & Collins S. (1997). Low plasma leptin in response to dietary fat in diabetes- and obesity-prone mice. *Diabetes* 46:1516-1520.

Tartaglia L.A. (1997). The leptin receptor. *J. Biol. Chem.* 272:6093-6096.

Tartaglia L.A., Dembski M., Weng X., Deng N., Culpepper J., Devos R., Richards G.J., Campfield L.A., Clark F.T., Deeds J., Muir C., Sanker S., Moriarty A., Moore K.J., Smutko J.S., Mays G.G., Woolf E.A., Monroe C.A. & Tepper R.I. (1995). Identification and expression cloning of a leptin receptor, OB-R. *Cell* 83:1263-1271.

Taylor S.I., Barr V. & Reitman M. (1996). Does leptin contribute to diabetes caused by obesity? *Science* 274:1151-1152.

Trayhurn P. (1996). New insights into the development of obesity: obese genes and the leptin system. *Proc. Nutr. Soc.* 55:783-791.

Turpeinen A.K., Haffner S.M., Louheranta A.M., Niskanen L.K., Miettinen H. & Uusitupa M.I.J. (1997). Serum leptin in subjects with impaired glucose tolerance in relation to insulin sensitivity and first-phase insulin response. *Intl. J. Obesity* 21:284-287.

Wang Q., Bing C., Al-Barazanji K., Mossakowaska D.E., Wang X.-M., McBay D.L., Neville W.A., Taddayon M., Pickavance L., Dryden S., Thomas M.E.A., McHale M.T., Gloyer I.S., Wilson S., Buckingham R., Arch J.R.S., Trayhurn P. & Williams G. (1997). Interactions between leptin and hypothalamic neuropeptide Y neurons in the control of food intake and energy homeostasis in the rat. *Diabetes* 46:335-341.

Weigle D.S., Bukowski T.R., Foster D.C., Holderman S., Kramer J.M., Lasser G., Lofton-Day C.E., Prunkard D.E., Raymond C. & Kuijper J.L. (1995). Recombinant ob protein reduces feeding and body weight in the ob/ob mouse. *J. Clin. Invest.* 96:2065-2070.

Widdowson P.S. Upton R., Buckingham R., Arch J. & Williams G. (1997). Inhibition of food response to intracerebroventricular injection of leptin is attenuated in rats with diet-induced obesity. *Diabetes* 46:1782-1785.

Yu W.H., Kimura M., Walczewska A., Karanth S. & McCann S.M. (1997). Role of leptin in hypothalamic-pituitary function. *Proc. Natl. Acad. Sci.* USA 94:1023-1028.

Zachow R.J. & Magoffin D.A. (1997). Direct intraovarian effects of leptin: impairment of the synergistic action of insulin-like growth factor-I on follicle-stimulating hormone-dependent estradiol-17 production by rat ovarian granulosa cells. *Endocrinol.* 138:847-850.

Zachwieja J.J., Hendry S.L., Smith S.R. & Harris R.B.S. (1997). Voluntary wheel running decreases adipose tissue mass and expression of leptin mRNA in Osborne-Mendel rats. *Diabetes* 46:1159-1166.

Zakrzewska K.E., Cusin I., Sainsbury A., Rohner-Jeanrenaud F. & Jeanrenaud B. (1997). Glucocorticoids as counterregulatory hormones of leptin: toward an understanding of leptin resistance. *Diabetes* 46:717-719.

Zhang Y., Proenca R., Maffei M., Barone M., Leopold L. & Friedman J.M. (1994). Positional cloning of the mouse obese gene and its human homologue. *Nature* 372:425-432.

Zhou Y.-T., Shimabukuro M., Koyama K., Lee Y., Wang M.-Y., Trieu F., Newgard C.B. & Unger R.H. (1997). Induction by leptin of coupling protein-2 and enzymes of fatty acid oxidation. *Proc. Natl. Acad. Sci.* USA 94:6386-6390.

SESSION 1

ORGAN AND TISSUE METABOLISM

THE EFFECT OF DIETARY ASYNCHRONY ON CIRCADIAN CHANGES IN NET NUTRIENT FLUXES ACROSS THE PORTAL-DRAINED VISCERA IN CALVES

E.J. Kim[1], M.S. Dhanoa[1], J.M. Gooden[2], J.M. Dawson[3], P.J. Buttery [3] and N.D. Scollan [1]

[1]Institute of Grassland and Environmental Research, Plas Gogerddan, Aberystwyth, SY23 3EB, U.K. [2]University of Sydney, PMB 3, Camden NSW 2570, Australia and [3]Faculty of Agriculture and Food Science, University of Nottingham, Loughborough LE12 5RD, U.K.

Summary

The effect of feeding diets which were either asynchronous (A) or synchronous (S) for the supply of energy and nitrogen to the rumen microbes on portal drained viscera (PDV) metabolism were investigated in four Holstein-Friesian bull calves (150 kg LWt). Animals were fed twice daily and arterial concentrations and net PDV fluxes were measured over a 24 h period using arterio-venous techniques. A double Fourier model was employed to describe circadian patterns. Diet did not affect arterial or net PDV flux of oxygen, glucose, β-OH butyrate, ammonia or insulin. PDV acetate absorption was higher on diet A ($P<0.05$). PDV urea flux was negative on diet A ($P<0.05$). Portal blood flow and PDV ammonia flux increased with feeding and these responses were described well by the double Fourier model. Similar, but less well defined patterns were observed for arterial acetate, glucose and β-OH butyrate and insulin. It is concluded that portal vein blood flow is responsive to feeding and that urea recycling to the rumen is important on asynchronous diets. The circadian patterns noted for arterial acetate, glucose and β-OH butyrate concentrations may have important implications for peripheral tissue metabolism.
Key words: Dietary asynchrony, nutrient flux, PDV, calves, energy, N.

Introduction

Forage nitrogen use by ruminants is poor and one of the major factors contributing to such inefficiencies is loss of forage nitrogen absorbed as ammonia from the rumen. This is related to an asynchrony between the rapid release of non-protein nitrogen from the breakdown of the highly soluble plant proteins and the slower rate of release of energy, mainly from the forage structural carbohydrates. This lack of synchronisation between energy and nitrogen release in the rumen contributes towards lower amounts and efficiencies of rumen microbial protein synthesis (Sinclair *et al.*, 1995; Chamberlain and Choung, 1995). It is uncertain what further effects dietary energy and nitrogen synchrony may have, for example, either on metabolism of nutrients by portal drained viscera (PDV) or on metabolism by peripheral tissues such as skeletal muscle. This experiment examined the impact of feeding two diets, formulated to be asynchronous or synchronous with respect to the potential hourly supply of energy and nitrogen to the microbial fraction of the rumen, on net flux of selected nutrients across the PDV. Many studies on PDV metabolism are conducted under steady state conditions, but this experiment has adopted non-steady state conditions, using a twice-daily feeding regime, to enable circadian (or diurnal) patterns of nutrient fluxes across the PDV to be described.

Materials and Methods

Four Holstein-Friesian bull calves (approx. 150 kg LWt) were used in a two period changeover design, with 21 d periods. Each animal was prepared with an indwelling catheter in the portal vein, mesenteric vein and carotid artery and a transit-time ultrasound probe (size 24US, Transonic Systems Inc., Cornell, NY) on the portal vein. Two diets were formulated to be either asynchronous (A) or synchronous (S) and were based on wheat straw, barley, urea and/or rapeseed, respectively. Organic matter, total nitrogen and soluble nitrogen were 946 and 939, 18.6 and 19.2 and 8.1 and 4.3 g/kg DM for diets A and S, respectively. The diets were offered daily at 22 g DM/kg LWt, in 2 equal feeds at 12 h intervals at 09:00 and 21:00. All animals consumed their feed within 20 minutes. On day 14, animals were put into metabolism crates and on day 18, blood was sampled continuously from the portal vein and carotid artery as follows. A heparin bolus (5.5×10^4 IU) was administered at 07:15 and a heparin infusion commenced into the mesenteric vein at 150 IU/min. Blood collection commenced at 08:00 (12 ml/h) with the aid of a peristaltic pump at a rate of 12 ml/h. Blood was collected into plastic syringes bedded in ice. Two 30 min samples were acquired in the one hour period before each feed followed by a series of 20 min samples in the two hour period following feeding. At all other times one sample was collected per hour. Continuous estimates of portal blood flow were available via the flow probes. Blood samples were immediately analysed for blood gases. Plasma was prepared and later analysed for glucose, acetate, β-OH butyrate, ammonia, urea and insulin. Net nutrient fluxes across the PDV were calculated as ($[\text{nutrient}]_{\text{portal}}$ - $[\text{nutrient}]_{\text{artery}}$) x (blood flow). The data generated have been analysed using several approaches. First, a double Fourier model was fitted to describe the 24 h profile. Output from this model was a baseline value across the 24 h period with peak responses post AM and PM feeding, described as amplitudes 1 and 2, respectively. Then auto-correlation was used to test the extent to which the net flux of a nutrient depends on the previous value.

Results

The circadian pattern of portal vein blood flow was adequately described by the double Fourier model with, on average, approximately 75% of the variance accounted for. Blood flow increased in response to feeding (Table 1, Figure 1), but the degree and time of response differed between animals, ranging between 10 and 30% and 1.5 and 5 h post feeding, respectively.

Diet did not affect blood flow, arterial concentration or net PDV flux of oxygen, β-OH butyrate or glucose. Mean values were 2.88 and 0.74 mM and 14.6 and 0.63 mmol/min for arterial concentration and net PDV flux of oxygen and β-OH butyrate, respectively, across both diets. Arterial glucose concentration was approximately 5 mM and, although not significantly different, mean net PDV glucose flux was negative on diet A and positive on diet S (-0.44 and 0.41 mmol/min respectively). Arterial acetate concentration was not different between diets but net flux was higher on diet A ($P<0.05$). Ammonia absorption was similar for the two diets and the mean baseline value from the double Fourier model was 0.61 mmol/min. However, mean arterial urea concentrations were higher on diet A compared to diet S ($P = 0.06$); 1.96 vs. 1.68, respectively. Urea was extracted by the PDV on diet A but absorbed on diet S ($P<0.02$); mean fluxes of -0.15 and 0.20 mmol/min, respectively).

Considerable circadian variation was observed in both arterial concentration and net PDV flux of the nutrients reported. Only in some cases did the double Fourier model adequately explain this variation. Arterial acetate and glucose tended to increase after feeding but only approximately 40% of the variance could be explained by the model.

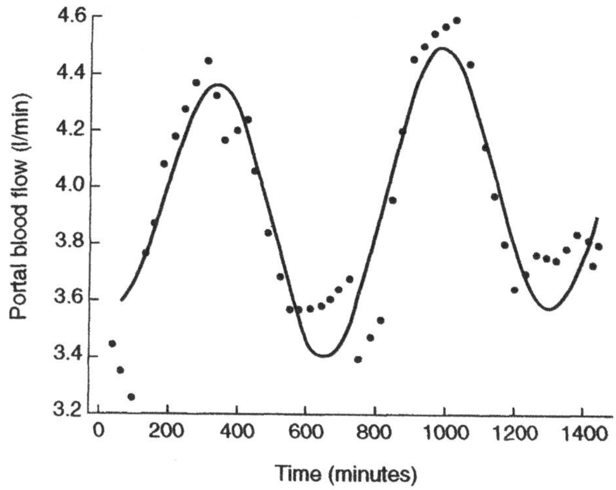

Fig. 1. A typical circadian pattern for portal vein blood flow (feed offered at 60 and 780 minutes).

In contrast, a well defined circadian pattern was found for β-OH butyrate, with peak response some 4-5 h post feeding. Approximately 80% of the variance was accounted for by the double Fourier model. No obvious pattern was observed for net PDV flux of acetate, glucose or β-OH butyrate. The model also appropriately described net PDV ammonia flux (80% of variance accounted for). On diet A, ammonia flux increased quite rapidly post feeding, in comparison to diet S, reflecting differences in the rate of feed nitrogen degradation in the rumen. Arterial urea reflected changes in PDV ammonia, but was not quite as well described by the model (approximately 50% of variance accounted for). Net PDV urea flux responses tended to be more linear across the day and were more negative on diet A than S.

In general, first order auto-correlations were high for both arterial concentrations and net PDV flux of all the measured nutrients with correlation values ranging between 0.65 and 0.90.

Discussion

There was little effect of diet on blood flow, arterial concentration or net PDV flux of oxygen, acetate, glucose or β-OH butyrate, although acetate absorption was lower on diet S ($P<0.05$), which may be related to the lower levels of straw and barley in this diet. In general, these nutrients tend to be more related to ME intake. Ammonia absorption was similar between the diets, but on diet A, net extraction of urea occurred compared with net absorption on diet S, suggesting that urea recycling is dominant on the asynchronous diet (Holder *et al.*, 1995).

Post-prandial increases in portal vein blood flow were in agreement with previous observations in both sheep and cattle (Barnes *et al.*, 1986). Considerable variation was observed across the 24 h sampling period for the measured parameters. Post-prandial increases were observed especially for arterial acetate, glucose, β-OH butyrate and insulin, but they were poorly described by the double Fourier model. These changes may have important implications for the pattern of nutrient delivery to peripheral tissues such as skeletal muscle. The large post-prandial responses (20-35%) in PDV ammonia absorption are generally a reflection of rumen ammonia patterns. No clear relationships were established between changes in pattern of nutrient availability to the PDV and their energy metabolism.

Table 1. Selected nutrient concentrations and net fluxes across the PDV of calves fed on diets designed to be either asynchronous (A) or synchronous (S) for the supply of energy and nitrogen to rumen microbes.

	Asynchronous	Synchronous	SEM	P
Blood flow (l/min)*				
base	5.89	5.68	0.201	NS
amplitude 1	0.77	0.49	0.661	NS
amplitude 2	0.54	0.59	0.193	NS
Arterial concentration (mM)				
Oxygen	5.48	5.26	0.264	NS
β-OH Butyrate	0.70	0.77	0.171	NS
Acetate	1.78	2.18	0.259	NS
Glucose	5.01	4.99	0.400	NS
Ammonia	ND	ND		
Urea	1.96	1.68	0.072	0.06
Insulin (μU/ml)	7.10	7.10	3.33	NS
Portal absorption (mmol/min)				
Oxygen	-8.28	-9.24	0.832	NS
β-OH Butyrate	0.76	0.50	0.452	NS
Acetate	3.86	1.66	0.478	0.05
Glucose	- 0.44	0.41	1.534	NS
Urea	- 0.15	0.20	0.050	0.02
Ammonia *				
base	0.54	0.68	0.166	NS
amplitude 1	0.18	0.20	0.087	NS
amplitude 2	0.17	0.14	0.030	NS

* Output from Double Fourier model; ND = not detected; NS = not significant.

References

Barnes R.J., Comline R.S. & Dobson A. (1986). The control of splanchnic blood flow. In: L.P. Milligan, W.L. Grovum and A. Dobson (Eds) Control of digestion and metabolism in ruminants. Prentice-Hall, Englewood Cliffs, NJ, pp. 41-59.

Chamberlain D.G. & Choung J.J. (1995). The importance of rate of ruminal fermentation of energy sources in diets for dairy cows. In: P.C. Garnsworthy and D.J.A. Cole (Eds) Recent advances in animal nutrition. Nottingham University Press, pp. 3-27.

Holder P., Buttery P.J. & Garnsworthy P.C. (1995). The effect of dietary asynchrony on rumen nitrogen recycling in sheep. *Animal Sci.* 60:528.

Sinclair L.A., Garnsworthy P.C., Newbold J.R. & Buttery P.J. (1995) Effects of synchronizing the rate of dietary energy and nitrogen release in diets with a similar carbohydrate composition on rumen fermentation and microbial protein synthesis in sheep. *J. agric. Sci. Cambridge* 124:463-472.

OXYGEN CONSUMPTION BY HINDQUARTERS IN COLD-EXPOSED NEW-BORN PIGS

G. Lossec, Y. Lebreton, J.C. Hulin, M. Fillault & P. Herpin

INRA, Station de Recherches Porcines, 35590 Saint Gilles, France

Summary

Hindquarters (HQ), i.e. skeletal muscle, and whole-body oxygen consumption (Vo_2) were measured in 1 and 5 day-old piglets at thermal neutrality and during intense shivering. The cold challenge was designed to induce the same increase (+90%) in whole-body Vo_2 at both ages. The cold-induced increase in HQ blood flow amounted to 25% at 1 d and 65% at 5 d of age. Oxygen extraction by HQ rose from 30-35% at thermal neutrality to 65-70% in the cold, at both ages. Then, HQ Vo_2 increased by 140% at 1 d and 230% at 5 d in the cold. Skeletal muscles contributed to 50% and 70-80% of total Vo_2 at thermal neutrality and in the cold, respectively, at both ages. Moreover, 82% and 118% of cold-induced thermogenesis could be attributed to skeletal muscles at 1 d and 5 d of age, respectively. These results indicate that the enhancement of blood supply to the muscles plays a key role in the improvement of the thermoregulatory response of the piglet to cold between 1 and 5 d of life.

Key words: Newborn piglets, oxygen consumption, hindquarters, muscles, cold exposure.

Introduction

At birth, the piglet usually experiences a 15°C decrease in its thermal environment. It shivers vigorously and is assumed to rely essentially on muscular thermogenesis for thermoregulatory purposes (Berthon *et al.*, 1994). Physiological mechanisms involved in the development of shivering thermogenesis should therefore be of utmost importance for successful adaptation to extra-uterine life. Further, we have shown recently that in cold conditions, the thermogenic efficiency of shivering increases during the first 5 days of life (Berthon *et al.*, 1994), but there is a complete lack of data to explain this postnatal improvement. This study was designed to provide qualitative and quantitative information on skeletal muscles, i.e. hindquarters, oxygen consumption at 1 d and 5 d of life during intense shivering. Calculations were performed to estimate the contribution of skeletal muscles to whole body Vo_2 at thermal neutrality and under cold conditions, as well as its contribution to the whole thermogenic response of the piglet.

Materials and Methods

At 2 h or 4 d of age, piglets were placed under halothane anaesthesia and surgically fitted with catheters in the caudal vena cava (via the left femoral vein) and in the right carotid artery. An ultrasonic transit-time flow probe was implanted around the caudal aorta between the renal and iliac vessels. Animals were placed in thermoneutral conditions during recovery and were fed reconstituted milk at regular intervals (30 g/kg at 2 h and 40 g/kg at 4 d) before the experiment. Measurements of heat production at various times after surgery in experimental and in control piglets, confirmed that recovery was complete in less than one day. Twenty hours later, i.e. at 1 d (n=6; 1.600 $\pm$ 0.056 kg) and 5 d (n=6; 2.022 $\pm$ 0.093 kg) of age, piglets were placed in a respiration chamber for 1 hour at thermal neutrality (33°C at 1 d and 30°C at 5 d of age) and for 90 minutes in the cold. The cold temperature was set at 23°C at 1 d and 15°C at 5 d of age, to provide the same increase in total Vo_2 at both ages (+90%). During each experiment, the energy metabolizm of one control piglet from the same litter (same age, same weight, surgically intact and fed the same amount of milk) was measured to confirm that the experimental animals exhibited a normal thermoregulatory response during cold stress.

Whole-body Vo_2 and muscle blood flow were measured continuously, and serial blood samples were taken from both catheters at thermal neutrality, and at 15, 30, 60 and 90 minutes after the beginning of cold exposure. Blood oxygen content was determined. Hindquarters Vo_2 (HQ Vo_2) was calculated from the averaged HQ blood flow at each sampling time while the animal was lying quietly and the corresponding arterio-venous (A-V) difference in O_2. At the end of the experiment, body temperature and weight were recorded. Piglets were killed and HQ were dissected to determine the actual composition (% of skeletal muscle, bone and skin) of the drained tissue compartment. Whole-muscle Vo_2 was estimated, assuming that skeletal muscles represent 30% of body mass in newborn pigs and that HQ comprises a wide range of muscle fibre types and may therefore be regarded as reasonably representative of the entire skeletal muscle mass. The contribution of skeletal muscles to total Vo_2 and to the increase in Vo_2 in the cold were calculated.

Data are presented as means ± SEM. Variations of all parameters were analysed (ANOVA) in order to determine the statistical effects of age and cold exposure.

Results

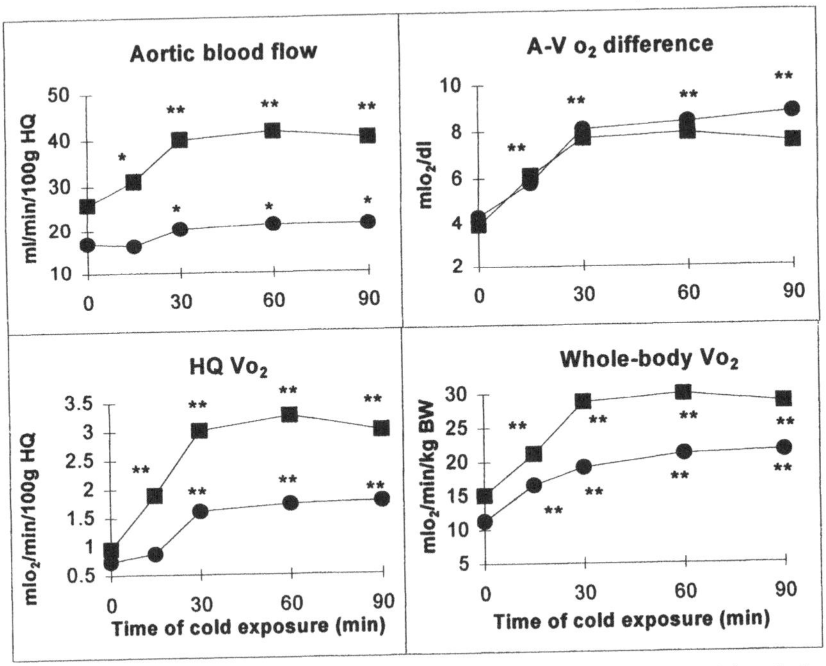

Fig.1. Aortic blood flow, A-V o_2 difference, HQ Vo_2 and whole-body Vo_2 during cold exposure at 1 (●) and 5 d (■) of age in piglets ; effect of cold (*, $P<0.01$; **, $P<0.001$).

During cold exposure, experimental piglets were able to maintain body temperature just as well as control piglets (38.6°C vs. 38.5°C, 39.3°C vs. 39.1°C at 1 and 5 d, respectively) and whole-body Vo_2 increased similarly (+90%) in both groups. At thermal neutrality, HQ blood flow was 50% higher at 5 d than at 1 d of life (Figure 1). The cold-induced increase in HQ blood flow amounted to 25% at 1 d and 65% at 5 d of age. In addition, the A-V difference in O_2 rose from 30-35% at thermal neutrality to 65-70% in the cold, at both ages.

As a consequence, HQ Vo_2 increased by 140% at 1 d and 230% at 5 d in the cold. Changes in blood flow, A-V difference in O_2, HQ Vo_2 and whole-body Vo_2 were parallel: all parameters increased during the first 30 min of cold exposure and plateaued afterwards. The calculation of skeletal muscle Vo_2 and its contribution to whole-body Vo_2 and to cold-induced regulatory thermogenesis are shown in Table 1.

Table 1. Contribution of skeletal muscles to the cold-induced increase in Vo_2 at 1 day and 5 days of age in lying pigs (values are means ± SEM; n=6).

	1 day			5 days		
	TN[1]	C	Variation	TN	C	Variation
HQ heat production [2] (kJ/d/kg HQ)	210.9 ±14.4	502.7*** ±26.0		274.4[b] ±14.4	895.6***b ±98.2	
Estimated total muscle[3] Vo_2 (ml/min/kg BW)	6.0 ±0.4	14.3*** ±0.7	+8.3	7.1[a] ±0.4	23.2***a ±2.5	+16.1
Whole-body Vo_2 (ml/min/kg BW)	11.3 ±0.3	21.4*** ±0.6	+10.1	15.1[b] ±0.5	28.7***b ±1.5	+13.6
Muscle Vo_2 (% whole body Vo_2)	53.5 ±3.8	67.1** ±3.8		47.6 ±3.9	77.2** ±6.6	
Estimated % regulatory Vo_2 attributable to muscle			82%			118%

[1] : TN : thermal neutrality, C : cold exposure.
[2] : Hindquarters were composed of 63.8% of bone + skin and 36.2% of muscle at 1 d, corresponding values are 60.7% and 39.3% at 5 d of age.
[3] : Effect of cold: **, P<0.001 ; ***, P<0.0001. Effect of age: [a], P<0.001, [b], P<0.0001.

Discussion

Working on newborn multicatheterized piglets means that recovery from surgery should be effective and rapid. It is not possible to state, unequivocally, that recovery was complete by 20 h after surgery but there is clear evidence that experimental piglets exhibited the same thermoregulatory responses as the control littermates: Vo_2 was similar at thermal neutrality and in the cold, as were rectal temperature and growth. Because the resistance of the newborn pig to cold is rapidly enhanced during the first days of life, different cold temperatures were set at both ages. As expected this was necessary to induce a comparable thermogenic demand and increase in Vo_2 (+90%) at both ages. All the measurements were obtained on lying animals because physical activity induces major changes in HQ blood flow (22-90% at 1 d and 13-44% at 5 d) and in energy expenditure which could interfere with the treatment.

Our data showed that in HQ (i.e. skeletal muscle), Vo_2 increases considerably during cold-induced shivering thermogenesis as previously demonstrated in rats (Adan *et al.*, 1995), young oxen (Bell *et al.*, 1974) or ducklings (Duchamp & Barré, 1993). This rise is explained by both an increase in blood flow and in O_2 extraction. However, whereas O_2 extraction was similarly enhanced at both ages, changes in blood flow were 3-times greater at 5 d than at 1 d of life suggesting that blood supply to the shivering muscles was considerably improved. These changes occurred during the first 30 min of cold exposure and probably contributed to the postnatal enhancement of the thermogenic efficiency of shivering. The limited cardiovascular adaptations observed at 1 d of life could be due to the relative immaturity of the sympatho-adrenal system in newborn pigs or to a limited development of skeletal muscle capillary beds. At 5 d of life, the cold-induced increase in blood flow was in agreement with results obtained on 3-4 d old piglets (Mayfield *et al.*, 1986). Conversely, the rise of O_2 extraction by HQ is not a general characteristic of cold adaptation and is not found in all species. Indeed, in ducklings (Duchamp & Barré, 1993) the rise of skeletal muscle Vo_2 is entirely explained by an increase in muscle blood flow with no change in the extraction of O_2. In addition, during more prolonged treatments, changes in O_2 extraction rates sometimes compensate for opposing changes in blood flow.

The calculated contribution of skeletal muscles to whole-body Vo_2 (~50%) at thermal neutrality is close to the results reported for rats (40%, Field *et al.*, 1939), higher than in ducklings (27%, Duchamp & Barré, 1993) and much higher than in ruminants (8-16%, Ortigues & Visseiche, 1995). One contribution to explain this difference could be the size of the animal because specific metabolic rates of muscular tissues are known to vary with animal body mass$^{-0.25}$ (Schmidt-Neilsen, 1984) and must increase with decreasing body size. This contribution could have been slightly overestimated because oxygen consumption by skin and bones was considered to be negligible in the HQ. Finally, the contribution of skeletal muscle to the cold-induced increase in total Vo_2 is estimated to be 82% and 118% in 1 d and 5 d old piglets, respectively. The results: 1) confirm the major role of skeletal muscle in the thermogenic response of the newborn pig and 2) indicate that the priority of the thermoregulatory function could lead to a reduction of the oxygen supply to non-muscular tissues at 5 d of life.

References

Adan C., Ardevol A., Closa D., Remesar X., Alemany M. & Fernandez-Lopez J.A. (1995). Hind-leg heat losses in cold-exposed rats. *J. Therm. Biol.* 20: 343-348.

Bell A.W., Thompson G.E. & Findlay J.D. (1974). The contribution of the shivering hind-leg to the metabolic response to cold of the young ox. *Pflügers Arch.* 346:341-350.

Berthon D., Herpin P. & Le Dividich J. (1994). Shivering thermogenesis in neonatal pig. *J. Therm. Biol.* 19: 413-418.

Duchamp C. & Barré H. (1993). Skeletal muscle as major site of non-shivering thermogenesis in cold-acclimated ducklings. *Am. J. Physiol.* 265: R1076-R1083.

Field J., Belding H.S. & Martin A.W. (1939). An analysis of the relation between basal metabolizm and summated tissues respiration in the rat. *J. Cell. Comp. Physiol.* 17:R415-R421.

Mayfield S.R., Stonestreet B.S., Brubakk A.M., Shaul P.W. & Oh W. (1986). Regional blood flow in newborn piglets during environmental cold stress. *Am. J. Physiol.* 251: G308-G313.

Ortigues I. & Visseiche A.L. (1995). Whole-body fuel selection in ruminants : nutrient supply and utilization by major tissues. *Proc. Nutr. Soc.* 54:235-251.

Schmidt- Nielsen K. (1984). Scaling: Why is animal size so important? Cambridge University Press. Cambridge.

DIETARY PROTEIN IS THE MAJOR SOURCE OF ENERGY FOR THE PORTAL-DRAINED VISCERA OF FED PIGLETS

B. Stoll, D.G. Burrin, F. Jahoor, J. Henry, H. Yu and P.J. Reeds

USDA/ARS Children's Nutrition Research Center, Department of Pediatrics, Baylor College of Medicine, Houston, TX, USA

Summary

Seven female pigs (24 days old; 8.0 kg) were implanted with arterial, gastric, portal and peripheral venous catheters and an ultrasonic portal blood flow probe. After 5 days of recovery, they were fed a high protein milk-based diet once-hourly and infused intragastrically with [U-^{13}C]algal protein. On average, 60% of the intake of essential amino acids was recovered in the portal blood. The fractional portal tracer balances of lysine, leucine and phenylalanine, but not of threonine, exceeded their portal mass balances. No more than 20% of the measured first-pass metabolism of tracer threonine, leucine, lysine and phenylalanine was recovered in mucosal protein. The sum of ammonia, arginine and alanine production by the intestinal tissues accounted for 23-26% of the dietary protein nitrogen intake and was equivalent to oxidation of 3800 μmol amino acid carbon/kg/h. This accounted for 80% of the CO_2 production of the portal-drained viscera.

Key words: Piglets, stable isotopes, intestinal energy metabolism, mucosal protein synthesis.

Introduction

The intestinal tissues make a disproportionate contribution to whole body energy expenditure and variations in mucosal mass have a measurable effect on maintenance energy needs (Burrin *et al.*, 1990). *In vitro* evidence demonstrates the wide metabolic capabilities of these tissues including their ability to extensively metabolize glutamine, aspartate and glutamate (e.g. Windmueller & Spaeth, 1980). However there is little information on the substrates used by the mucosa *in vivo*.

Research based on the prolonged intravenous infusion of ^{15}N-amino acids (de Lange *et al.*, 1990) suggests that many dietary proteins are <90% digested. However, measurements of the portal outflow of amino acids in piglets suggest that less than 100% of the amino acids that have been removed from the lumen appear as free amino acids in the portal circulation (Ebner *et al.*, 1994). The difference between the luminal disappearance and the portal appearance of dietary amino acids implies their substantial metabolism by the mucosal cells. The objectives of this study were to quantify the first-pass intestinal metabolism of dietary amino acids and to measure the contributions of mucosal protein synthesis and amino acid catabolism.

Materials and Methods

Seven, 2 week old, female piglets (Large White x Duroc x Hampshire) were fed a milk-replacer diet that supplied 1300 kJ and 22 g crude protein/kg$^{0.75}$/d. After 10 days, they were implanted with gastric, portal, arterial and peripheral venous catheters (Ebner *et al.*, 1994). An ultrasonic portal blood flow probe was also implanted. The animals were studied after a 5 d recovery period, over which they gained approximately 200 g body weight per d. The pigs were fasted from 1800 h to 0700 h. Baseline blood samples were taken and the animals were offered a meal supplying 1/12 of their preceding day's intake. For the next 6 h, the animals consumed hourly meals of a complete diet, each providing 600 mg protein/kg/h, and were infused intragastrically with [U-^{13}C] *Spirulina platensis* (Martek, Malvern, MA; 60 mg [U-^{13}C]protein/kg/h). Portal and arterial blood samples were taken each hour and at 15 min intervals over the last hour. After sacrifice, the proximal 2 m of small intestine were ligated, excised and frozen. Mucosa was subsequently isolated from this sample. The remainder of the intestine was removed and weighed.

Free amino acids (a.a.) were isolated from whole blood by ultrafiltration. Mucosal, dietary and *Spirulina* protein samples were hydrolysed (5/M HCl at 110°C). Amino acids were quantified by reverse phase HPLC of their phenylisothiocyanate derivatives. Amino acid labelling was measured by negative chemical ionization mass spectrometry of the n-propyl ester of their heptafluorobutyramide derivatives. The crude ion spectra were converted to tracer:tracee ratios (mol [U^{13}C]a.a. per mol [U-^{12}C]a.a.).

Calculations

Portal mass balance (μmol/kg/h) = $[C_{PV} - C_A]$
where C = a.a. concentration (μmol/l); PBF = portal blood flow rate (l/kg/h);
A = arterial; PV = portal.

Portal tracer balance (μmol/kg/h) = $[(C_{PV} \times t/T_{PV}) - (C_A \times t/T_A)] \times PBF$
where t/T is the tracer:tracee ratio of the a.a. (mol [U-^{13}C] a.a./mol [^{12}C] a.a.)

Fractional balance (percent) = (Balance/Input) x 100
for mass balance, input is rate of a.a. intake;
for tracer balance, input is rate of tracer infusion.

Tracer incorporation into mucosal protein (μmol/kg/h) =
$t/T_{\text{protein bound a.a.}}$ x mucosal protein bound a.a. content (μmol/kg body weight) x 1/6

Estimated a.a. catabolism
(a) a.a. intake - [(fractional tracer balance - fractional tracer incorporation) x a.a. intake]
(b) Portal ammonia balance + net synthesis of arginine and alanine

Results

Fasting portal blood flow (3.0 ± 0.3 l/kg/h) rose rapidly to 3.6 ± 0.4 l/kg/h within an hour of the first feed. There was also a rapid increase in portal vein ammonia concentration. Steady-state absorption of amino acids was achieved by 3 h of infusion. Table 1 shows the mass balance (± between animal SD).

Table 1. Portal mass balance of selected amino acids.

Amino acid	Mass balance	
	(μmol/kg/h)	(percent of intake)
Essential amino acids	1100 ± 400	60 ± 22
Threonine	157 ± 97	43 ± 22
Methionine	33 ± 18	48 ± 23
Glutamate	37 ± 55	7 ± 10
Arginine	84 ± 14	140 ± 20
Alanine	901 ± 96	205 ± 28
Ammonia	760 ± 190	NA

On average, 60% of the essential amino acid intake appeared in the portal vein, although a lower ($P<0.05$) proportion of the intake of threonine and methionine was absorbed. The portal balance of glutamate, aspartate and cystine was not significantly different from zero. The portal mass balance of arginine and alanine exceeded their intakes by 40 and 105%, respectively.

The metabolism of enteral tracer threonine, leucine, lysine and phenylalanine is shown in Table 2. With the exception of threonine, the fractional tracer balance was significantly ($P<0.05$) higher than the fractional mass balance. Incorporation into mucosal protein (Table 2) accounted for no more than 20% of the total mucosal metabolism.

Table 2. Intestinal metabolism of selected enteral tracer amino acids.

Amino acid	Tracer infused	Tracer balance	Tracer metabolism	Protein incorporation	Total metabolism - protein incorporation (percentage of infusion)
	(μmol/kg/h)				
Threonine	33	12 ± 5	21 ± 5	2.2 ± 0.4	56 ± 11
Leucine	48	31 ± 4	17 ± 5	3.5 ± 0.6	28 ± 5
Lysine	24	15 ± 3	9 ± 3	1.8 ± 0.2	30 ± 8
Phenylalanine	16	10 ± 2	6 ± 2	1.2 ± 0.2	30 ± 8

Estimates of amino acid catabolism, based either on amino acid tracer metabolism plus glutamate and aspartate catabolism, or on ammonia, alanine and arginine production, are shown in Table 3.

Table 3. Estimates of amino acid catabolism in the portal-drained viscera (μmol N/kg/h).

First-pass metabolism - protein synthesis (essential amino acids)	532 ± 186
Glutamate and aspartate catabolism	686 ± 178
Total amino acid catabolism	*1218 ± 371*
Net arginine and alanine synthesis	544 ± 75
Portal ammonia production	760 ± 190
Total amino acid catabolism	*1304 ± 252*

The results suggested that 23-26% of the dietary amino acid N was catabolized in first pass by the intestine. Assuming that the "average" amino acid contains 5 atoms of carbon, then after accounting for the carbon incorporated into arginine and alanine, the catabolism of amino acids would have generated 3800 mol/kg/h of CO_2 or 80% of the CO_2 output by the portal-drained viscera (4700 ± 740).

Discussion

The present results suggest that first-pass splanchnic metabolism of enteral amino acids (e.g. Matthews *et al.*, 1993), may reflect intestinal rather than hepatic metabolism. Furthermore, although the intestine utilized approximately 35% of dietary protein, the fractional portal tracer balance exceeded the fractional mass balance, suggesting that the portal-drained viscera continued to utilize arterial amino acids. In a separate study, we (Stoll *et al.*, 1997) have shown that about 50% of the uptake of phenylalanine from the mesenteric artery is incorporated into mucosal protein.

The metabolic fate of dietary amino acids in the mucosa has considerable nutritional significance. In designing this study, we expected to find that protein synthesis dominated the metabolism of enteral amino acids, but our results showed that no more than 20% of the measured first-pass metabolism was recovered in mucosal constitutive protein. It should be emphasized that tracer incorporation into constitutive protein probably underestimates the true impact of mucosal protein synthesis, because secretory protein synthesis is underestimated. It was noteworthy that less threonine (an amino acid that is a major component of intestinal mucins) was incorporated into mucosal protein than other essential amino acids. Nevertheless, even though isotopic measurements probably underestimate mucosal protein synthesis, catabolism (and presumably oxidation) was still the predominant metabolic fate of the enteral amino acids in the mucosa. This conclusion was supported by measurements of ammonia, arginine and alanine production. Taken together, the results imply that amino acid catabolism accounts for approximately 80% of the total CO_2 production by the portal-drained viscera and of this about 50% could be ascribed to the oxidation of glutamate and aspartate.

It is not certain if the results reflect an obligatory utilization of amino acids. It should be noted that *in vitro* studies (Kight & Fleming, 1993) imply that the selection of substrates for oxidation by mucosal enterocytes appears to depend on their relative concentrations rather than on extracellular regulators. The present results might, therefore, reflect an effect of high protein intake and it is possible that mucosal amino acid catabolism may be less active in animals consuming low protein diets.

Acknowledgements

Funding for this work has been provided in part from the USDA/ARS under Co-operative Agreement No. 58-6258-6100. The contents of this publication do not necessarily reflect the views or policies of the USDA. Mention of trade names or commercial products does not imply endorsement by the US Government. We thank L. Loddeke for her editorial advice.

References

Burrin D.G., Ferrell C.L., Britton R.A. & Bauer M. (1990). Level of nutrition, visceral organ size and metabolic activity in sheep. *Br. J. Nutr.* 64:439-448.

de Lange C.F.M., Souffrant W.B. & Sauer W.C. (1990). Real ileal and amino acid digestibilities in feedstuffs for growing pigs as determined with the ^{15}N-isotope dilution technique. *J. Anim. Sci.* 68:409-418.

Ebner S., Schoknecht P., Reeds P.J. & Burrin D.G. (1994). Growth and metabolism of gastrointestinal and skeletal muscle tissues in protein-malnourished neonatal pigs. *Am. J. Physiol.* 266:R1736-R1743.

Kight C.E. & Fleming S.E. (1993). Nutrient oxidation by rat intestinal epithelial cells is concentration dependent. *J. Nutr.* 123:876-882.

Matthews D.E., Marano M.A. & Campbell R.G. (1993). Splanchnic bed utilization of leucine and phenylalanine in humans. *Am. J. Physiol.* 264:E109-E118.

Stoll B., Burrin D.G., Henry J., Jansor F. & Reids P.J. (1997). Phenylalanine metabolizims in intestinal mucosa and liver measured with intravenous and intragastric tracers in pigs. *Am. J. Physiol.* In press.

Windmueller H.G. & Spaeth A.E. (1980). Respiratory fuels and nitrogen metabolism *in vivo* in small intestine of fed rats. Quantitative importance of glutamine, glutamate, and aspartate. *J. Biol. Chem.* 255:107-112.

ENERGY EXPENDITURE AND MITOCHONDRIAL PROTON LEAK IN LEAN AND OBESE STRAINS OF RATS

J.J. Ramsey[1], D.E. Johnson[2], and K.A. Johnson[3]

[1]Wisconsin Regional Primate Research Centre, University of Wisconsin, Madison, WI, USA; [2]Department of Animal Sciences, Colorado State University, Fort Collins, CO, USA; [3]Department of Animal Sciences, Washington State University, Pullman, WA, USA

Summary

The purpose of these experiments was to determine if differences in energy expenditure with obesity or cold exposure were associated with differences in mitochondrial proton leak. In the first experiment, energy expenditure was measured in obese and lean rats. Adjusted to a common food intake, obese rats had, at least, 21% lower ($P<0.05$) energy expenditures ($kJ/W^{0.75}$) than lean rats. Respiration rates and membrane potential were measured in hepatic mitochondria. Proton leak kinetics were visualized by plotting membrane potential against proton leak. Lean rats had higher proton leak rates than obese rats. The same measurements performed in the first experiment were completed in a second study using obese and lean Zucker rats housed at 24°C or following five days at 10°C. The obese animals had lower ($P<0.05$) energy expenditures ($kJ/W^{0.75}$) than the lean animals at 24°C while energy expenditure was not different between either group at 10°C. Cold exposure had no significant effect on proton leak rate in the obese rats while proton leak was decreased in the lean rats. The results suggest that a low proton leak rate may contribute to the low energy expenditure observed in obese Zucker rats. Increases in hepatic proton leak do not appear to play a role in adaptation to cold exposure.
Key words: Energy expenditure, mitochondria, proton leak, obesity, cold exposure, rats.

Introduction

Proton leak across the inner mitochondrial membrane is a major contributor to whole animal resting energy expenditure. This inefficiency in oxidative phosphorylation may be responsible for as much as 20-40% of the basal respiration rate of hepatocytes (Brand, 1990). Standard metabolic rate differences between mammals and reptiles (Brand *et al.*, 1991) and rats of different thyroid status (Harper & Brand, 1993) have been correlated with differences in mitochondrial proton leakage. There is increasing evidence that mitochondrial proton leak is a major determinant of metabolic rate under a variety of physiological conditions.

Two experiments examined the relationship between energy expenditure and proton leak. The purpose of the first study was to determine if energy expenditure differences between lean and obese rats are associated with differences in mitochondrial proton leak while the second study determined if differences in energy expenditure with cold exposure are associated with differences in hepatic mitochondrial proton permeability.

Materials and Methods

Indirect respiration calorimetry measurements were made in obese Zucker (n=5), Sprague-Dawley (n=5), Fisher 344 (n=5), and lean Zucker (n=3) rats in the first experiment. The second experiment used obese (n=8) and lean (n=9) Zucker rats. Half of the animals in the second experiment were housed at 24°C while the other half were housed at 10°C for 7 days with calorimetric measurements on days 5 and 6 of cold exposure. All other procedures were identical between the two experiments. Oxygen consumption was analysed over a 24-hour period and energy expenditure was calculated using 21.5 kJ/l O_2.

After the calorimetry measurements, the rats were euthanized and internal organs were removed, cleaned of fat, and weighed. The liver was placed into ice cold media (250 mM sucrose, 10 mM Tris/HCl, 1 mM (K)EGTA, pH 7.4) and homogenized. Mitochondria were isolated by differential centrifugation. The mitochondrial pellet was resuspended in medium containing 120 mM KCl, 5 mM HEPES, and 1 mM (K)EGTA and protein content was determined by the Bradford method.

Mitochondrial respiration measurements involved mixing equal volumes of medium containing the suspended mitochondria and of respiration medium (120 mM KCl, 5 mM HEPES, 1 mM (K)EGTA, 10 μM rotenone, 4 μg of oligomycin/ml, 0.8 μg of nigericin/ml, 20 mM succinate, and one of the following: 0, 2, 6, 10, or 20 mM malonate, pH 7.2). Oxygen consumption was then measured by placing the mixture in a 37°C chamber containing a Clarke-type electrode.

Mitochondrial matrix volume was determined by adding ^3H_2O (1.0 μCi/ml) and [^{14}C] sucrose (0.14 μCi/ml) to the respiring mixture. Matrix volume was equal to ^3H_2O space minus [^{14}C] sucrose space. Membrane potential was determined by adding [^{3}H]methyltriphenylphosphonium (TPMP) bromide (0.1 μCi/ml) to a separate, respiring, mitochondrial mixture. Membrane potential was calculated as equal to 59 x log (calibration factor x accumulation ratio for $TPMP^+$). Proton leak was calculated assuming six protons are leaked per oxygen atom consumed (Brand *et al.*, 1991).

Body composition was determined from body water by lyophilizing carcasses to constant weight. Fat-free mass was calculated as body water divided by 0.73. All statistical analyses were performed using the glm procedure in SAS.

Results

Body composition and calorimetry data are summarized in Table 1 for the first experiment and in Table 2 for the second experiment.

Table 1. Body composition and energy expenditure in lean and obese strains of rats.

	Obese Zucker (n=5)	Sprague-Dawley (n=5)	Fisher 344 (n=5)	Lean Zucker (n=3)	SEM
Body mass (kg)	0.742[a]	0.485[b]	0.345[c]	0.404[d]	0.042
Lean body mass (kg)	0.362[a]	0.421[b]	0.294[c]	0.369[ab]	0.039
Organ mass (% of lean mass)	12.94[a]	7.51[b]	7.31[b]	6.78[b]	0.61
Energy expenditure at common intake:					
Energy expenditure (kJ/d)	239.2[a]	221.4[ac]	174.3[b]	208.4[c]	19.6
Energy expenditure (kJ/W$^{0.75}$)	299.4[a]	381.0[b]	386.7[b]	410.6[b]	25.8

[1]Values in a row not sharing a common superscript differ (P<0.05).

As expected, obese Zucker rats were at least 46% heavier than strains in either experiment. This body weight difference was due almost exclusively to increased fat mass in the obese Zucker rats. Lean body mass was similar between the different strains. Obese animals also had larger (P<0.05) livers, stomachs, kidneys and hearts than lean rats. As a percentage of lean mass, organ mass was at least 44% larger in the obese Zucker rats. Cold exposure did not result in any significant changes in body composition or organ mass.

Table 2. Energy expenditure in lean and obese Zucker rats at 24°C or 10°C.

	Obese 24°C (n=4)	Obese10°C (n=4)	Lean 24°C (n=4)	Lean 10°C (n=5)	SEM
Energy expenditure (kJ/d)	321.6[b]	414.4[a]	213.0[c]	324.4[b]	27.1
Energy expenditure ($kJ/W^{0.75}$)	425.9[b]	594.2[a]	427.7[b]	654.0[a]	40.1

[1]Values in a row not sharing a common superscript differ (P<0.05).

Energy expenditure (kJ/d) was higher (P<0.05) in the obese Zucker rats than the lean animals at 24°C. In the first experiment, adjustment of energy expenditure for body weight and food intake resulted in the obese Zucker rats having energy expenditures ($kJ/W^{0.75}$) at least 21% lower than the other strains. Cold exposure resulted in a significant increase (P<0.05) in energy expenditure in both the lean and obese rats.

Mitochondrial proton leak kinetics are presented in Figure 1 (experiment 1) and Figure 2 (experiment 2). In the first experiment, mitochondrial respiration rate was not different between strains of rats. Membrane potential, however, was higher (P<0.05) in the obese rats than in the lean strains. These results were confirmed in experiment 2 and show that the 24°C housed lean rats had at least 25% higher proton leak rates than the obese Zucker rats. Mitochondrial respiration, membrane potential, and proton leak were not altered by cold exposure in the obese animals. The lean Zucker rats housed at 10°C, however, had lower mitochondrial respiration rates and higher membrane potentials than the lean animals at 24°C. These results show that proton leak rate was either unchanged (obese) or decreased (lean) by cold exposure.

Discussion

Obese Zucker rats had higher energy expenditures ($kJ/W^{0.75}$) than lean rats at 24°C. Expressed relative to metabolic body size, obese Zucker rats had similar or lower energy expenditures than lean strains. Obese animals, however, consumed approximately 50% more feed per day than lean animals. Mathematical adjustment of energy expenditure to a common intake resulted in obese animals having significantly lower (P<0.05) energy expenditures than lean animals. This agrees with previous experiments that also showed a lower energy expenditure in obese Zucker rats than lean rats at the same intake (Pullar & Webster, 1974).

Organ masses were measured for all of the rats to determine if energy expenditure corresponded with changes in organ mass. Liver and gastrointestinal tract are responsible for approximately 50% of the whole animal's energy expenditure (Johnson *et al.*, 1990), so differences in organ mass can have a substantial effect on whole animal energy expenditure. Expressed as a percentage of lean body mass, obese animals had at least 58% larger internal organs than lean animals. This difference was greater than any observed differences in energy expenditure. This finding, along with the calorimetry results, suggests that the obese animals were energetically more efficient and had lower energy expenditures than the lean animals. Cold exposure resulted in a 40-50% increase in energy expenditure in both lean and obese rats. The increases in energy expenditure, and absence of hypothermia, observed for both lean and obese animals in this study suggest no major defects in temperature regulation.

Comparison between lean and obese rats at 24°C showed that hepatic mitochondrial proton leak was at least 25% lower in obese rats than lean rats at comparable membrane potentials. Cold exposure caused no change in proton leak in the obese animals while proton leak was decreased in the lean animals. Proton leak has been shown to account for 20-30% of resting oxygen consumption of hepatocytes (Brand, 1990). Consequently changes in this process could have a substantial effect on whole animal energetics if the observed changes are also present in other tissues.

The mechanism of proton leak was not investigated in this work, although the recent discovery (Fleury *et al.*, 1997) of an uncoupling protein (UCP2) that is widely expressed in most tissues may explain the current findings. In agreement with the proton leak measurements in this study, UCP2 has been linked with obesity. Exposure to cold did not influence UCP2 mRNA levels (Fleury *et al.*, 1997). In addition to UCP2, thyroid hormones (Harper & Brand, 1993) and membrane fatty acid composition (Brand *et al.*, 1991) have been proposed to influence proton leak. Both processes may be altered in obesity.

The results suggest that a low mitochondrial proton leak rate may contribute to a low energy expenditure in obese Zucker rats. Increases in proton leak, however, do not appear to play a role in increasing energy expenditure with cold exposure.

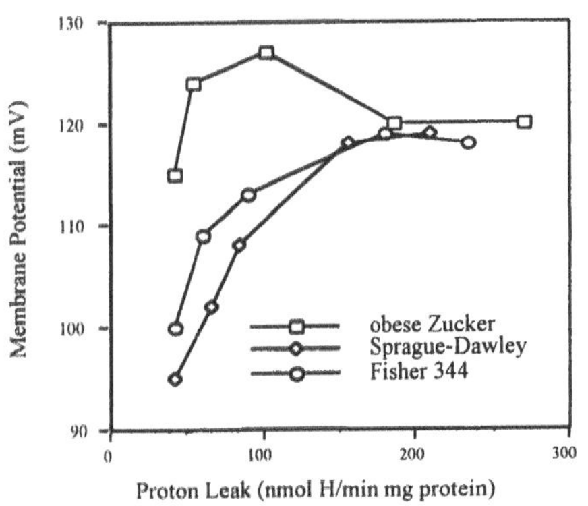

Fig. 1. Mitochondrial proton leak kinetics in liver mitochondria from lean and obese strains of rats.

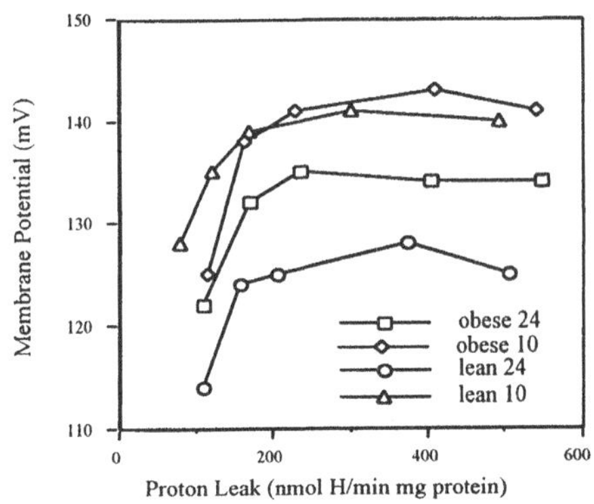

Fig. 2. Mitochondrial proton leak kinetics in liver mitochondria from rats adapted to 24°C or 10°C.

References

Brand M.D. (1990). The contribution of the leak of protons across the mitochondrial inner membrane to standard metabolic rate. *J. Theor. Biol.* 145:267-286.

Brand M.D., Couture P., Else P.L., Withers K.W. & Hulbert A.J. (1991). Proton permeability of the inner membrane of liver mitochondria is greater in a mammal than a reptile. *Biochem. J.* 275:81-86.

Fleury C., Neverova M., Collins S., Raimbault S., Champigny O., Levi-Meyrueis C., Bouillaud F., Seldin M.F., Surwit R.S. & Warden C.H. (1997). Uncoupling proton-2: a novel gene linked to obesity and hyperinsulinemia. *Nature Gen.* 15:269-272.

Harper M.E. & Brand M.D. (1993). The quantitative contributions of mitochondrial proton leak and ATP turnover reactions to the changed respiration rates of hepatocytes from rats of different thyroid status. *J. Biol. Chem.* 268:14850-14860.

Johnson D.E., Johnson K.A. & Baldwin R.L. (1990). Changes in liver and gastrointestinal tract energy demands in response to physiological work load in ruminants. *J. Nutr.* 120:649-655.

Pullar J.D. & Webster A.J.F. (1974). Heat loss and energy retention during growth in congenitally obese and lean rats. *Br. J. Nutr.* 31:377-392.

INFLUENCE OF ENERGY DENSITY AND METABOLIZABLE ENERGY INTAKE ON VISCERAL ORGAN GROWTH IN SHEEP

K.R. McLeod and R.L. Baldwin,VI

Nutrient Conservation and Metabolism Laboratory, Livestock and Poultry Sciences Institute, Agricultural Research Service, U.S. Department of Agriculture, Beltsville, MD 20705, USA

Summary

Twenty-eight crossbred lambs were fed diets containing either 75% forage or 75% concentrate at either 0.416 or 0.757 MJ ME/kg $BW^{0.75}$ per d for 52 d. Lambs were slaughtered and visceral organs removed, weighed, sectioned, and epithelial cells isolated from the rumen and small intestine. Visceral organ mass (**VOM**) and organ mass as a percentage of empty body weight (**VOME**) increased in response to increasing ME intake. Conversely, increasing dietary energy density decreased VOM and VOME. Oxygen consumption by isolated epithelial cells from the rumen and small intestine was unaffected by either ME intake or dietary energy density. Thus, greater energy use by visceral tissues associated with increasing ME intake, as well as enhanced efficiency for animals fed predominantly concentrate diets, may be attributable to changes in visceral organ mass.
Key words: Viscera, dietary energy density, ME intake, sheep, oxygen consumption.

Introduction

The relationships between dietary energy intake and visceral organ mass and oxygen consumption have been well established (Seal & Reynolds, 1993). However, the regression equation developed by Seal & Reynolds (1993), generated using data from the literature, consistently over-predicts portal-drained drained viscera (**PDV**) heat production associated with concentrate diets and under predicts PDV heat production associated with roughage diets. Therefore, the objectives of this experiment were to identify the components of ruminant gut tissue oxygen consumption and nutrient use, i.e. visceral organ mass and (or) mass specific metabolism, which are sensitive to ME intake and dietary energy density.

Materials and Methods

Twenty-eight crossbred lambs (20.1± 0.3 kg BW) were assigned randomly to a factorial arrangement of treatments. Treatments included pelleted diets containing either 75% orchardgrass hay and 25% concentrate (**FORG**; 9.56 MJ ME/kg DM) or 75% concentrate and 25% orchardgrass hay (**CONC**; 11.74 MJ ME/kg DM) fed once daily at either 0.42 or 0.76 MJ ME/kg $BW^{0.75}$ per d. Diets were formulated to provide an equal N to ME ratio. During the 52 d feeding period, lambs were weighed twice weekly and the amount of feed offered was adjusted weekly.

Lambs were exsanguinated and the visceral organs were removed. Wet weights of the rumen, reticulum, abomasum, omasum, small and large intestine and liver were recorded. Sections (1 m) from the duodenum, jejunum and ileum were filled with buffer solution (Krebs salts plus 25 mM HEPES, pH 7.4, 37°C) and transported to the laboratory for isolation of cells. Likewise, ruminal epithelium was taken from the ventral cranial region of the rumen and immediately placed in buffer.

Intestinal and ruminal epithelial cells were isolated according to the procedures of Watford *et al.* (1979) and Baldwin & Jesse (1991), respectively. Cell concentrations were adjusted to approximately 1 x 10^8 viable cells/ml. Oxygen uptake was determined as described by McBride *et al.* (1985). Homogenized epithelial tissue was used to determine NA^+, K^+ ATP-ase activity (Caperna & Gavelek, 1994).

Results

As expected, DM, ME and N intakes and average daily gains were greater ($P<0.001$) in lambs fed higher ME intake (Table 1). Average daily gain of lambs fed low ME intake did not differ from zero ($P>0.20$), indicating that they were at, or near, maintenance. Lambs fed CONC consumed less DM ($P<0.001$) and gained at a faster rate ($P<0.01$) than lambs fed FORG reflecting greater efficiency of ME use for tissue gain from the CONC diet.

Table 1. Effects of dietary energy density and ME intake on lamb performance.

	Forage		Concentrate			Significance of effect[a]		
Item	Low Intake	High Intake	Low Intake	High Intake	SEM	Diet	Intake	DxI
Initial wt., kg	29.1	29.6	29.7	30.2	1.3			
Final wt., kg	28.6	41.7	30.0	44.9	1.6		***	
Days on feed	52.3	52.4	51.3	52.4	0.9			
Gain[b], kg/d	0.00	0.22	0.02	0.26	0.01	**	***	
DM intake, kg/d	0.54	1.13	0.45	0.97	0.03	***	***	
ME intake, MJ/d	5.15	10.84	5.27	11.42	0.29		***	
N intake, g/d	13.1	27.7	13.7	29.8	0.1		***	

[a] Probability of a larger F statistic; * $P<0.05$, ** $P<0.01$, *** $P<0.001$.

[b] Calculated over the final 49 days only to minimize initial effects of gut fill.

Wet weights of rumen, omasum, abomasum, large intestine, and liver were greater ($P<0.001$) in lambs fed high ME intake (Table 2). Lambs fed CONC had lower ($P<0.001$) omasum and higher ($P<0.01$) liver wet weights than lambs fed FORG. Wet weights of reticulum and small intestine were greater ($P<0.001$) in lambs fed FORG compared with those fed CONC at high ME intake. However, wet weights of these tissues were similar for both diets at low ME intake (diet × intake interaction, $P<0.05$).

As percentages of empty body weight (**EBW**), rumen and liver weights were greater ($P<0.001$) in lambs fed higher ME intake. Weights of omasum ($P<0.001$) and large intestine ($P<0.05$) were greater in lambs fed FORG than in those fed CONC. Weights of reticulum, abomasum and small intestine increased ($P<0.02$) with greater intake of FORG, but were unaffected by greater intake of CONC (diet × intake interaction, $P<0.05$).

Oxygen uptake by cells isolated from the rumen and small intestine was not affected by ME intake or diet (Table 4). In general, oxygen consumption by intestinal epithelial cells was greater than that by rumen epithelial cells. Na^+, K^+ ATP-ase activity was higher in ruminal ($P<0.05$) and ileal ($P<0.001$) tissue from lambs fed CONC than those fed FORG. Greater ME intake increased ($P<0.05$) Na^+, K^+ ATPase activity in ileal tissue.

Discussion

The positive relationship between ME intake and visceral organ mass is consistent with previous data for lambs (Ferrell *et al.*, 1986). Total digestive tract organ mass (**DTM**), as a percentage of EBW, increased from 8.4 to 10.8% as ME intake increased from approximately 1 to 2 times maintenance. The lack of change in *in vitro* oxidative capacity in the current study indicates that a greater proportion of the total maintenance energy requirement is associated with the digestive tract of lambs fed above maintenance.

An inverse relationship between dietary energy density and visceral mass in the current study is consistent with the data of Reynolds *et al.* (1991) who showed an increase in PDV oxygen use by heifers fed a 75% forage diet compared with heifers fed a 75% concentrate diet at equal ME intake. Also, Reynolds *et al.* (1991) demonstrated that a lower proportion of whole animal oxygen consumption was accounted for by PDV tissues in concentrate-fed heifers. In the current experiment, DTM, as a percentage of EBW, was 19% lower in concentrate-fed lambs at high energy intake than in forage-fed lambs. Because *in vitro* oxygen use did not differ between diets, the shift in the proportion of oxygen use by the PDV noted by Reynolds *et al.* (1991) can be explained, at least in part, by changes in DTM.

Table 2. Effects of dietary energy density and ME intake on visceral organ mass in lambs.

	Forage		Concentrate			Significance of effect[a]		
Item	Low Intake	High Intake	Low Intake	High Intake	SEM	Diet	Intake	DxI
Rumen								
Wet wt, g	260.7	530.4	266.6	567.0	25.4		***	
% EBW[b]	1.36	2.06	1.36	1.85	0.11		***	
Reticulum								
Wet wt., g	81.5	139.6	71.8	98.5	6.0	***	***	*
% EBW[b]	0.34	0.45	0.28	0.25	0.02	***	*	**
Omasum								
Wet wt., g	83.8	130.4	55.0	92.5	7.6	***	***	
% EBW[b]	0.34	0.43	0.22	0.25	0.04	***		
Abomasum								
Wet wt., g	135.5	219.7	131.4	182.8	13.1		***	
% EBW[b]	0.55	0.72	0.51	0.48	0.05	**		*
Small intestine								
Wet wt., g	409.2	705.7	392.7	593.7	20.7	**	***	*
% EBW[b]	1.69	2.30	1.56	1.61	0.12	**	**	*
Large intestine								
Wet wt., g	335.8	523.1	335.9	506.5	21.1		***	
% EBW[b]	1.38	1.70	1.32	1.35	0.10	*		
Liver								
Wet wt., g	340.6	600.7	377.4	747.8	28.7	**	***	
% EBW[b]	1.40	1.93	1.48	2.00	0.12		***	

[a] Probability of a larger F statistic; * $P<0.05$, ** $P<0.01$, *** $P<0.001$.
[b] Wet organ weight expressed as a percentage of empty body weight.

Changes in liver mass associated with ME intake are likely due to differences in nutrient inputs Changes in DTM appear to be more complex. Although the diets in the current study were pelleted (to minimize differences in physical form) comparisons between ME intake and diet were confounded by DM intake and it is difficult to determine whether the observed changes in DTM were due to differences in diet composition or in DM intake. Rompala *et al.* (1988) reported that increasing dietary bulk resulted in increased visceral organ mass. However, many factors (i.e. humoral, local and luminal) are involved in the regulation of ruminal (Galfi *et al.*, 1991) and intestinal epithelial growth

(Johnson, 1987). Additional work is needed to delineate the mechanisms responsible for the differential effects of diet composition on digestive tract organ growth.

Table 3. Effects of dietary energy density and ME intake on oxygen uptake by isolated cells and Na^+,K^+ ATPase activity of epithelial homogenates.

Item	Forage		Concentrate			Significance of effect[a]		
	Low Intake	High Intake	Low Intake	High Intake	SEM	Diet	Intake	DxI
Rumen								
Oxygen uptake[b]	1.37	1.05	0.96	1.42	0.19			
Na^+,K^+ ATPase[c]	3.66	3.34	4.38	4.54	0.46	*		
Duodenum								
Oxygen uptake[b]	2.05	2.92	3.20	3.08	0.59			
Na^+,K^+ ATPase[c]	9.83	13.37	12.20	14.47	1.82			
Jejunum								
Oxygen uptake[b]	2.27	2.97	2.20	2.34	0.36			
Na^+,K^+ ATPase[c]	12.57	16.47	12.02	15.55	2.82			
Ileum								
Oxygen uptake[b]	1.35	1.95	1.39	1.14	0.27			
Na^+,K^+ ATPase[c]	5.53	7.86	9.64	14.31	1.42	***	*	

[a] Probability of a larger F statistic; * $P<0.05$, ** $P<0.01$, *** $P<0.001$.
[b] $\mu l\ O_2/10^6$ viable cells/h [c] nmole PO_4/mg protein/min.

References

Baldwin R.L., VI & Jesse B.W. (1991). Technical note: Isolation and characterization of sheep ruminal epithelial cells. *J. Anim. Sci.* 69:3603-3609.

Caperna T.J. & Gavelek D. (1994). Tissue ATP-ase activity in somatotropin-treated pigs. In J.F. Aguilera (Ed) Energy Metabolism of Farm Animals. *EAAP Publ. No.* 76. pp. 125-128.

Ferrell C.L., Koong L.J. & Nienaber J.A. (1986). Effect of previous plane of nutrition on body composition and maintenance energy costs of growing lambs. *Br. J. Nutr.* 56:595-605.

Galfi P., Neogrady S. & Sakata T. (1991). Effects of volatile fatty acids on the epithelial cell proliferation of the digestive tract and its hormonal mediation. In T. Tsuda, Y. Sasaki and R. Kawashima (Eds) Physiological Aspects of Digestion and Metabolism in Ruminants. Academic Press, San Diego, CA. pp. 49-59.

Johnson L.R. (1987). Regulation of gastrointestinal growth. In L.R. Johnson (Ed) Physiology of the Gastrointestinal Tract. Raven Press, New York, NY. pp. 301-333.

McBride B.W. & Milligan L.P. (1985). Influence of feed intake and starvation on the magnitude of Na^+, K^+-ATPase (EC 3.6.1.3)-dependent respiration in duodenal mucosa of sheep. *Br. J. Nutr.* 53:605-614.

Reynolds C.K., Tyrrell H.F. & Reynolds P.J. (1991). Effects of diet forage-to-concentrate ratio and intake on energy metabolism in growing beef heifers: Whole body energy and nitrogen balance and visceral heat production. *J. Nutr.* 121:994-1003.

Rompala R.E., Hoagland T.A. & Meister J.A. (1988). Effect of dietary bulk on organ mass, fasting heat production and metabolism of the small and large intestines in sheep. *J. Nutr.* 118:1553-1557.

Seal C.J. & Reynolds C.K. (1993). Nutritional implications of gastrointestinal and liver metabolism in ruminants. *Nutr. Res. Rev.* 6:185-208.

Watford M., Lund P. & Krebs H.A. (1979). Isolation and metabolic characterizacteristics of rat and chicken enterocytes. *Biochem. J.* 178:589-596.

CHANGES IN GUT AND LIVER GLUCOSE RELEASE AND OXYGEN CONSUMPTION DURING GESTATION AND LACTATION IN THE EWE

H.C. Freetly and C.L. Ferrell

USDA, ARS, Roman L.Hruska, U.S. Meat Animal Research Center, Clay Center, NE, USA

Summary

Oxygen consumption and net glucose flux across the portal-drained viscera (PDV) and liver of ewes prepared with chronic indwelling catheters in the abdominal aorta, the portal vein and a hepatic vein were determined during pregnancy and lactation. Oxygen consumption by the PDV and liver increased in late pregnancy and remained elevated through mid-lactation compared to non-pregnant ewes. Net hepatic glucose release increased in late pregnancy and remained elevated through mid-lactation. This study supports the hypothesis that the increased energy requirement during pregnancy and lactation partially results from an increased energy expenditure by the gut and liver.
Key words: Gut, liver, pregnancy, lactation.

Introduction

Energy requirements of the ewe increase 50-120% over maintenance at the end of pregnancy and remain elevated during lactation. The increase in energy demand during pregnancy is required to support conceptus growth and growth of non-reproductive tissues, and to meet maintenance requirements of the conceptus and ewe. During lactation, increased energy is required for growth and maintenance of the mammary gland, and for synthesis of milk. At 90 d of gestation, about 32% of the increase in heat production is attributable to maternal tissues other than the gravid uterus (Bell, 1986). Limited information is available on the relative contribution of different maternal tissues to the increase in heat production. During lactation, the PDV and liver are primary contributors to animal oxygen consumption (Huntington & Tyrrell, 1985; Reynolds *et al.*, 1986). The liver supports pregnancy and milk production by being a net producer of glucose ; whereas the gravid uterus and mammary gland are net users of glucose (Battaglia & Meschia, 1981; Miller *et al.*, 1991). During both late pregnancy and lactation, glucose requirements are high. The objectives of this study were to determine the pattern of net glucose release and oxygen consumption by the PDV and liver during gestation and lactation.

Material and Methods

Seven multiparous polled Dorset ewes were individually penned and provided with water and a pelleted diet *ad libitum* (57% dehydrated alfalfa, 28% corn cobs, and 15% corn as fed; ME = 9.61 MJ/kg dry matter). Catheters were surgically placed in the portal vein, a branch of the hepatic vein, a mesenteric vein and the abdominal aorta. Fifty-four days after surgery, ewes were bred.

Forty-seven days after surgery (-154 ± 4 d from parturition), blood flow and net nutrient flux across the PDV and liver were determined by continuously infusing *p*-aminohippuric acid (PAH; 0.15 M) into the mesenteric vein. Blood samples were drawn into heparinized syringes (10 ml) from the aortic, portal venous, and hepatic venous catheters. Blood samples for determination of oxygen concentration were anaerobically drawn into 1 ml syringes. Samples were collected at 30-min intervals to provide a total of seven sets of samples per period (aortic, portal venous, and hepatic venous). Additional sets of samples were collected at -105, -84, -63, -42, -21, -7, 7, 21, 42, and 70 (SD± 3) days from parturition. After parturition, lambs were kept with the ewes.

Blood and plasma flows were calculated using an indicator-dilution technique (Katz & Bergman, 1969) and net release of metabolites was calculated by multiplying blood or plasma flow by the concentration difference in the vessels (Katz & Bergman, 1969). In this technique a negative release rate represents a net uptake by the tissue. Data were analysed by ANOVA using a model in which collection day was a discrete effect.

Results

Before breeding, ewes mean body weight was 76.6 ± 4 kg. Body weight increased during pregnancy (Table 1). Average feed intake (1959 ± 308 g/d) during gestation (periods 2-7) did not differ between collection periods (P = 0.32, Table 1), but it was lower during gestation than before breeding (2452 ± 237 g/d; P<0.001). Feed intakes on days 42 and 70 are combined intakes for the ewe and lambs.

Table 1. Mean ± SE body weight and feed intake.

	Pre-parturition					
	105	84	63	42	21	7
BW, kg	84.2 ± 4	87.8 ± 4	91.8 ± 4	93.6 ± 5	95.9 ± 4	98.5 ± 4
Feed, g/d	2049 ± 76	1924 ± 105	2021 ± 110	1848 ± 131	2125 ± 108	1788 ± 154

	Post-parturition			
	7	21	42	70
BW, kg	85.7 ± 4	85.4 ± 4	83.5 ± 4	81.8 ± 5
Feed, g/d	2497 ± 211	3281 ± 172	3983 ± 399	3684 ± 361

Portal venous blood flow was decreased in mid-pregnancy (between days 84 and 42 pre-parturition) but did not differ from non-pregnant blood flow rates in late pregnancy and during lactation (P>0.05; Fig. 1). Hepatic venous blood flow followed the same general pattern as portal venous blood flow but differed in that blood flow at -21, 7, and 21 d from parturition was higher than when animals were not pregnant (P<0.05).

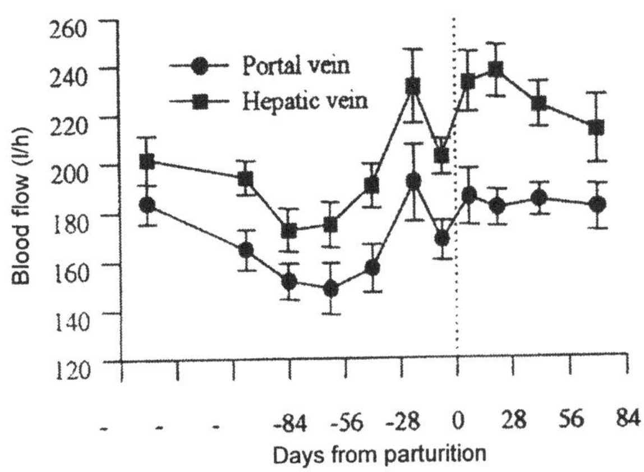

Fig. 1. Portal venous and hepatic venous blood flow (mean ± SE) of mature ewes (n = 7) during gestation and lactation.

With the exception of day -7, hepatic arterial blood flow in late pregnancy through mid-lactation (-21 through 42 d) was higher than in early- and mid-pregnancy ($P<0.05$).

Both hepatic and PDV oxygen consumption increased as pregnancy progressed. Hepatic and PDV oxygen consumption remained high through early lactation before decreasing in late lactation (Fig. 2).

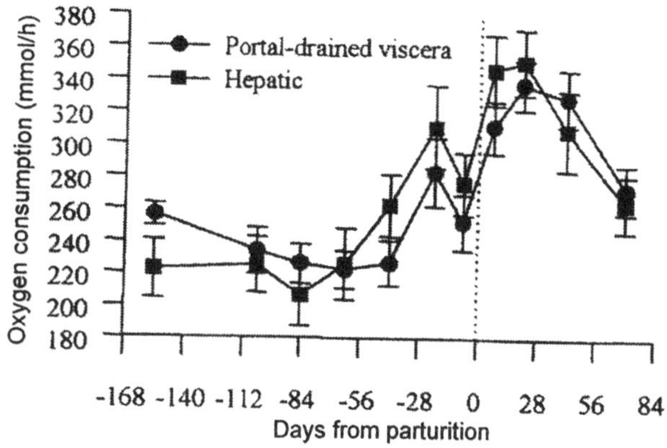

Fig. 2. Portal-drained viscera and hepatic oxygen consumption (mean ± SE) of mature ewes (n = 7) during gestation and lactation.

Net hepatic glucose release followed a pattern similar to that of hepatic oxygen consumption (Fig. 3). With the exception of -21 and 42 d, net PDV glucose release did not differ from zero (Fig. 3). Arterial plasma glucose concentration did not differ ($P = 0.64$) with time from parturition (3.69 mM).

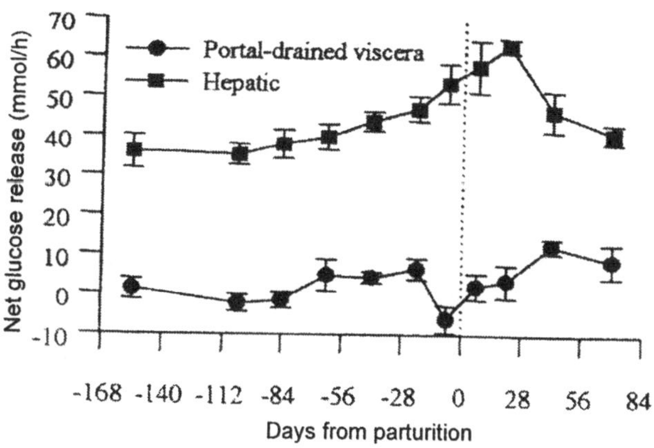

Fig. 3. Net portal-drained viscera and hepatic glucose release (mean ± SE) of mature ewes (n = 7) during gestation and lactation.

Discussion

Calculated hepatic arterial blood flow increased in late pregnancy independently of increases in portal venous blood flow. In late pregnancy, circulating levels of oestrogens and progesterone increase. The liver is the primary site of steroid metabolism in the ewe. Previous studies have demonstrated that hepatic arterial flow increases in the presence of elevated estradiol (Freetly & Nienaber, 1994; Rosenfeld *et al.*, 1976), suggesting that the increase in hepatic arterial flow in late pregnancy may facilitate steroid metabolism. Portal venous blood flow increases with increased feed intake (Freetly *et al.*, 1995), suggesting that the increase in portal venous blood flow after parturition is partially due to an increase in feed intake during lactation.

The transition between pregnancy and lactation represents a unique period. In late pregnancy the metabolic demands of uterine tissues and the mammary gland are increasing and after parturition metabolic demand of the mammary gland continues to increase while that of uterine tissues decreases. Energy requirements in the animal increase rapidly in late pregnancy (Moe & Tyrrell, 1971). Hepatic oxygen consumption follows a similar pattern, suggesting that as pregnancy progresses, metabolic demand on the liver increases. In the lactating animal, heat production increases as milk energy secretion increases (Flatt *et al.*, 1965). In the current study, hepatic oxygen consumption followed a pattern similar to that of the lactation curve. These data support the hypothesis that part of the increase in energy demand during pregnancy and lactation is due to an increase in hepatic energy metabolism.

In this study, net glucose release during pregnancy changed dynamically, and the pattern of release followed a pattern similar to fetal growth (Koong *et al.*, 1975). In late gestation, the gravid uterus has a net glucose uptake of approximately 16.8 mmol/h (Battaglia & Meschia, 1981), which is consistent with the 17.5 mmol/h increase in net hepatic glucose release observed in this study. The mammary gland requires glucose for synthesis of lactose, triglycerides and redox units. Net hepatic glucose release follows the same pattern as milk production during lactation. At peak lactation (21 d), net hepatic glucose production increases 21.1 mmol/h compared to the non-pregnant ewe. The lactose and glycerol secreted in the milk account for 91% of the increase in hepatic glucose production. These data suggest that hepatic glucose production is regulated to meet the glucose requirements of the body.

References

Battaglia F.C. & Meschia G. (1981). Foetal and placental metabolism: Their interrelationship and impact upon maternal metabolism. *Proc. Nutr. Soc.* 40:99-113.

Bell A.W. (1986). Efficiency of prenatal growth and the heat increment of pregnancy. Proceedings Cornell Nutrition Conference of Feed Manufacturers, Syracuse, NY, p 64.

Flatt W.C., Coppock C. E. & Moore L.A. (1965). Energy balance studies with lactating, non-pregnant dairy cows consuming rations with varying hay to grain ratios. The 3rd Symposium on Energy Metabolism. Troon, Scotland. *EAAP Publ. No.* 11:121-130.

Freetly H.C., Ferrell C.L., Jenkins T.G. & Goetsch A.L. (1995). Visceral oxygen consumption during chronic feed restriction and realimentation in sheep. *J. Anim. Sci.* 73:843-852.

Freetly H.C. & Nienaber J.A. (1994). Gut and liver oxygen consumption following chronic exposure to estradiol-17b in the mature ewe. The 13th Symposium on Energy Metabolism of Farm Animals, Mojacar, Spain, *EAAP Publ. No.* 76:93-96.

Huntington G.B. & Tyrrell H.F. (1985). Oxygen consumption by portal-drained viscera of cattle: Comparison of analytical methods and relationship to whole body oxygen consumption. *J. Dairy Sci.* 68:2727-2731.

Katz M.L. & Bergman E.N. (1969). Simultaneous measurements of hepatic and portal venous blood flow in the sheep and dog. *Am. J. Physiol.* 216:946-952.

Koong L.J., Garrett W.N. & Rattray P.V. (1975). A description of the dynamics of fetal growth in sheep. *J. Anim. Sci.* 41:1065-1068.

Miller P.S., Reis B.L., Calvert C.C., de Peters E.J. & Baldwin R.L. (1991). Patterns of nutrient uptake by the mammary glands of lactating dairy cows. *J. Dairy Sci.* 74:3791-3799.

Moe P.W. & Tyrrell H.F. (1971). Metabolizable energy requirements of pregnant dairy cows. *J. Dairy Sci.* 55:480-483.

Reynolds C.K., Huntington G.B., Tyrrell H.F. & Reynolds P.J. (1986). Splanchnic tissue and whole animal oxygen consumption by lactating Holstein cows. *J. Dairy Sci.* 69 (Suppl. 1):193 (Abstr.)

Rosenfeld C.R., Morriss F.H. Jr., Battaglia F.C., Makowski E.L. & Meschia G. (1976). Effect of estradiol-17b on blood flow to reproductive and non-reproductive tissues in pregnant ewes. *Am. J. Obstet. Gynaecol.* 124:618-629.

EFFECTS OF ABOMASAL WHEAT STARCH INFUSION ON SPLANCHNIC METABOLISM AND ENERGY BALANCE OF LACTATING DAIRY COWS

C.K. Reynolds, D.J. Humphries, S.B. Cammell, J.Benson, J.D. Sutton and D.E. Beever

Centre for Dairy Research, Department of Agriculture, The University of Reading, Earley Gate, Reading RG6 6AT, U.K.

Summary

In lactating dairy cows infused abomasally with 1200 g wheat starch per d for 8 days the recovery of wheat starch, as increased net glucose absorption was 39%, whilst the recovery (as increased liver glucose production) of starch infused into the rumen, was 26%. Starch infusion into either site decreased ammonia absorption, which probably reflects an increase in N use for rumen or hindgut fermentation. In separate cows, 14 day abomasal starch infusions decreased N digestion but tended to increase tissue N retention. Of the starch energy infused, 59% was recovered as metabolizable energy (ME), whilst 74% of the increased ME was retained as tissue energy.
Key words: Starch, glucose, lactation.

Introduction

It is well established that ME from cereals is used for production in ruminants more efficiently than ME from forages (Armsby, 1903). In addition, it is often assumed that starch ME is used more efficiently if it is digested to glucose in the small intestine rather than being digested in the rumen to volatile fatty acids, of which propionate reaching the liver can be converted to glucose. This assumption is supported by data from maintenance fed sheep infused with glucose (Armstrong *et al*, 1960), but there is little evidence that post-ruminal starch digestion is more efficient than ruminal starch digestion in lactating dairy cows (Reynolds *et al*, 1994). In addition, studies with non-lactating ruminants catheterized for measurements of nutrient absorption by the portal-drained viscera (PDV) have found that net PDV glucose absorption accounts for only a third of the glucose in starch digested in the small intestine. Thus the present experiment was conducted to determine the effect of site of starch digestion on nutrient absorption and metabolism by the PDV and liver and the effect of post-ruminal starch supplementation on ME supply and energy balance in lactating dairy cows.

Materials and Methods

Two experiments were conducted. In the first, 6 lactating Holstein-Friesian cows, prepared with rumen fistulae during gestation and catheterized for measurements of net PDV and liver nutrient flux at 3 weeks postpartum, were fed (in equal, hourly meals) a 50:50 mixture (dry matter [DM] basis) of forage (60% dehydrated lucerne and 40% grass silage) and pelleted concentrates. Intakes were set at 95% of *ad libitum* to minimise refusals during sampling. Cows averaged 211 days postpartum, 129 days of gestation and 686 kg body weight (BW) at the start of the experiment. Six hourly measurements of blood flow and nutrient flux across the PDV and liver and 4 daily measurements of milk production were obtained at the end of a 6 d infusion of water (9 l/d into the abomasum, 9 l/d into the rumen) and an 8 d infusion of wheat starch (1200 g in 9 l water/d) into either the rumen or the abomasum in a split-plot design with 5 week periods. During starch infusion water infusion (9 l/d) continued into the site not receiving starch. For abomasal infusions, tubing was inserted via the rumen and reticulo-omasal orifice. Experiment 2 was conducted using

4 rumen-cannulated cows in late lactation fed the same ration as used in experiment 1 except that the forage was fed at 8 h intervals and the concentrate was fed hourly. Offered DM was set at 90% of *ad libitum* intake before the experiment to minimise variation in basal DM intake between periods. Energy balance was measured using 6-d digestion trials and simultaneous measurements of respiratory exchange at the end of 14-d abomasal infusions of water (18 l/d) or wheat starch (1200 g in 18 l H_2O/d) in a switch-back design with three 6-week periods. Cows averaged 192 d postpartum, 109 d of gestation and 631 kg BW during the first measurement and 681 kg BW during the third measurement.

Results

Experiment 1. Dry matter intake was not significantly affected by starch infusion (Table 1). Milk yield was increased by starch infusion into the abomasum (P<0.07), but not the rumen, whilst milk protein (P<0.05) and fat (P<0.01) concentration were increased by starch infusion into the rumen, but not the abomasum. Treatments had no effect on blood flow or oxygen use

Table 1. Dry matter intake (DMI), milk yield and milk composition during the last 4 days of water (H_2O) or starch infusions (1200 g/d) into the rumen or abomasum.

	Rumen			Abomasum		
	H_2O	Starch	P<	H_2O	Starch	P<
DMI, kg/d	20.1	19.8	0.50	20.6	20.1	0.17
Milk yield, kg/d	24.8	24.5	0.23	23.8	24.3	0.07
Milk protein, g/kg	36.0	36.8	0.05	36.7	37.2	0.13
Milk fat, g/kg	41.1	42.9	0.01	42.5	42.7	0.59

Table 2. Blood plasma flow and net nutrient metabolism by splanchnic tissues of lactating dairy cows during water (H_2O) or starch infusions (1200 g/d) into the rumen or abomasum.

	Rumen			Abomasum		
	H_2O	Starch	P<	H_2O	Starch	P<
Plasma flow, l/h						
PDV	1241	1304	0.20	1287	1335	0.31
Liver	1520	1616	0.27	1572	1606	0.68
Arterial concentration, mM						
Glucose	3.72	3.82	0.05	3.71	3.80	0.07
Lactate	0.419	0.501	0.02	0.464	0.439	0.31
Ammonia	0.131	0.107	0.21	0.153	0.125	0.17
Oxygen	6.1	5.7	0.28	5.9	5.8	0.68
Packed cell volume, %	29.8	30.1	0.13	30.9	30.4	0.44
PDV flux, mmol/h						
Glucose	1	-1	0.90	-7	93	0.01
Lactate	153	168	0.13	162	167	0.58
Ammonia	618	524	0.03	645	577	0.06
Oxygen	-3728	-3568	0.60	-3613	-3563	0.87
Liver flux, mmol/h						
Glucose	630	696	0.10	644	629	0.64
Lactate	-82	-15	0.01	-43	-67	0.15
Ammonia	-665	-563	0.02	-657	-638	0.50
Oxygen	-3468	-3106	0.35	-3283	-3573	0.45

by PDV or liver (Table 2). Starch infusion into the abomasum increased net glucose absorption by 100 mmol/h, but did not affect liver glucose production. Starch infusion into the rumen increased liver glucose production by 66 mmol/h, but did not affect net PDV glucose flux. Thus both treatments increased arterial concentration of glucose ($P<0.07$), but via different metabolic responses. Net lactate absorption by the PDV was not affected by starch infusions, but liver removal of lactate was decreased by ruminal starch infusion ($P<0.01$) and thus arterial concentration of lactate was increased ($P<0.02$). Starch infusion into either site decreased ($P<0.06$) net PDV absorption of ammonia.

Experiment 2. Feed DM intakes remained fairly constant during the experiment and were not affected ($P>0.74$) by abomasal starch infusion. Refusals were less than 1% of DM offered except for one cow infused with starch which refused 13% of DM offered in period 3. On average, total DM intake was increased 0.9 kg/d by infusion of 1.05 kg starch DM/d ($P<0.09$; Table 3). Milk yield and composition were not affected by starch infusion. Starch infusions had no affect ($P>0.36$) on faecal starch concentration (22 g/kg) or output (117 g/d), thus all the starch infused was digested and total starch digested increased from 3.3 to 4.3 kg/d ($P<0.001$). Nitrogen intake and milk N output were not affected by starch infusion. Faecal N concentration was increased (30.8 vs. 33.0 g/kg, $P<0.03$), total N digested ($P<0.05$) and urinary N excretion ($P<0.02$) were decreased, whilst tissue N retention tended to increase ($P<0.18$) when starch was infused. This increase in tissue N retention was significant (15 vs. 59 g/d, $P<0.01$) when the measurement with large feed refusals was excluded from the data set. Starch infusion supplied 18.3 MJ/d, which increased average energy intake by 16.3 MJ/d (Table 4). This tended to increase both digested ($P<0.16$) and metabolizable ($P<0.11$) energy by 13 and 14 MJ/d, respectively. Methane energy was not affected by abomasal starch infusion, but urinary energy was decreased ($P<0.01$). Milk, heat and tissue energy were not significantly affected by starch infusion, but energy balance was increased ($P<0.07$) due to an increase in tissue energy retention ($P<0.27$) which accounted for 74% of the increase in ME. A slight increase in heat energy accounted for a further 22% of the ME from starch infusion.

Table 3. Dry matter intake (DMI), milk yield and nitrogen balance (g/d) of lactating dairy cows during the last week of 14 day abomasal infusions of water (H_2O) or starch (1200 g/d).

	H_2O	Starch	SEM	P<
DMI, kg/d [1]	18.0	18.9	0.3	0.09
Milk yield, kg/d	21.8	21.4	1.0	0.78
Nitrogen balance				
Intake	529	526	7	0.82
Digested	370	352	5	0.05
Urine	224	184	7	0.02
Milk	127	127	5	0.99
Tissue	19	41	10	0.18

[1] Includes 1.048 ± 0.004 kg starch dry matter.

Discussion

The recovery of glucose from starch infused into the abomasum as net glucose absorption by PDV was 39%, which is similar to previous measurements in non-lactating animals using shorter adaptation periods and maize starch infusion. This low recovery may in part reflect an increase in PDV use of arterial glucose. In the present study we used wheat starch to maximise digestibility of infused starch and in these cows, later in lactation, the recovery of maize starch infused into the abomasum as increased net PDV glucose absorption was 25%.

Table 4. Energy metabolism (MJ/d) of lactating dairy cows during the last week of 14 day abomasal infusions of water (H_2O) or starch (1200 g/d).

	H_2O	Starch	SEM	P<
Feed energy intake	344	342	2	0.80
Starch energy infused	0	18.3	0.05	-
Digested	245	258	5	0.16
Methane	24.2	24.4	0.4	0.85
Urine	11.6	9.9	0.3	0.01
Metabolizable	210	224	5	0.11
Heat	128	131	2	0.39
Milk	72.1	72.6	3.6	0.94
Tissue	9.8	20.1	5.6	0.27
Energy balance	81.9	92.7	3.1	0.07

When starch was infused into the rumen the decrease in liver lactate removal and increase in liver glucose production may have been attributable to increased propionate absorption. The recovery of starch infused, as increased liver glucose production, was 26%. The decrease in apparent N digestion, when starch was infused abomasally, likely reflects increased starch fermentation in the hindgut (Ørskov, 1986) and use of N for microbial protein synthesis may explain the decrease in PDV ammonia absorption when starch was infused into either site in experiment 1. Of the starch energy infused, 71 and 77% were recovered as digested energy and ME, respectively. The infused energy not recovered as digested energy in part reflects the slight decrease in feed energy intake, whilst the remaining energy not recovered may reflect losses attributable to fermentation in the hindgut. The increase in ME was greater than the increase in digested energy due to the decrease in urine energy output which accompanied decreased urine N output. This decrease in urine N output was due to both a decrease in N digestion (45%) and a numerical increase in tissue N retention (55%). Abomasal starch infusion had only a small, non-significant effect on body heat production and there was little effect of abomasal starch infusion on milk energy, thus the increase in energy balance measured was a consequence of increased tissue energy. Of the increase in tissue energy due to abomasal starch infusion, 68% was deposited in body fat. As the recovery of infused starch as net glucose absorption was relatively low, a part of the increase in tissue energy deposition may have occurred in PDV adipose. In terms of energy balance, the utilization of total and metabolizable energy from abomasally infused starch was highly efficient (59 and 77%, respectively).

Acknowledgements

Funded in part by the U.K. Ministry of Agriculture, Fisheries and Food.

References

Armsby H.P. (1903). The Principles of Animal Nutrition. John Wiley and Sons, New York, NY.

Armstrong D.A., Blaxter K.L. & Graham N. McC. (1960). Fat synthesis from glucose by sheep. *Proc. Nutr. Soc.* 19:xxxi-xxxii.

Reynolds C.K., Harmon D.L. & Cecava M.J. (1994). Absorption and delivery of nutrients for milk protein synthesis. *J. Dairy Sci.* 77:2787-2808.

Ørskov E.R. (1986). Starch digestion and utilization in ruminants. *J. Anim. Sci.* 63:1624-1633.

FATTY ACID OXIDATION IN SKELETAL MUSCLE MITOCHONDRIA FROM NEWBORN PIGS

I. Schmidt, M. Fillault, J-C. Hulin and P. Herpin

INRA, Station de Recherches Porcines, 35 590 Saint Gilles, France

Summary

Changes in respiration rate, mitochondrial mass, fatty acid oxidation and carnitine palmitoyltransferase I (CPT I) activity were followed in intermyofibrillar mitochondria isolated from pig *longissimus dorsi* and *rhomboideus* muscles between birth and 5 days of life. Mitochondrial protein mass increased by 30% during the first 5 days of life ($P<0.05$). Mitochondria had high respiration rates, respiratory control ratios, fatty acid oxidation potentials and CPT I activities from birth. $C_{16:0}$, $C_{18:1}$ and $C_{18:2}$ were oxidized at a higher rate than $C_{18:0}$ (-37%) and C_8 (-55%). The sensitivity of CPT I to malonyl-CoA decreased by 65% ($P<0.05$) and malonyl-CoA level fell by 33% ($P<0.05$), in both muscles, with age. In conclusion, isolated intermyofibrillar mitochondria are functional and able to oxidize fatty acids at birth, but the rise in mitochondrial protein mass and the lowering of the inhibition of CPT I activity by malonyl-CoA are probably involved, at least in part, in the postnatal development of fatty acid oxidation in pig muscle.

Key words: Piglet, muscle, age, mitochondria, carnitine palmitoyltransferase I, malonyl-CoA.

Introduction

At birth, the newborn pig experiences a dramatic and sudden 15 to 20°C decrease in its thermal environment. Since it has no brown adipose tissue, it relies essentially on muscular shivering thermogenesis for thermoregulatory purposes (Herpin & Le Dividich, 1995). Body and colostral carbohydrate are the main energy substrates used by the piglet early after birth, but lipid utilisation develops progressively during the first postnatal days. As part of an ongoing program on the effect of age and cold exposure on skeletal muscle energy metabolism in the piglet, this study was performed to determine whether the low oxidation rates of fatty acids at birth was due to a limited quantity of mitochondria or to an immaturity of mitochondrial oxidative pathways. In this paper, the amount of mitochondrial proteins, respiration rates, fatty acid oxidation rates, and CPT I activity and regulation in intermyofibrillar mitochondria isolated from a slow-oxidative muscle, (*rhomboideus* RH), and a fast-glycolytic muscle (*longissimus dorsi*; LD), are compared in newborn and 5 d old piglets.

Materials and Methods

At birth (n=10) and 5 days of age (n=10), intermyofibrillar mitochondria were isolated from fresh samples of LD and RH muscles from Pietrain x Large White crossbred piglets. Mitochondrial protein mass was determined by the Biuret method. Respiratory control ratio (RCR) and fatty acid oxidation rates were measured polarographically. Malate, pyruvate and ADP were added successively to initiate state IV and state III respiration and to enable calculation of the RCR (state III/state IV). Fatty acid oxidation potential was measured in the presence of CoASH, ATP, GDP and malate, which are co-factors necessary to optimize the successive steps of fatty acid utilization and, under uncoupled conditions, with 100 mM 2,4-dinitrophenol, in order to remove the control exerted by the respiration chain upon the overall oxidative process. CPT I activity and sensitivity to malonyl-CoA inhibition (IC_{50}, concentration of malonyl-CoA required for 50% inhibition of the enzyme activity) were assayed on fresh isolated mitochondria according to Herbin *et al.* (1987). Malonyl-CoA concentration was determined by reversed-phase high performance liquid chromatography (HPLC) of muscle homogenates (King *et al.*, 1988).

Results

There was no effect of muscle type on yield of mitochondrial protein and RCR ratio (Figure 1). At 5 days, state III respiration rate was 26% ($P<0.05$) higher in RH than in LD muscle. The amount of isolated mitochondria increased by 30% in LD and RH muscles between birth and 5 days ($P<0.05$). The RCR ratio was close to 4 in both muscles and did not change with age, whereas respiration rate decreased by 24% ($P<0.05$) in LD muscle. These results indicate that mitochondria were well coupled and functional from birth.

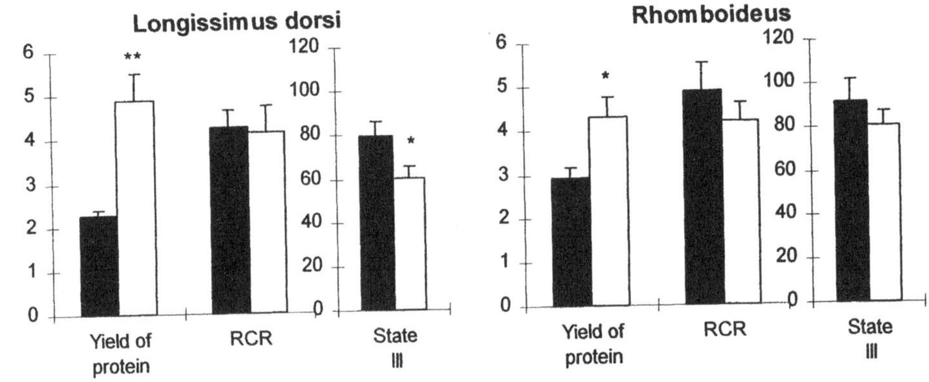

Fig. 1. Yield of mitochondrial protein, respiratory control ratio (RCR) and state III respiration in intermyofibrillar mitochondria from *longissimus dorsi* and *rhomboideus* muscles at birth (■) and 5 d (□) of life. Means±SEM (n=10). The yield of mitochondrial protein is expressed in mg of mitochondrial proteins/g muscle and state III respiration in no. of atoms oxygen/min/mg mitochondrial protein. Age effect : *, P<0.05 ; ***, P<0.001.

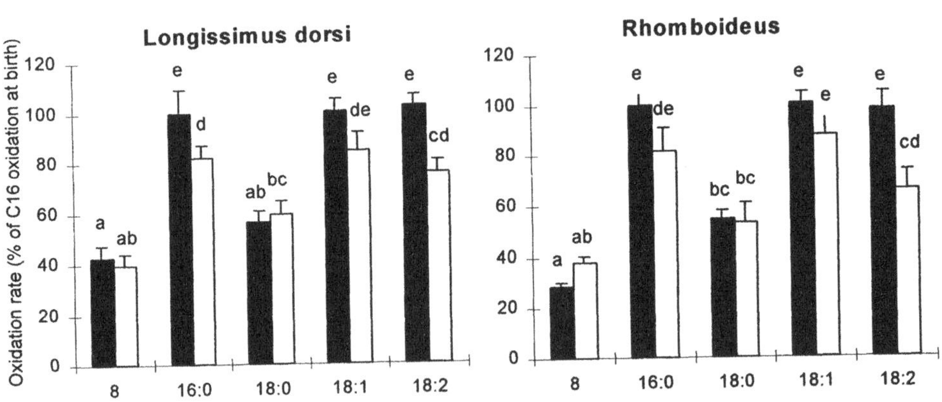

Fig. 2. Fatty acid oxidation rate in mitochondria from *longissimus dorsi* and *rhomboideus* muscle at birth (■) and 5 d (□) of life. Means±SEM (n=6). Effect of fatty acid chain length and saturation : values with different superscript letters are significantly different (P<0.05).

Mitochondrial oxidation of fatty acids of different carbon chain length and degrees of saturation was determined by measuring oxygen consumption under uncoupled conditions. There was no difference between LD and RH muscles (Figure 2). At birth, $C_{18:1}$, $C_{16:0}$ and $C_{18:2}$were oxidized more rapidly (P<0.05) than $C_{18:0}$ and more rapidly than $C_{8:0}$. This medium-chain fatty acid was the most poorly oxidized of the five fatty acid studied, its oxidation rate being only 37% of that of $C_{16:0}$. At 5 days, the same order of oxidation existed but $C_{16:0}$ and $C_{18:2}$ were 18 and 29% less rapidly oxidized, respectively, than at birth (P<0.05). The highest rate of oxidation was obtained when $C_{18:1}$ was added to the medium.

As shown in Table 1, there was no effect of muscle type on CPT I activity, IC_{50} and malonyl-CoA level. CPT I activity remained constant with age. From birth to 5 days of life, the concentration of malonyl-CoA required for 50% inhibition of CPT I (IC_{50}) increased by about 90% (P<0.05) in both muscles. Conversely, the concentration of malonyl-CoA was high at birth and decreased by 27 (P<0.01) and 42% (P<0.01) within 5 days in LD and RH muscle, respectively.

Table 1. Carnitine palmitoyltransferase I activity, sensitivity to malonyl-CoA inhibition (IC_{50}) and malonyl-CoA levels in *longissimus dorsi* and *rhomboideus* muscles at birth and 5 d of age.

	Longissimus dorsi		Rhomboideus	
	Birth	5 days	Birth	5 days
CPT I activity	1.51±0.15	1.57±0.11	1.67±0.14	1.58±0.12
IC_{50}	2.92±0.49	4.76±0.82*	3.32±0.60	5.48±1.10
Malonyl-CoA level	11.6±0.7	7.8±0.8*	13.1±0.9	8.8±0.6*

Means±SEM ; n=10 for CPT I activity and n=5 for IC_{50} and malonyl-CoA level. CPT I activity is expressed in nmoles of palmitoylcarnitine formed/min/mg mitochondrial protein, IC_{50} in mM of malonyl-CoA and malonyl-CoA levels in nmoles/g muscle. Effect of age : *, P<0.05.

Discussion

Our data show an increase in the mass of intermyofibrillar mitochondrial proteins in muscle between birth and 5 days of age. Similar results are obtained when the amount of subsarcolemmal mitochondria and the yield of recovery of mitochondria extraction are taken into account. Muscle mitochondrial mass increased by 85% with age and was 21% higher in RH than in LD muscle at 5 days of life (our unpublished observations). This is in good agreement with studies showing a considerable rise in mitochondrial volume density within 3 to 5 days after birth in both oxidative and glycolytic muscles of the pig (Handle & Stickland, 1987), and a rapid increase in mitochondrial number per unit tissue volume between birth and 2 weeks postpartum in pig liver (Mersman *et al.*, 1972). However, the nature of the postnatal rise in mitochondrial mass, i.e. an increase in mitochondrial size or of the surface area of the inner mitochondrial membrane and (or) an enhanced mitochondrial division and proliferation, cannot be assessed directly from the present study. Furthermore, our data show that there is essentially no change in the efficiency of ATP synthesis (RCR) in either muscle within 5 days indicating that pig muscle mitochondria are functional from birth. On the contrary, pig liver mitochondria exhibited functional differences within the first 2 days of life (Mersman *et al.*, 1972).

Oxidation of various fatty acids well-represented in sow colostrum was measured in our mitochondrial preparations and compared with that of octanoic acid. Firstly, the hierarchy observed in the oxidation rate of those fatty acids, i.e. $C_{18:1} \geq C_{18:2} \geq C_{16:0} > C_{18:0}$ was fairly consistent with the results obtained *in vivo* by Boyd *et al.* (1982). Oleic, linoleic and palmitic acids, which constitute about 80% of the fatty acids found in sow colostrum are readily oxidized by the newborn pig and its skeletal muscle mitochondria. This confirms that measuring fatty acid oxidation on isolated mitochondria might be an adequate procedure to get information on the utilisation of fatty acids for energy purposes. Secondly, our results show that octanoic acid is much less rapidly oxidized than palmitic acid, which suggests that octanoate oxidation might be limiting in muscle mitochondria. This is confirmed by the fact that when these measurements were performed on liver mitochondria from the same piglets no difference in oxidation rate was found between either of these fatty acids (data not shown). Thirdly, fatty acid oxidation and CPT I activity are high, and probably not limiting at birth in isolated IM mitochondria. However, there was a rise in IC_{50} value between birth and 5 days of life in LD and RH muscles. Interestingly, this was associated with a decrease in tissue levels of malonyl-CoA.

Similarly, in rat liver, an increased IC_{50} value has been observed between foetal and newborn animals (Peñas & Benito, 1986). Since the intracellular water space in skeletal muscle is approximately 0.8 ml/g wet weight, it follows that the concentration of malonyl-CoA in skeletal muscle ranges between approximately 8 to 16 mM. These concentrations exceed the IC_{50} of IM mitochondria in pig skeletal muscle (3-6 mM) which strengthens the biological significance of this result. If the response of CPT I to malonyl-CoA in the intact muscle is similar to the response obtained on isolated mitochondria, it would be strongly inhibited by malonyl-CoA as previously shown by Awan & Saggerson (1993) on myocytes.

It is unclear if muscle CPT I is exposed *in vivo* to the concentrations of malonyl-CoA calculated from the above-mentioned tissue content, but our results suggest that malonyl-CoA is present in skeletal muscle from the newborn pig and that its inhibitory effect upon CPT I activity is weakened during the early neonatal period thereby promoting the enhancement of fatty acid utilization. Together with the proliferation of skeletal muscle mitochondria, this mechanism is in good agreement with, and is likely to contribute to, the postnatal enhancement of lipid utilisation by the newborn pig to cover its energy needs. The mechanisms underlying the postnatal changes in CPT I sensitivity to malonyl-CoA cannot, however, be determined from the present study.

References

Awan M.M. & Saggerson E.D. (1993). Malonyl-CoA metabolism in cardiac myocytes and its relevance to the control of fatty acid oxidation. *Biochem. J.* 295:61-66.

Handle S.E. & Stickland N.C. (1987). The effects of low birthweight on the ultrastructural development of two myofibre types in the pig. *J. Anat.* 150:129-143.

Herbin C., Pégorier J.P., Duée P.H., Kohl C. & Girard J. (1987). Regulation of fatty acid oxidation in isolated hepatocytes and liver mitochondria from newborn rabbits. *Eur. J. Biochem.* 165:201-207.

Herpin P. & Le Dividich J. (1995). Thermoregulation and the environment. In: Varley M.A. (Ed) The neonatal pig. Development and survival. CAB International, Wallingford, U.K. pp. 57-95.

King M.T., Reiss P.D. & Cornell N.W. (1988). Determination of short-chain co-enzyme A compounds by reversed phase high performance liquid chromatography. *Methods Enzymol.* 166:70-79.

Peñas M. & Benito M. (1986). Regulation of carnitine palmitoyltransferase activity in the liver and brown adipose tissue in the newborn rat: effect of starvation and hypothermia. *Biochem. Biophys. Res. Commun.* 135:589-536.

BROWN ADIPOSE TISSUE THERMOGENESIS AND ONTOGENY IN *BOS INDICUS* AND *BOS TAURUS* CATTLE

G.E. Carstens[1], E.H. McPhail[1], S. Kahl[2], and S.B. Smith[1]

[1]Department of Animal Science, Texas A & M University, College Station, TX, USA;
[2]U.S. Department of Agriculture, Agricultural Research Service, Beltsville, MD, USA

Summary

Newborn *Bos indicus* calves are less cold tolerant, and appear to have less active brown adipose tissue (BAT) thermogenic activity than *Bos taurus* calves. This study sought to determine if differences between Angus and Brahman exist in BAT composition, uncoupling protein (UCP) gene expression and iodothyronine 5' deiodinase (5'D) activity in newborn calves (Experiment 1) and in BAT composition and 5'D activity during late foetal development (Experiment 2). In experiment 1, heat production (HP) was measured before and after norepinephrine (NE) infusion to assess BAT thermogenic activity *in vivo* in newborn Angus and Brahman calves. NE-induced peak metabolic rates were 19% lower ($P<0.01$) in Brahman than in Angus calves. No differences were found in BAT composition or UCP mRNA but BAT type I and II 5'D activities were higher in Brahman than Angus calves. In experiment 2, Angus and Brahman foetuses were collected at 6, 14, 24, 48 and 96 d prior to expected parturition (n = 3). Perirenal BAT from foetuses and newborn calves was weighed and analysed for lipid, DNA and 5'D activities. Brahman foetuses tended to have more perirenal BAT at 180 d of gestation than Angus foetuses, but this difference diminished as foetal age increased. Angus foetuses had a higher BAT lipid content than Brahman foetuses throughout late foetal development, but no differences were found in BAT DNA. Type I and II 5'D activities increased quadratically as foetal age increased, peaked at day 230 of gestation, and declined thereafter until term. No breed differences were found in type I and II 5'D activities during foetal BAT development. The data demonstrate that lower BAT thermogenic rates in newborn Brahman calves are not due to less UCP gene expression or lower BAT 5'D activities.

Key words: Brown adipose tissue, thermogenesis, cattle, uncoupling protein.

Introduction

Newborn calves have well-developed thermogenic mechanisms, including shivering thermogenesis in muscle and non-shivering thermogenesis in BAT, to support thermal balance. Brown fat is capable of generating heat by uncoupling oxidative phosphorylation from mitochondrial respiration via an UCP located in BAT mitochondria.

Calf mortality related to cold intolerance appears to be genetically influenced. Mortality rates of 3, 3, 8 and 21% for calves with 0, 25, 50 and 75% *Bos indicus* inheritance, respectively, suggest that *Bos indicus* calves are less cold tolerant than *Bos taurus* calves (Josey *et al.*, 1987). Data from our laboratory suggest that calves born to Brahman dams have less thermogenically active BAT than calves born to Angus dams.

Adrenergic stimulation and local production of BAT T_3 are essential in regulating synthesiz of UCP in BAT. Production of T_3 in BAT is a critical mechanism regulating BAT thermogenic capacity in newborn calves. There are two major outer-ring 5'D that convert T_4 to T_3 in peripheral tissues. Wu *et al.* (1991) demonstrated that newborn lamb BAT contains Type I and II 5'D isozymes, whereas, Giralt *et al.* (1989) found only Type I 5'D activity in newborn calf BAT. Objectives of this study were to determine if breed differences in foetal development of BAT 5'D activities, and UCP gene expression would explain the apparent lower BAT thermogenic capacity of Brahman calves.

Materials and Methods

Expt I*:* Newborn Angus (n=7) and Brahman (n=10) calves were obtained prior to suckling and fitted (at 6 h old) with indwelling jugular catheters and placed in a thermostatted (37°C) water-immersion bath. Gaseous exchange data was obtained by face mask to determine HP by indirect calorimetry. After 20 min of stable measurements to determine thermoneutral metabolic rates (TMR), NE was infused for 10 min at 20 μg/kg/min. NE-induced peak metabolic rates (PMR) were derived from the maximal average of 3 consecutive 1 min HP measurements following NE infusion (Carstens *et al.*, 1997). At 12 h of age, calves were killed, perirenal fat (primary BAT depot) excised, and samples collected for analysis of lipid, protein, DNA, UCP mRNA, and 5'D activity. RNA was isolated from BAT, and UCP mRNA determined by slot blot analysis. Band intensity was quantified by laser densitometry and corrected for ribosomal RNA (Martin *et al.*, 1997). Deiodinating activity was determined as ^{125}I release from ^{125}I-labelled rT_3 using assay conditions for type I and II (Leonard & Rosenberg, 1980). Blood samples taken prior to slaughter were analysed for T_3 and T_4 by radioimmunoassay.

Expt II*:* Pregnant Angus and Brahman cows (n=36) were randomly assigned, within breed, to full-term or one of five foetal age groups (n=3). Foetuses were obtained by Caesarean section at 6, 14, 24, 48 and 96 d prior to expected parturition at 282 and 292 d for Angus and Brahman, respectively. Full-term calves were killed within 6 h of birth. Perirenal adipose tissue was excised, weighed and analysed for lipid, DNA and type I and II 5'D activity as described above. Data were analysed by least squares procedures. Models included breed type, sex of foetus or calf, and breed x foetal age nonlinear functions.

Results and Discussion

Weight-specific NE-induced PMR was 19% lower ($P<0.01$) in newborn Brahman calves than Angus calves (Table 1). Carstens *et al.* (1997) reported that Angus-, Brahman- and Tuli-sired calves from Brahman dams had lower weight-specific PMR than calves from Angus dams (213 vs. 264 J/kg/min). The results suggest that increased susceptibility of newborn Brahman calves to hypothermia during cold exposure may be due to less thermogenically active, or less functional, BAT.

Table 1. Effects of breed type on thermoneutral (TMR) and NE-induced peak metabolic rates (PMR), and BAT composition, uncoupling protein (UCP) gene expression and type I (5'D-I) and II (5'D-II) iodothyronine 5' deiodinase activity in newborn calves.

	Angus	Brahman	SE	P-value
TMR, J/kg/min	124	113	5.4	0.15
PMR, J/kg/min	268	218	13.1	0.01
Brown fat composition:				
lipid, mg/g	683	665	17	0.42
protein, mg/g	36.7	39.2	3.6	0.61
DNA, μg/g	921	1,090	188	0.50
mRNA UCP:28S x 100	23.4	31.6	7.8	0.32
5'D-I, nmol I/h/mg protein	0.60	0.94	0.12	0.04
5'D-II, pmol I/h/mg protein	1.89	4.85	0.51	0.001

Brahman calves had higher ($P<0.05$) plasma T_3 (521 vs. 339 ± 62 ng/dl) and T_4 (24.3 vs. 16.3 ± 2.6 μg/dl) than Angus calves. Plasma T_4 tended to be negatively correlated ($r=-0.41$; $P=0.10$) with NE-induced PMR, suggesting that elevated T_4 may suppress BAT thermogenesis.

Lipid, DNA and protein concentrations of BAT were similar in Brahman and Angus newborn calves (Table 1). No breed differences were found in the level of UCP gene expression in BAT but both type I and II 5'D activities were higher in Brahman than Angus calves. NE infusion appeared to increase 5'D activities. Catheters of 2 Brahman calves failed which prevented NE infusion but type I and II 5'D activities of these calves (0.19 ± 0.02 nmol I^-/h/mg protein and 0.50 ± 0.12 pmol/h/mg protein, respectively) were substantially lower than in NE-infused Brahman calves (Table 1).

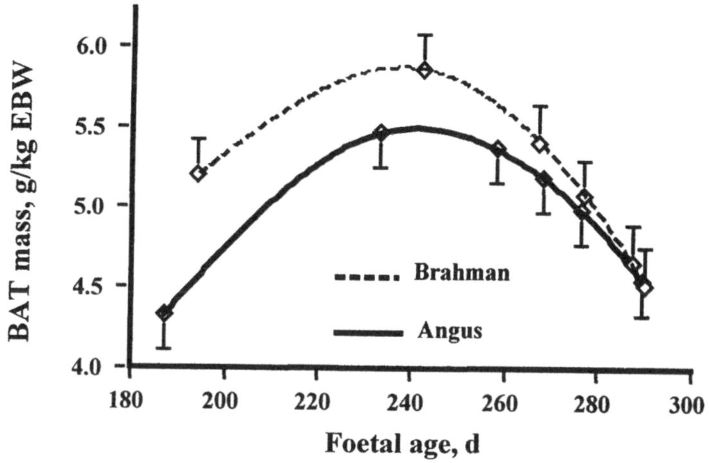

Fig. 1. Effect of breed on relative growth of perirenal BAT during foetal development (R^2=0.46).

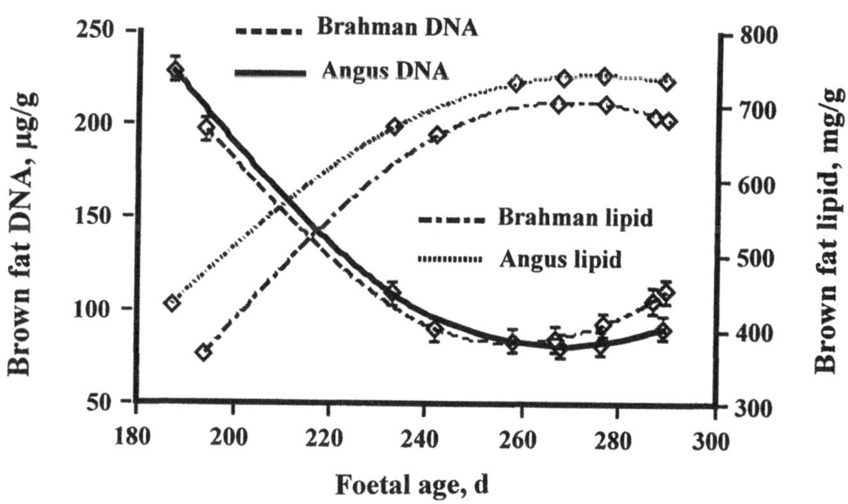

Fig. 2. Effect of breed on DNA and lipid content of perirenal BAT during foetal development. (R^2 =0.79 and 0 .92, respectively).

The growth of perirenal BAT relative to empty foetal weight (EBW) increased quadratically with foetal age, reaching maximal values at 234 and 241 d for Brahman and Angus foetuses (Figure 1). Brahman foetuses tended (P<0.10) to have more perirenal BAT at 200 d of gestation than Angus foetuses (5.4 vs. 4.8 g/kg EBW), but not at older foetal ages. This breed difference in relative BAT mass decreased with age (breed x foetal age; $P<0.04$), and BAT mass in Angus and Brahman foetuses was similar at the end of gestation. During foetal development, BAT lipid content increased and DNA content decreased quadratically with foetal age (Figure 2). Angus foetuses had higher ($P<0.05$) BAT lipid content than Brahman foetuses throughout the last trimester of gestation (518 vs. 420; 693 vs. 655 and 741 and 707 mg/g at 200, 240 and 280 d of gestation, respectively). In contrast, the DNA content of BAT was similar for Angus and Brahman foetuses throughout gestation.

Substantial type I and II 5'D activities were already present in foetal BAT at the start of the third trimester of gestation (Figure 3). Type I and II 5'D activities peaked at 230 d of gestation and declined thereafter. Giralt *et al.* (1989) examined changes in 5'D activity in bovine foetal BAT from day 100 of gestation to full term and found that type I BAT 5'D activity increased markedly from day 150, peaked at 220 to 230 d of gestation, and declined thereafter to term. Differences between breeds were not detected in type I or II 5'D activities during foetal BAT ontogenesis. Lower BAT thermogenic rates in newborn Brahman calves cannot be attributed to lower BAT 5'D activities or less UCP gene expression.

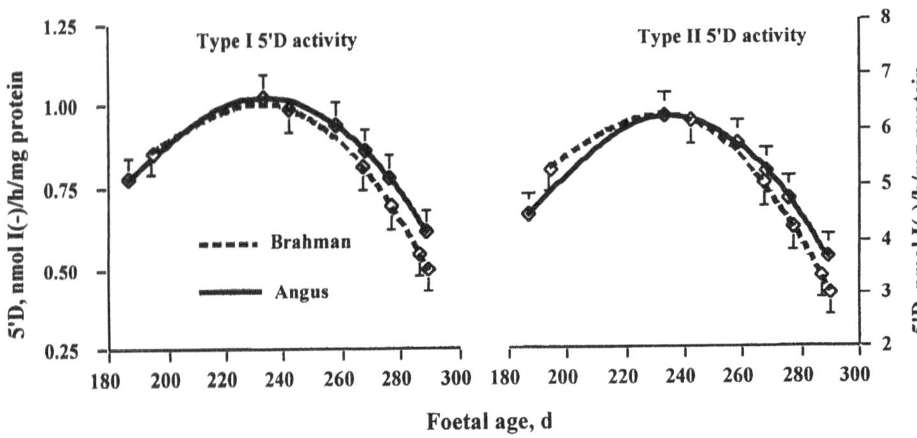

Fig. 3. Effect of breed on type I and II iodothyronine 5' deiodinase (5'D) activity in BAT during foetal development (R^2 were 0.43 and 0.42, respectively).

References

Carstens G.E., Mostyn P.M., Lammoglia M.A., Vann R.C., Apter R.C. & Randel R.D. (1997). Genotypic effects on norepinephrine-induced changes in thermogenesis, metabolic hormones, and metabolites in newborn calves. *J. Anim. Sci.* 75:1746-1755.

Giralt M., Casteilla L., Vinas O., Mampel T., Iglesias R., Robelin J. & Villarroya F. (1989). Iodothyronine 5'-deiodinase activity as an early event of prenatal brown-fat differentiation in bovine development. *Biochem. J.* 555-559.

Leonard J.L. & Rosenberg I.N. (1980). *Endocrinol.* 107:1376-1383.

Martin G.S., Carstens G.E., Taylor T.L., Sweatt C.R., Eli A.G. & Smith S.B. (1997). Prepartum protein restriction does not alter norepinephrine-induced thermogenesis or brown adipose tissue function in newborn calves. *J. Nutr.* (In press).

Wu S.Y., Kim J.K., Chopra I.J. & Fisher D.A. (1991). Postnatal changes in lambs of two pathways for thyroxine 5'-monodeiodination in brown fat. *Am. J. Physiol.* 261:E257.

EFFECTS OF DIETARY FATS (MEDIUM-CHAIN vs. LONG-CHAIN FATTY ACIDS) ON THE OXIDATIVE CAPACITY OF TISSUES IN THE PRERUMINANT CALF

C. Piot[1], J.H. Veerkamp[2], D. Durand[1], D. Bauchart[1] and J.F. Hocquette[1]

[1] INRA, Laboratoire Croissance et Métabolismes des Herbivores, Centre de Recherches Theix, 63122 St-Genès Champanelle, Clermont-Ferrand, France. [2] Department of Biochemistry, University of Nijmegen, P.O. Box 9101, 6500 HB Nijmegen, The Netherlands

Summary

The influence of dietary fatty acids on the oxidative capacity of liver, heart and two skeletal muscles was studied, *in vitro*, in two groups of five preruminant calves fed a milk replacer rich in tallow (TA) or in coconut oil (CO). Cytochrome oxidase (COX) and citrate synthase (CS) activities did not differ significantly between TA and CO calves. However, the COX/CS ratio was 1.7-fold higher in the liver of the TA calves ($P<0.05$) than in that of the CO calves. Activities of these two enzymes were higher in the heart than in other tissues. Oxidation rates of lauric, palmitic and oleic acids did not significantly differ between the two groups of calves whatever the tissues. The oxidative capacities of liver and heart muscle were higher than those of skeletal muscles. In addition, the oxidation rate of laurate, the major fatty acid of the CO diet, was higher than that of palmitate and oleate in the liver but not in other tissues.
Key words: Fatty acid, oxidative capacity, preruminant calves, liver, muscles.

Introduction

Energy metabolizm in muscles is of prime importance for anabolic processes such as protein accretion. In previous studies, coconut oil (CO), a natural source of medium-chain fatty acids (MCFA), was shown to allow increased nitrogen balance, and higher growth rate, than tallow (TA) when given in the milk diet of preruminant calves during the first month of life (Aurousseau *et al.*, 1984). For a better understanding of the biological mechanisms involved in this better nutrient utilization, the oxidative metabolizm of MCFA or long-chain fatty acids (LCFA) was studied in calves fed a milk replacer containing either TA or CO (224 g/kg of dietary DM) for 3 weeks. Possible underlying mechanisms were investigated by measuring the oxidation of the major fatty acids (FA) from CO or TA in homogenates of heart, liver or skeletal muscles. The oxidative capacity of tissues was also assessed by measuring mitochondrial enzymes.

Materials and Methods

Experiments were performed using two groups of 5 preruminant Holstein-Friesian male calves. The TA group was fed a conventional milk replacer providing LCFA where the saturated/unsaturated FA ratio equalled 1 while the CO group received the same diet rich in saturated MCFA. Animals were slaughtered at the same age (5 weeks) and at the same body weight (68 kg). Samples of liver, heart and of two muscles of the carcass (*rectus abdominis*, RA, *longissimus dorsi*, LD) were taken at the time of slaughter. Activities of citrate synthase (CS) and cytochrome oxidase (COX) were determined in tissue homogenates as described by Glatz *et al.* (1984). One unit of CS or COX activity per g of tissue wet weight represents 1 µmol CoA liberated per min and 1 µmol cytochrome c oxidized per min, respectively.

Total FA oxidation capacity of muscles was determined by measuring oxidation rates of [1-^{14}C]-palmitate, [1-^{14}C]-oleate or [1-^{14}C]-laurate in homogenates of fresh tissue according to the procedure of Veerkamp & Van Moerkerk (1985). Laurate (C12:0), palmitate (C16:0) and oleate (C18:1) were used as substrates since they were abundant either in CO diet (C12:0 = 42% of total

FA), or in TA diet (C18:1 = 40% of total FA) or in both diets (C16:0 = 22 and 13% of total FA for TA and CO respectively). Oxidation rates were expressed in nmol of FA oxidized to $^{14}CO_2$ and ^{14}C-labelled perchloric acid-soluble products per min and per g tissue wet weight or relative to COX activity.

Analysis of variance of the data was made using the GLM procedure of SAS (1987). For CS and COX activities and oxidation rates per FA (Table 1), the effects tested in the model included group (G) of calves (TA, CO), calf (C) tested within the treatment group and tissue (T). For the oxidation rates in the liver (Table 2), the tested effects included group (G) of calves (TA, CO), calf (C) tested within the treatment group and FA. The group factor was tested against calves within groups. The residual mean square was used as the error term for other effects.

Results and Discussion

Growth and body composition

Average daily gain did not significantly differ between the two groups of calves (0.93 ± 0.23 and 0.86 ± 0.10 kg per day for TA and CO groups respectively). In addition, the weights of liver, heart, RA and LD muscles were similar for TA and CO groups.

Citrate synthase and cytochrome oxidase activities

Whatever the diet, CS activity was significantly higher in the heart (33.6 units) than in the liver, LD and RA muscles (3.6, 6.1 and 4.8 units respectively, $P<0.0001$) but did not differ between liver and skeletal muscles (Table 1). Whatever the diet, COX activity was significantly higher in the heart (222 units) than in the liver (69.5 units), and higher in the liver than in LD and RA muscles (21.5 and 24.7 units respectively, $P<0.003$) (Table 1). These results confirmed that heart and liver have a higher oxidative capacity than skeletal muscles as previously shown in other species (Veerkamp & Van Moerkerk, 1985). The ratio of COX to CS activities also differed between tissues: it was highest in the liver (20.33), intermediate in heart (7.96) and lowest in RA and LD (5.91 and 3.94 respectively). This may indicate tissue specificity.

COX and CS activities did not significantly differ between the calves of the two groups. However, the ratio of COX to CS activities was significantly higher in the liver of the TA calves than in that of the CO calves (25.5 vs. 15.2 respectively). This may indicate an adaptation of the oxidative metabolizm of the liver to the diet.

Total oxidative capacity

Oxidation rates of three different FA (C12:0, C16:0 and C18:1) were measured in four tissues (heart, RA, LD and liver) from calves fed a milk replacer diet rich in TA or in CO.

Effect of tissue

Irrespective of whatever the FA used (Table 1), total oxidation rates of FA were higher in the liver and in the heart than in skeletal muscles ($P<0.002$) which is in agreement with the higher specific metabolic activities of these tissues. In addition, oxidation rates of C12:0 and C16:0 were 2.2- and 1.5-fold higher respectively in the liver than in the heart ($P<0.0001$). These results are in accordance with previous studies in human and rat tissues (Veerkamp & Van Moerkerk, 1985).

Normalization of FA oxidation to mitochondrial marker enzymes allows comparisons among tissues. On average, for the three tested FA, rates of FA oxidation relative to COX activity were 5.7-fold higher ($P<0.002$) in the liver than in the heart due to higher activity of this enzyme in the heart. Values were 3-fold higher ($P<0.0001$) in the liver than in skeletal muscles due to a much higher mitochondrial content in the liver. However, values did not differ significantly between heart and skeletal muscles (Table 1).

Table 1. Citrate synthase (CS) and cytochrome oxidase (COX) activities and oxidation rates of laurate ($C_{12:0}$), palmitate ($C_{16:0}$) and oleate ($C_{18:1}$) in tissue homogenates from preruminant calves fed either tallow (TA) or coconut oil (CO).

Group	Tallow (n = 5)				Coconut oil (n = 5)					Effect of (P<)		
Tissue :	L	H	RA	LD	L	H	RA	LD	SEM	G	C	T
CS activity	3.45^{b}	28.20^{a}	4.71^{b}	4.80^{b}	3.77^{B}	39.00^{A}	4.90^{B}	7.38^{B}	3.118	0.28	0.11	0.001
COX activity	83.0^{b}	231.6^{a}	24.4^{c}	19.9^{c}	56.1^{B}	211.4^{A}	25.0^{B}	23.1^{B}	13.84	0.43	0.13	0.001
COX / CS	25.48^{a}	9.64^{b}	6.53^{bc}	4.46^{c}	15.18^{A}*	6.28^{B}	5.29^{B}	3.42^{B}	1.336	0.11	0.002	0.001
$C_{12:0}$ oxidation rate	432^{a}	187^{b}	28^{c}	26^{c}	363^{A}	176^{B}	27^{C}	21^{C}	42. 7	0.58	0.06	0.001
$C_{16:0}$ oxidation rate	284^{a}	167^{b}	31^{c}	24^{c}	268^{A}	190^{B}	19^{C}	25^{C}	21.4	0.96	0.06	0.001
$C_{18:1}$ oxidation rate	180^{a}	147^{a}	21^{b}	19^{b}	163^{A}	170^{A}	15^{B}	15^{B}	19.8	0.97	0.02	0.001
C_{12} oxidation rate / COX	5.01^{a}	0.86^{a}	1.62^{a}	2.39^{a}	8.14^{A}	0.82^{B}	1.09^{B}	0.96^{B}	1.438	0.77	0.34	0.001
C_{16} oxidation rate / COX	3.56^{a}	0.85^{b}	1.53^{b}	1.80^{b}	5.47^{A}	0.90^{B}	0.77^{B}	1.22^{B}	0.627	0.66	0.76	0.001
$C_{18:1}$ oxidation rate / COX	2.09^{a}	0.69^{a}	0.99^{a}	1.11^{a}	3.87^{A}	0.80^{B}	0.62^{B}	0.67^{B}	0.665	0.62	0.30	0.005

Citrate synthase (CS) and cytochrome oxidase (COX) activities were measured in whole homogenates; results are expressed in µmol CoA liberated per min and per g tissue wet weight and in µmol cytochrome c oxidized per min and per g tissue wet weight, respectively. Oxidation rates of fatty acids are expressed in nmol of fatty acid oxidized per min and per g tissue wet weight or relative to the COX activity.

L, liver; H, heart ; RA, *rectus abdominis* ; LD, *longissimus dorsi* ; G, C, T, significant effects of group (G) tested against calves within group (TA or CO) of calf (C) and of tissue (T).

$^{A, B, C, a, b, c}$ Means within a group (TA or CO) with different superscript letters were significantly different ($P<0.05$).

* $P<0.05$, difference between TA and CO calves per tissue.

Table 2. Oxidation rates of laurate, palmitate and oleate in liver homogenates from preruminant calves fed tallow (TA) or coconut oil (CO).

Group	Tallow (n = 5)			Coconut oil (n = 5)				Significance of effect of (P<)		
FA :	C12:0	C16:0	C18:1	C12:0	C16:0	C18:1	SEM	G	C	FA
Oxidation rate	432^{a}	284^{ab}	180^{b}	363^{A}	268^{AB}	163^{B}	52.7	0.62	0.06	0.002
Oxid. rate/COX	5.01^{a}	3.56^{a}	2.09^{a}	8.14^{A}	5.47^{AB}	3.87^{B}	1.136	0.32	0.001	0.02

For details of units, abbreviations and symbols, see Table 1.

Effect of fatty acid

Irrespective of the diet, oxidation rates expressed either as absolute values or relative to COX activity in the heart and in skeletal muscles did not differ significantly between C12:0, C16:0 or C18:1 (Table 1). However, the oxidation rate of C12:0 in the liver was 1.4- and 2.3-fold higher than those of C16:0 and C18:1 respectively ($P<0.05$, Table 2). Similar differences were observed when results were expressed per unit of COX activity (Table 2), but the difference was significant only between C12:0 and C18:1 (x 2.2 ; $P<0.05$).

Effect of diet

Irrespective of the tissue and the FA used in the experiment, oxidation rates did not significantly differ between the animals of the 2 groups (Tables 1 and 2). However, data in the liver were lower for the CO diet when expressed in nmol of FA oxidized per min and per g tissue wet weight but higher when expressed per unit of COX activity (Table 2).

Conclusions

Our results show that changes in dietary FA did not affect oxidation rates of C12:0, C16:0 and C18:1 either in the liver or in muscles of preruminant calves. Nevertheless, large differences in the ability of the liver to oxidize FA were shown since MCFA (C12:0) were better catabolized than mono-unsaturated LCFA (C18:1). Consequently, it can be hypothesized that the amount of FA oxidized in the liver *in vivo* could be enhanced with the CO diet since the laurate content of the CO diet was 15-fold higher than that of the TA diet. These specificities could enhance the production of ketone bodies by the liver and their subsequent catabolism in muscles.

References

Aurousseau B., Thivend P. & Vermorel M. (1984). Influence du remplacement d'une partie du suif d'un aliment d'allaitement par de la tricaproïne ou de la tricapryline en association à de l'huile de coprah sur la croissance du jeune veau préruminant. *Ann. Zootech.* 33:219-234.

Glatz J.F.C., Jacobs A.E.M. & Veerkamp J.H. (1984). Fatty acid oxidation in human and rat heart. Comparison of cell-free and cellular systems. *Biochim. Biophys. Acta.* 794:454-465.

Statistical Analysis Systems (1987). *SAS/STAT®* Guide for personal computers. Version 6, SAS Institute Inc., Cary, NC.

Veerkamp J.H. & Van Moerkerk H.T.B. (1985). Effect of various agents and conditions on palmitate oxidation by homogenates of rat liver and rat and human skeletal muscle. *Int. J. Biochem.* 11:1163-1169.

INFLUENCE OF ENERGY DENSITY AND METABOLIZABLE ENERGY INTAKE ON SUBSTRATE METABOLISM BY RUMINAL EPITHELIAL CELLS ISOLATED FROM SHEEP

R.L. Baldwin, VI and K.R. McLeod

Nutrient Conservation and Metabolism Laboratory, LPSI, ARS, U.S. Department of Agriculture Beltsville, MD 20705, USA

Summary

Twenty-eight lambs (20.1±0.3 kg BW) were fed diets containing either 75% forage or 75% concentrate at either 0.416 or 0.757 MJ ME/kg $BW^{0.75}$ per d for 52 d. Isolated ruminal epithelial cells were incubated in the presence of multiple concentrations of oxidizable substrates. Oxidation rates of butyrate and propionate were greater in cells isolated from concentrate fed lambs as compared to those fed forage, while rates of acetate oxidation were enhanced with increasing ME intake. Changes in oxidation rates did not affect the overall predicted Michaelis-Menten parameter estimates. The estimated K_{ox} values are consistent with the contention that VFA are the primary oxidizable fuels used by ruminal epithelial cells. In conclusion, while subtle changes in VFA oxidation do occur, overall oxidative capacity of ruminal epithelial cells is largely unaffected by diet or ME intake.
Key words: Dietary energy density, ME intake, rumen epithelium, sheep.

Introduction

Effects of dietary manipulation on ruminal epithelial metabolism has received minimal attention despite its importance to the energy metabolism of the animal. Previous studies *in vitro* focusing on ruminal epithelial metabolism have been conducted using tissue pieces (Harmon, 1986). However, ruminal epithelial morphology is influenced to a large degree by alterations in luminal environment (Weigand *et al.*, 1975). Therefore, studies using papillae pieces to compare the effects of forage and concentrate diets on ruminal epithelial metabolism may be confounded by changes in morphology. Furthermore, few studies have investigated the effect of substrate concentration on substrate metabolism within ruminal epithelium. The objectives of this experiment were to determine metabolic characteristics of ruminal epithelium using an isolated ruminal epithelial cell system and to assess the sensitivity of these characteristics to ME intake and dietary energy density.

Materials and Methods

Lambs (n=28, 20.1±0.3 kg BW) were assigned randomly to a factorial arrangement of treatments including: pelleted diets containing either 75% orchard grass hay and 25% concentrate (FORG; 9.56 MJ ME/kg DM or 75% concentrate and 25% orchard grass hay (CONC; 11.74 MJ ME/kg DM fed, once daily, at either 0.42 or 0.76 MJ ME/kg $BW^{0.75}$ per d which approximated to 1 and 2 times maintenance, respectively. Diets were formulated to provide an equal N to ME ratio. During the 52-d feeding period, lambs were weighed twice-weekly and the quantity of feed offered was adjusted weekly.

Following slaughter, rumen epithelium from the ventral cranial region was removed and epithelial cells isolated and incubated as described by Baldwin & Jesse (1996) with ^{14}C substrates added to produce final concentrations of 0 to 50 mM butyrate, propionate, acetate, glucose, glutamine and glutamate to assess dose response kinetic parameters.

Estimates of V_{max} (maximal rate of substrate oxidation) and K_{ox} (concentration of substrate at which half maximal oxidation rate is achieved) were determined. The equation used was $Y = V_{max}$ (conc/ (K_{ox}+conc)) in which Y= metabolite oxidized to CO_2 (nmoles / 90 min) and conc= initial metabolite concentration. Parameter estimates determined for individual lambs with $r^2 > 0.90$ (n=5/treatment) were used to compare the effect of energy density of diet and ME intake. Oxidation and kinetic parameter data were analyzed with diet, intake and diet×intake and time of slaughter included in the model.

Results

Ruminal pH of CONC lambs was lower ($P<0.05$) compared with FORG lambs (Table 1). Likewise, ruminal pH tended to decrease ($P<0.10$) with increasing ME intake. CONC lambs had lower ($P<0.05$) ruminal concentrations of acetate, propionate, iso-valerate, and total VFA compared with FORG. Conversely, acetate, propionate, n-butyrate, n-valerate, and total VFA concentrations were higher ($P<0.05$) at high ME intake.

Table 1. Effects of dietary energy density and ME intake on ruminal VFA concentrations.

Concentration, mM	Forage		Concentrate			Significance[a]		
	Low intake	High intake	Low intake	High intake	SEM	Diet	Intake	D×I
pH	6.67	6.29	6.25	6.13	0.08	0.04	0.084	---
acetate	32.47	41.13	22.41	35.77	1.89	0.02	0.002	---
propionate	9.82	15.73	5.75	13.45	0.95	0.05	0.001	---
n-butyrate	6.09	8.11	5.03	6.82	0.40	---	0.017	---
iso-butyrate	1.14	1.26	0.95	1.14	0.07	---	---	---
n-valerate	1.19	1.53	0.63	1.69	0.16	---	0.030	---
iso-valerate	1.79	1.99	1.34	1.66	0.09	0.04	---	---
Total VFA	52.48	69.69	36.10	60.51	2.79	0.00	0.001	---

[a] Probability of a larger F statistic; ---, P>0.10.

Rates of glucose, glutamate, glutamine and acetate oxidation to CO_2 were unaffected by energy density of the diet (Table 2). Conversely, rates of substrate oxidation to CO_2 were higher ($P<0.05$) for propionate, at 10 and 50 mM, or tended to be higher ($P<0.08$) for butyrate, from 1 to 10 mM in cells isolated from lambs fed CONC. Rates of glucose, propionate, and butyrate oxidation to CO_2 were unaffected by ME intake. Ruminal epithelial cells from lambs fed at high ME intake tended ($P<0.09$) to oxidize glutamate (10 mM) and glutamine (1 mM) at higher rates than cells isolated from lambs fed a low ME intake. Rates of acetate oxidation to CO_2 were higher at 1 and 5 mM and tended to be higher at 50 mM in cells isolated from lambs fed a high ME intake.

There were no significant effects of energy density of the diet or ME intake on the predicted V_{max} and K_{ox} for any of the substrates (Table 3). For glutamate and glutamine, the predicted V_{max} over-estimated the observed data (Table 2) thus, V_{max} was not attained. Across treatments, mean predicted K_{ox} values were 0.40, 0.55, 2.70, 6.42, 9.85, and 16.81 for butyrate, propionate, glutamate, glutamine, acetate and glucose, respectively.

Discussion

Rates of butyrate oxidation to CO_2 in the present experiment are consistent with previously reported rates using isolated cell systems (Baldwin & Jesse, 1996) and higher than rates reported using papillae (Goosen, 1976). Similarly, rates of propionate and glucose oxidation to CO_2 are higher than those previously reported for papillae (Weekes, 1974). The higher rates of substrate oxidation observed with isolated cells compared to papillae is probably a result of increased availability of both substrate and oxygen.

Table 2. Effects of dietary energy density and ME intake on substrate oxidation to CO_2 by isolated ruminal epithelial cells[a].

	Forage		Concentrate			Significance		
Substrate (mmoles/l)	Low intake	High intake	Low intake	High intake	SEM	Diet	Intake	DxI
Glucose								
1	31.5	43.1	42.3	54.3	11.6	---	---	---
5	77.5	114.5	113.1	122.6	22.3	---	---	---
10	75.7	102.2	87.2	116.3	21.7	---	---	---
50	260.7	287.6	239.5	281.1	51.5	---	---	---
Glutamate								
1	110.4	152.5	128.7	123.0	31.9	---	---	---
5	210.0	319.6	257.0	362.3	75.8	---	---	---
10	276.2	444.7	328.8	476.8	90.8	---	0.086	---
Glutamine								
1	82.9	265.2	209.4	227.7	65.9	---	0.071	---
5	270.5	506.7	483.5	656.6	157.9	---	---	---
10	346.2	582.5	474.4	590.1	170.0	---	---	---
50	376.9	638.4	781.7	831.3	283.8	---	---	---
Acetate								
1	57.6	135.3	64.5	120.9	23.2	---	0.001	---
5	138.9	214.0	145.6	215.7	39.8	---	0.033	---
10	228.9	326.8	253.6	286.6	52.5	---	---	---
50	494.9	549.1	473.5	778.2	111.9	---	0.068	---
Propionate								
0.1	92.5	70.7	62.4	99.1	30.7	---	---	---
1	342.2	361.9	370.1	472.2	73.2	---	---	---
5	406.2	411.9	490.1	449.3	74.1	---	---	---
10	312.6	332.8	408.9	431.7	65.2	0.074	---	---
50	267.2	352.3	396.4	642.0	144.2	0.077	---	---
Butyrate								
0.1	37.7	28.2	31.0	33.0	4.3	---	---	---
1	110.2	101.3	134.5	144.1	14.9	0.013	---	---
5	131.5	146.9	169.7	187.6	21.2	0.039	---	---
10	142.3	144.1	166.7	189.3	21.5	0.071	---	---

[a] Oxidation rates expressed as nmoles substrate oxidized to CO_2/1 x 10^6 cells/90 min.
[b] Probability of a larger F statistic; ---, P>0.05.

Previous experiments designed to compare effects of forage and concentrate diets on ruminal epithelial metabolism failed to demonstrate altered VFA metabolism (Weigand *et al.*,1975) or enzyme activities (Young *et al.*, 1969). In the current experiment, rates of substrate oxidation to CO_2 were altered by both energy density (butyrate and propionate) and ME intake (acetate). Although these changes did not alter V_{max} and K_{ox}, it does indicate that changes in VFA metabolism occur in response to dietary manipulation. Because a clear relationship between VFA concentrations and substrate oxidation was not demonstrated, the factors responsible for shifts in ruminal epithelial VFA oxidation rates are not apparent.

Harmon (1986) reported that papillae from steers fed concentrate at 6.3 MJ ME/d oxidized both glucose (5 mM) and glutamine (1 mM) to CO_2 at higher rates than from steers fed forage at 3.4 MJ ME/d. In the current study, although no differences were detected in glucose oxidation rate, at the concentrations of substrates used by Harmon (1986), the same pattern is observed for glucose and glutamine oxidation. These substrate concentrations are below the estimated K_{ox} predicted by the current data set and may reflect a change in kinetics of glucose and glutamine use rather than absolute capacity (i.e. V_{max}).

Table 3. Effects of dietary energy density and ME intake on predicted V_{max} and K_{ox} for substrate oxidation to carbon dioxide by isolated ruminal epithelial cells.

		Forage		Concentrate	
Substrate	Parameter	Low intake	High intake	Low intake	High intake
Glucose	V_{max}[a]	309.9	302.8	179.6	369.3
	K_{ox}[b]	20.8	26.3	20.6	18.7
Glutamate	V_{max}	430.5	704.6	530.3	531.1
	K_{ox}	4.3	2.6	3.5	3.8
Glutamine	V_{max}	559.5	1256.0	967.0	1418.3
	K_{ox}	6.8	4.7	2.6	4.9
Acetate	V_{max}	656.5	676.5	480.9	526.8
	K_{ox}	18.2	5.9	13.1	12.1
Propionate	V_{max}	542.3	599.4	670.1	575.7
	K_{ox}	0.4	1.2	0.7	0.9
Butyrate	V_{max}	145.6	152.6	173.3	203.9
	K_{ox}	0.3	0.4	0.4	0.5

[a] V_{max} = maximal rate of substrate oxidation, nmoles oxidized to CO_2 /1 x 10^6 cells/90 min.
[b] K_{ox} = concentration of substrate mM at which half V_{max} oxidation rate is achieved.

Predicted V_{max} values reflect previous data with the notable exceptions of glutamine and glutamate which failed to achieve V_{max}. Kight & Fleming (1993) reported similar problems in predicting glutamine oxidation by rat enterocytes. The estimated K_{ox} values reported here reflect the expected hierarchy of substrate use *in vivo*. Based on derived K_{ox} estimates, VFA oxidation would be expected to operate at or near maximal rates continuously *in vivo*. In contrast, predicted K_{ox} values for glucose, glutamine, and glutamate are well above ruminal and arterial blood concentrations, indicating that these substrates are not primary fuel sources for ruminal epithelial cells.

References

Baldwin R.L., VI & Jesse, B.W. (1996). Propionate modulation of ruminal ketogenesis. *J. Anim. Sci.* 74:1694-1700.

Goosen P.C.M. (1976). Metabolism in rumen epithelium. Oxidation of substrates and formation of ketone bodies by pieces of rumen epithelium. Z. *Tierphysiol. Tierernahr. Futtermitelk.* 37:14.

Harmon D.L. (1986). Influence of dietary energy intake and substrate addition on the *in vitro* metabolism of glucose and glutamine in rumen epithelial tissue. *Comp. Biochem. Physiol.* 85B:643-647.

Kight C.E. & Fleming S.E. (1993). Nutrient oxidation by rat intestinal epithelial cell is concentration dependent. *J. Nutr.* 123:876-882.

Weekes T.E.C. (1974). The *in vitro* metabolism of propionate and glucose by the rumen epithelium. *Comp. Biochem. Physiol.* 49B:393-406.

Weigand E., Young J.W. & McGilliard A.D. (1975). Volatile fatty acid metabolism by rumen mucosa from cattle fed hay or grain. *J. Dairy Sci.* 58:1294-1300.

Young J.W., Thorp S.L. & Lumen H.Z. De (1969). Activity of selected gluconeogenic and lipogenic enzymes in bovine rumen mucosa, liver and adipose tissue. *Biochem. J.* 114:83-88.

ENERGY AND PROTEIN REQUIREMENTS OF GROWING FEMALE ETAWAH CROSSBREED GOATS DETERMINED FROM METABOLIC PARAMETERS

D.A. Astuti[1], D. Sastradipradja[1], T. Sutardi[2] and B. Haryanto[3]

[1]Department of Physiology and Pharmacology, Faculty of Veterinary Medicine, Institut Pertanian Bogor (IPB-Bogor Agriculture University), Jalan Taman Kencana 1 no. 3, Bogor 16151, Indonesia; [2]Department of Animal Nutrition IPB; [3]Centre for Animal Research, Agency for Agricultural Research and Development (AARD) - Ciawi-Bogor, Indonesia

Summary

Twenty female, weaned, Etawah crossbreed goats (ave. BW 14 kg) were randomly allocated into five dietary treatment groups (R1 to R5) anticipated to consume respectively 100%, 90%, 80%, 70% and 60% of *ad libitum* intake. The pellet fed ration contained 21% CP and 19.31 MJ GE/kg DM. A one-month adaptation period was allowed before a two-week measurement of energy and N balances. Average daily gain (ADG), daily heat production (HP), energy expenditure of, and volatile fatty acid (VFA) absorption by the splanchnic bed, glucose flux, cardiac minute volume, retained energy (RE), microbial protein synthesis (MPS) and urinary allantoin and nitrogen were measured and multiple regression equations were developed for the estimation of metabolizable energy (ME) and protein requirements of growing goats. Using stepwise multiple regression analysis of independently measured metabolic parameters, equations for ME and protein requirements were formulated as below:

$$ME\ (MJ/d) = 2.12 + 0.0007\ ADG - 0.02153\ kg^{0.75} + 0.0263\ glucose\ flux + 0.5965\ HP + 0.5836\ HP_{splanchnic\ bed} + 0.7913\ RE\ (R^2 = 0.98);$$

$$Protein\ (g/d) = 20.592 + 0.1533\ ADG + 9.6934\ kg^{0.75} + 0.0931\ allantoin - 0.440\ catab.\ prot + 0.813\ RP - 24.596\ MPS\ (R^2 = 0.99).$$

Key words: Goats, ME requirements, protein requirements, growth, digestion.

Introduction

Quantitative approaches have been used to elucidate mechanisms of ruminant digestion, metabolism and related areas. This information is important to relate performance to metabolic activity and environmental conditions. In a system with several components, the model provides a way of bringing together knowledge about the parts and to give a coherent view of the behaviour of the whole system. Estimation of nutrient requirements should ideally integrate those parameters into prediction equations. The present study was undertaken to investigate which digestion, metabolic and performance parameters are related to ME and protein requirements of female growing goats.

Materials and Methods

Twenty, female, weaned Etawah crossbreed goats (6-8 months old; ave. BW 14 kg) were used in a two-months experiment. The animals were randomly allocated to five dietary treatment groups R1 to R5, anticipated to consume respectively 100%, 90%, 80%, 70% and 60% of *ad libitum* intake. The rations offered were pellets containing 21% CP (crude protein) and 19.31 MJ GE (gross energy)/kg. A one-month adaptation period was allowed before energy and N balances were measured in a two-weeks trial (4 d acclimatization to cages followed by 10 d total collection of excreta and unconsumed feed). Animals were weighed once weekly and metabolic body size (MBS) calculated. A completely randomized design with one way classification was used. ADG (average daily gain) was obtained from body weight changes during the 2 months experimental period. Daily heat production (HP), glucose kinetics (glu. flux) and cardiac minute volume (CMV) were measured by the carbon dioxide entry rate technique (CERT) combined with single pulse labelling of glucose-2-^{3}H as previously described (Sastradipradja, 1992). The CERT measurements were completed within one week after termination of the balance trials. For the calculation of energy balance (EB), energy losses as ruminal methane production were estimated to be 8% of GE intake and urinary energy (UE) loss as urinary-N x 34.0 kJ/gN. Retained energy was estimated as the difference between ME (metabolizable energy) and HP. During the following week, portal blood flow ($BF_{splanchnic\ bed}$), heat production, and VFA absorption by the splanchnic bed at peak digestive activity 2 h postfeeding were measured using an acute surgical technique on anaesthetized animals (xylapex 0.05 ml/kg BW followed by ketamine 0.11 ml/kg BW i.m. injection) according to Katz & Bergman (1969a, 1969b). Labelled PAH-^{3}H was infused instead of common *p*-aminohippuricacid (PAH) for the $BF_{splanchnic\ bed}$ measurement. Urinary allantoin and nitrogen were measured on urine samples and related to rumen microbial protein synthesis (MPS) which was approximated by the *in vitro* batch technique involving $NaH_2{}^{32}PO_4$ labelling of samples of rumen contents (Suwandyastuti, 1986) measured prior to the balance trials.

Differences between means were compared using Duncans Multiple Range Test after ANOVA for one way classified data (Steel & Torrie, 1986). Using a stepwise multiple regression analysis of the metabolic data which were independently measured one from another, equations for energy and protein requirements were developed. MINITAB/SPSS release 6.1 (1988) was used for the multiple regression procedures.

Results and Discussion

The results of the digestibility and metabolizm studies are presented in Table 1. The average intake of the *ad libitum* group (R1) was 466 g DM/d containing 8.53 MJ GE/d and 102g CP/d. Groups R2, R3, R4 and R5 were offered the respective calculated percentages of *ad libitum* feeding. Although given lesser amounts of feed than that given to R1, part of the offered amounts were not always consumed. Thus, the actual average consumptions of R2, R3, R4 and R5 were respectively 83.7, 73.5, 65.4 and 55.2% of *ad libitum* feeding. These percentage levels deviated further if based on DE or ME intakes. Results of the present investigation showed that the digestion and subsequent metabolizm of absorbed nutrients were influenced by gastrointestinal fill. Thus, when DE for R1 was taken as 100% *ad libitum*, then the percentages for R2, R3, R4 and R5 were lowered respectively to 75, 71.2, 71 and 54.4%. Body protein catabolism was much affected by level of feeding with high values for R1 and R5 ($P<0.01$). Decrease in protein intake was parallel with lowered energy intake, but inadequate energy intake (R5) would have caused more protein to be used for energy needs. There were no differences in HP between treatment groups. The goats were very excitable and noisy when given inadequate feed, causing the HP values to be high (3.25 to 3.60 MJ/d). These values were 1.5-1.7 times basal metabolic rate and as a consequence, the calculated values for ME were not different between treatments R2, R3 and R4. The experiment would, however, still give information on relationships between performance parameters and ME intakes.

Due to excitable behaviour, the tropical Etawah crossbreed goat may show higher metabolic rates, e.g. HP of the overall individual as well as HP of the splanchnic bed, glucose flux etc. Empirical equations of best fit to relate these data to ME were as follows:

$HP = 2.84 + 0.167ME$ ($r=0.60$); Glu. Flux $= -15.37 + 8.065ME$ ($r=0.94$);

$HP_{splanchnic\ bed} = -0.890 + 0.553ME$ ($r=0.96$); $VFA_{absorption} = 13.59e^{0.468ME}$ ($r=0.91$);

$RE = [0.305ME^{1.52} - 2]$ ($r=0.96$); $RP = 16.89ME^{0.308}$ ($r=0.90$).

Allantoin excretion in urine was intimately related to MPS which reflected the portion of protein supply to the animal from rumen microbes. Allantoin is a product of purine metabolism which originates mainly from the degradation of rumen microbial nucleic acids (Chen *et al.*, 1992). Our results demonstrate that urinary allantoin excretion relates to MPS according to: allantoin excretion (gld) = -336.37 + 190.25MPS (r=0.83).

Table 1. Digestibility and metabolism of nutrients by growing, Etawah crossbreed, female goats.

Parameter	R1	R2	R3	R4	R5	P level
Intakes & digestibilities:	(n=4)	(n=4)	(n=4)	(n=4)	(n=4)	
DM (g/d)	466	390	342	305	258	
GE (MJ/d)	8.53	7.14	6.27	5.58	4.71	
DE (MJ/d)	5.22^{a}	3.84^{b}	3.71^{bc}	3.66^{bc}	2.83^{c}	< 0.01
CP (g/d)	102	85	75	66	56	
Energy/nutrient balances:	(n=4)	(n=4)	(n=4)	(n=4)	(n=4)	
ME/DE (%)	86	84	85	86	85	NS
ME (MJ/d)	4.47^{a}	3.23^{b}	3.17^{bc}	3.18^{bc}	2.40^{c}	< 0.01
E-CH_4 (MJ/d)	0.68^{a}	0.57^{b}	0.50^{c}	0.45^{d}	0.38^{e}	< 0.01
EU (MJ/d)	0.064^{a}	0.057^{ab}	0.043^{b}	0.038^{b}	0.056^{b}	< 0.05
HP (MJ/d)	3.62	3.60	3.25	3.17	3.31	NS
HP/GE (%)	43	50	51	57	70	
RE (MJ/d)	0.85^{a}	-0.37^{ab}	0.085^{ab}	0.005^{ab}	-0.91^{b}	< 0.05
MBS ($kg^{0.75}$)	7.75	7.63	7.12	6.76	6.48	NS
ADG (g/d)	147.0^{a}	107.0^{a}	95.0^{ab}	57.0^{b}	30.0^{c}	= 0.07
Catprot(UNx6.25g/d)	12.68^{a}	7.48^{b}	7.55^{b}	7.68^{b}	11.72^{a}	< 0.01
MPS (g/d)	3.14^{a}	2.83^{ab}	2.65^{ab}	2.48^{ab}	2.29^{ab}	< 0.05
RProt (g/d)	67.14^{a}	50.88^{b}	48.50^{b}	44.55^{b}	31.01^{c}	< 0.01
Glu. Flux (mg/min)	21.02^{a}	11.28^{bc}	12.20^{b}	6.68^{cd}	4.63^{d}	< 0.01
Cardiac min.vol.(ml/min)	3160	3590	3160	3290	2880	NS
Organ metabolism (at 2 h post feeding):	(n=2)	(n=2)	(n=2)	(n=2)	(n=2)	
$BF_{splanchnic\ bed}$(ml/min)	1032	624	625	486	394	NS
$HP_{splanchnic\ bed}$/HP (%)	45	28	24	21	16	
$VFA_{absorption}$ (mmol/min)	22.8^{a}	13.6^{ab}	15.4^{ab}	9.9^{ab}	8.6^{b}	< 0.05
Urin. Allantoin (g/d)	0.27^{a}	0.25^{a}	0.14^{b}	0.11^{b}	0.11^{b}	< 0.01

* Urinary energy values were low due to technical difficulties in the determination of urinary N

Nutrient requirements are the minimum requirements of an animal to meet the needs for survival, health and production. To understand causal relationships of the metabolic system, a mechanistic model was constructed by viewing the key parameters and analysing the behaviour of the system in terms of its individual components and their interactions with one another. Nutrient utilization for growth might consider components at the individual level as well as their partition to lower levels, like dynamics of body fat, protein, plasma metabolites and gastrointestinal fill, which include interactions such as protein and fat turnover, gluconeogenesis, nutrient supply and absorption. The present research tried to incorporate the measured metabolic parameters into a mechanistic description of nutrient utilization for growth. Using a stepwise multiple regression analysis, the following equations for ME and protein requirements were derived:

1. Using ADG and MBS only,

$$ME\ (MJ/d) = 0.5836 + 0.0034\ ADG + 6.3336\ MBS\ (R^2 = 0.52);$$

$$Protein\ (g/d) = -11.288 + 0.0494\ ADG + 11.6912\ MBS\ (R^2 = 0.44)$$

The confidence limits for each variable for the ME requirement equation were P = 0.70 for the intercept, P = 0.25 for ADG and P = 0.15 for MBS. For the protein equation, the values were P = 0.69 for the intercept, P = 0.38 for ADG and P = 0.01 for MBS.

2. Using also other metabolic parameters, the R^2 values were very much improved.

$$ME\ (MJ/d) = 2.12 + 0.0007\ ADG - 0.02153\ MBS + 0.0263\ glu.flux + 0.5965\ HP + 0.5836\ HP_{splanchn.ic\ bed} + 0.7913\ RE\ (R^2 = 0.98);$$

$$Protein\ (g/d) = 20.592 + 0.1533\ ADG + 9.6934\ MBS + 0.0931\ allantoin - 0.440\ catab.\ prot + 0.813\ RP - 24.596\ MPS\ (R^2 = 0.99).$$

The confidence limits for each variable for the ME requirement equation were P = 0.42 for the intercept, P = 0.45 for MBS, P = 0.51 for glucose flux, P = 0.56 for ADG, P = 0.20 for HP, P = 0.12 for $HP_{splanchnic\ bed}$ and P = 0.02 for RE. For the protein equation, the values were P = 0.12 for the intercept, P = 0.53 for ADG, P = 0.05 for MBS, P = 0.01 for allantoin excretion, P = 0.31 for catabolized protein, P = 0.007 for retained protein and P = 0.10 for MPS.

Acknowledgements

Supported by Research Grants 021/P4M/DPPM/PHB I/2/1993 and 061/P4M/DPPM/94/-PHB I/3/1994 from the Directorate General of Higher Education of the Department of Education and Culture of the Republic of Indonesia.

References

Chen X.B., Franklin M.F., Ørskov E.R. & Shand W.J. (1992). The effect of feed intake and body weight on purine derivative excretion and microbial protein supply in sheep. *J. Anim. Sci.* 70:1534-1542.

Katz M.L. & Bergman E.N. (1969a). Simultaneous measurements of hepatic and portal venous blood flow in the sheep and dog. *Am. J. Physiol.* 216:946-952.

Katz M.L. & Bergman E.N. (1969b). Hepatic and portal metabolizm of glucose, free fatty acids, and ketone bodies in the sheep. *Am. J. Physiol.* 216:953-960.

Minitab Statistical Software. (1988). Minitab reference manual. PC version. Release 6.1. Minitab Inc., 3081 Enterprise Drive, State College, PA 16801-3008, USA.

Sastradipradja D. (1992). Tracer techniques for measuring energy expenditure and related metabolic parameters. Proceedings 5th Asia and Oceania Congr. Nucl. Med. and Biology, vol. 2, PKBNI Jakarta. pp. 451-457.

Steel R.G.D. & Torrie J.H. (1986). Principles and procedures of statistics. McGraw-Hill Book Co. Inc., Singapore-New York.

Suwandyastuti S.N.O. (1986). Peningkatan mutu jerami padi ditinjau dari neraca mineral esensial pada sapi perah. PhD dissertation, Bogor Agriculture University, Indonesia.

NUTRIENT UTILIZATION OF GROWING FEMALE ETAWAH CROSSBREED GOATS FED DIETS CONTAINING DIFFERENT PROTEIN/ENERGY RATIOS

I.G.A. Arta Putra[2], D. Sastradipradja[1], D.A. Astuti[1], B. Kiranadi[1] and L.A. Sofyan[3]

[1]Department of Physiology and Pharmacology, Faculty of Veterinary Medicine, Institut Pertanian Bogor (IPB-Bogor Agriculture University), Jalan Taman Kencana 1 no. 3, Bogor 16151, Indonesia; [2]Department of Animal Production, Faculty of Animal Husbandry, Udayana University, Denpasar-Bali, Indonesia; [3]Department of Animal Nutrition IPB

Summary

Nutrient utilization by 12 growing, female, Etawah crossbreed goats receiving pelleted diets (I, II, III) formulated to contain the same energy (18.6 MJ/kgDM) but different CP levels (18.46, 16.24 and 14.0%DM), was studied. Daily feed intakes, energy and N balances, average daily gain, *in vivo* body composition, VFA production rates, portal blood flow, heat production of and VFA absorption by the splanchnic bed at peak digestive activity (postfeeding) were measured. The results indicated that all animals received adequate energy, but different protein intakes. Rumen fermentation patterns did not differ among treatments, but protein digestibility and VFA absorption rates differed significantly among treatments. Higher protein:energy ratio stimulated gut activity and increased retained protein. An adequate supply of energy would provide readily available energy resources for body catabolic processes causing a sparing effect of protein to be directed towards body protein retention.
Key words: Goats, protein:energy ratio, VFA production rates, splanchnic bed, protein retention.

Introduction

Protein and energy metabolism are intimately linked. In ruminants, dietary protein is vigorously subjected to ruminal microbial attack and amino acids are fermented to provide energy for microbes while, simultaneously, ammonia and carbon skeletons are used for amino acid synthesis. Amino acids absorbed from the small intestine are not only used for body protein synthesis, but also as energy sources. Protein synthesis has been established as a major energy-requiring process both at the cellular level and in the whole animal. Protein digestion and metabolism are subject to internal and external regulation affected by a number of factors including animal species and breeds. The purpose of the present investigation was to document energy and protein utilization of growing female Etawah crossbreed goats fed diets containing different protein:energy (P:E) ratios.

Material and Methods

Twelve growing, female, Etawah crossbreed (PE) goats (5 months old, BW 16.4 $\pm$ 1.79 kg) were randomly allocated to 3 groups receiving pelleted diets (I, II, III) formulated to contain the same energy (18.6 MJ/kgDM) but different crude protein levels (18.46, 16.24 and 14.0%DM). A completely randomized design with one-way classification was used. The animals were housed in individual pens allowing restricted movement. Goats consumed about 700 g of feed per day and water was given *ad libitum*. Half of the feed was divided into 3 equal parts and given at 06.00, 10.00 and 14.00, while the remaining half was given at 18.00. All goats were adapted to diets for 2 weeks before they were subjected to energy and N balance trials (4 d adaptation to the metabolic cages and 10 d total collection of excreta and unconsumed feed). For estimation of metabolizable energy

(ME), energy losses as ruminal methane production was estimated to be 8% of gross energy (GE) intake and urinary energy (UE) loss as urinary-N x 34.0 kJ/gN.

Body weights were recorded at weekly intervals. Goats were twice subjected to *in vivo* body composition measurements by the urea space (US) technique to enable calculations of energy and protein retentions (RE and RProt. respectively). The first measurement was conducted at the beginning of the experiment (2nd week of the feed adaptation period) and the second measurement during the week after the balance trials. US was measured on all animals by injecting into the jugular vein 0.65 ml of a saline solution of 20% urea per kg MBS (metabolic body size). Several blood samples were drawn thereafter for determination of the US according to Rule *et al.* (1986) and bodywater was calculated using the equation EBW (empty body water as %BW) = 59.1 + 0.22*US - 0.04*BW. *In vivo* body composition could then be calculated using an equation for fat content (Panaretto & Till, 1963): Body fat (%EBW) = 98.0 - 1.32* EBW, and for protein content (Panaretto, 1963): Empty Body Protein (kg) = 0.34*EBW - 0.52. Validation was carried out in a separate experiment on 4 animals by comparing *in vivo* results with slaughter technique (ST) analysis data on the same goats. RE and RProt. were estimated from the changes of body fat and protein contents measured from the first and second US trials. Conversion factors of 37.67 kJ/g fat and 16.74 kJ/g protein were used to calculate energy contents (Powers & Howley, 1990). Assuming minimal physical activity, heat production (HP) of individual goats was calculated as the difference between ME and RE.

The week after the balance trials (after finishing the 2nd US measurements) the same goats were also used to measure portal blood flow (PBF), heat production (HP) of splanchnic bed and VFA absorption by the splanchnic bed ($VFA_{absorption}$) at peak digestive activity 2h postfeeding. PBF was measured by an acute surgical method on anaesthetized animals (xylapex 0.05 ml/kg BW followed by ketamine 0.11 ml/kg BW i.m. injection) using the approaches described by Katz & Bergman (1969a, 1969b) but using labelled (^{3}H) instead of unlabelled *p*-aminohippuric acid (PAH).

The following (i.e. final) week of the experiment was used to estimate ruminal VFA production rates ($VFA_{production}$) by the zero-time *in vitro* method according to Whitelaw *et al.* (1970). Due to the small size of PE goats, only two samples of rumen contents were taken, via rumen fistulae, after each feeding during the day-time with different time procedures. Data for each animal were pooled to produce a composite time-course curve of VFA concentration and daily VFA production rates were calculated as the area under the day time curve multiplied by 2x rumen volume to cover the night period also. Rumen volume was estimated by the ^{51}Cr-EDTA technique (IAEA, 1985).

Differences between means were compared using least significance difference after ANOVA for one way classified data (Steel & Torrie, 1986). MINITAB software release 8.2 (1991) was used for the analysis.

Results and Discussion

Validation of the *in vivo* body composition method using US was done separately on 4 goats using ST analysis in addition to the *in vivo* calculations on the same animals. EBW, estimated according to Rule *et al.* (1986), resulted in underestimation by 4.2%. Therefore, the use of US required a correction factor of 1.044. For use with small ruminants, the Panaretto (1963) equation for body protein and the Panaretto & Till (1963) equation for body fat are based on EBW. The use of corrected values in the aforesaid equations resulted in good agreement with our ST results.

Table 1. Body composition (% BW) according to slaughter technique (ST) analysis and *in vivo* equations estimated from 4 female growing goats.

	Body water		Body protein		Body fat	
	ST	Eq. 1	ST	Eq. 2	ST	Eq. 3
Mean ± SD	60.9±5.94	60.9±0.06	20.2±0.39	20.2±0.02	16.2±5.74	17.6± 0.08

Eq. 1 = EBW according to Rule *et al.* (1986) US equation multiplied by 1.044,
Eq. 2 = Panaretto (1963) body protein equation using EBW value from Eq. 1,
Eq. 3 = Panaretto & Till (1963) body fat equation using EBW value from Eq. 1.

The results are presented in Table 1. The digestibilities of nutrients and metabolic parameters are presented in Table 2. The results showed no differences in DM, GE or DE intakes although reductions were observed for protein digestibility and digestible protein (DProt.) in the groups with lower protein intakes. There was no significant difference of total VFA production ($VFA_{production}$) rates observed among treatments with daily acetate production rates for the relatively small PE goats varying from 0.97 to 1.35 moles and daily propionate production rates from 0.425 to 0.615 moles. These values are comparable to the values reported by Bergman (1990) for temperate-climate sheep. Instant total VFA absorption by the splanchnic bed at peak digestive activity 2 h postfeeding was significantly lowered due to lower P/E ratio, but the values do not reflect total daily VFA absorption.

Table 2. Metabolic parameters of growing female crossbreed Etawah goats.

Metabolic parameter	I	II	III	P sig. level
Intakes and digestibilities:	n = 4	n = 4	n = 4	
DM intake (g/d)	634.41	634.06	637.91	
GE intake (MJ/d)	11.83	11.81	11.92	
CP intake (g/d)	117.11	102.97	89.46	
Energy digestibility (%)	59.57	59.44	59.59	
Protein digestibility (%)	71.00^{a}	68.32^{ab}	65.19^{bc}	< 0.05
Ruminal nutrient supply:	n = 2	n = 2	n = 2	
Total $VFA_{production}$ rate (mmol/h)	64.59	77.10	97.12	NS
$Propionate_{production}$ rate (mmol/h)	17.71	19.69	25.63	NS
Rumen volume (l)	4.33	4.11	4.11	NS
*$VFA_{absorption}$ (mmol/min)	13.44^{a}	7.62^{ab}	5.22^{bc}	< 0.05
*$HP_{splanchnic}$ bed (MJ/d)	0.528	0.563	0.342	= 0.14
Energy and N balance:	n = 4	n = 4	n = 4	
ME (MJ/d)	5.93	5.96	6.0	NS
UE (MJ/d)	0.20	0.14	0.16	NS
HP (MJ/d)	4.97	4.73	4.72	NS
RE (MJ/d)	0.935	1.032	1.057	NS
Rprotein (g/d)	43.86^{a}	41.78^{a}	26.75^{b}	< 0.05
ADG (g/d)	95.5	101.5	102	NS

* measured at 2 h post feeding.

The number of replications was small (n = 2 for each treatment group) but the results do however suggest that a large amount of VFA had been metabolized by the rumen epithelium to support the active metabolism of the portal-drained viscera, which included the energetic cost of digestive and absorptive processes. Indeed, $HP_{splanchnic\ bed}$ values in the present study varied from 0.34 to 0.56 MJ/h which accounted for 7.20 to 11.84% of the total HP of the whole body. The data on retentions of energy and of protein revealed that no difference was obtained between treatments for energy retention, but for protein retention the value was lower with a lower P:E ratio.

In conclusion, the results indicated that all animals received adequate energy, but different protein intakes. Although rumen fermentation characteristics did not differ among treatments, protein digestibility differed significantly and there were large differences in VFA absorption rates. A higher P:E ratio would stimulate gut activity and this is supported by the data on $HP_{splanchnic\ bed}$, which showed a tendency to increase ($P = 0.14$). Considering that an adequate P:E ratio would increase retained protein, a rational explanation would be that the adequate supply of energy in the present study would provide readily available energy resources for body catabolic processes causing a sparing of protein to be diverted towards body protein retention.

Acknowledgements

Supported by Research Grant no. 004/P4M/DPPM/95/PHB I/4/1995 from the Directorate General of Higher Education of the Department of Education and Culture of the Republic of Indonesia.

References

Bergman E.N. (1990). Energy contributions of volatile fatty acids from the gastrointestinal tract in various species. *Physiol. Rev.* 70:567-590.

IAEA (1985). Laboratory training manual on the use of nuclear techniques in animal nutrition, Vienna. pp. 191-192.

Katz M.L. & Bergman E.N. (1969a). Simultaneous measurements of hepatic and portal venous blood flow in the sheep and dog. *Am. J. Physiol.* 216:946-952.

Katz M.L. & Bergman E.N. (1969b). Hepatic and portal metabolism of glucose, free fatty acids, and ketone bodies in the sheep. *Am. J. Physiol.* 216:953-960.

Minitab Statistical Software (1991). Minitab reference manual. PC version. Release 8. Minitab Inc., 3081 Enterprise Drive, State College, PA 16801-3008, USA.

Panaretto B.A. & Till A.R. (1963). Body composition *in vivo*. II. The composition of mature goats and its relationship to the antipyrine, tritiated water and N-acetyl-4-aminoantipyrine spaces. *Austr. J. Agric. Res.* 14:926-943.

Panaretto B.A. (1963). Body composition *in vivo*. III. The composition of living ruminants and its relation to the tritiated water spaces. *Austr. J. Agric. Res.* 14:944-952.

Powers S.K. & Howley E.T. (1990). Exercise Physiology. WCB Publ., USA. p 539.

Rule D.C., Arnold R.N., Hentges E.J. & Beitz D.C. (1986). Evaluation of urea dilution as a technique for estimating body composition of beef steers *in vivo*: validation of published equations and comparison with chemical composition. *J. Anim. Sci.* 63:1935-1948.

Steel R.G.D. & Torrie J.H. (1986). Principles and procedures of statistics. McGraw-Hill Book Co. Inc., New York.

Whitelaw F.G., Hyldgaard-Jensen J., Seid R.S. & Kay M.G. (1970). Volatile fatty acid production in the rumen of cattle given an all concentrate diet. *Br. J. Nutr.* 24:179-195.

MID-TERM EFFECT OF PHOTOPERIOD ON ADIPOSE TISSUE LIPOGENESIS, PLASMA METABOLITES, INSULIN, SOMATOTROPIN AND PROLACTIN IN UNDERFED-REFED EWES

Y. Chilliard, Y. Faulconnier, M. Bonnet and F. Bocquier

Laboratoire Ruminant Undernutrition, INRA-Theix, 63122 St-Genès-Champanelle, France

Summary

Twenty-five dry non-pregnant ewes were divided into a control group (n=5; slaughtered after 3 weeks on a diet providing 125% of the maintenance energy requirement, MER) and the remainder (n=20) divided into 2 groups and subjected to either 8 or 16 h/d light, and underfed to 25% MER for 7 days. Half of the ewes in each of the non-control groups were slaughtered and the remaining ewes were refed at 200% MER for a further 14 days, until slaughtering. Underfeeding reduced plasma insulin, prolactin, acetate, triglycerides and urea, increased free fatty acid (NEFA) and covered the activity of the lipogenic enzymes fatty acid synthase, glucose-6-phosphate dehydrogenase, malic enzyme and glycerol-3-phosphate dehydrogenase. Refeeding partially or totally reversed these effects. Difference in daylength did not modify any of the measured parameters, except for plasma prolactin (which was higher in ewes exposed to longer daylight) and NEFA (which were higher in underfed ewes with a shorter light period).
Key words: Ewe, underfeeding, photoperiod, adipose tissue, lipogenic enzymes, metabolites, insulin, prolactin, somatotropin, NEFA.

Introduction

Adipose tissue (AT) metabolism in sheep responds to seasonal changes (Vernon & Flint, 1984), and photoperiod is involved in seasonal variations of body lipid mobilization and deposition in lactating ewes (Bocquier *et al.*, 1986, 1990). Variation in daylight plays a role in voluntary feed intake regulation (Kay, 1985). As feed intake is an important factor in the variation of AT metabolism, it is important to understand whether photoperiod acts directly on AT metabolism, or indirectly through nutrient flow and/or metabolic hormones. The aim of the present study was to measure the effects of two different photoperiod patterns on AT lipogenic enzyme activities, and endocrine and metabolic profiles, of pair-fed ewes during an underfeeding-refeeding experiment.

Materials and Methods

Twenty-five multiparous, dry, non-pregnant, ovariectomized Lacaune ewes were divided into 5 groups containing 5 ewes each. The experiment took place in January and February. The control group (C) was slaughtered after 3 weeks on a hay-concentrate (70-30) diet providing 125% of the maintenance energy requirement (MER). The remaining 20 ewes were divided into 2 groups subjected either to a short (S; 8 h/d light - fluorescent bulb - 500 lux) or to a long (L; 16 h/d light) constant photoperiod. They received the control diet for 3 weeks and then were underfed to 25% MER for 7 days on a hay-straw (60-40) diet. Five S-ewes and five L-ewes were slaughtered at the end of the underfeeding period (groups S-U and L-U), and the remaining ewes (groups S-R and L-R) were refed until slaughtering, at 200% MER for 14 days on a hay-concentrate (45-55) diet.

Jugular blood was sampled on the day before slaughtering at 14.00 (4 h after feeding) and every 20 min from 09.00 to 15.00, to determine mean prolactin and somatotropin concentrations. Insulin (INSI-PR kit, ORIS group), prolactin and somatotropin (Bocquier *et al.*, 1990) were determined by radio-immunoassay. Plasma concentrations of glucose, acetate, lactate, triglycerides, urea, 3-hydroxybutyrate (3-OHB), free glycerol and NEFA were determined enzymatically by autoanalyser.

Perirenal AT (PAT) was taken after slaughter, placed in liquid nitrogen and stored at -80°C pending analyses. Mean adipose cell volume and the activities of a number of lipogenic enzymes : fatty acid synthase (FAS), glucose-6-phosphate dehydrogenase (G6PDH), malic enzyme (ME) and glycerol-3-phosphate dehydrogenase (G3PDH) were measured as described by Chilliard *et al.* (1991).

Results

Underfeeding significantly decreased insulin (-63%), acetate (-85%), triglycerides (-76%) and urea (-55%), and increased NEFA (+1504%) (Table 1). These parameters returned or tended to return to control levels during refeeding, except for a rise of urea above control level. Underfeeding tended to decrease postprandial plasma 3-OHB (-13%) but increased the preprandial level (+44%) (data not shown).

PAT weight and cellularity were not affected by the treatments. Underfeeding decreased the activity of lipogenic enzymes: -62% for FAS ($P<0.10$) and G6PDH ($P<0.05$), -41% for ME ($P<0.05$) and -50% for G3PDH ($P<0.10$ S and 0.05 L). Refeeding increased ($P<0.05$ vs. underfed ewes) the FAS (+381%), G6PDH (+280%) and ME (+120%) activities and ($P<0.10$) G3PDH (+99%) activity. FAS activity was higher in refed than control ewes. Results were similar on a per cell (Table 1) or per gram basis (data not shown).

Daylength did not affect any of the measured parameters in underfed or refed ewes, except for plasma prolactin and, to a lesser extent, NEFA. Plasma prolactin was increased by long light. When considering the 2 x 2 factorial design (feeding level x photoperiod) composed only of the 4 groups of S- and L-ewes, prolactinaemia was significantly ($P<0.05$) increased not only by photoperiod, but also by level of feeding. There was a significant ($P<0.10$) interaction, with a higher prolactin response to photoperiod in refed groups (Table 1). Conversely, postprandial plasma NEFA was significantly higher in the underfed group with the shorter photoperiod (Table 1). This effect of photoperiod on preprandial plasma NEFA was, however, not significant (data not shown).

Discussion

Underfeeding decreased plasma levels of exogenous nutrients, increased NEFA mobilized from PAT and decreased insulinaemia and prolactinaemia, in agreement with previous data (Chilliard *et al.*, 1995). However, there was no trend towards subclinical ketosis, as indicated by the low level of plasma 3-0HB.

Changes in FAS, G6PDH and ME activities of PAT were in agreement with published data in underfed-refed growing sheep (Ingle *et al.*, 1973). We are not aware of published data on G3PDH activity. Responses to refeeding were more marked in the present trial, probably because animals were refed at a higher level of intake, for a longer period, and had a lower relative energy requirement than growing sheep. FAS activity showed a higher response both to underfeeding and refeeding, in keeping with its regulatory role in PAT lipogenesis. In this situation we showed that regulation of FAS activity in ruminant PAT partly occurs at a pre-translational level (Bonnet *et al.*, 1996).

Changes in photoperiod were shown to modify prolactin level after a few days, milk yield after 2 weeks (Bocquier *et al.*, 1986), and reproductive activity after about 6 weeks (Chemineau *et al.*, 1996), while voluntary feed intake (Kay, 1985) and basal energy expenditure (Walker *et al.*, 1991) started to change after 2 to 4 months. In the present experiment, prolactin responded to changes in photoperiod within 3 weeks, i.e. before the beginning of the underfeeding period (results not shown).

Table 1. Effects of underfeeding, refeeding and photoperiod on plasma prolactin, insulin and metabolites, and adipose tissue lipogenic enzymes (5 ewes per group).

Feeding level[1]	125	25		200		SEM
Photoperiod[2]	Natural	S	L	S	L	
Dry matter intake (g/d)[3]	1296[b]	290[c]	302[c]	1588[a]	1589[a]	50
Plasma hormones						
Prolactin (ng/ml)	12.8[bc]	7.0[c]	19.3[b]	13.2[bc]	34.4[a]	2.4
Somatotropin (ng/ml)	0.81	0.55	0.96	0.92	0.65	0.10
Insulin (μU/ml)	41.7[a]	14.6[b]	16.0[b]	38.9[a]	44.9[a]	6.5
Plasma metabolites (μM)						
NEFA	37[c]	694[a]	493[b]	51[c]	41[c]	55
Free glycerol	121[ab]	140[a]	124[ab]	114[b]	122[ab]	6
3-OH-butyrate	521	435	474	549	598	53
Glucose	3663[ab]	3243[b]	3181[b]	3880[a]	3508[ab]	154
Lactate	1170[a]	715[ab]	566[b]	1148[a]	789[ab]	123
Acetate	1067[a]	116[c]	203[c]	645[b]	727[b]	60
Triglycerides	274[a]	83[b]	49[b]	122[b]	113[b]	32
Urea	5764[b]	2567[c]	2668[c]	8175[a]	8194[a]	558
Body weight (kg)	75.4	71.1	71.6	76.1	74.9	2.9
Perirenal adipose tissue						
Weight (kg)	2.12	2.31	2.14	2.40	2.26	0.28
Adipocyte volume (pl)	958	935	843	712	867	99
Fatty acid synthase[4]	121[b]	49[b]	44[b]	206[a]	241[a]	21
G6PDH[4,5]	388[a]	149[b]	147[b]	548[a]	576[a]	60
Malic enzyme[4]	136[b]	74[c]	87[c]	147[b]	208[a]	13
G3PDH[4,6]	7338[a]	4169[ab]	3217[b]	6625[ab]	8059[a]	996

[1]: Percent of maintenance. [2]: S = 8 h/d light ; L = 16 h/d light. [3]: Recorded during the 2 days before slaughtering. [4]: Unit = nmol/min/10^6adipocytes. [5]: Glucose-6-phosphate dehydrogenase. [6]: Glycerol-3- phosphate dehydrogenase.
[a, b, c]: figures with different letters differ significantly (P<0.05).

A higher postprandial NEFA concentration in underfed ewes subjected to a short photoperiod, i.e. with lower prolactinaemia despite the same feeding level and plasma somatotropin (Table 1), has not been reported before and could reflect an enhanced ability of ewes to mobilize PAT during winter (short days) when less food is available.

In this mid-term experiment, there was no effect of photoperiod changes (over 6 weeks in refed ewes) on AT lipogenic enzymes. This suggests that any putative changes in AT lipogenesis were due to indirect effects of photoperiod, e.g. through a change in milk yield or feed intake but, alternatively, it could be that the duration of the photoperiodic treatments was not long enough, or that a constant daily light duration did not mimic all the effects of the natural change in seasonal daily light.

In conclusion, an increase in the daily light period over 4-6 weeks increased prolactinaemia in underfed-refed ewes and decreased plasma NEFA in underfed animals. However there was no change in insulinaemia nor in PT lipogenic activities.

Acknowledgements

This work was supported by the "Action Incentive Programme : Lipogenesis in species of zootechnical interest" (INRA).

References

Bocquier F., Thériez M., Kann G. & Delouis C. (1986). Influence de la photopériode sur la répartition de l'énergie nette entre la production laitière et les réserves corporelles chez la brebis traite. *Reprod. Nutr. Develop.* 26:389-390.

Bocquier F., Kann G. & Thériez M. (1990). Relationships between secretory patterns of growth hormone, prolactin, and body reserves on milk yield in dairy ewes under different photoperiod and feeding conditions. *Anim. Prod.* 51:115-125.

Bonnet M., Faulconnier Y., Hocquette J.F., Leroux C., Langin D., Martin P. & Chilliard Y. (1996). mRNAs encoding lipogenic and lipolytic enzymes in the adipose tissue of underfed-refed ewes and cows. *Int. J. Obesity* 20:148.

Chemineau P., Malpaux B., Pelletier J., Leboeuf P., Delgadillo J.A., Deletang F., Pobel T. & Brice G. (1996). Emploi des implants de mélatonine et des traitements photopériodiques pour maîtriser la reproduction saisonnière chez les ovins et les caprins. *INRA Prod. Anim.* 9:45-60.

Chilliard Y., Gagliostro G., Fléchet J., Lefaivre R. & Sebastian I. (1991). Duodenal rapeseed oil infusion in early and midlactation cows. 5. Milk fatty acids and adipose tissue lipogenic activities. *J. Dairy Sci.* 74:1844-1854.

Chilliard Y., Doreau M., Bocquier F. & Lobley G.E. (1995). Digestive and metabolic adaptations of ruminants to variations in food supply. In: M. Journet, E. Grenet, M-H. Farce, M. Thériez and C. Demarquilly (Eds) Recent developments in the nutrition of herbivores. *Proceedings of the IVth International Symposium on Nutrition of Herbivores.* INRA, Paris, pp. 329-360.

Ingle D.L., Bauman D.E., Mellenberger R.W. & Johnson D.E. (1973). Lipogenesis in the ruminant : effect of fasting and refeeding on fatty acid synthesis and enzymatic activity of sheep adipose tissue. *J. Nutr.* 103:1479-1488.

Kay R.N.B. (1985). Seasonal variation of appetite in ruminant. In: Haresign W. and Cole D.J.A. (Eds) Recent Developments in Ruminant Nutrition, Butterworths, London, pp. 199-210.

Vernon R.G. & Flint D.J. (1984). Adipose tissue: metabolic adaptation during lactation. *Symp. Zool. Soc. Lond.* 51:119-145.

Walker V.A., Young B.A. & Walker B. (1991). Does seasonal photoperiod directly influence energy metabolism. In: C. Week and M. Boessinger (Eds) Energy Metabolism of Farm Animals. ETH - Zürich, Switzerland. *EAAP Publ. No.* 58:372-375.

SUMMARY OF THE DISCUSSION

Discussion leader: D.S. Parker

Secretaries: M. Klein & A. Wylie

Introduction

The importance of integrating data from different systems used in the study of animal metabolism cannot be overemphasized. Molecular biology provides an important tool in developing our understanding of cellular control mechanisms but we must not ignore the complexities of the whole animal which represents the ultimate 'model' which we are trying to understand. An example of the strength of evaluating data from different approaches would be the use of data for VFA metabolism by rumen epithelial cells isolated from animals on different diets (paper of Baldwin & McLeod) and those from whole animal studies evaluating the relationship between rumen VFA production rates and portal vein absorption. Both these approaches identify the extent of butyrate metabolism in the gut wall and the 'sparing' of propionate for gluconeogenesis in the liver.

The data presented by Stoll *et al.* was particularly interesting, demonstrating the extensive metabolism of amino acids by the tissues of the small intestine. The original observations of Tagari and Bergman, that amino acid disappearance from the gut lumen of sheep was significantly higher than amino acid appearance into the portal vein, has been the basis of considerable research in this area. Recently this work has been re-evaluated in sheep (MacRae *et al.*, 1997) and the results indicated that some 60% of lumenal loss can be accounted for in portal blood. The difference between these two studies may reflect differences between the requirements of gut tissues in young pigs when compared with those of sheep. Both studies provide evidence that lumenal amino acids only contribute some 20% of the amino acids required for protein synthesis in gut tissues, the majority arising from arterial supply. However, it may not bc possible to generalize, given the earlier observations of Alpers (1972) which showed that the gut cell populations may differ in their source of essential nutrients such as amino acids.

Further studies in this area using stable isotope technology should provide essential quantitative information on the fate of both carbon and nitrogen moieties of amino acids within gut cells providing a clearer picture of the nutrient requirements of this highly metabolically active tissue.

General Discussion

In response to a question by Reeds, as to whether substrate use by the gut is regulated or the mucosa simply uses what is there, Parker replied that he was not aware of any specific data but if the rate of gut cell turnover is altered by use of gut active growth promoters the overall cost of nutrient usage changes. It is not known if the pattern of nutrient use is altered.

In reply to a query about peptide uptake from algal protein, Reeds commented that they could not find labelled peptides in the algal protein hydrolysate. Once moieties of >5000 molecular weight were excluded there was no evidence of peptide uptake across the gut. It was pointed out that since there are peptide transporters on the basolateral membrane of the enterocyte cell some transfer would be anticipated and Reeds replied that there may be some transfer of proline-rich peptides which are not hydrolysed extensively by the brush border and cytoplasmic peptidase enzymes. Reynolds stated that the Reading group have looked at A-V peptide concentrations across the gut and, if anything, find uptake rather than release by the gut tissues and it was suggested that Tagari now believes that there is absorption of short peptides possibly via the omasal transporter identified by Webb. Parker stated that, following infusion of labelled peptides into the rumen in large doses, some portal uptake occurred but the overall nutritional importance of peptides would be low.

In relation to attempts to reduce the protein content of pig diets the question was raised about the requirement of the gut for amino acids rather than fatty acids for energy generation? Reeds replied that, although the figure of 30% utilization of amino acids by the gut does appear high, others have reported first pass metabolism of 40%. He was not sure if the usage recorded is an obligatory demand or just a reflection of excessive protein in the diet. Amino acid metabolism to provide energy is inefficient and certain amino acids (such as threonine which is extensively used for mucus synthesis) may become limiting. Data from the Houston laboratory suggested low efficiency of recycling of amino acids in the gut.

Further discussion centred on the apparent differences in metabolism of mitochondria in animals of different ages and from different tissue sources. In reply to a question on the effects of cold adaptation on UCP2 in muscle, Herpin stated that he had no data but that it is known that UCP3 is expressed in muscle which could provide some interesting new regulatory mechanisms. Carstens observed that he had not looked at UCP3 but that in brown fat, UCP1 takes 7-14 days to disappear, and UCP2 and 3 are probably present post-natally.

Several papers had investigated changes in gut tissue function following changes in intake or physiological 'load'. There are some discussions as to whether changes in mass of tissues such as the liver were due to hypertrophy or hyperplasia. Reeds was of the opinion that, when the liver is forced to 'work' harder, the increase in liver mass is largely through hyperplasia and it was reported that Burrin's data showed an increase in cell number rather than in cell size.

Parker commented that studies on changes in the ruminant tract with intake seem to focus on the reticulorumen and wondered what is the 'signal' with respect to the remainder of the tract. McLeod stated that he was not surprised by changes in the rumen mass between forage and concentrate diets. Forage diets stimulate reticulum development. He continued that, in the post-ruminal tract, it is changes in the epithelial cells rather than muscular tissues which are apparent. The role of butyrate as a key nutrient in colonocyte metabolism and in regulating hypertrophy of the rumen epithelium was raised. It was pointed out that butyrate is cytotoxic to cells in culture but at low levels may be stimulatory with other growth factors i.e. it may have a permissive role with EGF and IGF-1. In this way butyrate may alter the sensitivity/responsiveness of rumen epithelial cells to growth factors. There may even be a butyrate binding protein in mice.

Parker commented that the IGER Double Fourier model seemed to fit absorption data for β-OH butyrate but not for acetate or propionate. He asked whether this was because β-OH butyrate production is directly linked to butyrate production in the rumen whereas, for acetate, there are a number of sources (endogenous acetate for example) which are not directly linked to rumen events and whether, given such constraints, the model helps us to understand the non-steady state. Scollan replied that the model was used to try and describe the problems of asynchrony of nutrient supply in the rumen associated with feeding grass silage diets in the UK. The model was tested using the diets developed at

Nottingham to provide synchronous and asynchronous release of nutrients within the rumen. It was suggested that a key question is the extent to which asynchronous release of nutrients may affect PDV metabolism and also what influence the timing of amino acid supply to muscle may have upon protein synthesis. The main recorded differences appear to be in energy supply and insulin release. Sutton pointed out that in studies on causes of milk fat depression they had moved from 2x to 6x daily feeding to try and establish a steady-state situation. The IGER model only fitted some of the data and not all of it.

In closing the discussion Parker drew attention to a remark, many years ago, by John Rooke that the mammary gland appears to synthesize milk continuously even though nutrient supply may be discontinuous. However it was possible that the modern high-yielding dairy cow is in fact closer to steady-state than we think, given the levels of intake achieved and the continuous high rumen throughput.

SESSION 2

LACTATION AND REPRODUCTION

THE INFLUENCE OF COW GENOTYPE AND CONCENTRATE PROPORTION IN THE DIET ON ENERGY UTILIZATION BY LACTATING DAIRY COWS

C.P. Ferris, F.J. Gordon, D.C. Patterson and M.G. Porter

Agricultural Research Institute of Northern Ireland, Hillsborough, Co. Down, Northern Ireland BT26 6DR, U.K.

Summary

Twenty-eight high genetic merit and 32 medium genetic merit Holstein/Friesian dairy cows, were used in a continuous design, 2 (cow genotypes) x 4 (forage:concentrate ratios) factorial experiment consisting of eight treatments. The four forage:concentrate DM ratios were 63:37, 52:48, 41:59 and 30:70. During this study 12 animals of each genotype, three from each concentrate treatment, were subject to nutrient utilization studies. With the exception of the higher methane energy outputs of the medium merit cows and the reduction in urinary energy outputs with increasing proportion of concentrate in the diet, treatment had no effect on any of the other energy utilization parameters measured. Most significantly, k_l was found to be unaffected by genotype, suggesting that increased nutrient intakes are required to sustain the higher milk yields of high merit cows.
Key words: Dairy cows, genetic merit, concentrate level, energy utilization.

Introduction

The genetic merit of the UK and Irish dairy herds is currently increasing rapidly. A recent comparison confirmed the higher milk output potential of high merit cows, their daily milk yield being 6.6 kg/day higher than for animals of medium genetic merit (Gordon *et al.*, 1995). An understanding of whether these higher yields can be attributed to differences between genotypes in the efficiency of nutrient utilization or in nutrient partitioning, would appear to be a prerequisite to deciding if management changes are necessary to sustain them. In addition, the trend towards offering high merit cows higher levels of concentrates may also have an effect on the efficiency of nutrient utilization. This study was designed to examine the effects of genotype and level of concentrate inclusion in the diet on nutrient utilization.

Materials and Methods

Twenty-eight high genetic merit and 32 medium genetic merit Holstein/Friesian dairy cows, with Predicted Transmitting Abilities for milk fat plus protein, calculated using 1990 as a baseline ($PTA_{(90)}$ fat + protein), of 10.9 and 53.2 kg respectively, were used in this trial. This was a continuous design, 2 (cow genotypes) x 4 (forage:concentrate ratios) factorial experiment consisting of eight treatments. The four forage:concentrate DM ratios were 63:37, 52:48, 41:59 and 30:70. The forage component comprised a good quality first cut grass silage, while the concentrate offered had a crude protein concentration of 222 g/kg DM. Diets were offered *ad libitum* in the form of a complete diet. Animals commenced experimental treatments after a mean post-calving interval of 10.3 days (SD 2.42), and remained on the treatments for, on average, 84 days.

During this milk production trial, 24 animals, 12 of high and 12 of medium genetic merit, representing three animals from each concentrate level within each genotype, were used in nutrient utilization studies. These studies commenced after a mean of 63 days (SD 5.6) on the production trial. Animals were placed in indirect open-circuit calorimeters for 72 hours, with measurements of gaseous exchange during the final 48 hour period in the chambers being used in the calculation of heat production. On completion of the three day period in the respiration chambers, animals were tied in individual standings in a cowshed and a six day ration digestibility study undertaken.

Results

There were no significant genotype x concentrate ratio interactions for any of the parameters examined ($P>0.05$), consequently only main effects are presented here.

During both the main production study and the period in the respiration chambers, high merit cows had higher intakes and milk yields (Table 1) than the medium merit cows ($P<0.01$). In addition, both milk yields and intakes increased significantly with increasing levels of concentrate offered during the production study ($P<0.001$), although no such effects were observed with the animals selected for use in the nutrient utilization study (Table 2). Values for both parameters tended to be lower with the animals in the respiration chambers.

Table 1. Effect of cow genotype on animal performance and energy utilization.

	Cow genotype			
	High	Medium	SEM	Sig.
Total DMI (kg/day)				
Main study	21.4	20.2	0.30	**
Respiration chamber	20.4	17.9	0.61	**
Milk output (kg/day)				
Main study	37.0	32.9	0.42	***
Respiration chamber	34.9	29.2	1.20	**
Energy utilization data				
Energy digestibility	0.789	0.785	0.0043	NS
Methane E/GE intake	0.063	0.070	0.0022	*
Urine E/GE intake	0.045	0.043	0.0010	NS
Milk E/GE intake	0.313	0.295	0.0128	NS
ME/GE (q)	0.681	0.672	0.0054	NS
ME concentration (MJ/kg DM)	12.43	12.47	0.131	NS
Heat E/MEI	0.587	0.627	0.0169	NS
Milk E/MEI	0.460	0.441	0.0212	NS
Retained E/MEI	-0.047	-0.067	0.0344	NS
k_l (AFRC, 1990)	0.55	0.52	0.012	NS
k_l (Yan *et al.*, 1997)	0.63	0.60	0.013	NS

Faecal and urinary energy losses for the chamber period have been calculated using the data from the digestibility study by assuming that these losses remained as a constant proportion of the gross energy intake within each animal. The digestibility of the energy component of the diet was unaffected by treatment. While neither urine nor milk energy outputs, as a proportion of gross energy intake, were influenced by genotype, medium merit cows had a higher methane energy output than high merit animals. In addition, urinary energy output, as a proportion of GE intake, decreased with increasing proportion of concentrate in the diet.

When expressed as a proportion of ME intake, neither genotype nor concentrate proportion in the diet had a significant effect ($P>0.05$) on heat energy, milk energy or retained energy. In calculating the efficiency of ME utilization for lactation (k_l), energy available for production has been calculated as net energy intake less a maintenance component, the latter having been calculated according to two different methods, AFRC

(1990) and Yan *et al.* (1997). Irrespective of the maintenance component used, neither genotype nor concentrate proportion in the diet had a significant effect on k_l (P>0.05).

Table 2. Effect of concentrate proportion in the diet on animal performance and energy utilization.

	Concentrate proportion in the diet					
	0.37	0.48	0.59	0.70	SEM	Sig.
Total DMI (kg/day)						
Main study	18.8	20.4	21.8	22.2	0.42	***
Respiration chamber	18.5	19.9	18.0	20.4	0.86	NS
Milk output (kg/day)						
Main study	33.3	33.1	35.9	37.0	0.59	***
Respiration chamber	33.1	31.2	32.8	31.0	1.70	NS
Energy utilization data						
Energy digestibility	0.789	0.781	0.782	0.797	0.0060	NS
Methane E/GE intake	0.071	0.067	0.067	0.062	0.0032	NS
Urine E/GE intake	0.048	0.044	0.041	0.042	0.0014	*
Milk E/GE intake	0.336	0.284	0.319	0.276	0.0182	NS
ME/GE (q)	0.670	0.671	0.674	0.692	0.0077	NS
ME concentration (MJ/kg DM)	12.36	12.42	12.46	12.55	0.184	NS
Heat E/MEI	0.618	0.603	0.626	0.579	0.0239	NS
Milk E/MEI	0.505	0.425	0.474	0.399	0.0300	NS
Retained E/MEI	-0.123	-0.028	-0.101	0.022	0.0487	NS
k_l (AFRC, 1990)	0.54	0.53	0.52	0.54	0.018	NS
k_l (Yan *et al.*, 1997)	0.62	0.62	0.61	0.63	0.018	NS

Discussion

The higher intakes and milk yields of the high merit cows in this study agree with the findings of Gordon *et al.* (1995). That intakes and milk yields recorded in the respiration chambers were lower than for the main production study, may indicate that animals selected for the chambers were not representative of all animals on the production study, while in addition, the chamber environment may have altered performance.

That the digestibility of the energy component of the diet was unaffected by genotype agrees with the observations of Gordon *et al.* (1995) and suggests that animals of high and medium genetic merit digest food with similar efficiency. The higher methane energy output of the medium compared to the high merit animals is difficult to explain and conflicts with the findings of Gordon *et al.* (1995), who found no such effect. Although a reduced methane output may be associated with an increased rate of passage through the rumen, the fact that ration digestibility was unaffected by genotype in this study suggests that rate of passage was unaffected by genotype. The trend towards reduced methane losses with increased concentrate levels is in line with previous work and may be due to an increased rate of passage of digesta at the higher concentrate feed levels. The fall in urinary energy outputs with increasing concentrate levels reflects the fact that actual urinary energy losses remained relatively constant irrespective of treatment.

When k_l was calculated according to first principles from AFRC, values obtained were considerably below the value of 0.66, predicted from the relationship between k_e and metabolizability (q) in AFRC (1990). That low k_l values are common with grass silage based diets has been noted previously (Unsworth *et al.*, 1994), with Yan *et al.* (1996) suggesting that the maintenance requirements of lactating dairy cows offered grass silage based diets may in fact be 40% higher than those calculated using the equations of AFRC (1990). Consequently when maintenance energy requirements in the current study were calculated as 0.67 MJ/kg $LWT^{0.75}$, the value calculated by Yan *et al.* (1997) a mean k_l value of 0.62 was obtained, this value being similar to that predicted on the basis of the metabolizability (q) of the diets offered.

That similar k_l values were obtained for both high and medium merit cows in this study, irrespective of the maintenance component used, suggests that genetic merit has little effect on the partial efficiency of ME use for milk production. This is in line with the findings of Gordon *et al.* (1995) who recorded k_l values of 0.58 for both high and medium merit cows. However that k_l was unaffected by concentrate treatment is in disagreement with the equation presented by Offer *et al.* (1996) which indicates an increase in k_l with increasing proportion of concentrate in the diet.

Conclusions

The findings of this study suggest that neither genotype nor level of concentrate inclusion in the diet had a significant effect on the overall efficiency of energy utilization by dairy cows. This would appear to confirm the suggestion that the higher milk yields of high merit cows are largely due to their ability to mobilize more of their own body reserves in early lactation, while replenishing less of their own body reserves in later lactation, compared to animals of medium genetic merit. Thus as the higher milk yields can not be attributed to differences in the efficiency of nutrient utilization, it is likely that extra nutrients are required to support these higher milk outputs. Consequently management practices must be altered to ensure that higher intakes are achieved. If higher levels of concentrates are offered to satisfy these higher nutrient demands, the results of this study would suggest that this should have little effect on the efficiency of energy utilization for lactation.

References

Agricultural and Food Research Council (1990). Technical Committee on Responses to Nutrients, Report No. 5, Nutritive Requirements of Ruminant Animals : Energy. *Nutr. Abs. Rev. (Series B)*, 60:729-804.

Gordon F.J., Patterson D.C., Yan T., Porter M.G., Mayne C.S. & Unsworth E.F. (1995). The influence of genetic index for milk production on the response to complete diet feeding and the utilization of energy and nitrogen. *Anim. Sci.* 61:199-210.

Offer N.W., Cottril B.R. & Thomas C. (1996). The relationship between silage evaluation and animal responses. Proceedings of the 11th International Silage Conference, Aberystwyth, pp. 26-39.

Unsworth E.F., Mayne C.S., Cushnahan A. & Gordon F.J. (1994). The energy utilization of grass silage diets by lactating dairy cows. Proceedings of the 13th Symposium on Energy Metabolizm of Farm Animals. *EAAP Publ. No.*76:179-182.

Yan T., Gordon F.J., Agnew R.E., Porter M.G. & Patterson D.C. (1997). The metabolizable energy requirement for maintenance and the efficiency of utilization of metabolizable energy for lactation by dairy cows offered grass silage-based diets. *Livest. Prod. Sci.* 51:143-153.

UTILIZATION OF ENERGY AND PROTEIN BY SOWS UNDERFED DURING LACTATION

H. Fandrejewski, S. Raj and L. Lassota

The Kielanowski Institute of Animal Physiology and Nutrition, 05-110 Jablonna, Poland

Summary

Primiparous sows fed during 4 weeks lactation on a protein-free diet or control diet (12% d.p.) at low ME daily intakes (30 MJ ME), mobilized large amounts of protein and fat to maintain milk production. Sows fed a protein-free diet reduced milk yield and percentage of protein in the milk and body. Sows fed the control diet lost more fat from the body. Nitrogen requirement for maintenance amounted to 0.37 $g/kg^{0.75}/d$. Efficiency of utilization of digestible protein and body protein reserves for milk were 0.66 and 0.90, respectively. Energy utilization was affected by milk composition.
Key words: Sows, lactation, energy utilization, protein utilization, body composition.

Introduction

High-producing lactating sows exhibit a large weight loss which was usually assumed to be body fat. However, numerous studies (e.g. Kotarbinska, 1983; Everts, 1994) showed that considerable amounts of protein were also involved. Composition of weight loss may vary depending on which dietary component, energy or protein, is deficient and which component of body, fat or protein, is mobilized. Information concerning the energy and protein metabolism by lactating sows in various nutritional states is still limited (Burlacu *et al.*,1986; Noblet & Etienne, 1988; Everts,1994).

In the present study the dietary energy for lactating sows was in deficit. In addition, some of these animals were fed a protein-free diet to gain precise information on utilization of body protein alone for milk production.

Materials and Methods

Thirty-eight primiparous Landrace sows after parturition were allotted to one of three treatments P, E or C. Ten sows (treatment P) were offered a protein-free diet (12.9 MJ ME per kg) to appetite based on maize starch (87%). The remaining sows (treatments E and C.) were fed with a commercial feed compound (maize, soybean meal and premix) containing 151 g crude protein, 120 g digestible protein, 7.8 g lysine, 6.4 g methionine + cystine and 13.0 MJ ME per kg. The C sows (n=10) were fed near *ad libitum* level, whereas the Esows (n=18) received a similar daily amount of ME as the P sows (Table 1).

Shortly after birth the size of most litters was equalized to 10 piglets. Milk yield was determined by weight-suckle-weight method on d 4, 11, 18 and d 25 of lactation (6 measurements per day with suckling intervals of 1 hour). On the same days, milk samples (n=152) were obtained from all sows' teats by manual expression after an intramuscular injection of oxytocin, and then were analysed chemically. Fat percentage was determined by the Gerber method. Lactose content of the milk was obtained by subtraction of fat protein (N*6.38) and ash from dry matter. Energy content in milk was estimated using factors of 24.8, 35.8 and 15.9 kJ per g respectively for protein, fat and lactose.

Sows were kept in separate pens and fed twice daily. The underfed sows remained in the experiment until their production decreased markedly. The C sows were slaughtered on day 28 (±0.3) of lactation. All sows removed from litters were immediately slaughtered and (after autoclaving) their bodies were analysed chemically for nutrient and energy content. The comparative slaughter method was used to measure protein and energy loss during lactation. Initial body composition was predicted from data on live weight and ultrasonic backfat thickness obtained from another study with 63 sows slaughtered at 112 day of pregnancy.

Results and Discussion

Although a protein-free diet was offered to appetite, the P sows consumed only 2.3 kg of food daily. As expected, all animals lost weight during the lactation period. Sow weight loss was greater in the E. and P. group (1.32 and 1.78 kg/d) than in the C group (0.56 kg/d). The mean daily milk yields differed between treatments (Table 1). Yields were more dependent on intake of protein than energy. The differences in milk output were markedly greater at the end than at the beginning of the suckling period. At day 25 of lactation, the C, E and P sows produced respectively 7.1, 6.0 and 3.7 kg of milk per day (P<0.01).

Table 1. Milk yield and composition.

	C	E	P	SEM
No. of sows	10	18	10	-
ME intake, MJ/d	55.3	30.4	29.4	0.97
DP intake, g/d	511	280	-	11.0
Days of lactation	28.4	30.0	29.6	0.52
Litter size	10.2	10.0	9.9	0.06
Initial live weight, kg	152	163	171	2.1
Loss of live weight, kg	15.9^{A}	40.8^{B}	51.9^{C}	1.5
Loss of fat thickness, mm	2.5^{A}	7.6^{B}	5.1	0.4
Milk yield, kg/d	6.40^{Aa}	5.82^{b}	4.66^{Bb}	0.1
Milk DM. %	18.56	18.88	18.47	0.1
in DM, g/kg, protein	291^{a}	281^{a}	248^{b}	2.7
fat	407	432	431	4.6
lactose	260	243^{a}	280^{b}	4.0
ash	42	44	41	0.5
energy, MJ	26.0	26.4	26.1	0.2
Protein in milk, g/d	340^{A}	303^{B}	210^{C}	5.0

^{a,b} - P<0.05 ^{A,B,C} - P<0.01.

The P sows had less protein in milk than their analogues, and more lactose than the E sows (Table 1). In consequence, at similar intakes of ME, the E sows fed the control diet produced much more protein in milk than those fed the protein-free diet. Different amounts of the control diet (30 or 56 MJ/d) did not change the chemical composition of milk. There was no effect of treatment on energy content of the milk.

Table 2. Chemical body composition of sows.

	C	E	P	SEM
No	10	18	10	-
Empty body weight, kg	128^{A}	113^{B}	113^{B}	1.87
Protein, g in the EBW	153^{A}	156^{A}	135^{B}	1.64
Fat, g in the EBW	202^{A}	206^{A}	292^{B}	6.82
Ash, g in the EBW	32	34	34	0.66
LBM/protein ratio	5.23	5.09	5.24	0.23
Ash/protein ratio	0.21^{A}	0.22^{A}	0.25^{B}	0.01

^{A,B} - P<0.01

Feeding levels were without effect on the body composition of the C and E sows (Table 2). In contrast, the P sows contained less protein and more fat (P<0.01). They lost 8.6±0.47kg of protein from the body (36% of the initial content). However, treatment had no significant influence on the LBM/protein ratio in the body. In contrast, the ash/protein ratio was higher (P<0.05) in the P sows. This indicated that, during food restriction, ash is a more stable component in the body than water.

All protein and energy balances were negative, even those measured in the control group. The C sows were omitted from further analysis due to energy mobilization being too low in applying the slaughter procedure. Protein and energy balances of underfed sows are presented in Table 3.

Table 3. Protein and energy balance in lactating sows underfed in terms of energy (E) or protein (P).

		E	P	SEM
	Intake (d.p.)	280	0	11.5
	Loss (Nx6.38)	174^A	292^B	7.68
Protein, g	In milk	303^A	210^B	6.25
	Efficiency*	0.67	0.72	0.12
	Intake	29.4	30.4	1.01
	Loss, total	30.4	25.6	1.39
Energy, MJ	In fat	26.2^A	18.6^B	1.29
	In protein	4.2^A	7.0^B	1.83
	In milk	28.8^A	22.4^B	0.80
	Efficiency*	0.75^a	0.66^b	0.02

* Production in milk /(intake + body loss), LWm = 41.1 (E), 41.7 $kg^{0.75}$ (P)

^{a,b} - P<0.05, A,B - P<0.01.

Sows fed the protein-free diet mobilized significantly (67%) more protein from the body, and secreted in milk 45% less protein than E group. Gross efficiency of protein (milk /(intake + body loss)) tended to be higher on treatment P than treatment E. Partial efficiency of nitrogen was measured in the E and P groups separately by linear regression related to metabolic weight ($kg^{0.75}$). In the P group (n=10), the relationship between nitrogen loss from the body (N_{body}) and nitrogen secreted in milk (N_{milk}) was expressed by the equation:

$$N_{body} = (1.11\ 0.095)N_{milk} + (0.243 \pm 0.077) \quad R^2=0.94,\ RSD=0.041 \qquad (1)$$

From the equation it can be calculated that the efficiency of conversion of body protein into milk was 0.90 (1/1.11). No such direct measurements are available in the literature. In Equation 1, the intercept (0.243 g/$kg^{0.75}$) may express the total obligatory losses (via faecal, urine and skin). The estimated OL were 36% higher than those found for growing pigs fed diets extremely low in protein (Carr *et al.*,1977).

Assuming the E sows (n=18) obligatory losses of 307mg dN (243/0.79), the efficiency of milk production from body (N_{milk}) and digestible nitrogen (dN) was calculated as:

$$N_{milk} = (0.91 \pm 0.12)N_{body} + (0.66 \pm 0.14)(dN - 0.307) \quad RSD=0.14,\ R^2=0.98 \qquad (2)$$

This equation confirms the utilization of protein body for milk production estimated by Equation 1 (0.90), and shows that digestible protein is used with an efficiency of 0.66. The Equations 1 and 2 allow an estimate of lactational maintenance requirement. The estimated maintenance is 0.37 g digestible N/kg$^{0.75}$. This value is only 2% lower than that proposed by Mullan *et al.* (1993)

At similar rates of daily ME intake the E sows lost more energy from the body due to higher fat mobilization, and produced more energy in milk (Table 3). The gross efficiency of energy [milk/(intake + body loss)] was significantly higher ($P<0.05$) in sows fed the control than the protein-free diet (0.75 vs. 0.66). This could be partly due to higher protein content (and lower content of lactose) in milk of the E sows, as the theoretical efficiency of milk synthesis is higher for protein than non-protein components (Baldwin, 1968). Approximate estimates of partial efficiency of utilization of food (k_l) and body reserves (k_r) for milk production were 0.56±14 and 0.77±0.04 respectively (by assumption that ME_m=500 kJ). Although our estimates are lower than that obtained by others (e.g. Burlacu *et al.*,1983; Verstegen *et al.,* 1985), the difference between k_l and k_r agrees with the value obtained by Noblet & Etienne (1987).

Conclusions

It may be stated that sows, restricted in both protein and energy, can mobilize large amounts of protein and fat to maintain the milk production, but the composition of milk and weight loss of lactating sows changes depending on which dietary component, energy or protein, is more deficient. The estimated maintenance requirement for protein in lactating sows was 0.37g/kg$^{0.75}$. Net efficiency of protein utilization from body reserves and digestible protein were 0.90 and 0.66 respectively. Change of milk composition affected energy utilization.

References

Baldwin R.L. (1968). Estimation of theoretical calorific relationship as a teaching technique. A review. *J. Dairy Sci.* 51:104-111.

Burlacu G., Iliescu M. & Carmida P. (1986). Efficiency of food utilization by pregnant and lactating sows. *Archiv. Anim. Nutr.* 36:803-825.

Carr J.R., Boorman K.N. & Cole D.J.A. (1977). Nitrogen retention in the pig. *Br. J. Nutr.* 37:143-155.

Everts H. (1994). Nitrogen and energy metabolism of sows during several reproductive cycles in relation to nitrogen intake. Thesis.

Kotarbinska M. (1983). Chemical body composition of gilts and sows. *Pig News and Information.* 4:275-278.

Mullan B.P., Close W.H. & Cole D.J.A. (1993). Predicting nutrient responses of the lactating sow. In: D.J.A. Cole, W. Haresign and P.C. Garnsworthy (Eds) Recent developments in pig nutrition 2. Nottingham, pp. 332-346.

Noblet J. & Etienne M. (1987). Metabolic utilization of energy and maintenance requirements in lactating sows. *J. Anim. Sci.* 64:774-781.

ENERGY METABOLISM OF LACTATING FIRST LITTER SOWS AS AFFECTED BY FEEDING LEVEL AND DIET COMPOSITION

H. van den Brand, J.W. Schrama, M.J.W. Heetkamp and B. Kemp

Department of Animal Sciences, Wageningen Institute of Animal Sciences, P.O. Box 338, 6700 AH, Wageningen, The Netherlands

Summary

Twenty-four primiparous sows were assigned to a 2 x 2 factorial design of treatments during a three week period of lactation. Treatments were diet composition: fat rich (F) or carbohydrate rich (C) and feeding level: 44 MJ of NE (H) or 33 MJ of NE (L) per day. From day 6 until day 20 two, one week energy balances were determined in climate respiration chambers. Milk production was higher on C-diet compared to F-diet, but growth of the piglets was not affected by diet composition. Piglets grew faster when sows were fed a high feeding level compared to a low feeding level. Body weight loss of sows was not affected by diet composition, but sows lost more weight when fed a low feeding level. Heat production and net energy for gain of sow and piglets were affected by feeding level but not by diet composition. Respiration quotient was higher in C-diet and high feeding level compared to F-diet and low feeding level.
Key words: Pigs, lactation, energy balance, diet composition, feeding level, milk production.

Introduction

In primiparous sows, an increased body weight loss and back fat loss during lactation results in a prolonged weaning to oestrus interval (WOI) (Kirkwood *et al.*, 1987; Baidoo *et al.*, 1992; Vesseur *et al.*, 1994). Ovulation rate seems not to be affected by body weight loss and back fat loss during lactation (Kirkwood *et al.*, 1987; Baidoo *et al.*, 1992). Effect of body weight loss and backfat loss on embryo mortality is not yet clear. Baidoo *et al.* (1992) found no effect, whereas Kirkwood *et al.* (1987) found a lower embryo survival in sows with a high body weight loss and backfat loss.

Restricted energy and protein intake combined with high demands for milk production are the major factors responsible for body weight loss and backfat loss. However the effect of energy source in this case is hardly investigated. Nelssen *et al.* (1985) concluded that primiparous sows fed a fat rich diet lost more weight during lactation compared to corn starch fed sows. Also backfat loss tended to be greater in sows fed tallow compared to isocalorically corn starch fed sows. Schoenherr *et al.* (1989) and Babinszky (1992) found no effects of energy source during lactation on body weight loss and backfat loss.

In the present study the effect of energy source at two feeding levels during lactation on milk production, growth of piglets, body weight loss of the sows, heat production of sows and piglets and reproduction parameters in subsequent pregnancy were investigated.

Materials and Methods

Twenty-four Great Yorkshire x Dutch Landrace primiparous sows, with 9 piglets each, were assigned to a 2 x 2 factorial arrangement of treatments. Calculated feeding levels during a three week period of lactation were 44 MJ of NE (H) or 33 MJ of NE (L) per day. At each feeding level, 50% of the sows was assigned to a carbohydrate rich diet (C: 380.0 g/kg carbohydrate, 33.4 g/kg fat) and the other half to a fat rich diet (F: 196.7 g/kg carbohydrate, 135.0 g/kg fat). Diets within each feeding level were fed isocalorically and were iso-nitrogenous.

From day 3 of lactation onwards, eight groups of 3 sows with 9 piglets each were housed in one of two large climate respiration chambers (Verstegen *et al.*, 1987). From day 6 onwards energy and nitrogen balances were measured during two 7 d balance periods. Heat production of sows and piglets together was measured during 4 days a week.

Milk production was measured at day 10 and 17 by weigh-suckling-weigh method for 8 consecutive measurements at one-hour intervals. Multiplying by 24 estimated the daily milk production. For the calculations of milk yield, corrections of weight loss of piglets due to metabolic rate, evaporations, urinations and defecations during suckling were applied.

The following model was used for analysing body weight loss, WOI, ovulation rate and embryo survival:

$$Y_{ijk} = m + L_i + D_j + L_i{*}D_j + e_{ijk}$$

Milk production and growth of the piglets were analysed using the model:

$$Y_{ijkl} = m + L_i + D_j + L_i{*}D_j + S_k(L_i{*}D_j) + W_l + e_{ijkl}$$

For energy intake, HP, RQ and NE gain for sow + piglets next model was used:

$$Y_{ijlm} = m + L_i + D_j + L_i{*}D_j + G_m(L_i{*}D_j) + W_l + e_{ijlm};$$

where $Y_{ijk(lm)}$ = dependent variable; m = mean; L_i = feeding level (i=H,L); D_j = diet (j=F,C); $L_i{*}D_j$ = interaction feeding level and diet; W_l = week (l=1,2); $S_k(L_i{*}D_j)$ = sow variation between interaction feeding level and diet; $G_m(L_i{*}D_j)$ = group variation between interaction feeding level and diet; $e_{ijk(lm)}$ = residual error.

Results

Results of body weight loss of the sow during lactation and some reproduction parameters in subsequent pregnancy are shown in Table 1.

Table 1. Body weight loss between day 6 and day 20 and reproduction results in subsequent litter (mean±SEM).

Diet	Fat		Carbohydrate	
Feeding level	High	Low	High	Low
Inseminated/pregnant	6/6	6/5	6/5	5/5
Body weight loss (kg)	9.8 (1.6)	13.7 (2.7)	6.2 (1.6)	12.7 (2.1)
WOI (hour)	115 (5.6)^a	158 (15.1)^b	116 (5.7)^a	129 (15.7)^{ab}
Ovulation rate	18.7 (1.3)	16.0 (1.3)	19.0 (1.1)	17.4 (1.0)
Embryo survival (%)	72.5 (6.3)	60.0 (5.4)	63.5 (5.4)	69.3 (6.6)

^{a,b} Means in a row with different superscripts differ significantly (P<0.05).

Body weight loss between day 6 and day 20 was significantly affected by feeding level (P<0.05), but not by diet composition. Low feeding level resulted in a prolonged weaning to oestrus interval, especially in the fat rich diet, resulting in a feeding level * diet interaction (P=0.06). Body weight loss between day 6 and day 20 was not significantly correlated with any reproduction parameter in the subsequent litter (P>0.10).

Average milk production and growth of piglets in both balance periods is shown in Table 2.

Table 2. Average milk production (kg/sow/d) and growth of piglets (g/d) in each balance period (mean±SEM).

Diet	Fat		Carbohydrate	
Feeding level	High	Low	High	Low
Number of observations	6	6	6	6
Milk production (week 1)	7.0 (0.25)	6.6 (0.14)	7.3 (0.19)	7.2 (0.21)
Milk production (week 2)	7.4 (0.33)	7.2 (0.17)	9.2 (0.23)	7.9 (0.29)
Growth of piglet (week 1)	251 (6.7)	199 (10.2)	261 (6.4)	207 (8.4)
Growth of piglet (week 2)	258 (6.7)	229 (9.6)	284 (4.5)	245 (7.9)

Milk production was significantly higher in sows fed a carbohydrate rich diet compared to a fat rich diet (7.0 vs. 7.8 kg/d; $P<0.05$) and was increased in week 2 compared to week 1 (7.0 vs. 7.9 kg/d; $P<0.001$). Feeding level did not affect milk production (7.7 vs. 7.2 kg/d for high and low feeding level respectively), but growth of piglets was significantly higher in sows fed a high feeding level compared to a low feeding level (263 vs. 221 g/d, $P<0.01$). Diet composition and week had no effect on growth of piglets.

In Table 3 the energy intake of the sow, total heat production of sow and piglets, respiration quotient and net energy for gain of sow and piglets is shown.

Table 3. Balance parameters for sows and piglets (d6-d20; LS means).

Diet	Fat		Carbohydrate		
Feeding level	High	Low	High	Low	SEM
Sow (kJ/kg$^{0.75}$/d)					
Gross energy intake	1767	1297	1784	1295	30
Sow+piglets					
Total HP (kJ/kg$^{0.75}$/d)	707	663	750	668	12
RQ (CO_2/O_2)	0.83	0.81	0.92	0.87	0.01
NE gain (kJ/kg$^{0.75}$/d)	110	-64	136	-34	30

Energy intake by sows was affected by feeding level ($P<0.001$), but not by diet composition. Intake during the first week tended to be lower compared to the second week ($P=0.09$). Heat production of sows and piglets together was higher at high feeding level (729 vs. 666 kJ/kg$^{0.75}$/d; $P<0.01$), where diet and week had no effect. NE gain for sows and piglets together was higher in high feeding level and in week 1 compared to low feeding level ($P<0.01$) and week 2 ($P<0.05$) but was not affected by diet. Respiration quotient was significantly higher in the carbohydrate rich diet, high feeding level and first week compared to the fat rich diet, low feeding level and second week ($P<0.05$).

Discussion

Body weight loss of primiparous sows during lactation was affected by feed intake. Kirkwood *et al.* (1987) and Baidoo *et al.* (1992) found comparable results. Diet composition did not significantly affect body weight loss. This is in agreement with Schoenherr *et al.* (1989) and Babinszky (1992). Nelssen *et al.* (1985) found in sows fed a diet with tallow a higher body weight loss compared to a corn starch rich diet.

Although the body weight loss of the sows was not affected by diet composition, the milk yield was higher in the sows fed a carbohydrate rich diet and also the growth of the piglets (not significant) was higher. Schoenherr *et al.* (1989) and Babinszky (1992) found no differences in milk production between a starch rich and a fat rich diet. However in the present study milk production per gram growth of the piglets was similar for both diets (3.4 vs. 3.5 g milk/g growth for fat and carbohydrate diet respectively). This indicates that total energy secretion in the milk per day is comparable for both diets. Future determination of milk composition will provide this information.

Total heat production of sows and piglets was not affected by diet composition as well as the net energy for gain of sows and piglets. Heat production at the low feeding level was comparable for both diets, whereas at the high feeding level heat production was higher in the carbohydrate rich diet, although not significant. This is in accordance with the results of Babinszky (1992), who also found a higher heat production in starch rich diets compared to fat rich diets (695 vs. 674 $kJ/kg^{0.75}/d$).

The higher respiration quotient for carbohydrate rich diet is in accordance with Babinszky (1992). The high RQ for the C-diet indicates that milk fat has been synthesized mainly from carbohydrates.

In conclusion these preliminary results show that WOI and possibly other reproduction parameters are affected by treatment. A carbohydrate rich diet can partly compensate for the negative impact of a low energy intake during lactation on WOI. Although diet composition does not seem to affect body weight loss, milk production was (not significantly) increased in sows fed a carbohydrate rich diet.

Acknowledgements

The financial support of the Dutch Organization of Scientific Research is gratefully acknowledged as well as the technical assistance of J.M. van der Linden.

References

Babinszky L. (1992). Energy metabolism and lactation performance of primiparous sows as affected by dietary fat and vitamin E. PhD thesis. Wageningen Agricultural University.

Baidoo S.K., Aherne F.X., Kirkwood R.N. & Foxcroft G.R. (1992). Effect of feed intake during lactation and after weaning on sow reproductive performance. *Can. J. Anim. Sci.* 72:911-917.

Kirkwood R.N., Baidoo S.K., Aherne F.X. & Sather A.P. (1987). The influence of feeding level during lactation on the occurrence and endocrinology of the post weaning estrus in sows. *Can. J. Anim. Sci.* 67:405-415.

Schoenherr W.D., Stahly T.S. & Cromwell G.L. (1989). The effects of dietary fat or fibre addition on yield and composition of milk from sows housed in a warm or hot environment. *J. Anim. Sci.* 67:482-495.

Verstegen M.W.A., van der Hel W., Brandsma H.A.,. Henken A.M & Bransen A.M. (1987). The Wageningen respiration unit for animal production: a description of the equipment and its possibilities. In: M.W.A. Verstegen and A.M. Henken (Eds) Energy metabolism in farm animals. Martinus Nijhoff Publishers, Dordrecht, pp. 21-48.

Vesseur P.C., Kemp B. & den Hartog L.A. (1994). Factors affecting the weaning to oestrus interval in the sow. *J. Anim. Phys. Anim. Nutr.* 72:225-233.

MODELLING METABOLIZABLE ENERGY UTILIZATION IN LAYING HENS

B. Zhang and C.N. Coon

Department of Animal Science, University of Minnesota, St Paul, MN 55108, USA

Summary

Experiments were conducted to investigate the energy maintenance requirement (ME_m) as affected by temperature (T °C), and utilization efficiency for egg synthesis (EUEP) using either a new technique with hens treated with Tamoxifen (TAM) or restriction feeding technique (RFT). The estimated ME_m was 117.3 ± 2.0 kcal/kg $W^{0.75}$, and the EUEP was 60.4 ± 2.0% using TAM at 19.7°C. Using TAM technique in another experiment at temperatures of 7.5, 15.2, 21.4, 25.8, 32.4, 36.5°C, the relationship of ME_m and T was well described by: ME_m (K=kcal/$W^{0.75}$) = $0.0905T^2$ - 6.5873T + 209.4 for a non-laying period, and ME_m (kcal/$W^{0.75}$) = 143.7 - 1.612T for a laying period when assuming a constant 63% EUEP. The ME_m determined during the non-laying period appeared to overestimate ME_m during the laying period for hens housed at 7.5 and 15.2°C. The data indicates that dietary energy not retained during egg formation may be used to maintain body temperature for layers housed in cold temperatures. Using RFT, EUEP was 0.684, 0.716, and ME_m 120 and 123 kcal/kg $W^{0.75}$ from two regression equations. From the coefficients determined using TAM, two ME intake prediction models were developed, and tested against other models in the literature. Model 2 [MEI=$W^{0.75}$(143.7 -1.612T) + 5DW + EM*EEC/0.63, where MEI = energy intake (kcal/h), W = BW (kg), DW = BW change (g/h/d), EM = egg mass (g/h/d), EEC = egg energy concentration (kcal/g EM)] produced the nearest predictions among the models tested.
Key words: Energy, prediction, model, efficiency, utilization.

Introduction

Prediction models for ME intake are of importance since dietary concentration of other nutrients depends on ME intake in laying hens. The accuracy of energy partition models relies on the estimation of maintenance (ME_m), and energy efficiency utilization (EU) for production. Significant variations exist in the estimation of ME_m and EU in the literature, thus the prediction models, which are mainly a result of difficulties in the estimation methodology of the coefficients (Chwalibog, 1991).

Therefore the objective of this study was to determine the ME_m, and EU for egg production.

Materials and Methods

Experiment 1. A total of 50 laying hens, 46 week of age were individually caged at 19.7°C and fed *ad libitum*. Hens were intramuscularly injected with 10 mg TAM/kg body weight at days 1, 4, 7, and 10. Body weight (W, kg) and feed intake (FI) were recorded every third day during the 9 week experiment. Egg production (EP) was recorded daily. Zero EP was reached one week after the first TAM administration. The ME intake (MEI, kcal/h/d) was regressed against the $W^{0.75}$ and weight change (DW, g/d) using the data collected during the zero EP period for each hen to estimate ME_m and energy required for unit weight gain. These estimated values were then used to calculate the EU for production after the hen rejuvenated lay. Thus the ME_m and EU for production were estimated independently for each individual.

Experiment 2. A total of 120 individually caged laying hens, 62 weeks of age, were assigned into 6 rooms of 20 each (7.5, 15.2, 21.4, 25.8, 32.4, 36.5°C). The same technique as in experiment 1 was used in this experiment to determine ME_m as affected by temperature (T, °C), and EU for production.

Experiment 3. A total of 90 laying hens, 84 weeks of age, were assigned into 9 groups with 10 hens each. One group (10 hens) was sacrificed at beginning of the experiment for body composition analysis. The remaining groups were provided daily FI of 30, 40, 50, 60, 70, 80, 90 g/h, and *ad libitum* for 16 days, and sacrificed at the end of the trial and analysed for body composition. Eggs laid during the experiment were weighed and analysed for egg composition. The MEI was regressed against body energy change (BEC), and energy produced from egg mass (EME).

Experiment 4. From the estimates obtained in Experiments 1-3, two ME intake (MEI, kcal/d) prediction models were developed: $MEI=W^{0.75}(209.4-6.5873T+0.0905T^2)+5DW + EM*EEC/0.63$ (Equation 1); $MEI=W^{0.75}(143.7 -1.612T) + 5DW + EM*EEC/0.63$ (Equation 2), where EM = egg mass (g/h d), EEC = egg energy concentration (kcal/g EM). The two models were tested against the models in the literature (NRC, 1994; Emmans, 1974; Peguri and Coon, equation 3, 1988; Pesti *et al.*, 1992) using the following experimental data.

A total of 480 individually caged hens, 26 weeks of age, were assigned into a 4 x 4 (temperature x strain: 10.0, 18.3, 26.7, 35°C; A,B,C,D) factorial arrangement. Four diets were formulated for each temperature based on the FI during a 2 week acclimatization period, and provided to hens *ad libitum* for 3 periods each of 4 weeks. Individual performance (FI, W, EM) was measured.

Results and Discussion

Experiment 1. The estimated average daily ME_m was 117.3 ± 2.0 kcal/$W^{0.75}$. The EU for egg production was 60.4 ± 1.78% (Table 1).

Table 1. Estimated ME_m (a), energy required for unit weight change (c and c') using two regression models[1] and the estimated EU for production (EUEP) in Expt. 1.

	Model 1			Model 2		EUEP
	a	c	Adjusted R^2	c'	Adjusted $R^{2'}$	(%)
n	46	46	46	46	46	38
Mean	117.3	3.764	0.977	3.742	0.768	60.45
SE	2.03	0.101	0.003	0.098	0.023	1.776

[1]Model 1: MEI = a$W^{0.75}$ + cWC, where WC = weight change (g/d).
Model 2: MEI = a' + c'WC (without forcing the line through the origin).

Experiment 2. The EU for production was estimated to be 112.8, 81.4, 61.4, 66.7, 63.0 and 69.8% at 7.5, 15.2, 21.4, 25.8, 32.4, 36.5°C, respectively (Table 2). The relationship of ME_m and T was well described by the equation: ME_m (kcal/$BW^{0.75}$) = $0.0905T^2 - 6.5873T + 209.4$ ($R^2 = 0.9805$, $P<0.005$, MSR=28.4) during the non-laying period. ME_m determined during non-laying period appeared to be over-estimates of ME_m during laying period as evidenced by the over-estimated EU for production at the cold Ts (7.5 and 15.2°C). By assuming a constant EUEP of 63% for the laying period across all temperatures the relationship of ME_m and T was well described by a linear regression line: ME_m (kcal/$BW^{0.75}$) = 143.7 - 1.612T (R^2=0.78). Figure 1 illustrates the differences in ME_m predicted by the two equations for the six temperatures.

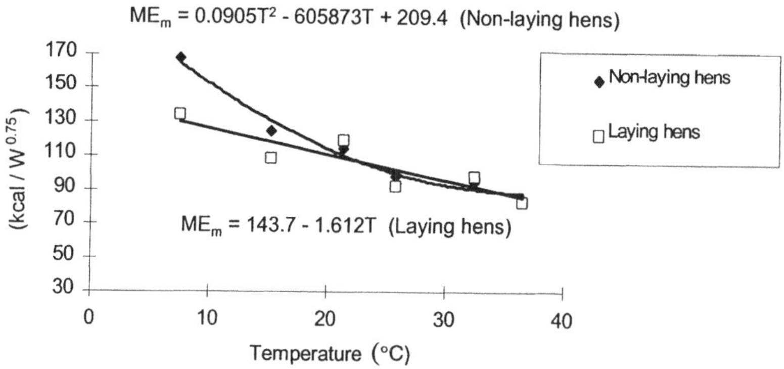

Fig. 1. The effect of temperature on energy maintenance requirement.

Table 2. Production parameters, and calculated coefficients in Expt 2.

T (°C)	7.5	15.2	21.4	25.8	32.4	36.5
n	18	14	19	14	14	9
EM (g/h/d)	48.15	48.08	48.48	52.29	49.57	31.85
Egg GE (kcal/gEM)	1.65	1.61	1.64	1.66	1.65	1.60
BW (kg)	1.64	1.73	1.66	1.91	1.63	1.53
W change (g/h/d)	-0.19	-1.71	0.91	-0.06	-4.76	-4.05
FI (g/h/d)[1]	108.1	94.3	103.0	97.3	83.8	59.3
EU for production (%)	112.8	81.4	61.4	66.7	63.0	69.8
ME_m (kcal/$W^{0.75}$/d)	167.2	124.2	114.5	97.8	94.6	86.9
ME_m Corrected[2]	134.2	108.5	118.9	92.4	98.2	83.1

[1]Feed contained 2954 kcal/kg .AME_n
[2]Corrected maintenance was calculated by assuming a 63% EU for production and 5 kcal increase or decrease in energy requirement per g weight gain or loss.

Experiment 3. Three regression equations were derived from the data: MEI = 161.6 + 1.026BEC + 1.4623 EME (R^2=91.51) (equation 1); MEI = 120.1$W^{0.75}$ + 0.968 BEC + 1.397 EME (R2=0.9905) (equation 2); and using data from hens with negative weight change: MEI = 123.5$W^{0.75}$ + 0.985BEC + 1.348EME (R2=0.989) (equation 3). The data showed that body energy was used at an equivalent value compared to dietary energy. EUEP was 0.684, 0.716, 0.725 from equations 1-3, and ME_m was 120 and 123 kcal/kg $W^{0.75}$ from equations 1 and 2, respectively at 19.7°C.
Experiment 4. Equation 2 produced the closest predictions to the actual ME intake. The NRC (1994) equation produced the second best predictions among the models, but appeared to overestimate the intake at cold Ts. The empirical model (Pesti *et al.*, 1992) produced the most unreliable predictions (Table 3).

Table 3. Actual ME intake (MEI; kcal/d) and predicted ME intake (kcal/d) using prediction models in Expt 4.

Temp	Strain	MEI	Model 1	Model 2	Emmans 1974	NRC 1994	Pesti 1992	Peguri and Coon
10		324	361	324	357	338	364	354
18.3		297	319	311	333	322	333	342
26.7		282	290	294	305	301	301	324
35		223	219	215	214	219	328	227
	A	277	288	278	290	286	338	296
	B	296	314	302	326	313	335	341
	C	291	305	293	309	302	293	321
	D	264	284	273	287	282	362	292

References

Pesti G.M., Dorfman J.H. & Gonzalez-A M.J. (1992). Model for predicting egg production and metabolizable energy intake of laying pullets. *Br. Poult. Sci.* 33:543-552.

Peguri A. & Coon C.N. (1988). Development and evaluation of prediction equations for metabolizable and true metabolizable energy intake for the DeKalb XL-link White Leghorn hen, In: 49th Minnesota Nutrition Conference and Degussa Technical Symposium, Bloomington, Minnesota, pp. 199-211

National Research Council (NRC) (1994). Nutrient Requirements of Poultry. 9th ed., National Academy Press, Washington, DC.

Emmans G.C. (1974). The effect of temperatures on the performance of laying hens, In: T. R. Morris and B. M. Freeman (Eds) Energy requirements of poultry, British Poultry Science Limited, Edinburgh, Scotland, pp. 79-90

Chwalibog A. (1991). Energetics of Animal Production. *Acta Agric. Scand.* 41:147-160.

SOME ASPECTS OF THE ENERGY METABOLISM IN LACTATING GOATS

L. Rapetti, G.M. Crovetto, A. Tamburini, G. Galassi, A. Sandrucci and G. Succi

Istituto di Zootecnia Generale, Facoltà di Agraria, via Celoria 2, 20133 Milano, Italy

Summary

Data from 3 trials (27 energy balance experiments) on lactating Saanen goats fed 9 different diets typical of an intensive feeding system were considered to evaluate energy requirements for maintenance and production. The estimated daily ME_m was 484 kJ/kg $W^{0.75}$ and the efficiency of utilization of metabolizable energy for milk production (k_l) was 0.619. The relationship between RE and LE showed that the reduction of 1 kJ of milk energy yield was associated with an increase in body energy retention equal to 0.986 kJ. The results obtained show that ME_m and k_l in the lactating goats are similar to those of dairy cows, while the energy content of goat 4% FCM (2889 kJ/kg) is lower.

Key words: Lactating goats, energy metabolism, energy requirements.

Introduction

Armstrong & Blaxter (1965) and Aguilera *et al.* (1990) have measured energy requirements for maintenance and production of lactating goats, estimated from measurements of energy balance. These experiments were performed on goats characterized by a milk energy yield lower than that produced by actual high yielding dairy goats. The present study evaluated the energy metabolism of lactating Saanen goats in an intensive-type feeding system.

Materials and Methods

Twenty-seven energy balance experiments were carried out in three different trials on 18 pluriparous Saanen goats in mid-lactation. In each trial six animals were paired and fed *ad libitum* the different diets, in a 3 x 3 Latin square design.

The rations tested were typical of an intensive feeding system. The differences between diets were based on: in the first trial the base forage: permanent pasture hay (H), ryegrass silage (R) or maize silage (M) (Crovetto *et al.*, 1994); in the second trial different forage to concentrate ratio: 30:70 (30), 50:50 (50) and 70:30 (70) (Rapetti *et al.*, 1997); in the third trial different energy sources: maize starch (MS), whey lactose (W) and rumen protected fat (F) (Rapetti *et al.*, 1995).

Individual apparent digestibilities were determined on the actual ingested diets (8 days for each collection period), while respiratory exchanges were recorded by indirect calorimetry for each pair of goats (3 or 4 x 24 h cycles) during digestibility measurement. Total heat production (HP) was determined by indirect calorimetry from the respiration exchanges using the Brouwer (1958) equation.

The goats were fed and milked twice a day. Samples of feedstuffs, feed refusals, faeces and urine were taken daily; milk was sampled twice a day. Gross energy of all samples was determined in an adiabatic calorimeter (IKA 4000). Milk energy secretion (LE) was also corrected as a function of retained energy (RE) as follows: corrected milk energy secretion (LE_c) = LE + (0.84 x negative retained energy RE) + (1.0142 x positive RE), where 0.84 is derived from AFRC (1980) for balances with RE<0, while 1.0142 originates from the present studies on the balances with RE>0.

Regression analysis was utilized to determine: 1) metabolizable energy for maintenance (ME_m), by regression of LE plus RE versus metabolizable energy intake (MEI); 2) the relationship between body energy retention (RE) and milk energy secretion (LE), by regression of RE on LE both expressed as a percentage of MEI above maintenance (ME_p), according to Armstrong & Blaxter (1965), using for ME_m the value obtained by Equation (1); 3) the efficiency of utilization of metabolizable energy for milk production (k_l) by regression of LE_C versus MEI; and 4) the energy content of goat milk.

Results

The daily energy balances of the goats fed the different diets tested in the three trials are summarized in Table 1.

Table 1. Daily energy utilization and partition of the diets considered in the study.

Diet	H	R	M	30	50	70	MS	W	F
kJ/kg W$^{0.75}$									
Intake (IE)	2265	2135	2422	2528	2303	2071	2325	2376	2351
Faeces	688	567	733	777	738	718	642	612	605
DE	1577	1568	1689	1751	1565	1353	1683	1764	1746
Urine	69	67	69	87	76	87	66	60	65
Methane	166	141	161	140	151	141	173	194	159
ME	1342	1360	1460	1524	1338	1125	1443	1510	1522
Heat	785	790	825	851	785	750	834	872	837
Milk	426	456	490	455	410	353	445	452	506
RE	131	114	145	218	143	22	163	185	178
Partition of IE (%)									
Faeces	30.4	26.5	30.3	30.7	32.0	34.6	27.6	25.8	25.7
DE	69.6	73.5	69.7	69.3	68.0	65.4	72.4	74.3	74.3
Urine	3.1	3.2	2.8	3.5	3.3	4.2	2.8	2.5	2.8
Methane	7.4	6.6	6.6	5.5	6.6	6.8	7.5	8.2	6.8
ME	59.1	63.7	60.3	60.3	58.1	54.4	62.1	63.5	64.7
Heat	34.6	37.0	34.0	33.8	34.2	36.4	36.0	36.8	35.7
Milk	18.7	21.4	20.2	18.0	17.8	17.1	19.1	19.0	21.4
RE	5.8	5.3	6.1	8.5	6.1	0.9	7.0	7.7	7.6
Partition of ME (%)									
Heat	58.5	58.1	56.4	56.0	58.9	66.9	58.0	57.9	55.2
Milk	31.6	33.5	33.6	29.8	30.6	31.3	30.7	29.9	33.1
RE	9.9	8.4	10.0	14.2	10.5	1.8	11.3	12.2	11.7

The metabolizable energy for maintenance was computed using the following equation which we have developed:

$$LE + RE = 0.634\ MEI - 307\ (n=27;\ r^2 = 0.93;\ RSD = 28.5;\ P<0.001) \qquad (1)$$

where LE, RE and the metabolizable energy intake (MEI) are expressed as kJ/kg $BW^{0.75}$ (see Fig. 1a). Assuming LE + RE = 0, ME_m resulted to be 484 kJ/kg $BW^{0.75}$.

Energy balances were always positive except in one case. The regression of RE on LE (see Fig. 1b) (both expressed as percentage of ME above maintenance, using for ME_m the value of 484 kJ/kg $BW^{0.75}$) calculated on the positive balances, showed that the reduction of 1 kJ of milk energy yield was associated with an increase in body energy retention equal to 0.986 kJ. In fact the equation obtained was:

$$RE = -\ 0.986\ LE + 62.92\ (n=26;\ r^2 = 0.67;\ RSD = 3.00;\ P<0.001) \qquad (2)$$

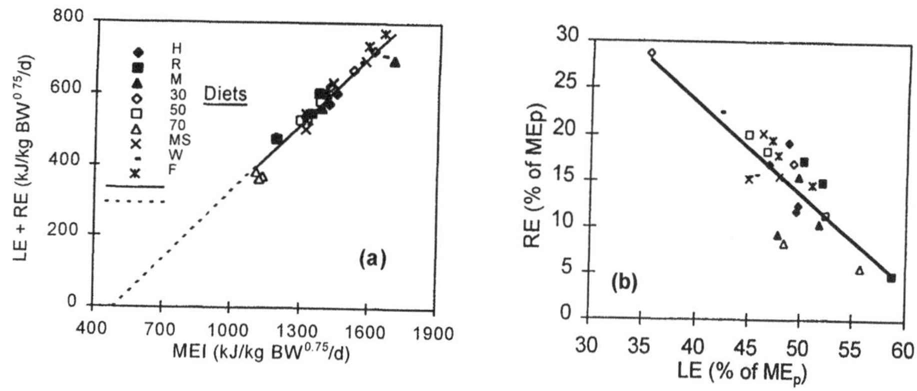

Fig.1. Lactating energy plus retained energy versus metabolizable energy intake (a), and relationship between RE and LE (both expressed as percentage of ME above maintenance) determined on the positive balances (b), of lactating goats fed different diets (see text materials and methods for the legends).

The regression of LE_c on MEI (with the data expressed as kJ/kg $BW^{0.75}$) indicates a k_l of 0.619.

$$LE_c = 0.619 \times MEI - 281 \; (n=27; r^2 = 0.94; RSD = 27.4; P<0.001) \qquad (3)$$

The energy content (E, kJ/kg) and the fat content (F, g/kg) of milk were determined. The developed equation was:

$$E = 43.238 \times F + 1159.6 \; (n=54; r^2 = 0.94; RSD = 66.9; P<0.001) \qquad (4)$$

Discussion

Considering energy utilization data, diets with higher proportion of concentrate and/or with high energy concentration gave better digestibility and metabolizability (q). In comparison with other studies (Armstrong & Blaxter, 1965; Aguilera *et al.*, 1990), the animals in our trials had higher energy intake and milk energy yield. Metabolizability in our study was higher than Aguilera's but lower if compared to the q values reported by Armstrong & Blaxter.

The daily ME_m estimated in the present study was equal to 484 kJ/kg $BW^{0.75}$. This value is higher than that obtained by Aguilera *et al.* (1990) with Granadina goats (401 kJ/kg $BW^{0.75}$), by Armstrong & Blaxter (1965) (319 kJ/kg $BW^{0.73}$) and by Sauvant and Morand-Fehr (1991) that reported a mean value (445 kJ/kg $BW^{0.75}$) derived from feeding trials and calorimetric experiments. Moe *et al.* (1970), in a study on lactating cows characterized by an average milk energy yield (419 kJ/kg $BW^{0.75}$) similar to that registered in our trials (444 kJ/kg $BW^{0.75}$), estimated that the daily ME_m was equal to 510 kJ/kg $BW^{0.75}$. Van Es (1975), evaluating more than 1000 balance trials performed with lactating cows, estimated a ME_m of 488 kJ/kg $BW^{0.75}$ and showed that 60% of ME above maintenance was converted into milk and body tissue energy; the value of ME_m is almost identical to our result while the k_p value we obtained (63.4%) is slightly higher.

Considering the relationship between RE and LE, Equation 2 suggests that in the lactating goats the ME_p is utilized for energy retention with almost the same efficiency as for lactation. The value we obtained (0.986) is very close to 0.96 (Armstrong & Blaxter, 1965) and slightly higher than the value of 0.907 determined by Aguilera *et al.* (1990).

The efficiency of utilization of ME for lactation (0.619) obtained in the present study is lower than the values reported in previous researches on lactating goats by Armstrong & Blaxter (1965) and Aguilera *et al.* (1990): 0.691 and 0.667, respectively. The discrepancy between 0.619 and 0.667 can be partially explained by the different coefficient utilized for the correction of milk energy secretion as a function of positive RE: in fact, utilizing the value (1.05) proposed by AFRC (1980) and used by Aguilera *et al.* (1990) we would have obtained a k_l of 0.645. Also, in comparison with Armstrong & Blaxter, determining the k_l value as they did (from Equation 2, assuming RE=0) we would have obtained a k_l value of 0.638.

The mean energy value of 1 kg of 4% fat corrected milk (FCM) obtained from Equation 4 is equal to 2889 kJ. This is quite similar to the average value of 2971 kJ derived from the calorimetric studies reported by Sauvant and Morand-Fehr (1991) but lower than the value of 3096 kJ suggested for cows milk (Vermorel, 1988).

Conclusions

The results obtained in the present study show that ME_m and k_l in the lactating goat are similar to values for the dairy cow. Furthermore, the efficiency of utilization of ME for body energy retention during mid lactation is very close to that for milk production. Lastly, it has been confirmed that the energy value of milk in high yielding goats (e.g. Saanan) is lower than that of cows milk.

References

AFRC (1980). The nutrient requirements of ruminant livestock. Commonwealth Agricultural Bureaux, Slough, pp. 90-91.

Aguilera J.F., Prieto C. & Fonolla J. (1990). Protein and energy metabolism of lactating Granadina goats. *Br. J. Nutr.* 63:165-175.

Armstrong D.G. & Blaxter K.L. (1965). Effects of acetic and propionic acids on energy retention and milk secretion in goats. In: K.L. Blaxter (Ed) Symposium on Energy Metabolism. Academic Press, London, pp. 59-70.

Brouwer E. (1958). On simple formulae for calculating the heat expenditure and the quantities of carbohydrate and fat metabolized in ruminants from data on gaseous exchange and urine-N. In: G. Thorbek and H. Aersøe (Eds) Symposium on Energy Metabolism. Statens Husdyrbrugsudvalg, Copenhagen V. Denmark, pp. 182-194.

Crovetto G.M., Rapetti L., Tamburini A., Succi G. & Galassi G. (1994). Energy utilization of diets with hay, maize silage or italian ryegrass silage as main forage in lactating goats. In: J.F. Aguilera (Ed) Energy metabolism of farm animals. *EAAP Publ. No.*76:175-178.

Es A.J.H. van (1975). Feed evaluation for dairy cows. *Livest. Prod. Sci.* 2:95-107.

Moe P.W., Tyrrell H.F. & Flatt W.P. (1970). Partial efficency of energy use for maintenance, lactation, body gain and gestation in the dairy cow. In: A. Schurch and C. Wenk (Eds) Energy Metabolism of Farm Animals. *EAAP Publ. No.* 13:65-68.

Rapetti L., Crovetto G.M., Sandrucci A., Battelli G. & Succi G. (1995). Effects of rumen-protected fat, lactose or starch on milk yield and composition in lactating goats. In: J.A.M. van Arendonk (Ed) Abstracts of the 46th Annual Meeting of the European Association for Animal Production. Wageningen Pers, The Netherlands, pp. 251.

Rapetti L., Tamburini A., Crovetto G.M., Galassi G. & Succi G. (1997). Energy utilization of diets with different hay proportions in lactating goats. *Zootecnica e Nutrizione Animale* (in press).

Sauvant D. & Morand-Fehr P. (1991). Energy requirements and allowances of adult goats. In: P. Morand-Fehr (Ed) Goat nutrition. Wageningen, Pudoc, Netherlands, pp. 61-72.

Vermorel M. (1988). Nutrition énergétique. In: R. Jarrige (Ed) Alimentation des bovins, ovins et caprins. INRA, Paris, pp. 57-71.

THE METABOLIZABLE ENERGY REQUIREMENT FOR MAINTENANCE AND THE EFFICIENCIES OF UTILIZATION OF METABOLIZABLE ENERGY FOR LACTATION AND TISSUE RETENTION IN DAIRY COWS

T. Yan, F.J. Gordon, R.E. Agnew and M.G. Porter

The Agricultural Research Institute of Northern Ireland, Hillsborough BT26 6DR, UK

Summary

In Experiment 1 four regression techniques were used to estimate the metabolizable energy (ME) required for maintenance (ME_m) and the efficiency of ME use for lactation (k_l) in lactating dairy cows (n=221) given grass silage-based diets. The mean derived ME_m (0.67 MJ/kg$^{0.75}$) is 40% higher than, but the k_l (0.65) is the same as that predicted from AFRC (1990), respectively. The high ME_m accords with data obtained in Experiment 2, a fasting study using non-lactating dairy cows given grass-based diets at near *ad libitum* prior to fasting. Fasting heat production (FHP) (0.45 MJ/kg$^{0.75}$) obtained in Experiment 2 is 41% higher than the predicted fasting metabolism (AFRC, 1990). The effects of lactation states on the energetic efficiency in dairy cows were also examined in Experiment 2. The mean k_l during lactation was 9% higher than the efficiency of ME use for tissue retention (k_g) during the dry period.

Keywords: Dairy cows, maintenance energy, fasting heat production, energetic efficiency.

Introduction

The ME_m in the UK ME system is based on data from fasting metabolism studies with non-lactating cows and beef steers offered restricted nutrition prior to fasting (ARC, 1980). When using this ME_m the k_l has been reported to be lower than expected when lactating dairy cows were given grass silage-based diets (Unsworth *et al.,* 1994). This could be a true effect or might suggest that the ME_m for dairy cows offered grass silage-based diets has been underestimated. Fasting following a long period of restricted nutrition can result in a number of metabolic disorders in ruminant animals (Chowdhury & Ørskov, 1994). Fasting metabolism has also been found to be positively related to the plane of nutrition prior to fasting in sheep (Marston, 1948) and cattle (Birkelo *et al.,* 1991). The objectives of the present studies were to examine the maintenance energy requirement for dairy cows given grass-based diets using both regression techniques on calorimetric data of lactating dairy cows and by measuring the fasting metabolism of dry cows offered diets at near *ad libitum* prior to fasting. A further objective was to compare the k_l during lactation with the k_g during the dry period, when the same diets were given at similar ME intakes throughout the study.

Materials and Methods

In Experiment 1 a total of 221 Holstein/Friesian lactating cows, drawn from several feeding trials, were subjected to energy metabolism measurements in indirect open-circuit calorimeters at this Institute. The animals were drawn from a range of genetic merits, parities (1-9) and lactation stages (milk yield, 3.2-49.1 kg/d). The mean ratio of ME:gross energy (GE) in the diets was 0.65 (S.D. 0.033). The animals received either grass silage only (n=36) or a mixture of grass silages and concentrates (concentrate proportion (DM basis), 0.18-0.70) (n=185). The ME_m and k_l were estimated using both regression of milk energy output, adjusted to zero energy balance (AFRC, 1990) ($E_{l(0)}$), against ME intake

(MEI) and regression of MEI against metabolic live weight (MW), milk energy output (E_l) and energy balance (E_g).

In Experiment 2 a total of 24 Holstein/Friesian multiparous lactating cows were used. All cows were non-pregnant and with a mean milk yield of 11.6 kg/d (SD 2.51). Sixteen cattle were in group 1, with half of them offered dried ryegrass nuts only and the other half a mixture of dried ryegrass nuts and concentrate at a ratio of 0.80:0.20 (GE basis). The remaining 8 animals were in group 2 and given a complete diet of ryegrass silage and concentrate at a ratio of 0.80:0.20 (GE basis). Each animal received the diet at the level which met her ME requirements for maintenance and lactation (AFRC, 1990) at the start of the study and remained unchanged throughout the study. Three measurement periods were carried out in group 1, i.e. full lactation, half lactation (by drying off 2 teats) and dry period (by drying off the remaining 2 teats). Group 2 had only full lactation and dry period (by drying off all 4 teats) measurements. In each measurement period the animals were allowed 11 days for adaptation followed by 3 days in calorimeters for energy metabolism measurement. Immediately after the dry period, 12 cows, half for the dried grass only diet and half for the mixed diet in group 1, remained in calorimeters for 4 to 5 days for FHP measurements.

Results and Discussion

Within each regression method the model explored in Experiment 1 accounted for the following factors: trial, parity, cow genetic merit and silage composition; and continuous variables: milk energy output, lactation days and silage GE intake as a proportion of total GE intake (silage-GE/total-GE). In all the regression methods the only factor reaching significance ($P<0.05$) was trial and hence the regression equations presented in Table 1 relate to the model in which this factor is removed. Within the continuous variables silage-GE/total-GE was the only one to reach significance ($P<0.05$) and for this reason the whole data set (n=221) was divided into three sub-sets (silage-GE/total-GE: <0.50, 0.51-0.99 and 1.00, n=99, 86 and 36, respectively) as presented in Fig. 1. The derived ME_m values increased ($P<0.01$) with increasing silage proportion in the diets (0.59, 0.68 and 0.74 MJ/kg$^{0.75}$, respectively), while the derived k_l values were similar (0.62, 0.64 and 0.63, respectively)

For each equation the relationship was highly significant ($P<0.001$) and the S.E. for estimates of ME_m and k_l, were also relatively small. The estimated ME_m values (MJ/kg$^{0.75}$) ranged from 0.61 (equation 2) to 0.75 (equation 4), with a mean of 0.67 MJ/kg$^{0.75}$. Similarly, the derived k_l values ranged from 0.61 (equation 2) to 0.68 (equation 4), with a mean of 0.65. The derived ME_m is 40% higher than, but the k_l is the same as, that predicted from AFRC (1990) (0.48 MJ/kg$^{0.75}$ and 0.65, respectively). Two recent studies using regression techniques by Unsworth *et al.* (1994) and Hayasaka *et al.* (1995) also reported high ME_m values of 0.64 and 0.59 MJ/kg$^{0.75}$, respectively. Estimating the k_l from the present data for each of the 221 cows, using the ME_m obtained in the present study, indicated that the mean k_l was considerably higher ($P<0.001$) than that obtained by using the predicted ME_m (AFRC, 1990) (0.65 vs. 0.53).

Table 1. The regression equations and derived ME_m (MJ/kg$^{0.75}$) and k_l (the figures in brackets after values are S.E. of the coefficients).

Equations	R^2	ME_m	k_l	
$E_{l(0)} = 0.65\ (0.023)\ MEI - 0.435\ (0.048)$	0.90	0.67	0.65	(1)
$E_{l(0)} = 0.61\ (0.024)\ MEI + 0.93 \times (q_m - 0.65) - 0.372\ (0.049)$	0.90	0.61	0.61	(2)
$E_{l(0)} = -6.94\ [1 - EXP\ (-(-0.088)\ (MEI - 0.66\ (0.050)))]$	0.89	0.66		(3)
$MEI = 0.75\ (0.026)\ MW + 1.48\ (0.032)\ E_l + 1.09\ (0.037)\ E_g$	0.92	0.75	0.68	(4)

The units in equations (1) to (3) are MJ/kg$^{0.75}$ and in equation (4) MJ/d; q_m = ME/GE.

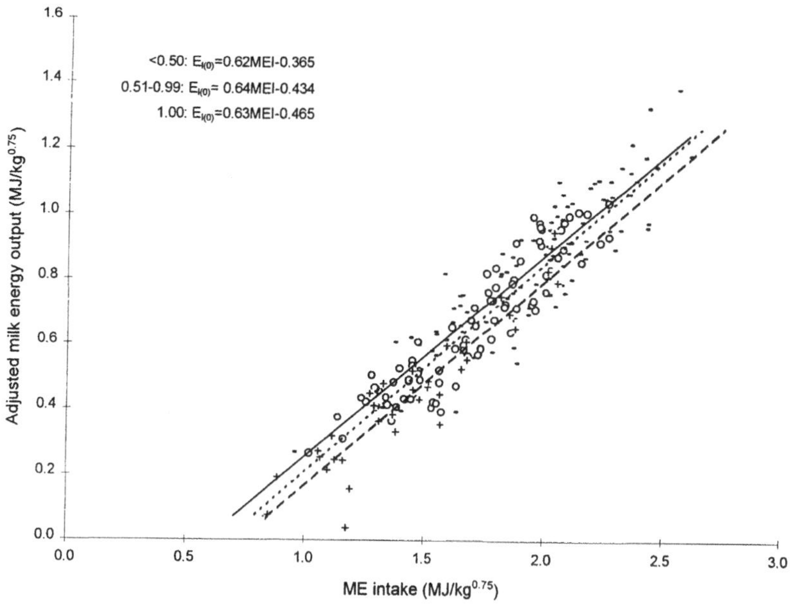

Fig. 1. Effects of silage-GE/total-GE on relationship between ME intake and milk energy output adjusted to zero energy balance (<0.50;- _____; 0.51-0.99; o -----; 1.00, + _ _ _).

In Experiment 2 the FHP values in group 1 for the dried grass only diet were respectively 0.442, 0.482, 0.412, 0.444, 0.439, 0.504 (mean 0.454) MJ/kg$^{0.75}$; and for the mixed diet 0.455, 0.412, 0.466, 0.511, 0.400, 0.470 (mean 0.452) MJ/kg$^{0.75}$, giving an overall mean of 0.453 (SD 0.0354) MJ/kg$^{0.75}$. This finding is higher than that obtained previously in any of the studies reported in the review of ARC (1980) when cattle received a long period of restricted nutrition prior to fasting, and is 41% higher than the fasting metabolizm (0.32 MJ/kg$^{0.75}$) predicted from AFRC (1990). However this result accords with that reported in Experiment 1.

The energetic efficiencies at the different lactation states are presented in Table 2. The ME intakes were similar in each group between lactation and the dry period and this allowed the efficiencies of ME utilization to be examined across lactation states without confounding effects of ME intake. The ratio of ME/GE in the dry period in group 1 was the same as, but in group 2 was lower ($P<0.05$) than that in lactation period(s). In group 2, heat production was increased in the dry period ($P<0.05$), resulting in a significantly lower k_g in the dry period than the k_l in lactation period, when the ME_m was based on the present study ($P<0.05$) or AFRC (1990) ($P<0.01$). Similar trends were obtained in group 1, although none of the differences were significant. A similar result was also reported by Moe *et al.* (1970) when using multiple regression techniques to analyse calorimetric data of dairy cows (n=350).

Conclusions

The maintenance energy requirement for dairy cows obtained in the present study, using both the regression techniques and FHP, has been shown to be considerably higher than predicted using the equations of AFRC (1990). However the k_l obtained in study 1 was in line with that predicted from AFRC (1990).

Table 2. Energy utilization between full lactation, half lactation and dry period in Experiment 2.

Lactation states	Group 1 (n=16) Full	Half	Dry	SEM	Group 2 (n=8) Full	Dry	SEM
Energy intake (MJ/d)							
Gross energy	265	264	259	3.0	213	218	2.7
Metabolizable energy	126	126	125	1.8	144	141	2.1
Energy output (MJ/d)							
Faeces	119	117	113	1.8*	44	48	1.4
Urine	8	9	9	0.3***	10	13	0.4**
Methane	12	12	12	0.3	15	16	0.3
Heat production	90	92	94	1.2	96	104	2.1*
Milk energy	35	20	-		45	-	
Energy balance (MJ/d)	1	14	31		3	37	
Productive NE (MJ/d)	36	34	31	1.7	48	37	2.3*
Energy utilization#							
ME/GE	0.48	0.48	0.48	0.005	0.67	0.65	0.005*
k_l or k_g (ARINI)	0.70	0.68	0.66	0.020	0.66	0.56	0.020*
k_l or k_g (AFRC)	0.53	0.51	0.48	0.016	0.54	0.44	0.018**

The k_l and k_g were calculated from the ME_m obtained in the present study (ARINI) or predicted from AFRC (1990).

References

Agricultural and Food Research Council (1990). Technical Committee on Responses to Nutrients, Report Number 5, Nutritive Requirements of Ruminant Animals: Energy. *Nutr. Abstr. Rev. (Series B)*, 60: 729-804.

Agricultural Research Council (1980). The Nutrient Requirements of Ruminant Livestock, Technical Review. CAB International, Farnham Royal.

Birkelo C.P., Johnson D.E. & Phetteplace H.P. (1991). Maintenance requirements of beef cattle as affected by season on different planes of nutrition. *J. Anim. Sci.*, 69:1214-1222.

Chowdhury S.A. & Ørskov E.R. (1994). Implications of fasting on the energy metabolism and feed evaluation in ruminants. *J. Anim. Feed Sci.*, 3:161-169.

Hayasaka K., Takusari N. & Yamagishi N. (1995). Energy metabolism in lactating Holstein cows (in Japanese, with English abstract). *Anim. Sci. Technol.*, 66:374-382.

Marston H.R. (1948). Energy transactions in sheep. 1. The basal heat production and heat increment. *Aust. J. Sci. Res.* B., 1:93-129.

Moe P.W., Tyrrell H.F. & Flatt W.P. (1970). Partial efficiency of energy use for maintenance, lactation, body gain and gestation in the dairy cows. In: A. Schürch and C. Wenk (Eds) Energy metabolism of farm animals. *EAAP Publ. No.* 13:65-68.

Unsworth E.F., Mayne C.S., Cushnahan A. & Gordon F.J. (1994). The energy utilization of grass silage diets by lactating dairy cows. In: J. F. Aguilera (Ed) Energy metabolism of farm animals. *EAAP Publ. No.* 76:179-181.

SUMMARY OF THE DISCUSSION

Discussion leader: C.K. Reynolds
Scientific secretaries C.P. Ferris & J.F. Hocquette

Introduction

Papers in this session used traditional respiration or slaughter balance calorimetry to measure maintenance energy requirements or the efficiency of energy and protein utilization for production. Many of these papers also addressed questions posed at previous sessions of this symposium about the effects of diet composition, level of nutrition and genotype on the efficiency of energy utilization for production or maintenance requirements. In all but one case production, in these papers, was synonymous with lactation. The exception was the paper by Zhang & Coon where egg production by hens represented the only direct measurement of the energy cost of reproduction in this session. Their studies involved a treatment used to switch egg production on and off. We might gain useful insight into the maintenance requirement of lactating dairy cows if we could turn milk production on and off so easily. Lacking such an approach, Yan *et al.* dried off half the udder in their studies of maintenance requirements of lactating dairy cows.

Effects of Genetic Potential on Energy Metabolism

The papers by Ferris *et al.* and Yan *et al.*, from their work on energy metabolism in dairy cows, posed two questions addressed by other research presented at previous symposia. The first question concerns the possible effect of genetic potential for milk yield on the efficiency of metabolizable energy use for milk production, (k_l), as well as maintenance energy requirements. Numerous studies presented at previous energy symposia have addressed effects of breeding on energy requirements for maintenance and production. The second question concerns the effect of diet 'quality' on maintenance energy requirements or k_l. First considering breed effects, work conducted at Beltsville, USA (Moe *et al.*, 1972; Tyrrell *et al.*, 1990; Reynolds *et al.*, 1990) comparing the use of metabolizable energy for lactation in Holstein, Jersey and Hereford-Angus cows fed 'high quality' diets found that when scaled to body weight raised to the 3/4 power the regression of milk energy output adjusted for tissue energy loss or gain on metabolizable energy intake was virtually the same in the 3 breeds. In these studies the lactating beef cows had a much lower genetic potential for milk production and only produced from 2 to 12 kg milk per day during measurements of energy balance. This means that the incremental use of metabolizable energy for lactation and tissue energy retention is virtually the same across these breeds. It is interesting to note that in the paper by Rapetti *et al.* the regression of energy balance data from lactating 'high genetic merit' goats is almost identical to that obtained in lactating cows. One reason these regressions differ between lactating goats and cows is that the authors used different adjustments for tissue energy. The predominant difference between breeds and genetic lines with differing propensity for milk yield is the partition of metabolizable energy between milk and tissue energy. Studies by Tyrrell *et al.* (1982) and Kirchgessner *et al.* (1989), reported at previous symposia, have shown that treating cows with somatotropin does not affect the efficiency of metabolizable energy use for milk production but shifts the partition of metabolizable energy away from body tissue and towards milk. Tyrrell *et al.* (1990) found that Jersey cows produce milk with a higher fat content by using more metabolizable energy for milk energy at the expense of tissue energy. Thus higher yielding cows, or cows producing milk with higher fat content, partition energy away from tissue and towards milk, but are not more efficient in their incremental use of metabolizable energy for net (milk and tissue) energy. Higher yielding cows are more efficient in terms of the cow maintenance cost required to produce a unit of milk. Another consideration when comparing regressions of net energy on metabolizable energy is the well documented difference between breeds in their maintenance requirements in the non-lactating state. These regressions suggest that, if breed differences in maintenance requirements are maintained in the lactating state, then those differences do not affect the incremental use of metabolizable energy for milk and tissue energy. They also suggest that either all of these animals were fed diets of similar 'quality' or that effects of diet composition on the amount of metabolizable energy lost as heat alters maintenance and not k_l, as was concluded by Yan *et al.* Regardless, the intercept of the linear regressions from the Beltsville studies cited was similar across breeds fed diets of similar 'quality'.

Effects of Diet Quality

One obvious conclusion from the paper by Ferris *et al.* is that, as higher yielding cows do not use metabolizable energy more efficiently, they have to consume more energy to produce more milk or else lose a greater amount of body tissue. This is a major concern in the UK and Ireland where conserved grass is a predominant forage for dairy cows and metabolizable energy intakes may be restricted in cows fed poorly conserved grass. In addition, the efficiency of utilization of metabolizable energy from grass may be lower than for other forages with more digestible or lower fibre contents. One major concern, with increasing rolling herd averages, for milk yield in grass-based production systems is the potential negative effects of prolonged tissue energy loss on health, reproduction and total productive life of the cow. Therefore much of the calorimetry work at Hillsborough, Northern Ireland has concerned the energy metabolism of dairy cows fed grass-based diets.

The paper by Yan *et al.* addresses the effects of feeding grass silage on maintenance energy requirements of the dairy cow. They conclude that feeding grass silage increases fasting heat production and that maintenace requirements of cows fed grass silage are higher than predicted by current feeding systems. These maintenance requirements were derived from analysis of their data using regression procedures. If the higher maintenance requirement is then used in calculating k_l then k_l is not decreased by feeding grass silage. The physiological basis of this elevation in maintenance requirement is intriguing and may in part be attributable to an increase in gut mass and heat production as shown in growing sheep fed grass compared to concentrate by Baldwin & McLeod and McLeod & Baldwin in their papers from the session on Organ and Tissue Metabolism.

References

Kirchgessner M., Schwab W. & Muller H.L. (1989). Effect of bovine growth hormone on energy metabolism in lactating cows in long-term administration. In: Energy Metabolism in Farm Animals. Proceedings of the 11th Symposium. Y. Van der Honing and W.H. Close (eds). *EAAP Publ.* No. 43:143-146.

Moe P.W., Flatt W.P. & Tyrrell H. F. (1972). Net energy value of feeds for lactation. *J. Dairy Sci.* 55:945-958.

Reynolds C.K. & Tyrrell H.F. (1990). Energy metabolism in lactating and dry beef cows. *J. Anim. Sci.* 68(Suppl. 1):523.

Tyrrell H.F., Reynolds C.K. & Baxter H.D. (1990). Energy metabolism of Jersey and Holstein cows fed total mixed diets with or without whole cottonseed. *J. Dairy Sci.* 73(Suppl. 1):192.

Tyrrell H.F., Brown A.C.G., Reynolds P.J., Haaland G.L., Peel C.J., Bauman D.E. & Steinhour W.C. (1982). Effect of growth hormone on utilization of energy by lactating Holstein cows. In:. A. Ekern and F. Sundstol (eds) Energy Metabolism of Farm Animals. Proceedings of the 9th Symposium. *EAAP Publ No.* 29:46-49.

General Discussion

Much of the general discussion concerned the paper by Ferris *et al.* on energy metabolism of lactating cows, of low and high genetic potential for milk yield, fed diets differing in forage-to-concentrate ratio in a continuous design experiment. Cows used for the energy balance measurements were taken from a larger group of animals used for a production study and one concern expressed was that milk yield and dry matter intake were lower during energy balance measurements than for cows in the production study. This was attributed to the individual cows selected for the respiration trials (it was suggested that random selection of cows had chosen cows of lower yield than the average yield for the production study) or that placing cows in the respiration chambers had restricted production. The potential negative effects of chamber confinement on intake and production has been a frequent topic of discussion at previous meetings and the importance of adequate adaptation of animals to chambers before measurements begin was again stressed. Another concern expressed was that the difference in yields achieved between genetic lines was too small to draw conclusions. It was pointed out that milk fat and protein concentrations were higher in the higher genetic merit cows. Another question raised concerned the lack of difference between diets and it was suggested that the grass silage fed may have been of such high quality that effects of forage to concentrate ratio were minimized. In this regard there was considerable discussion about the high digestibility (79%) of the diets fed in this study. This was attributed to the very immature state at which the grass was harvested.

Another line of discussion concerned the regulation of intake in animals selected for high milk yield. It was suggested that, in selecting animals for milk yield, the most important trait selected may be the ability to consume large amounts of feed. It was pointed out that in the short term, treatment of lactating cows with somatotropin increases milk yield at the expense of tissue energy but that, in the long term, intake increases to support higher milk yield and ultimately replenish tissue energy lost. This response was used to illustrate the concept that milk production was pulling or allowing higher intake. On the other hand, somatotropin may be increasing intake directly but with a lag in the response compared to effects on milk synthesis. An interesting observation discussed was that studies in France have shown that when milk yield is restricted in early lactation by only milking once a day, cows with restricted yields had higher intakes, but this was attributed to a reduction in lipid mobilization which is known to inhibit intake in early lactation.

The source of glucose produced by the liver of ruminants was discussed. Data from cattle and sheep were reviewed and the conclusion was that, on a net basis 50 to 60% comes from propionate, 20 to 30% comes from amino acids and the rest comes from lactate and glycerol. It was pointed out that lactate represented a point of metabolic flexibility and that, as the contribution of other precursors changed, the contribution of lactate was the most variable. The contribution of lactate balances the cytosolic pyruvate pool and also varies with redox state. On a net basis most of the lactate removed by the liver comes from the portal-drained viscera. The amount of lactate from other tissues is usually small except in early lactation and in fasted animals. The origins of lactate released by the portal-drained viscera would include fermentation (during ensiling of feeds but especially ruminal fermentation) and tissue metabolism of glucose and propionate. However, many studies have suggested that the contribution of propionate to lactate produced by the portal-drained viscera is small. Glycerol contributions may increase during periods of lipid mobilization and may be as high as 25% in fasted, non-lactating sheep, but 10 to 15% of the glycerol from lipolysis will be used for milk synthesis in lactating cows.

SESSION 3

REGULATION OF METABOLISM

BOVINE ADAPTIVE THERMOGENESIS AND EXPRESSION OF ATP-SENSITIVE K^+ CHANNELS IN MONOCYTES

B. Löhrke[1], M. Derno[2], H.-D. Matthes[2], W. Jentsch[2]

[1]Research Unit of Population Biology and Breeding Research; [2]Research Unit of Nutritional Physiology, "Oskar Kellner", Research Institute for the Biology of Farm Animals, Dummerstorf-Rostock, Germany

Summary

Resting metabolic rate and heat production, respond to sympathetic outflow and circulatory work, respectively, and also to the expression of sulphonylurea receptors which are components of ATP-sensitive K^+ channels (K_{ATP}), using mononuclear blood cells (monocytes) as indicators. Thyroid hormones and cortisol failed to have a direct thermoregulatory effect.

Key words: Cattle, metabolism, sympathetic outflow, sulphonylurea receptors, environment, leukocytes.

Introduction

Adaptive thermogenesis occurs in response to changes in feeding level, ambient temperature and locomotory activity and after feeding or activation of the immune system. In all cases, redistribution of the blood flow seems to be involved which raises the hypothesis that the major contributions to metabolic rate (MR) and, by corollary, to heat production (HP), arose from the circulatory system directly or by pathways directly dependent on perfusion, such as supply of oxygen and substrates. Sympathetic nervous activity is involved in the regulation of the circulatory and respiratory systems, and may be mainly responsible for both the resting and regulatory HP. However, blood flow through tissues is also important for the overall HP and is regulated at the level of tissue's capillaries, sensing ions and tissue-produced hormone-like substances, such as adenosine. A link between intracellular metabolism (increase of which requires changed blood supply) and systemic HP may be brought about by K_{ATP}. Since K_{ATP} and sulphonylurea receptors (SUR), respectively, are able to sense the intracellular ATP (and ADP) level, contributing to the stimulation of the cardiac output by K^+ release when a tissue's ATP level is low. Moreover, the K_{ATP} activity is under purinergic receptor control so that the vasoregulatory action of adenosine, produced by the tissue, can act on a tissue's blood supply and purinergic receptors may act via K_{ATP} on intracellular respiratory regulation. Expression of SUR may have a role, not only in cellular metabolism, but also in whole body HP, but studies on the relationship between SUR expression and HP seem to be lacking.

Materials and Methods

Young Black Pied dairy bulls (280±10 kg; 150±5 kg in cromakalim trials) were used in consecutive experiments, four animals in each case. MR and HP were monitored in four climatically controlled respiratory chambers, measuring the O_2 consumption and the production of CO_2, CH_4 and urinary nitrogen continuously over 24 h. HP was calculated as described by Brouwer (1965) and heart rate (HR) was measured by biotachometer (Polar, Finland). The animals were provided with a diet adequate for normal rumen function, (Jentsch *et al.*, 1994).

Feeding levels were 1.0-fold the metabolizable energy requirement for maintenance (ME_m) and 1.6-fold the ME_m. Ambient temperatures were 18°C and 4°C. Animals were familiarized with these conditions for two weeks (with one restorative week between the conditioning weeks) before starting the experiments. Sympathetic outflow was reduced by the mainly central action (Zoukos *et al.*, 1993) of the α_2-adrenergic agonist clonidine infused, once only (8-10 min) via a jugular vein catheter at a dose of 20 nmole/kg or 15 nmole/kg in abrogating the clonidine effect by the K_{ATP} opener cromakalim (400 nmole/kg). Controls: Administration of the α_2-adrenergic antagonist yohimbine (600 nmole/kg) or of the carrier of the drugs. Circulating levels of hormones (T_3, T_4, cortisol) and metabolites (non-esterified fatty acids (NEFA), glycerol, glucose) were measured by commercial kits using modifications as described (Löhrke *et al.*, 1994) in the case of metabolite determination. Mononuclear blood cells were prepared (Löhrke *et al.*, 1994) and incubated with fluorescent or fluorogenic probes for detecting the plasma transmembrane potential difference ($\Delta\psi_p$) with 125 nM $DiBaC_4(3)$ as described (Wilson *et al.*, 1985), the oxidative activity with DHR, 2µM or dose-dependently, the sulphonylurea receptors (SUR) with 40 nM glibenclamide-Bodipy (or dose-dependently) using plate fluorimetry or flow cytometry for the analysis of single cells and blood cell subpopulations as described (Löhrke *et al.*, 1994, 1995, 1996). SUR types were determined by the dose-dependent response of $\Delta\Psi_p$ to cromakalim, pinacidil and diazoxide in cells depolarized by 60 mM KCL which was the highest concentration in an almost linear relationship between $\Delta\Psi_p$ and $\Delta[KCl]_o$ (6-60 mM) using intact cells. Results are given by means (symbols) and standard deviations (bars) or confidence limits ($P < 0.05$).

Results and Discussion

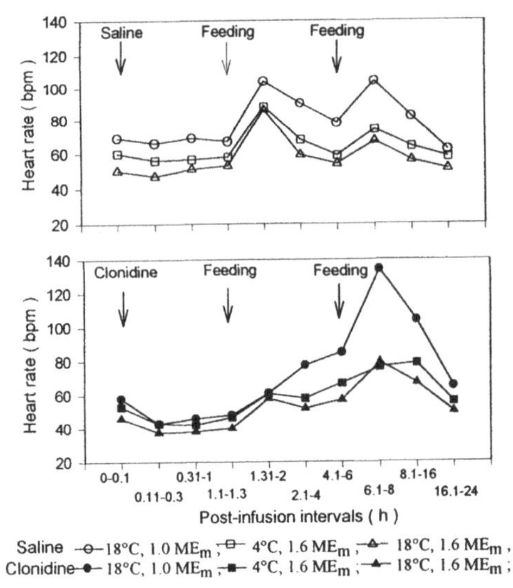

Fig. 1. Heart rate response to α_2-adrenergic agonist in different environmental conditions, SEM were within the symbols.

The HP at 18°C and a feeding level of 1.0- or 1.6-fold the ME_m or at 4°C and 1.6-fold the ME_m was 472±28, 539±30, and 614±29 kJ ($kg^{0.75}$/d) respectively.

Effects of the clonidine-reduced sympathetic outflow on HR (see Fig. 1) paralleled the HP response, except at a feeding level of 1.0-fold the ME_m. The clonidine-reduced HP at 4°C and 1.6-fold the ME_m; at 18°C and either 1.0-fold or 1.6-fold the ME_m was 0.66-, 0.77- and 0.79-fold the HP after vehicle infusion, suggesting different sympathetic sensitivity.

Yohimbine effects on MR, HP, and HR were not significant. Clonidine-reduced HR was nullified by cromakalim, indicating a role for K_{ATP} in stimulating the circulatory output. Environment-associated differences in the resting levels of NEFA, glucose (see Fig. 2) and cortisol (data not shown) were not significant whereas the glycerol level rose significantly at 4°C compared 18°C (0.20±0.06 mM vs. 0.14±0.02 mM), suggesting an increased NEFA mobilization in the cold. Levels of

T_3, T_4 and HP tended to be parallel whereas the clonidine-induced HP reduction was not reflected by T_3 or T_4 changes.

Clonidine-reduction of HR, NEFA (Figs 1 and 2) and glycerol levels was dependent on the environmental condition. In intact cells, mitochondrial activity and $\Delta\Psi_p$ are linked by ion perfusion and, in the case of K_{ATP}, by the intracellular ATP level. Thus, DHR oxidation and $\Delta\Psi_p$ in cells from the same cell suspension whereby the cell donors were exposed to the differing environmental conditions. Figure 3 shows a linear relationship between the response of $\Delta\Psi_p$ and DHR oxidating cells (percentage of mononuclear blood cells) to environmental changes. The correlation is inverse (the higher $\Delta\Psi_p$, indicated by low $DiBaC_4(3)$ uptake, the lower the rhodamine 123 production from DHR). This relationship occurs, perhaps, due to a high respiratory rate at 4°C, producing a K_{ATP} level that suppresses K_{ATP} activity. This conclusion is supported by the effect of the K_{ATP} opener cromakalim on both variables *in vivo* or *in vitro*, increasing the number of polarized cells and decreasing the portion of DHR oxidating cells (data not shown). *In vitro* cromakalim, pinacidil and diazoxide polarized blood cells dose-dependently (after depolarization by replacing [NaCl] of the medium by $[KCl]_o$ = 60 mM), suggesting expression of smooth muscle type SUR. HP at rest paralleled the number of monocytes made fluorescent by the glibenclamide Bodipy (see Fig. 4), indicating covariation between SUR expression and HP in changed environmental conditions.

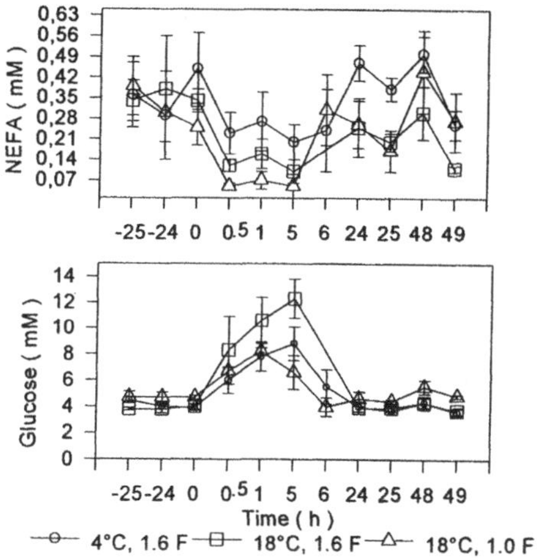

Fig. 2. Responses in circulating metabolites to carrier (- 25, - 24) or clonidine infusion. F = feeding level (ME_m)

In conclusion, the reduction of the sympathetic outflow by the α_2-adrenergic agonist, clonidine, revealed significantly different HP and HR responses to different ambient temperatures or feeding levels, suggesting adaptation is mediated by changed sensitivity of the α_2-adrenergic neurotransmission as well as strong involvement of the α_2-adrenergic component of the sympathetic outflow in the regulation of the HR and HP. At the cellular level, environmentally induced RMR differences correspond to different expression of SUR, indicating an important role for K_{ATP} for signalling demands of tissue metabolism for circulatory supply.

References

Brouwer E. (1965). Report of sub-committee on constants and factors. In: K.L. Blaxter (Ed) Energy Metabolism of Farm Animals. *EAAP Publ. No.* 11: pp. 441-443.

Jentsch W., Matthes H.-D., Derno M., Wegner J., Voigt J., Löhrke B., Nürnberg K. & Beyer M. (1994). Untersuchungen zum Soffwechsel, zur Wärmeproduktion, zum Verhalten und zur Morphologie von Jungbullen der Rassen Galloway und Schwarzbuntes Milchrind. *Arch. Tierz.* 37:363-375.

Löhrke B., Derno M., Jentsch W., Matthes H.-D. & Göbel D. (1994). Neurokrine und immunologische Gesichtspunkte zur Anpassung des Wärmehaushaltes an Umweltänderungen beim Rind. *Forschungsinstitut für die Biologie landw. Nutztiere Dummerstorf-Rostock Schriftenr.* 7:123-157.

Löhrke B., Renne R., Viergutz T., Ender K. & Krüger B. (1996). Effects of stress-related signal molecules on cell associated with muscle tissue. *Analyt. Quant. Cytol. Histol.* 18:383-388.
Löhrke B., Wegner J., Viergutz T., Dietl G. & Ender K. (1995). Flow-cytometric analysis of oxidative and proteolytical activities in tissue-associated phagocytes from normal and hypertrophic muscles. *Anal. Cell. Path.* 9:281-293.
Wilson H. & Chused T.M. (1985). Lymphocyte membrane potential and Ca^{2+}-sensitive potassium channels described by oxonol dye fluorescence measurement. *J. Cell. Physiol.* 125:72-81
Zoukos Y., Thomaides T., Pavitt D.V., Cuzner M.L. & Mathias C.J. (1993). Beta-adrenoceptor expression on circulating mononuclear cells of ideopathic Parkinson`s disease and autonomic failure patients before and after reduction of central sympathetic outflow by clonidine. *Neurology* 43:1181-1187.

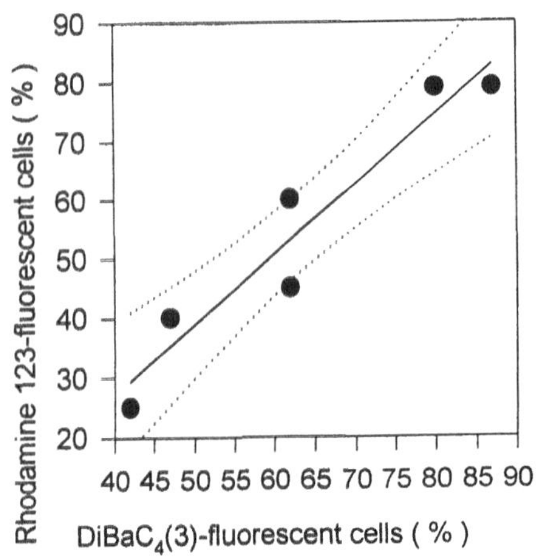

Fig. 3. ψ_p (DiBaC)-oxidative activity (R123) relationship in mononuclear blood cells.

Left, medium and right circles: 4 °C, 1.6 fold the ME_m, 18 °C, 1.6-fold the ME_m, 18 °C, 1.0-fold the ME_m.
Dashed lines indicate confidence limits ($P<0.05$) with $r^2 = 0.921$.

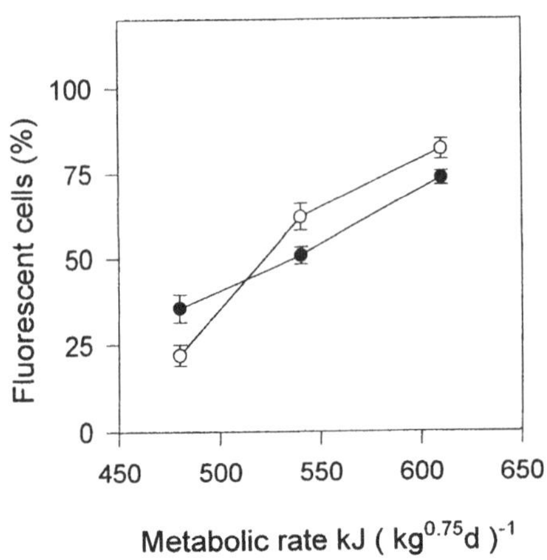

Fig. 4. Relationship between the metabolic rate and the portion of glibenclamide-Bodipy binding monocytes.

O Resting state, ● Yohimbin
Coefficient of determination $r^2 = 0.94$.

Acknowledgements

These investigations were supported by the Deutsche Forschungsgemeinschaft.

INSULIN RECEPTOR CHARACTERIZTICS IN RUMINANT TISSUES. EFFECTS OF BREED, SEX AND STAGE OF DEVELOPMENT

A.R.G. Wylie [1,2], P.D. McGrattan [2] and J. Nelson[3]

[1]Agricultural and Environmental Sciences Division, The Department of Agriculture for Northern Ireland, Newforge Lane, Malone Road, Belfast BT9 5PX, United Kingdom; [2]School of Agriculture and Food Science, The Queen's University of Belfast; [3]School of Biology and Biochemistry, The Queen's University of Belfast

Summary

Differences in binding-affinity and concentration of cell membrane insulin receptors were demonstrated between the principal insulin-sensitive target tissues of heavy Continental-type steers and of three breeds of growing lambs. In lambs, receptor affinity of individual tissues was unrelated to sex but both receptor concentration and specific insulin binding were breed and sex related. Insulin affinity of individual muscle and adipose depots did not change with stage of development in Holstein-Friesian steers. Specific binding of insulin to liver and adipose membrane receptors changed with age in Holstein-Friesian steers and these changes were associated with changes in receptor concentration. The results suggest that differences in insulin receptor affinity for insulin and in insulin receptor number exist between tissues in meat-producing ruminants. Such differences could have significance for partitioning of nutrients between tissues.

Key words: Insulin receptor, binding affinity, steers, lambs, ruminants, muscle, adipose.

Introduction

Insulin is implicated in the partitioning of nutrients to fat and muscle in both ruminants and non-ruminants. Insulin binding to its receptor causes receptor autophosphorylation, translocation of the insulin-responsive glucose transporter (GLUT4) and selective activation via intracellular kinases, of the pathways of fat, protein and carbohydrate ,etabolizm.

Although insulin is strongly associated with fat deposition, finishing cattle deposited more fat from silage than from concentrate diets at equal energy intakes (Steen, 1994) despite higher insulin levels with concentrate feeding (Thorp *et al.*, 1995). Much of the additional fat is abdominal (Steen; personal communication) reflecting higher lipogenesis in perirenal than subcutaneous fat of silage-fed steers (Greathead *et al.*, 1996) and also the higher GLUT4 levels in perirenal than in subcutaneous fat of steers (Hocquette *et al.*, 1996).

A 3-fold decrease in insulin receptor concentration of skeletal muscle was found between the 3-5 month old bovine foetus and 16 month old heifer, without apparent change in receptor affinity (Boge *et al.*, 1995) suggesting that receptor concentration, rather than receptor affinity, might be an important regulator of the metabolic effects of insulin.

Evidence, in humans and rats, of the existence of two, alternatively spliced, insulin receptor mRNA transcripts (Mosthaf *et al.*, 1990) and expression of two receptor isoforms differing in affinity for insulin, suggests that regulated isoform expression could result in different tissue-specific affinities for insulin. The current work sought to determine whether insulin receptor affinity and/or receptor concentration differs between ruminant tissues and if they are affected by sex, breed or by stage of development to maturity.

Materials and Methods

Study 1. Charolais-cross steers (n=5) were offered grass silage *ad libitum* from 18 months of age until slaughter at 701 ± 22.9 kg. Liver (L), two skeletal muscles (*m. pectineus*, A and *m. rectus capitis*, B) and subcutaneous (SCF), omental (OF) and perirenal (PF) fats were sampled as soon as possible after death, snap frozen in liquid N and stored at -70°C.

Study 2. Greyface (Scottish Blackface x Border Leicester), Dutch Texel and Rouge de l'Ouest lambs (3 male and 3 female) were fed grass nuts for 14 weeks prior to slaughter at 47 ± 2.1 kg. Intake was kept constant across breeds by adjusting allowances to the *ad libitum* intake of the Texel lambs (male and females separately). Liver (L), *m. rectus capitis* (B), *m. longissimus dorsi* (C) and SCF and PF fat were sampled as above.

Study 3. Holstein Freisian steers at 15 (n=5) and 36 months (n=5) were fed grass silage *ad libitum* for 8 weeks before sampling of *m. semimembranous* (D) and OF and PF fats by biopsy. Also, 5 Holstein-Freisian bull calves were fed reconstituted milk until slaughter (approx 10 d old) and *m. semimembranous* (D) and OF and PF fats were sampled as above.

Receptor preparation. Tissue was homogenized by Ultra-turrax (75% max. speed; 30 s) in 10-50 mM Tris-HCl buffer (pH 7.4) containing a cocktail of protease inhibitors and centrifuged (1500 g; 5 min; 4°C). Infranatant was ultra-centrifuged (75,600 g; 1 h; 4°C), the pellet solubilized by gentle end-over-end rotation and the suspension re-centrifuged (75,600 g, 1 h; 4°C). The supernatant was passed down a column of wheat-germ agglutinin on Sepharose and retained insulin receptor eluted by 300 mM N-acetyl glucosamine.

Insulin receptor binding assay. Insulin binding was assessed, in all tissues, using a tracer amount of human A-14 ^{125}I-insulin (Amersham, UK) and increasing concentrations of unlabelled bovine insulin (0-10 mM). Solubilized receptors were incubated (16 h; 4°) and precipitated using bovine gamma globulin and polyethylene glycol. K_d (dissociation constant) and receptor concentrations were determined, assuming a one-site model, and using 'LIGAND' (Munson & Rodbard, 1980). A one-way ANOVA was applied to all data and least significant differences were calculated using the Student's *t*-Test.

Results

In studies 1 and 2, liver had markedly lower insulin affinity (higher K_d) than the other tissues and was omitted from further ANOVAs comparing only muscle and adipose data.

In heavy Continental steers, one-site Scatchard analysis of insulin-binding revealed tissue-specific differences in receptor affinity (K_d) and receptor concentrations (Table 1).

Table 1. K_d and receptor concentration (R) in a variety of heavy weight steer tissue depots.

Receptor parameter		Tissue					
	n	[A]	[B]	[SCF]	[OF]	[PF]	SEM
K_d [nM]	5	0.173	0.231	0.315	0.170	0.076	0.035
R [fmol/mg protein]	5	2510	2836	1314	2982	4236	521.3

K_d of solubilized membrane receptors did not differ between *pectineus* (A) and *rectus capitis* (B) muscles but differences were found between muscle and fat depots (A vs. SCF, P<0.05; B vs. PF, P<0.05) and between fat depots SCF and OF (P<0.05) and SCF and PF (P<0.001). Receptor concentration of solubilized membranes differed between *m. pectineus* (A) and PF (P<0.05) and between SCF and OF (P<0.05) and SCF and PF

(P<0.01).

In lambs, insulin affinity of individual tissues showed no sex or breed effects overall but the K_d of *m. rectus capitis* (B) was lower (P<0.01) in Greyface and Texel compared to Rouge (tissue × breed interaction). Differences in K_d were found (Table 2) between muscle and adipose (*l.dorsi* (C) vs. PF and *rectus capitis* (B) vs. SCF, P<0.05 and B vs. PF, P<0.001).

Table 2. Mean K_d and receptor concentration (R) across 3 breeds of male and female lamb.

Receptor parameter				Tissue		
	n	[B]	[C]	[SCF]	[PF]	SEM
K_d [nM]	18	0.180	0.155	0.123	0.090	0.018
R [fmol/mg protein]	18	1135	1217	1999	2018	145.6

Overall, Greyface lambs had lower receptor concentrations (P<0.05) compared to Texel and Rouge lambs and males had higher receptor concentrations (P<0.001) than females in all tissues except *m. rectus capitis* (B) (Fig. 1). There were differences in receptor concentration between muscle and adipose tissues (B vs. SCF and PF, C vs. SCF and PF, P<0.001) but not between the two muscles or between the two adipose depots.

Insulin receptor affinity of individual muscle and adipose tissues did not change with stage of development in Holstein-Friesian steers. Receptor concentration in adipose depots OF and PF generally increased with age (Table 3) and these changes were associated with an increase in insulin binding capacity.

Table 3. Specific insulin binding and receptor concentration in bovine tissue depots at different developmental stages. Different superscripts within a column indicate a significant difference for specific binding (P<0.01) and receptor concentration (P<0.001) respectively.

Animal		Specific binding [fmol/mg protein]				Receptor concentration [fmol/mg protein]			
	n	[D]	[OF]	[PF]	SEM	[D]	[OF]	[PF]	SEM
Preweaned	5	207	239^a	439^a		861	1501^a	3488^a	
15 months	5	245	315^a	1028^b		1470	1918^a	7681^b	
36 months	5	125	893^b	1314^b	126.7	205	15667^b	5637^{ab}	806.9

Discussion

Differences in binding affinity for insulin demonstrated between insulin-sensitive tissues of humans (Kotzke *et al.*, 1995) and, in this study, of ruminants also, may be important in the partitioning of nutrients between tissues. In humans, two functionally distinct insulin receptor isoforms, differing by up to three-fold in insulin-affinity, are generated by the expression of two alternatively spliced mRNA transcripts in a tissue-specific manner (Mosthaf *et al.*, 1990). Expression of the alternative transcripts is altered by the chronic hyperinsulinaemia associated with insulinoma in humans (Sbraccia *et al.*, 1996). Furthermore, insulin receptor numbers are inversely related to insulin concentration through receptor down-regulation (Krupp & Lane, 1981). This raises the possibility that in ruminants, in which normally low levels of plasma insulin are markedly elevated for several hours by once daily feeding of high levels of concentrates (Thorp *et al.*, 1995), both the affinity and concentration of insulin receptors in a number of tissues may be altered by diet-induced hyperinsulinaemia with consequences for tissue responses including nutrient partitioning.

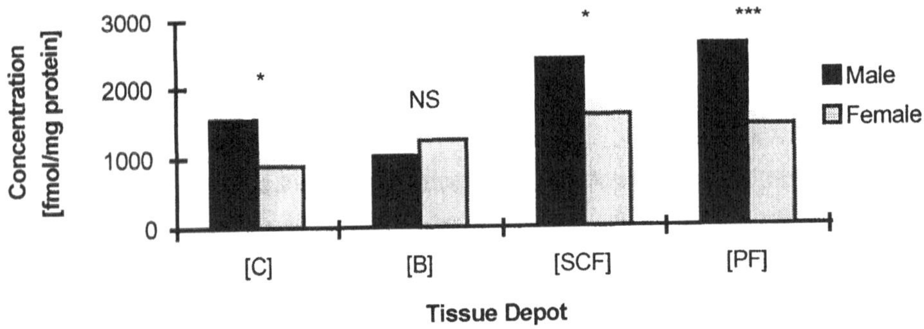

Fig. 1. Insulin receptor concentration in muscle and adipose depots from male and female lambs (n=9, SEM=201.8). [NS = not significant; * = P<0.01; *** = P<0.001 when comparing differences between male and female lambs for individual tissues.]

References

Boge A., Sauerwein H. & Meyer H.H.D. (1995). IGF-1 and insulin receptors in bovine skeletal muscle. Comparisons of different developmental ages, two different genotypes and various individual muscles. *Exp. Clin. Endocrinol. Diabetes* 103:99-104.

Greathead H.M.R., Dawson J.M., Sessions V.A., Tye F.T., Scollan N.D., McAllan A.B. & Buttery P.J. (1996). Fat synthesiz in different depots of cattle fed grass silage and dried grass diets. *Anim. Sci.* 62:656-657A.

Hocquette J-F., Castiglia C., Ferre P. & Vermorel M. (1996). Variations in GLUT4 protein content among bovine adipose tissues. *Proc. Nutr. Soc.* 55:21A.

Kotzke G., Schutt M., Missler U., Moller D.E., Fehm H.L. & Klein H.H. (1995). Binding of human, porcine and bovine insulin receptors from human brain, muscle and adipocytes and to expressed recombinant receptor isoforms. *Diabetologia* 38:757-763.

Krupp M. & Lane M.D. (1981). On the mechanism of ligand induced down-regulation of insulin receptor level in the liver cell. *J. Biol. Chem.* 256:1689-1694.

Mosthaf L., Grako K., Dull T.J., Coussens L., Ullrich A. & McClain D.A. (1990). Functionally distinct insulin receptors generated by tissue-specific alternative splicing. *EMBO J.* 9:2409-2413.

Munson P.J. & Rodbard D. (1980). LIGAND: A versatile computerised approach for characterization of ligand-binding systems. *Anal. Biochem.* 107:220-239.

Sbraccia P., D'Adamo M., Leonetti F., Caiola S., Iozzo P., Giaccari A., Buongiorno A. & Tamburrano G. (1996). Chronic primary hyperinsulinaemia is associated with altered insulin receptor mRNA splicing in muscle of patients with insulinoma. *Diabetologia* 39:220-225.

Steen R.W.J. (1994). Effects of forage:concentrate ratio in the diet and restricted dry matter intake on the performance and carcass composition of steers. *Anim. Prod.* 58:443.

Thorp C.L., Steen R.W.J., Wylie A.R.G., McEvoy J.D. & Shaw C. (1995). Effect of different forage:concentrate ratios and intake restriction on the rumen fermentation and hormone profiles of finishing beef steers. *Anim. Prod.* 58:444.

ROLE OF D,L-1, 2-PROPANEDIOL IN RUMINANT GLUCOSE METABOLISM

D. L. Palmquist[1] and H. Brunengraber[2]

[1]Department of Animal Sciences, OARDC/The Ohio State University, Wooster; [2]Department of Nutrition, Case Western Reserve University, Cleveland, Ohio, USA

Summary

The gluconeogenic agent, D,L-1,2-propanediol (PG) is an antiketogenic agent in lactating cows. The primary route of PG metabolism is believed to be by absorption intact from the rumen, with ruminal conversion to propionate as a secondary route. We observed rapid responses in early lactation after oral drenching of PG; plasma glucose and insulin increased, and non-esterified fatty acids decreased, within 30 min. When PG was orally drenched or intravenously administered in lactating cows the L-isomers of both PG and lactate disappeared from plasma more rapidly than the D-isomers. Plasma concentrations of PG peaked higher in orally-drenched than in i.v.-infused cows. Lactate concentrations remained within the normal physiological range in all cases. Peak plasma glucose (mM) and insulin (mU/ml) values after oral and i.v. doses were 4.1 and 30, and 3.5 and 16, respectively. Two lactating cows were infused continuously with [6,6-$^2H_2$] glucose for 4 hours. After 2 hours, cows were orally dosed with 500 ml PG. Glucose turnover was not changed by PG, but glucose concentration increased, reflecting decreased glucose disposal, probably resulting from peripheral utilization of lactate derived from PG. Our data support direct absorption and metabolism of PG as the primary route of utilization. Glucose may not be an intermediate.

Key words: Glucose metabolism, propylene glycol, lactation, stable isotopes.

Introduction

Propanediol, also called propylene glycol, is commonly used as a glucogenic agent to treat ketotic cows (Christensen *et al.*, 1997). Concentrations of glucose, insulin and non-esterified fatty acids (NEFA) in plasma respond rapidly (within 30 min) to an oral drench of PG (Table 1; see also Grummer *et al.*, 1994).

Table 1. Changes in plasma concentrations of glucose, insulin and NEFA after oral drenching with PG (Miyoshi, 1995).

	PG (n = 17)			Control (n = 16)			TRT x Time
	0 min	30 min	0 min	0 min	30 min	90 min	P <
Glucose (mg/dl)	55.8	59.5	60.8^a	53.5	53.0	52.5	0.01
Insulin (mU/ml)	6.6	11.4	1.8^a	6.3	6.4	6.9	0.01
NEFA (meq/l)	379	287	242^a	425	386	347^b	0.01

[a]Linear effect of time (P<0.01) after drenching.
[b]Linear effect of time (P<0.05) after drenching.

Emery *et al.* (1967) concluded that the primary route of PG metabolism is by absorption directly from the rumen. Conversion to propionate in the rumen was considered to be a secondary route of metabolism. The half-life of PG in the rumen was 1.7 hours and maximal labelling of blood glucose was achieved at 3 hours after administering ^{14}C-PG intraruminally. Furthermore, these authors concluded that PG is classically glucogenic, in that labelling patterns of glucose and glutamic acid in milk suggested conversion to glucose via pyruvate and oxalacetate.

The metabolism of D,L-1,2-propanediol first requires conversion to D,L-lactate, by action of alcohol and aldehyde dehydrogenases. Harmon *et al.* (1984) reported that bovine tissues could readily use both D- and L- lactate for oxidation and gluconeogenesis at low concentrations (< 1mM). However, maximal oxidation rates of the D (-) isomer were lower at substrate concentrations up to 50 mM, and gluconeogenic rates hardly increased at concentrations above 1mM.

Methods

To investigate the metabolism of PG in more detail, in experiment 1 we provided PG orally (520 g) or intravenously (76 g in 750 ml saline over 35 min) to two lactating cows. Blood samples were taken frequently and analysed for concentrations of glucose, insulin and D- and L- isomers of PG and lactate. Cows weighed 655 kg, were fed a diet of concentrates, maize silage and lucerne hay, and yielded 28 kg milk per day. In experiment 2, lactating cows in early lactation (610 kg body weight, 50 kg milk per day, 22-25 kg dry matter intake per day) were continuously infused with [6,6-$^2H_2$] glucose for 4 h (7.3 mg min^{-1} after a priming dose of 445 mg). After 120 min of infusion, the cows were orally dosed/intraruminally with 500 ml PG. Blood samples were taken for concentrations of glucose, and for $^2H_2$ enrichment of glucose.

Glucose enrichment was determined as described by Hazey *et al.* (1997). The glucose was isolated (it is important to remove all glycerol) by preparative HPLC and treated with sodium periodate, which oxidized C_6 of glucose to formaldehyde without loss or exchange of $^2H_2$. Excess periodate was destroyed and the sample was treated with NaOH and 15 M ammonium acetate, which converted formaldehyde to hexamethylenetetramine (HMT). The HMT was extracted twice with chloroform, dried, and analysed by gas chromatography-mass spectrometry (GC-MS). Conversion of the 6 carbon of glucose to HMT amplified enrichment by 6-fold, allowing very low enrichment of $^2H_2$ in plasma glucose to be measured accurately. Propanediol enantiomers were assayed according to Power *et al.* (1994).

Results

Experiment 1 Oral drenching caused PG to peak in plasma in a biphasic manner with approximately 5.5 mM of each isomer appearing at 45 min post-drenching, decreasing to 4 mM, and a second peak (6-7 mM) occurring at 120 min (Table 2). Peak concentrations of D- and L- isomers were similar; however, concentration of L- declined more rapidly. Concentration of PG in plasma peaked at the end of intravenous (i.v.) infusion. Concentration of D- declined more slowly than L-, as for oral drenching. Rates of decline of both isomers were more rapid for intravenous than for oral drenching, probably due to the smaller dose.

Table 2. Kinetic characteristics of PG metabolism after oral or intravenous dosing with PG.

Dose		Plasma peak (mM)		Utilization (k/h)	
Route	ml	D-	L-	D-	L-
Oral	520	7.0	6.0	0.28	0.45
i.v.	76	2.0	1.8	0.48	0.68

Experiment 2 The enrichment of plasma glucose achieved a plateau of 0.28 and 0.24%, corresponding to irreversible loss rates of glucose of 16.4 and 16.7 mmol/kg/min (2600 /d).

Infusion of PG increased plasma glucose from 3 to approximately 4 mM. However, irreversible disposal of glucose did not change when PG was infused.

Lactic acid concentrations in plasma (Table 3) were similar regardless of route of dosing. The biphasic peak observed for PG occurred also for lactate with the oral drench. Glucose concentration in plasma was higher when the oral dose was given. This was true also for insulin concentrations (30 and 16 mU/ml for oral and i.v., respectively).

Table 3. Concentration of plasma metabolites after oral or intravenous dosing with PG.

Route	Lactate (mM)		Glucose (mM)	Insulin (mU/ml)
	D	L		
Oral	0.2	0.6 - 1.2	4.2	30
i.v.	0.1 - 0.2	0.4 - 1.0	3.5	16

Discussion

The data from experiment 1 clearly documented the rapid absorption of PG and its metabolism via lactate. However, the data from experiment 2, in which PG dosing increased blood glucose concentration without changing glucose irreversible disposal, suggested that lactate was being used directly and sparing glucose utilization. This would be consistent with the ability to oxidize both D- and L- isomers, as shown by Harmon *et al.* (1984). Furthermore, our rates of disposal for PG were consistent with reported faster rates of utilization of L- lactate than the D- isomer, Harmon *et al.* (1984). Finally, our data are consistent with the conclusion of Emery *et al.* (1967) that direct absorption and metabolism of PG is the major route of utilization. However, our data suggested that PG carbon does not necessarily equilibrate with carbon in the glucose pool.

The methodology for amplifying the enrichment of glucose label is a powerful tool for metabolic studies in large animals. The same procedure may be adapted to amplify enrichment of N by four-fold.

Acknowledgements

Supported by the Ohio Agricultural Research and Development Centre/The Ohio State University, and the U.S. National Institutes of Health.

References

Christensen J.O., Grummer R.R., Rasmussen F.E. & Bertics S.J. (1997). Effect of method of delivery of propylene glycol on plasma metabolites of feed-restricted cattle. *J. Dairy Sci.* 80:563-568.

Emery R.S., Brown R.E. & Black A.L. (1967). Metabolism of DL-1, 2-propanediol -2-^{14}C in a lactating cow. *J. Nutr.* 92:348-356.

Grummer R.R., Winkler J.C., Bertics S.J. & Studer V.A. (1994). Effect of propylene glycol dosage during feed restriction on metabolites in blood of prepartum Holstein heifers. *J. Dairy Sci.* 77:3618-3623.

Harmon D.L., Britton R.A. & Prior R.L. (1984). *In vitro* rates of oxidation and gluconeogenesis from L(+)- and D(-) lactate in bovine tissues. *Comp. Biochem. Physiol.* 77B:365-368.

Hazey J.W., Yang D., Powers L., Previs S.F., David F., Beaulieu A.D., Puchowitz, M.A., Potter J.L., Palmquist D.L. & Brunengraber H. (1997). Tracing gluconeogenesis with deuterated water: measurement of low deuterium enrichments on carbons 6 and 2 of glucose. *Anal. Biochem.* 248:158-167.

Miyoshi S. (1995). Role of energy balance, plasma glucose and insulin on ovarian function and conception of dairy cows. M. S. Thesis. The Ohio State University, Columbus.

Powers L., Ciraolo S.T., Agarwal K.C., Kumar A., Bomont C., Soloviev M.V., David F., Desrochers S. & Brunengraber H. (1994). Assay of the enantiomers of 1,2-propanediol, 1,3-butanediol, 1,3 pentanediol, and the corresponding hydroxyacids by gas chromatography-mass spectrometry. *Anal. Biochem.* 221:323-328.

INFLUENCE OF CASEIN AND SOY PROTEIN ISOLATE ON THYROID HORMONE STATUS AND ON ENERGY AND PROTEIN METABOLISM IN GROWING PIGS

E. Saggau, M. Beyer, M. Klein, R. Schadereit, M. Derno, W. Jentsch , H. Scholze and H. Hagemeister

Research Institute for the Biology of Farm Animals, Research Unit of Nutritional Physiology, "Oskar Kellner", Dummerstorf-Rostock, Germany

Summary

Three experiments were carried out each using 12 growing, male, castrated pigs, which were fed semi-synthetic, isoenergetic diets based on either casein or soy protein isolate (3 × 650 kJ ME/kg $BW^{0,62} \times d^{-1}$). Three protein sources [(casein plus amino acid (AA) supplementation and soy protein isolate (SPI) without and with AA supplementation)] and two levels of protein supply (100% and 50% of the protein requirement) were tested. Both protein supply levels supported a lower protein deposition with SPI compared with casein+AA. At the 50% protein supply level, the decreased protein deposition was compensated by an increased fat deposition. At the 100% protein supply level there was no significantly higher fat deposition but a tendency of higher heat production. In our study, thyroid hormone levels were affected by the protein:energy ratio and the quality (AA pattern) of the dietary protein but higher thyroid hormone levels were not associated with an increased heat production.

Key words: Casein, soy protein isolate, energy metabolism, amino acid imbalance, thyroid hormones, pigs.

Introduction

Long-term feeding of diets which meet energy requirements, but not requirements for essential amino acids (AA) decreases growth and protein deposition in growing animals. An excess of dietary energy can either increase fat deposition or increase heat production. Both metabolic pathways were demonstrated when low-protein rations were fed while energy was supplied according to requirement (Fuller *et al.*, 1987; Tulp *et al.*, 1979).

Long-term feeding of low quality plant proteins (mostly soy protein isolate, SPI) compared with high quality animal proteins (mostly casein) as the sole protein source to monogastric animals induce higher plasma thyroid levels (Scholz-Ahrens *et al.*, 1990) but it is not known whether these increased levels are of physiological relevance. Increased thyroid hormone levels could indicate a shift towards increased heat production. In our study we intended to evaluate the influence of source and quality of dietary proteins on nutrient and energy metabolism of growing pigs as well as on thyroid hormone levels.

Materials and Methods

<u>Animals and diet</u>. Three separate experiments were carried out each with 12 male, castrated, German Landrace pigs, weighing between 30 and 90 kg. Pigs were housed individually at 22 ± 1°C. They were fed twice daily semi-synthetic, isoenergetic diets which provided three times the maintenance requirements of metabolizable energy (3 × 650 kJ ME/kg $BW^{0.62}$/d). Nutrient composition of the diets was similar to the composition of human diets in western industrial countries (protein, 8-18%; starch, 32-42%; sucrose, 20%; fat, 15%; cellulose, 7%; mineral and vitamin mix 8%, all wt/wt). Protein and starch

content was altered according to age and level of protein supply. Three protein sources were tested at two different levels. Protein sources were casein plus AA supplementation (Met, Thr, Trp), soy protein isolate (SPI) without and with AA supplementation (Lys, Met, Thr, Trp). Diet effects were compared at isoenergetic and isonitrogenous supply (kJ ME/kg $BW^{0.62}$/d and g N/kg $BW^{0.62}$/d). Casein plus AA supplement was adjusted to cover either 100% or 50% of the protein requirement. The SPI-diets provided equivalent amounts of nitrogen.

Experimental procedures and analyses. Animals were equipped with permanent jugular vein catheters at the initial weight of 30 kg for stress-free blood sampling. They were divided into groups of 4 to 6 animals and fed the experimental diets for an adaptation period of 21 days. Then followed 4 experimental periods; in each of the 4 periods energy metabolism was determined for a period of 1 week by indirect calorimetry. In the second period blood samples were taken five times a week for determination of thyroid hormones. Thyroxine (T_4), free T_4 (fT_4), triiodothyronine (T_3) and free T_3 (fT_3) were determined in serum by radioimmunoassays (Byk-Santec, Dietzenbach). Whole body protein synthesis was determined by means of an end product method with ^{15}N-glycine as tracer at a body weight of 40 kg and 100% protein supply level.

Statistical analysis was performed by t-tests. Differences were considered to be significant at $P \leq 0.05$.

Results

Average daily gains for casein+AA and SPI were 632 g and 533 g at 100% supply, respectively, and were 365 g and 318 g at 50% supply, respectively.

Protein energy deposition with SPI as compared with casein was significantly lower (by 19%) at the 100% protein supply level and by 31% at the 50% protein supply level (Table 1). At the 50% protein supply level, the decreased protein energy deposition was counterbalanced by a significantly increased fat energy deposition. There were no significant differences in the partition of metabolizable energy between energy deposition and heat production. At the 100% protein supply level, there was no significantly higher fat energy deposition but a tendency towards higher heat production. The energy deposition was decreased and the heat production was increased correspondingly on average by 32 kJ/kg $BW^{0.62}$/d (3% of total heat production) but the difference did not attain statistical significance. Though data on energy metabolism and thyroid hormones of

Table 1. Protein, fat and energy metabolism (per kg $BW^{0.62}$/d) as influenced by dietary protein source (casein plus AA, SPI or SPI plus AA) and level of protein supply.

Diet	Protein supply level	Intake		Deposition					
		ME[2] (kJ)	Prot.[3] (g)	Prot.[3] (g)	Fat (g)	PED[4] (kJ)	FED[5] (kJ)	ED[6] (kJ)	HP[1] (kJ)
Cas+AA	100%	1914	15.2	8.8[a]	14.1	210[a]	560	770	1144
SPI	100%	1914	15.3	7.1[b]	14.3	170[b]	568	738	1176
SPI+AA[7]	100%	1914	15.4	8.5[a]	13.5	203[a]	538	741	1173
Cas+AA	50%	1872	8.5	5.5[a]	15.4[a]	131[a]	613[a]	744	1128
SPI	50%	1872	9.0	3.8[b]	16.1[b]	91[b]	640[b]	731	1141

[1]Heat production, [2]Metabolizable energy, [3]Protein, [4]Protein energy deposition, [5]Fat energy deposition, [6]Energy deposition, [7]From a second, identical experiment. Different superscripts within a column and within the same protein supply level indicate significant differences.

SPI+AA diet originated from an experiment carried out at a different time, they clearly indicated that protein energy deposition was similar with supplementation by either protein source.

Preliminary results on whole body protein synthesis confirmed the decreased protein deposition of the SPI group as compared with the casein+AA and SPI+AA groups. The 100% protein supply level values expressed in terms of g per kg $BW^{0.62}$/d were 36.1 ± 4.0[a], 23.3 ± 4.5[b] and 34.1 ± 4.1[a] for casein+AA, SPI and SPI+AA, respectively, accompanied by protein depositions of 10.7 ± 0.3[a], 7.7 ± 0.9[b] and 10.4 ± 0.5[a] g per kg $BW^{0.62}$/d (different superscripts indicate significant differences).

A protein supply level of 100% caused significantly higher fT_3 values with SPI but not with casein+AA and SPI+AA (Table 2). At a protein supply level of 50%, when only SPI and casein+AA were tested, all parameters were distinctly affected by the protein source. SPI diet as compared to casein+AA caused a tendency towards higher T_4 levels and a significantly higher fT_4 level, but significantly lower T_3 and fT_3 levels. Irrespective of the protein source, the 50% protein supply level increased T_4 values on average by 30%, as compared with the 100% protein supply level.

Table 2. Thyroid hormone concentrations as influenced by the protein source and supply level.

Diet	Protein supply level	T_4 (nmol/l)	fT_4 (pmol/l)	T_3 (nmol/l)	fT_3 (pmol/l)
Cas+AA	100%	37.58 ± 2.19	13.55 ± 1.28	1.70 ± 0.14	0.44 ± 0.14[a]
SPI	100%	35.50 ± 3.18	12.58 ± 1.83	1.68 ± 0.29	0.74 ± 0.26[b]
SPI+AA[1]	100%	39.60 ± 10.58	14.41 ± 3.19	1.66 ± 0.40	0.30 ± 0.24[a]
Cas+AA	50%	50.79 ± 6.09	11.90 ± 1.58[a]	2.31 ± 0.29[a]	0.74 ± 0.26[a]
SPI	50%	55.68 ± 8.06	16.17 ± 1.41[b]	1.58 ± 0.24[b]	0.34 ± 0.22[b]

Data are means ± SD, n=6 animals/group, 15 analyses/animal. [1]From a second identical experiment. Different superscripts within a column and within the same protein supply level indicate significant differences.

Discussion

In the present study, the effects of source and quality of dietary proteins on nutrient and energy metabolism of growing pigs as well as on the thyroid hormone levels were investigated. Body weight gain and protein deposition decreased both when the concentration of protein in the diet was decreased and when non-supplemented SPI was used instead of supplemented casein. An excess of dietary energy, originating from limited protein deposition, can be used in two different ways. It can increase fat deposition or it can increase heat production, a metabolic pathway mediated by thyroid hormones (Tulp *et al*., 1979). Forsythe (1995) reviewed literature data on the influence of animal or plant proteins on the thyroid hormone status of monogastric animals and concluded that feeding of a plant protein with a lower biological quality (mostly soy protein isolate) as compared with an animal protein (casein) causes significantly higher T_4 levels, whereas effect on T_3, fT_3 and fT_4 varied.

Until now, there was no parallel measurement of energy metabolism. In the current study at 100% protein supply level, SPI as compared with AA-supplemented casein only tended to increase heat production. T_4, fT_4 and T_3 levels were not affected by the type of protein source. Only fT_3 was significantly higher following SPI as compared to AA-supplemented casein. Increased values of fT_3 were not clearly associated with significantly increased heat production.

At a 50% protein supply level, there was lower protein deposition with SPI as compared with AA-supplemented casein, which was compensated for by a significantly increased fat deposition. Heat production was not affected but there was a (not significantly) higher T_4 level in the SPI group. In several studies, significant effects were demonstrated when comparing plant and animal proteins. In this study fT_4 levels were significantly higher with SPI as compared with casein +AA.

A protein supply level of 50% produced a 30% higher T_4 concentration as compared with the 100% protein supply level, which was not affected by the protein source. It can be assumed that this effect was a consequence of the lower ratio of protein energy to total energy. Similar results were obtained by Schoppe (1988), who observed higher T_4 levels when feeding growing rats at a constant protein level and varying higher energy levels. In the previously mentioned study, energy supply might have affected the response, as energy-deficient diets were fed whereas, in the present study, diets exceeded energy requirements.

It is concluded that all thyroid hormone levels were affected differently by the protein:energy ratio of the diets as well as by the quality (AA pattern) of the dietary protein. In the partition of metabolizable energy to protein and fat deposition there were significant differences between the casein+AA-diets and SPI-diets, but only small or no differences in total energy deposition and heat production. Higher thyroid hormone levels within the physiological range were not associated with an increased heat production.

Acknowledgements

This project was supported by the Deutsche Forschungsgemeinschaft, DFG (German Research Foundation). Attendance at the 14th Symposium on Energy Metabolism of Farm Animals in September, 1997, Newcastle, Northern Ireland was sponsored by H. Wilhelm Schaumann Stiftung, Hamburg.

References

Forsythe W.A. (1995). Soy protein, thyroid regulation and cholesterol metabolism. *J. Nutr.* 125:619S-623S.

Fuller M.F., Cadenhead A., Mollison G., & Seve B. (1987). Effects of the amount and quality of dietary protein on nitrogen metabolism and heat production in growing pigs. *Br. J. Nutr.* 58:277-285.

Scholz-Ahrens K.E., Hagemeister H., Unshelm J., Agergaard N. & Barth C. (1990). Response of hormones modulating plasma cholesterol to dietary casein or soy protein in minipigs. *J. Nutr.* 120:1387-1392.

Schoppe I. (1988). Die Reaktion des Stickstoffhaushaltes und der Schilddrüsenhormone auf Casein- und Sojadiäten mit reduziertem Verhältnis von Proteinenergie zu Ge-samtenergie. Dissertation, Agricultural Faculty, Christian-Albrechts-University , Kiel, Germany.

Tulp O.L., Krupp P.P., Danforth Jr. E. & Horton E.S. (1979). Characteriztics of thyroid function in experimental protein malnutrition. *J. Nutr.* 109:1321-1332.

EFFECT OF SALBUTAMOL ON BODY COMPOSITION AND ENERGY UTILIZATION IN GROWING PIGS FED TWO LEVELS OF PROTEIN

J.A. Fernández, N. Oksbjerg and H. Jørgensen

Danish Institute of Agricultural Sciences, Research Centre Foulum, PO Box 50, DK 8830 Tjele, Denmark

Summary

The experiment was conducted using 7 litters of 5 female pigs each. Initially, 4 pigs from each litter (25 kg body weight) were allocated to 4 treatments: 2 levels of dietary protein (standard or high) supplemented with either 0 or 3 ppm of the β-adrenergic agonist salbutamol. The standard diet contained the level of protein recommended by Danish standards while the high protein diet contained approximately 14% in excess. At 35 and 65 kg body weight, digestibility and balance experiments were carried out. At 90 kg the pigs were slaughtered and dissected into: skeletal muscles, fat+skin, bones and gastrointestinal tract and internal organs. The anatomical fractions were minced and sampled for analysis of dry matter, nitrogen, lipid, amino acids and energy. The fifth littermate was slaughtered at the beginning of the experiment to provide the initial values for the experimental pigs. Growth rate and feed conversion were not affected by dietary protein level or by salbutamol. However dietary inclusion of salbutamol increased muscle growth and decreased the growth of fat+skin. Protein in muscle was increased especially for the pigs fed the high protein level, while the opposite was the case for fat content. The composition of empty body growth as a whole responded in a similar way. Fat deposition and consequently energy retention was decreased by the high protein level. Even though salbutamol enhanced protein deposition, the utilization of the absorbed protein was lower in the high protein diets, which is also reflected in a slightly higher heat production from these diets. These results indicate that salbutamol *per se* does not increase heat production and that maximal muscle response is not obtained unless the dietary level of protein is elevated beyond that required for optimal muscle growth on a standard diet.
Key words: β-agonist, salbutamol, body composition, energy metabolism, pigs.

Introduction

Several studies have demonstrated that the β-agonist salbutamol (SALB) promotes protein deposition and productive efficiency in slaughter pigs (Oksbjerg *et al.*, 1990; Warris *et al.*, 1990) The extent to which the β-agonist enhances protein deposition however, is dependent on, among other factors, dietary protein level. Dunshea *et al.* (1993) reported that dietary levels of protein have to be increased by 14% above recommendations in order to obtain maximal effects on protein deposition.

The objectives of this experiment were to study the effects of salbutamol administration, and of two levels of dietary protein, on the anatomical and chemical composition of pigs at slaughter and the concomitant impact on energy metabolism.

Material and Methods

Seven litters of 5 females each were used in this experiment. At an initial body weight (BW) of 25 kg, 4 pigs from each litter were allocated to 4 treatment groups, arranged as 2 different levels of dietary protein (S or H) supplemented with either 0 or 3 ppm of salbutamol (SALB), in a 2 x 2 factorial arrangement. The S diets contained a level of protein that was considered to be adequate for optimal growth performance by Danish standards. The H diets were formulated to contain approximately 14% in excess of the level in diet S. Dietary protein level was reduced in all finishing diets (50 to 90 kg BW). The pigs were scale fed twice a day and BW was recorded weekly throughout the experiment.

The fifth littermate was slaughtered at the beginning of the experiment to provide initial values of body composition. At 90 kg BW, the experimental pigs were stunned, bled and dissected (along with the 25 kg BW littermates) into the following anatomical fractions: skeletal muscles, subcutaneous fat + skin (fat+skin) and skeleton and entrails (gastrointestinal tract and internal organs). The different fractions were weighed, minced and sampled for analysis of dry matter, ash, nitrogen, lipid, energy and amino acids. Digestibility balances were carried out at 35 and 65 kg BW.

Results and Discussion

Gain of empty body EBWg (EBW = weight at slaughter - content of the GI-tract) and feed conversion were not influenced by SALB or by dietary protein level (Table 1). Protein deposition was increased by SALB especially in the pigs receiving diet H+SALB. However, the utilization of dietary protein was decreased by protein level, while SALB inclusion tended to improve utilization. The combined effect of SALB and increased protein level thus resulted in an increased protein oxidation, which suggested that H diets contained more protein than was necessary for a maximal response to SALB.

Muscle content (% of EBW) and muscle growth (g/kg EBWg) was increased by SALB. Muscle gain however, was highest on diet H+SALB. Skeletal growth and growth of entrails were not affected by treatments. Fat+skin growth on the contrary, was significantly reduced by SALB and, similarly to muscle growth, the effect was greatest in pigs receiving the H+SALB diet. Similar results were achieved by Dunshea *et al.* (1993) with ractopamine treated finishing gilts. Since the biological value of dietary protein was kept equal (level differences were obtained by dilution with N-free maize starch) these results suggest also, that SALB-treated and feed-restricted female pigs have higher protein requirement than untreated pigs, although this higher requirement appears to be less than the 14% applied in this experiment.

The protein content of muscle was significantly increased by SALB and protein level. In addition, the amino acid composition of muscle protein was unchanged by protein level but tended to be affected by SALB. Thus, glutamic acid tended to be increased but only when diet H+SALB was fed, whereas proline was decreased by diet H+SALB. In agreement with the results of Warris *et al.* (1991), this study demonstrates the protein-anabolic effect of SALB. Moreover, the results indicate that the effect of SALB is restricted to skeletal muscles in a tissue-selective manner as suggested by the change in amino acid composition of muscle protein.

Digestibility of protein was not influenced by SALB addition or by protein level. The content of GE was slightly higher in H diets due to the higher content of crude protein.The higher content of GE in H diets was ameliorated by a lower digestibility induced by the lower content of the highly digestible maize starch. ME content of H diets was further reduced by increased urinary N-excretion.

The increased protein deposition and concomitantly lower fat deposition resulted in decreased energy retention, both in absolute terms and in relation to ME. Again, the highest response, although non-significant, was shown by diet H+SALB, as evidenced by the altered partition of the retained energy as protein and fat, respectively.

Table 1. Performance, anatomical and chemical composition, and energy metabolism of pigs treated with salbutamol (SALB: 3 ppm) and fed two levels of protein (S & H).

	S	S +SALB	H	H +SALB	Significance of effects: SALB	Protein	SALB X Protein	SD
Production performance								
Feed intake (g/d)	1836	1832	1829	1808	ns	ns	ns	17
Digest. protein (g/d)	237	237	272	268	ns	***	ns	3
EBW (kg)	85	84	83	84	ns	ns	ns	1
EBW-gain (g/d)	665	667	654	679	ns	ns	ns	14
kg feed/kg EBWg	2.75	2.74	2.82	2.70	ns	ns	ns	0.09
Dep. protein (g/d)	118	123	119	131	*	ns	*	2
Dep. protein (% of intake)	42	43	36	40	ns	**	ns	1
Muscle (% of EBW)	53	55	53	56	**	ns	ns	0.7
Growth of anatomical fractions (g/kg EBWg)								
Muscle	554	578	557	598	**	ns	ns	15
Bones	123	122	125	118	ns	ns	ns	9
Fat	181	163	174	152	*	ns	ns	5
Entrails	90	91	91	89	ns	ns	ns	4
Chemical composition of anatomical fractions (g/kg DM)								
Muscle protein	641	668	656	714	**	*	*	16
Muscle fat	275	287	257	241	ns	*	ns	17
Muscle glut. acid (g/16 g N)	14.6	14.6	14.2	15.0	ns	ns	ns	0.6
Muscle lysine (g/16 g N)	8.2	8.2	8.1	8.2	ns	ns	ns	0.2
Muscle proline (g/16 g N)	4.4	4.5	4.5	4.3	ns	ns	*	0.1
Entrails protein	524	522	553	577	ns	**	ns	26
Entrails fat	416	413	395	366	ns	*	ns	26
Skeletal protein	400	401	398	411	ns	ns	*	6
Chemical composition of growth (g/kg EBWg)								
Dry matter	415	404	402	391	ns	ns	ns	13
Ash	30	29	30	29	ns	ns	ns	2
Protein	179	185	182	195	**	*	ns	4
Fat	186	182	174	154	ns	ns	ns	11
Energy (MJ)	11.37	11.18	10.90	10.22	ns	ns	ns	0.60
Energy metabolism								
GE (MJ/kg DM)	18.24	18.24	18.31	18.31	--	--	--	--
DE (MJ/kg DM)	14.92	14.85	14.72	14.64	ns	*	ns	0.17
DE (% of GE)	82	81	80	80	ns	*	ns	1
ME (MJ/kg DM)	14.46	14.42	14.20	14.16	ns	*	ns	0.18
ME (% of DE)	96.9	97.1	96.4	96.7	ns	*	ns	0.4
RE (MJ/kg DM)	4.88	4.85	4.57	4.51	ns	*	ns	0.23
RE (% of ME)	34	34	32	32	ns	*	ns	2
Heat production (MJ/kg DM)	9.58	9.57	9.63	9.65	ns	ns	ns	0.30
RE in protein (% of RE)	36.9	38.0	38.8	43.7	*	*	*	1.4
RE in fat (% of RE)	63.1	62.0	61.2	56.3	*	*	*	1.4

*, **, *** significant at P<0.05, P<0.01 and P<0.001, respectively; ns = non-significant.

Heat production (calculated as ME-RE) was not altered by SALB or protein level. The small differences detected between S and H diets, can be accounted for by the heat generated from the extra protein catabolized when on H diets (6.6 kJ/g catabolized protein: Just, 1982).

In conclusion, the results reported here and those published earlier (Oksbjerg *et al.*, 1996) from the same experiment indicate that the β-adrenergic agonist salbutamol promotes muscle growth and depresses fat deposition. The extent of these effects is dependent on the dietary protein level. Salbutamol *per se* did not alter energy metabolism. The detected differences in energy retention were a consequence of a modified protein and fat deposition pattern.

References

Dunshea F.R., King R.H. & Campbell R.G. (1993). Interrelationships between dietary protein and Ractopamine on protein and lipid deposition in finishing gilts. *J. Anim. Sci.* 71:2919-2930.

Just A. (1982). The net energy value of crude (catabolized) protein for growth in pigs. *Livestock Prod. Sci.* 9:349-360.

Oksbjerg N., Blackshaw A., Henckel P., Fernández J.A. & Agergaard N. (1990). Alterations in protein accretion and histochemical characteriztics of the *m. longissimus dorsi* in pigs caused by salbutamol (a β-Adrenergic Agonist). *Acta Agric. Scand.* 40:397-401.

Oksbjerg N., Fernández J.A., Jørgensen H., Olsen O.H., Rolph T. & Agergaard N. (1996). Effects of salbutamol on protein and fat deposition in pigs fed two levels of protein. *J. Anim. Physiol. and Anim. Nutr.* 75:1-12.

Warris P.D., Brown S.N., Rolph T.P. & Kestin S.C. (1990). Interactions between the beta-adrenergic agonist salbutamol and genotype on meat quality in pigs. *J. Anim. Sci.* 68:3669-3676.

METHANE REDUCTION AND ENERGY TURNOVER IN LAMBS RECEIVING DIFFERENT DIETARY FAT SUPPLEMENTS

A. Machmüller, D.A. Ossowski and M. Kreuzer

Institute of Animal Sciences, Animal Nutrition, Swiss Federal Institute of Technology (ETH), ETH centre/LFW, CH-8092 Zurich, Switzerland

Summary

The present data from growing lambs confirmed the nature and extent of the suppressing effects of certain lipids on methane release and on rumen protozoa as previously found *in vitro.* Various oilseeds and coconut oil were effective against methanogenesis. Digestibility and utilization of dietary energy were slightly reduced by the oilseeds but not by coconut oil.
Key words: Dietary fat, methane, energy turnover, utilization.

Introduction

Pure plant oils which are rich in polyenoic acids have long been known for their methane reducing activity (compiled by Van Nevel & Demeyer, 1996). Recent *in vitro* studies have also demonstrated considerable potential of whole oil seeds and of solid plant fat in this respect (Machmüller *et al.*, 1997). The methane reduction was accompanied by a decrease in the number of protozoa which are involved in methanogenesis. Fats which affect methane-producing microbes could in some way also influence ruminal fibre digestion. This might reduce dietary energy utilization and metabolic protein supply from microbial origin. The aim of the present study was to verify the methane reducing effects of the same diets *in vivo* as has been applied *in vitro* (Machmüller *et al.*, 1997) and to study the effects of fats on nutrient and energy turnover in fattening lambs.

Materials and Methods

Six diets were formulated to contain either no additional fat (control) or commercial rumen-inert fat, coconut oil, rapeseed, sunflower seed and linseed as supplements to maize silage-hay-concentrate diets (Table 1). Ether extract accounted for 31 g/kg consumed dietary dry matter in control, and for 56 g/kg (mean) of all fat supplemented diets. In 2 series with 6 growing lambs each, 3 subsequent periods of total balance procedure and open circuit respiratory technique for determination of gaseous exchange were carried out. Each lamb received a different sequence of the diets which, in total, provided 6 measurement periods per diet. Each period consisted of 14 days of adaptation, 7 days of excreta collection, and 2 days of respiratory measurements. At the end of each period, the lambs were inoculated with fresh rumen fluid and subsequently fed for 5 days with the control diet during the intermittent period. All diets were supplied to provide (according to current Swiss standards) the same daily amount of metabolizable protein and energy at each experimental period representing different growth stages. This resulted in moderately different organic matter and crude protein intakes. The data were evaluated by analysis of variance for diet, series and period within series.

Table 1. Composition, intake and apparent digestibility of the diets[1].

Group / fat component	Control	Inert fat	Coconut oil	Rapeseed	Sunflower seed	Linseed	SEM[2]	P[3]
Ingredients, g/kg DM								
Maize silage	583	677	684	684	675	690		
Hay	70	70	69	67	62	69		
Concentrate	346	254	246	250	263	241		
Fat component	-	28	25	59	57	67		
Barley	218	49	51	20	67	19		
Soybean meal	94	126	119	118	67	106		
Potato protein	15	30	31	34	53	29		
Minerals	20	20	20	20	19	20		
Analysed composition, per kg DM as consumed								
Gross energy, MJ	18.4	19.2	18.4	19.1	19.3	18.7		
Ether extract, g	31	57	54	59	56	55		
Crude protein, g	138	147	145	158	153	153		
NDF, g	367	371	371	370	370	376		
Intake, g/d								
Organic matter	889	859	856	886	888	916	16.8	
Coefficients of apparent digestibility								
Gross energy	0.67	0.67	0.66	0.61	0.62	0.63	0.035	0.03
Organic matter	0.71^{a}	0.70^{ab}	0.70^{a}	0.67^{ab}	0.65^{b}	0.68^{ab}	0.027	0.01
Ether extract	0.51	0.61	0.62	0.36	0.52	0.50	0.164	0.12
Crude protein	0.60^{b}	0.63^{ab}	0.67^{a}	0.64^{ab}	0.62^{b}	0.63^{ab}	2.672	0.01
NDF	0.53^{a}	0.52^{a}	0.51^{ab}	0.47^{ab}	0.42^{b}	0.48^{ab}	0.051	0.01

[1] Means carrying no common superscript are significantly different (P<0.05, Tukey)
[2] SEM, standard error of mean; [3] probability of dietary effect.

Results and Discussion

The average live-weight of the lambs during the experiment was 32 ± 4 kg, and feed dry matter intake was 0.94 ± 0.15 kg/d. Nutrient and energy digestibility was slightly different between diets (Table 1) being lowest for sunflower seed (significant for OM, CP and NDF) and, for ether extract only, for rapeseed.

The initial protozoal counts in rumen fluid were somewhat different between groups, but not in a systematic way (Table 2). Protozoal numbers were lower after 14 (P=0.12) and after 23 days of application of the diets in all fat-supplemented groups as compared with the controls. The lowest protozoal counts were observed using coconut oil (20% and 28% of control). This effect was of a similar extent to that reported for a comparable amount of coconut oil in the diet *in vitro* (34% of control; Machmüller *et al.*, 1997) and in wethers (13% of control; Machmüller & Kreuzer, 1997). Medium-chain saturated fatty acids are specifically toxic to protozoa (Matsumoto *et al.*, 1991). As in the *in vitro* study, with the whole crushed oil seeds (which contain unsaturated but partially rumen-inert oils) intermediate values occurred. All fat supplements, apart from the rumen-inert fat, reduced methane release (P=0.13 for g per day, P=0.08 for g per kg OM and P=0.04 for methane energy loss as % of gross energy intake) with the highest responses obtained with coconut oil (-25%) and sunflower seed (-23%). For coconut oil this was very close to the level of reduction found with wethers (-28%; Machmüller & Kreuzer, 1997) and somewhat weaker to that measured *in vitro* (-40%) whereas the reduction by the sunflower seed diet *in vitro* was 19% (Machmüller *et al.*, 1997). This was not predominantly a result of a reduced organic matter or fibre digestibility (Table 1) as can be seen from the methane formed per unit of organic matter digested (Table 2). Our previous *in vivo* investigation demonstrated that providing twice the amount of coconut oil by the use of a higher proportion of concentrate suppresses methane by more than 70% (Machmüller & Kreuzer, 1997) probably as a result of the reduction in fermentable cell wall carbohydrates and the direct adverse effects against rumen protozoa which are involved in methane formation (Kreuzer *et al.*, 1986). The intermediate effects of the oilseeds might be explained by their partly rumen-inert character.

Table 2. Protozoal counts and methane release.

Group / Fat component	Control	Inert fat	Coconut oil	Rape-seed	Sunflower seed	Lin-seed	SEM	P
Protozoal counts, 10^5/ml rumen fluid								
Initially	8.4	6.9	7.6	5.8	8.4	5.6	5.45	0.95
Day 14	10.3	5.1	2.1	2.8	4.4	4.0	5.02	0.12
Finally (day 23)	6.0	5.0	1.7	5.0	3.3	4.9	2.90	0.18
Methane								
g/d	15.5	14.9	11.6	12.5	11.9	14.2	2.85	0.13
g/kg OM	17.6	17.4	13.6	14.4	13.3	15.7	3.03	0.08
g/kg OM digested	24.6	25.0	19.5	21.5	20.4	23.2	4.39	0.19
% of energy intake	5.0	4.7	3.8	3.9	3.6	4.3	0.83	0.04

Retention and utilization of dietary energy and protein were not significantly affected by the dietary treatments (Table 3). Metabolizability tended to be lower for all the oilseeds with a slightly lower energy retention using sunflower seed and linseed (about 2 instead of about 3 MJ/d). This was accompanied by lower values for ME utilization. Heat production was quite similar in the treatments. The fat supplements reduced the respiratory quotient from 1.14 to 1.02 on average as would be expected from increased fat oxidation. As far as protein is concerned, the inert fat and rapeseed groups showed the highest retention and utilization. The average levels of utilization of ME and of dietary protein for growth fell within the normal range.

Table 3. Retention and utilization of dietary energy and protein.

Group / Fat component	Control	Inert fat	Coconut oil	Rape-seed	Sunflower seed	Lin-seed	SEM	P
Energy turnover								
Intake, (MJ/d)	17.3	17.5	16.7	18.0	18.2	18.2	0.91	
Heat, (MJ/d)	7.2	6.8	6.5	6.9	8.0	8.1	1.28	0.26
Retention, (MJ/d)	2.9	3.5	3.4	3.0	2.2	2.0	1.32	0.20
Metabolizability	0.59	0.59	0.59	0.54	0.55	0.55	0.033	0.06
Utilization of ME								
total (k_t)	0.28	0.31	0.35	0.29	0.19	0.20	0.125	0.17
for growth (k_g)[1]	0.57	0.60	0.55	0.59	0.36	0.41	0.257	0.41
Protein turnover								
Intake, (g/d)	130	134	132	149	144	149	3.2	
Retention, (g/d)	31	41	35	50	37	41	11.0	0.10
Retention/intake	0.24	0.31	0.26	0.33	0.26	0.28	0.071	0.27

[1] Assuming maintenance requirement of 0.380 MJME/d/kg metabolic live-weight (Thériez *et al.*, 1987).

In fattening cattle using similar diets (Ossowski *et al.*, 1997), growth effects resembled the present results in protein and energy retention, except for a lower growth with the coconut oil diet due to a reduced palatability. This was not the case in the lambs, whereas higher dietary proportions of coconut oil also decreased palatability in wethers (Machmüller & Kreuzer, 1997).

Conclusions

Although not statistically significant, the present study confirmed the nature and extent of the previous *in vitro* and *in vivo* results as far as the reduction of protozoal count and methane release by coconut oil and oilseeds is concerned. Overall, direct adverse effects of the quite moderate fat supplements on retention, digestion and utilization of energy and protein were low and not systematic. The slowly digestible hulls of the oilseeds were mainly responsible for the lower nutrient digestibility rather than their content of polyenoic fatty acid which widely differed between seeds. Regarding the particular side-effects, the use of coconut oil and oil seeds can be successfully applied in fattening ruminant diets as natural methane-reducing feed components which is not possible for pure plant oils because of their greater inhibition of fibre digestion.

References

Kreuzer M., Kirchgessner M. & Müller H.L. (1986). Effect of defaunation on the loss of energy in wethers fed different quantities of cellulose and normal or steamflaked maize starch. *Anim. Feed Sci. Technol.* 16:233-241.

Machmüller A. & Kreuzer M. (1997). Effect of dietary coconut oil on methane formation and energy turnover in sheep (in German). *Proc. Soc. Nutr. Physiol.* 6:94.

Machmüller A. Ossowski D.A., Wanner M. & Kreuzer M. (1997). Potential of various fatty feeds to reduce methane release from rumen fermentation *in vitro* (Rusitec). *Anim. Feed Sci. Technol.* (in press).

Matsumoto M., Kobayashi T., Takena A. & Itabashi H. (1991). Defaunation effects of medium-chain fatty acids and their derivates on goat rumen protozoa. *J. Gen. Appl. Microbiol.* 37:439-445.

Ossowski D.A., Sutter F. & Kreuzer M. (1997). Effects of rapeseed, sunflower seed and linseed on growth, carcass and meat quality of fattening bulls (in German). *Proc. Soc. Nutr. Physiol.* 6:118.

Thériez M., Bocquier F. & Brelurut A. (1987). Recommandations alimentaires pour les brebis à l'entretien et en gestation. *Bull. Tech. C.R.V.Z. Theix, INRA* 70, 185-197.

van Nevel C.J. & Demeyer D.I. (1996). Control of methanogenesis. *Environm. Monit. Assessm.* 42:73-97.

CRUDE PROTEIN AND SUPPLEMENTAL DIETARY TRYPTOPHAN EFFECTS ON GROWTH AND TISSUE NEUROTRANSMITTER LEVELS

R.W. Rosebrough

Growth Biology Laboratory, LPSI, USDA-ARS, BARC, Beltsville, MD 20705, USA

Summary

Indian River male broiler chickens growing from 7 to 28 d of age were fed diets containing 120, 210 and 300 g crude protein and 0, 1.7 or 16.7 g of added tryptophan (TRP)/kg diet. The hypothesis tested was that crude protein levels and TRP would affect both growth and neurotransmitter metabolism. Heart, brain and pancreatic neurotransmitter norepinephrine (NE), dopamine (DA), serotonin 5-HT and 5-hydroxy-indole-3-acetic acid (5-HIAA) concentrations were determined by HPLC separation and electrochemical detection. Cardiac NE concentrations were directly related to dietary crude protein levels while pancreatic levels were inversely related. An increase in dietary crude protein decreased both brain NE and DA. Supplemental dietary TRP increased both 5-HIAA and 5-HT. Changes in feed intake caused by different levels of both dietary crude protein and TRP are accompanied by altered levels of neurotransmitters. The present study indicates that much larger amounts of TRP are required to simultaneously change feed intake and neurotransmitters.
Key words: Diet, crude protein, intake, chickens, tryptophan, neurotransmitters, diet.

Introduction

The central nervous system (CNS) of rats will respond to changes in dietary regimens and will regulate eating behaviour. The addition of carbohydrates and fats to diets without changing the total energy level activates the sympathetic nervous system. Substitution of either sucrose or glucose for dietary protein also stimulates neural activity (Young & Landsberg, 1977). A more recent study indicated that dietary energy components might also change the rate of dopamine (DA) and norepinephrine (NE) turnover in rats (Vander Tuig & Romsos, 1984). The addition of casein raised urinary DA excretion while carbohydrate or fat supplementation decreased urinary DA.

A relationship also exists between consumption of protein and carbohydrate and the synthesis of serotonin (5-HT) by the brain. Serotonin is synthesized from the essential amino acid L-tryptophan (TRP) and may regulate food intake in rats. The initial step involves the hydroxylation of TRP to form 5-hydroxytryptophan (5-HTP) that is catalyzed by TRP hydroxylase. The major 5-HT metabolite in the brain is 5-hydroxy-indole-3-acetic acid (5-HIAA) and its relationship to 5-HT is indicative of 5-HT turnover (Fernstrom, 1979).

We have reported several times that broiler chickens eat less of a diet containing 120 g crude protein/kg diet than of one containing 180 g crude protein/kg. Birds fed diets containing low levels of crude protein also grow poorly and are proportionately fatter than are birds fed diets containing higher levels of crude protein. In this respect, feeding a diet containing a low level of crude protein produces a bird that is metabolically similar to the egg-laying bird. This experiment was designed to test the effects of crude protein and tryptophan on growth and neurotransmitter metabolism.

Materials and Methods

Animals - Male, Indian River broiler chicks growing from 7 to 28 d of age were fed three levels of crude protein (120, 210 and 300 g/kg) and three levels of added TRP (0, 1.67 or 16.7 g/kg diet) in a 3 x 3 factorial arrangement. At 28 d of age, one bird from each pen was randomly selected and killed and organs removed into liquid N_2 and stored at -70°C.

Neurotransmitters - Organs were homogenized in 0.1 M perchloric acid (PCA) and centrifuged at 12,000 x g for 15 min. Analyses of catecholamines were performed with a monochloroacetate buffer (150 mM) containing 1 mM EDTA and 1 mM sodium octyl sulphate at pH 3.0. The organic modifier was acetonitrile (50 ml/l). The electrochemical detectors were set at +0.65 V vs. Ag/AgCl reference electrode with ranges of 5 and 50 nA for NE, DHBA and DA. The mobile phase for indolamines was a monochloroacetate buffer (150 mM/l) containing 1 mM-EDTA and 1 mM-sodium octyl sulphate at pH 3.0. The organic modifiers were acetonitrile (40 ml/l) and tetrahydrofuran (10 ml/l), respectively. Detectors were set at +0.80 V and +0.65 V vs. a Ag/AgCl reference electrode and 5 and 10 nA for 5-HT and 5-HIAA. Data were analysed by standard ANOVA.

Results

Growth and Feed Consumption - Birds consuming the diets containing 120 g crude protein were smaller than birds consuming either of the two other levels of dietary crude protein (Table 1; P<0.0001). When analysing the effect of added TRP, it was noted that the addition of 16.7 g of TRP/kg of diet resulted in smaller birds than did the other levels of TRP (P<0.000). The interaction between crude protein and TRP was not statistically significant (P = 0.500). The same dietary effects were also noted on food consumption as well as efficiency of food utilization.

Table 1. Growth, feed intake and feed conversion efficiency of broilers.

	Dietary protein, (g/kg diet)									
	120			210			300			
	TRP			TRP			TRP			
Parameter	0	1.67	16.7	0	1.67	16.7	0	1.67	16.7	P<0.01
BWt	651	636	458	975	964	789	956	1391	1.72	Pro, TRP
Food	1147	1103	887	1384	1312	1291	911	1365	1.78	Pro, TRP
F/G	2.25	2.37	2.90	1.65	1.58	2.12	843	1312	1.88	Pro, TRP

Brain neurotransmitters - The effects of both dietary crude protein and TRP levels on brain neurochemical levels are presented in Table 2. An increase in dietary crude protein decreased both brain NE and DA (P = 0.006 and 0.014, respectively). The addition of TRP to the diets had no effect upon either NE or DA (P = 0.077 and 0.887, respectively). The ratio of DA:NE was not affected by either crude protein or TRP.

Table 2. Effect of diets on brain norepinephrine (NE), dopamine (DA) serotonin (5-HT) and 5-hydroxy-indole-3-acetic acid (5-HA) in broiler chickens (values are ng/g tissue).

	Dietary protein, g/kg diet									
	120			210			300			
	TRP			TRP			TRP			
Parameter	0	1.67	16.7	0	1.67	16.7	0	1.67	16.7	P<0.01
NE	398	366	409	368	345	401	336	352	289	Pro
DA	446	398	451	325	400	373	356	357	343	Pro
DA/NE 1.11	1.11	1.10	0.91	1.23	0.96	1.06	1.01	1.22		
5-HT	594	536	829	447	608	586	630	570	633	TRP
5-HA	225	237	498	224	288	405	223	236	416	TRP
5-HT/5-HA	2.69	2.49	1.72	2.03	2.36	1.76	2.87	2.53	1.52	Pro, TRP

Heart and pancreatic neurotransmitters - Dietary protein increased (P=0.033) cardiac NE, decreased (P<0.001) pancreatic NE, but had no effect upon cardiac DA (P=0.304; Table 3). Dietary TRP did not affect amines in either hearts or pancreases (P=0.296 and 0.960, respectively). Although both crude protein and TRP influenced 5-HT, these main treatment effects were confounded by a crude protein x TRP interaction (P=0.013). In contrast, only TRP influenced 5-HIAA (P=0.000). Both dietary crude protein and its interaction with TRP had little effect on 5-HIAA (P=0.661 and 0.368, respectively). Both crude protein and TRP influenced the ratio of 5-HT to 5-HIAA (P=0.01).

Table 3. Effect of diets on cardiac (H) and pancreatic (P) norepinephrine (NE) and dopamine (DA) in broiler chickens (all ng/g tissue).

	Dietary protein, g/kg diet									
	120			210			300			
	TRP			TRP			TRP			
Parameter	0	1.67	16.7	0	1.67	16.7	0	1.67	16.7	P<0.01
H NE	231	217	311	341	363	322	347	414	386	
H DA	1387	622	1219	969	1342	1286	1400	918	855	
H DA/NE	6.11	2.54	4.45	3.40	4.08	4.33	3.55	1.97	2.83	
P NE	446	481	430	328	322	266	271	260	363	Pro

Discussion

Kaufman *et al.* (1984) found that cardiac NE concentrations were 15 to 20% lower and NE turnover nearly two-fold greater in rats fed a diet containing a low level of crude protein. Furthermore, urinary NE excretion was 25% greater in the group fed the low protein diet. Although tyrosine availability could regulate NE turnover, supplementation of diets with tyrosine did not change turnover rates. Regretfully, these kinds of studies do not hint at the possible mechanisms involved in accelerated catecholamine metabolism in animals fed low-protein, high-carbohydrate diets. It should be noted that, in most studies, carbohydrates are often substituted for dietary protein sources during diet formulation and that the effect of a high-carbohydrate level cannot always be separated from that of a low protein level. Thus, the argument can always be made that effects are due to an increase in available dietary carbohydrate as well as a decrease in dietary protein. Indeed, Vander Tuig & Romsos (1984) stated that carbohydrate availability, rather than protein *per se*, is the dietary variable regulating CNS activity. This statement may not be all encompassing because both protein-free and high-protein diets reduce food intake. Depressed food intake of rats fed on either of these regimens is accompanied by changes in free, indispensable, amino acids but is not accompanied by any changes in either 5-HT or 5-HIAA. In addition, diets containing wide variations in amino acid quality and balance produce a competition among amino acids for uptake into the brain. This competition results in changes in feeding behaviour but does not change either 5-HT or 5-HIAA (Fernstrom, 1983)

It is possible that feed intake of chickens fed diets containing very low levels of crude protein may be affected by TRP through the influence of 5-HT. The value, to the chicken, of increased NE as a function of low-protein diets probably lies in the ability of these kinds of diets to stimulate fat synthesis. If pancreatic glucagon can be shown to be under the influence of the CNS, the lipolytic characteristics of this hormone may be a natural feedback on the production of excess fat in chickens fed high-carbohydrate diets. Unlike mammalian liver tissue, the avian liver may not respond to insulin but is more responsive to the catecholamine family of hormones.

References

Fernstrom J.D. (1979). Diet-induced changes in plasma amino acid pattern: effects on the brain uptake of large neutral amino acids, and on brain serotonin synthesis. *J. Neural Transmission* 15:217-219.

Fernstrom J.D. (1983). Role of precursor availability in control of monamine biosynthesiz in brain. *Phys. Rev.* 93:484-546.

Kaufman L.N., Young J.B. & Landsberg L. (1984). Protein stimulates sympathetic nervous (CNS) activity less than carbohydrate: evidence for nutrient-specific CNS responses. *Clin. Res.* 32:478-490.

Vander Tuig J.G. & Romsos D.R. (1984). Effects of dietary carbohydrate, fat, and protein on norepinephrine turnover in rats. *Metabolism* 33:26-33.

Young J.B. & Landsberg L. (1977b). Stimulation of the sympathetic nervous system during sucrose feeding. *Nature* 269:615-617.

DIETARY SEPARATION OF PROTEIN AND CARBOHYDRATE INTAKE

M. Kirchgessner and H.L. Müller

Institut für Ernährungsphysiologie, Technische Universität München 85350 Freising-Weihenstephan, Germany

Summary

The effect of a separation of intake of dietary protein and carbohydrates on thermogenesis and RE was studied in two experiments with adult sows. In the first experiment, dietary protein was offered once a day in the morning and, in the second experiment, at intervals of 33 h or 48 h. The trials were conducted as change-over designs with 3-week metabolism periods per treatment. Both experiments have clearly shown that the separated intake of dietary protein and carbohydrates by up to 48 h did not markedly change thermogenesis at the level of maintenance metabolism. Additional energy costs of regulation of substrate supply to the continuous cellular metabolism due to separate ingestion of dietary nutrients therefore can be neglected.
Key words: Protein, thermogenesis, dietary nutrient separation, sows.

Introduction

Despite the fact that food is taken up at regular or non-regular intervals, the biochemical pathways occur in continuous processes. There is incessant synthesis of body constituents, especially of proteins (about 20% of basal metabolism, Young *et al.*, 1992) which requires a steady flow of both amino acids and energy. Furthermore, under practical conditions all of the essential nutrients are not always supplied within each meal. In human nutrition, the principle of the HAY's diet even recommends a temporally separated intake of proteins and carbohydrates and promises favourable effects (Walb *et al.*, 1994). Therefore, in experiments with sows, we studied the effect of a separated intake of protein and carbohydrates (according to HAY's principle) on nitrogen and energy balance. In the first trial, protein was offered every 24 h and, in a second trial, at intervals of 33 and 48 h.

Materials and Methods

Experiment 1 was performed as a cross-over design with 8 adult, non-pregnant, sows. The daily ration consisted of 5% fish meal, 15% meat and bone meal, 22% wheat starch, 22% maize starch, 22% potato starch, 10.5% cellulose, 1% soybean oil and 2.5% of a mineral-vitamin premix. Feed was offered three times a day at 07.30, 11.30 and 16.30 h. In sows subjected to dietary separation all dietary protein was offered at the first feeding, while the carbohydrates and soybean oil were given in equal amounts at the remaining two feedings. Minerals and vitamins were spread evenly over the day. The control sows received the daily ration in three equal portions. The levels of all nutrients in the rations matched the maintenance requirements based on the initial bodyweight of the sows and remained unchanged throughout the experiment. In both 3-week metabolism periods, a complete energy balance was determined for each animal by the use of the collection technique (feed, faeces, urine) and a measurement of the gas exchange over 48 h in an open-circuit respiration chamber.

In experiment 2, a 3x3 latin square design was used with a total of nine sows. The composition of the ration was the same as in experiment I. Protein was added to each meal (control), or supplied at intervals of 33 h (treatment 1) or 48 h (treatment 2).

Table 1. Gaseous exchange, nitrogen and energy balance of Expt. I.

	Control	Treatment	SEM	P > F
Gaseous exchange (litre/animal/day)				
O_2	991	976	2.1	**
CO_2	981	964	1.9	**
CH_4	19.0	18.2	0.8	NS
RQ	0.990	0.988	0.004	NS
Nitrogen balance (g/animal/day)				
N intake	28.5	28.5	-	-
Faeces	9.8	9.4	0.3	NS
Urine	15.6	18.4	0.7	*
N retention	3.0	0.7	0.8	NS
Energy balance (MJ/animal/day)				
GE intake	25.21	25.21	-	-
Faeces	3.21	3.03	0.07	NS
DE	22.00	22.18	0.07	NS
Urine	0.70	0.77	0.01	*
CH_4	0.75	0.72	0.03	NS
ME	20.54	20.69	0.10	NS
HPcnrq	20.55	20.10	0.04	**
REcnrq	-0.01	0.59	0.11	**
DE/GE, %	87.2	88.0	0.3	NS
ME/GE, %	81.5	82.1	0.4	NS
HPcnrq/ME, %	100.1	97.2	0.5	**
REcnrq/ME, %	-0.2	2.8	0.5	**

HPcnrq, REcnrq: mean values from the CN and RQ method.

The rations also met the requirements for maintenance and were fed three times daily. With treatment 1, protein was offered in the morning and in the evening of the next day, with treatment 2 it was offered every other day in the morning. Complete energy balances were performed for each animal and experimental period using the same technique as in experiment 1. In both studies the calculation of heat production and energy retention was based on the CN balance method as well as on the respiratory quotient (RQ) method according to the Brouwer's formulae (Brouwer, 1965).

Results and Discussion

The separated intake of dietary protein and carbohydrates within a day had no effect on the bodyweight of the animals (178.3 vs. 178.6 kg with the control diet). Oxygen intake and carbon dioxide production were both reduced by 1.5% as a result of the separating feeding regime, so that RG remained unchanged (Table 1). With equal dietary supply of nitrogen

Table 2. Gaseous exchange, nitrogen and energy balance of Expt. 2.

	Control	33 h	48 h	SEM	P > F
Gaseous exchange (litre/animal/day)					
O_2	965	953	974	10.7	NS
CO_2	970	958	980	9.8	NS
CH_4	20.4^{a}	16.1^{b}	18.8^{ab}	0.9	**
RQ	1.007	1.006	1.007	0.009	NS
Nitrogen balance (g/animal/day)					
N intake	30.8	30.8	30.8	-	-
Faeces	11.7^{a}	9.2^{b}	9.9^{ab}	0.6	*
Urine	18.1	19.9	18.3	0.8	NS
N retention	0.9	1.7	2.6	0.9	NS
Mean energy exchange across 48 h (MJ/animal/day)					
GE intake	25.72	25.72	25.72	-	-
Faeces	3.15	2.61	2.74	0.17	NS
DE	22.57	23.11	22.98	0.17	NS
Urine	0.78	0.82	0.77	0.02	NS
CH_4	0.81^{a}	0.63^{b}	0.74^{ab}	0.04	**
ME	20.99^{a}	21.66^{b}	21.47^{ab}	0.18	*
HPcnrq	20.23	19.89	20.44	0.23	NS
REcnrq	0.76^{a}	1.77^{b}	1.03^{ab}	0.27	*
DE/GE, %	87.8	89.8	89.3	0.7	NS
ME/GE, %	81.6^{a}	84.2^{b}	83.5^{ab}	0.7	*
HPcnrq/ME, %	96.4^{a}	91.8^{b}	95.2^{ab}	1.2	*
REcnrq/ME, %	3.6^{a}	8.2^{b}	4.8^{ab}	1.2	*

there was no significant difference in faecal N. Urine N was higher with protein separation, which resulted in a slightly reduced N retention. The energy balance revealed no major differences up to the level of metabolizable energy. The significant difference in urinary energy of 0.07 MJ/d is too small in absolute terms to have an appreciable effect on ME. Heat production of the animals was 20.1 compared with 20.6 MJ/d on the control diet. Consequently energy retention was increased by 0.6 MJ/d in the treatment group. The small difference in heat production occurred mainly during the protein feeding phase (Kirchgessner & Müller, 1996).

In the second experiment, no systematic differences in the gaseous exchange between the feeding groups could be found (Table 2). Besides a lower faecal N in treatment 1 all the other variables of the nitrogen balance remained unchanged. In particular the tendencial decline in N retention, observed in the first experiment, could not be confirmed by the prolonged separation of protein and carbohydrates in the second experiment. The data of the energy balance showed a slightly reduced faecal energy and CH_4 production in both separation groups. Urine energy was not different between the groups. The metabolizability of energy declined by 2 percentage points on the rations with nutrient separation compared to the control group. Heat production was 20.2, 19.9 and 20.4 MJ/d respectively for the three treatment variants. There was no significant effect on the bodyweight of the sows which was 181.5, 180.7, and 180.5 kg for control, treatment 1 and treatment 2, respectively. The daily profile of heat production plotted at 5-minute intervals revealed no systematic patterns linked to the experimental variants (Müller & Kirchgessner, 1996).

In both experiments, the nitrogen and energy balances were not markedly affected by the separated intake of protein and carbohydrates. Although some minor differences were statistically significant, a consistent pattern could not be established when summarizing the two experiments and with respect to the increasing time intervals. In particular, the higher energy retention with nutrient separation relative to control in experiment II is mainly explained by corresponding differences in ME intake. On the basis of equal ME intake (assuming 75% efficiency of ME utilization with control feeding) the difference in energy retention between treatment 1 and the control group would decline from 1.01 to nonsignificant 0.5 MJ/d. We also have to realise, that in experiment I the variance (sem) of HP and RE was unusually small so that the statistical significance of both criteria should not be over-emphasized. The nitrogen balance of the two experiments did not differ between control groups and treatments (2.0 g vs. 1.7 g N retention , respectively).

There are many reports showing that administration of large amounts of proteins is followed by a considerable increase in thermogenesis (e.g. Hamilton, 1939; Blaxter, 1989; Robinson *et al.*, 1990). In contrast, the present results show that a thermogenic effect of proteins did not occur when the amount of digestible protein ranged between 0.7 and 1.4 g/kg bodyweight/meal, and the cumulative supply of protein did not exceed the maintenance requirements.

It can be concluded that separation of dietary protein and carbohydrates by up to 48 h has negligible effects on thermogenesis under conditions of maintenance. The energy expended for intermediary metabolism and regulation of the separated intake of protein and carbohydrates, if any, may not be recorded in a whole metabolism trial. Finally, it can be assumed that the supposed energetic effects of nutrient separation in human nutrition are not primarily due to the separation principle, but rather a consequence of change in food selection, fibre content and energy density of the chosen diets.

References

Blaxter K.L. (1989). Energy Metabolism in Animals and Man. Cambridge University Press, Cambridge.

Brouwer E. (1965). Report of Sub-committee on constants and factors. Proceedings of the 3rd Symposium on Energy Metabolism, *EAAP Publ. No.*11:441-443. Academic Press, London.

Hamilton T.S. (1939). The heat increments of diets balanced and unbalanced with respect to protein. *J. Nutr.* 17:583-599.

Kirchgessner M. & Müller H.L. (1996). Effect of Hay's separation principle on the energy metabolism. *Ann. Nutr. Metab.* 40:336-342.

Müller H.L. & Kirchgessner M. (1996). Effekt der zeitlichen Trennung von Protein und Kohlenhydraten bis zum Intevall von 48 Studen auf den Energiestoffwechsel. *J. Anim. Physiol. and Anim. Nutr.* 76:1-8.

Robinson S.M., Jaccard C., Persaud C. Jackson A.A., Jéquier E. & Schutz Y. (1990). Protein turnover and thermogenesis in response to high-protein and high-carbohydrate feeding in man. *Am. J. Clin. Nutr.* 52:72-80.

Young V.R., Yu Y.M. & Fukagawa N.K. (1992). Whole body energy and nitrogen (protein) relationships. In: J. M. Kinney and H. N. Tucker (Eds) Energy Metabolism: Tissue Determinants and Cellular Corollaries, pp. 139-161. Raven Press, New York.

Walb L., Walb I., Heintze T. & Heintze M. (1994). Original Haysche Trennkost. Haug, Heidelberg.

THERMIC EFFECTS OF CARBOHYDRATES AND UNSATURATED FATS FED TO SOWS BELOW AND ABOVE MAINTENANCE LEVEL

H.L. Müller and M. Kirchgessner

Institut für Ernährungsphysiologie, Technische Universität München, 85350 Freising-Weihenstephan, Germany

Summary

Three respiration studies were conducted with adult sows to investigate the heat production (HP) and energy retention (RE) of fats compared to carbohydrates. Fish oil reduced the HP in the total ration by about 4% compared to olive oil when supplemented to a basal ration up to maintenance level. Above maintenance this difference could not be reproduced and both fats resulted in the same RE. Starch and olive oil or a mixture of fats (50% olive oil, 25% soybean oil, 25% lard) behaved equivalently at maintenance level. With excess energy intake there was a difference of RE between starch and fats of 11%. The results confirm that the classical concept of k-values, together with the ATP approach to energetic efficiency, explains the efficiency of energy utilization of nutrients. However, specific effects of subclasses of the main nutrients should be allowed for when modelling energy metabolism under certain situations. Effects of the so-called non-obligatory thermogenesis could not be observed.

Key words: Fish oil, olive oil, fats, starch, thermogenesis, sows.

Introduction

There is still controversy about the concept of dietary-induced thermogenesis. Some earlier observations in human nutrition suggested a rise in heat production (HP) by high-fat diets. Unsaturated fatty acids are assumed to play a special role (Kasper *et al.*, 1973; James, 1992). On the other hand, recent data on substrate balances revealed that carbohydrates may be precisely balanced due to promotion of oxidation after carbohydrate intake whereas oxidation of fats is not influenced by fat intake (Jéquier, 1993; Flatt, 1995). We performed three experiments with sows to compare the thermic response to fats rich in unsaturated fatty acids with that to carbohydrates.

Materials and Methods

The first experiment was designed as a Latin Square with nine sows. The basal diet was composed mainly of barley, soybean meal and oat bran, and met 60% of the energy requirements. Isoenergetic supplements up to maintenance level were 176 kJ GE/kg $W^{0.75}$ as olive oil, fish oil (40% polyunsaturated FA with three or more double bonds), or wheat starch (control), respectively. During the experimental periods, complete balances for each animal were recorded, such that each period was divided into a preliminary phase of eight days, a collection period of six days and a two-day measurement of gaseous exchange in an open-circuit respiration chamber.

The second experiment was conducted with eight sows fed an excess of energy over eight periods of 16 days each. During the first and last experimental periods, the animals received a basal diet covering the maintenance requirements. From the second to the seventh period, supplements of olive oil, fish oil, or wheat starch were added to the basal diet so as to exceed the energy requirement for maintenance by approximately 20 and 40%, respectively. Each animal was given the six treatment combinations in random order.

During each period a complete energy balance was performed using the same technique as in experiment 1.

In a third, long-term (21weeks) experiment using a total of 14 adult non-pregnant sows, a carbohydrate group (50% maize starch, 50% wheat starch) was compared with an oil group (50% olive oil, 25% soybean oil, 25% lard). The basal diet was composed mainly of barley and soybean meals and supplied 60% of the ME requirement. Protein, minerals and vitamins were supplied in sufficient amounts to cover maintenance requirements. The daily intake of energy was kept constant during the whole experiment. Faeces and urine were collected during the first and last period of the experiment, while the gaseous exchange was measured over 48 h at intervals of three weeks on each of five occasions.

In all cases, heat production (HP) and energy balance (EB) were calculated according to the carbon-nitrogen (CN) and respiratory quotient (RQ) method according to Brouwer's formulae (Brouwer, 1965).

Results and Discussion

Experiment I. The dietary treatments did not influence the digestibility of the rations. The digestibility of energy of the total ration averaged 83.4% (Table 1) while the energy digestibility of the basal ration was calculated to be 76% according to the formula of Hoffmann *et al.* (1993). The data in Table 1 imply that all three supplements were almost completely digested. Fish oil increased urinary energy and decreased CH_4 production, but the shifts were, in absolute terms, very small. Consequently, the metabolizability of energy did not differ between the groups and averaged 79.3% in the total rations.

Table 1. Energy balance in Experiment I (kJ / kg $W^{0.75}$/day).

	Starch	Olive oil	Fish oil	SEM	F
Energy intake	531	529	531	-	-
Faeces	91	83	90	4	NS
Urine	16	16	18	1	*
CH_4	5.4	5.1	4.6	0.2	*
ME	419	424	419	4	NS
HPcn	412	411	394	5	NS
HPrq	416	410	401	4	NS
REcnrq	4	14	21	8	NS
DE/GE, %	82.9	84.3	83.0	0.7	NS
ME/GE, %	78.9	80.3	78.8	0.7	NS

RE is calculated as the mean value obtained by the CN and RQ methods.

The mean HPs were 414, 410, and 397 kJ/d when sows were fed with starch, olive oil or fish oil ($P<0.10$) respectively. When the HP of the starch diet was corrected for equal energy retention ($HP = HP_{starch} + \Delta RE/0.75*0.25$) the difference in thermogenesis between olive oil and starch was -7 kJ, and that between fish oil and starch was -23 kJ/kg $W^{0.75}$, respectively. These differences can be expressed in relation to starch:olive oil:fish oil = 1:0.95:0.86 on the basis of the supplementary energy. The relationship between starch and olive oil reflects exactly the theoretical expectation calculated from the ATP generated by oxidation of both nutrients (Müller & Kirchgessner, 1995). However, for fish oil, the daily HP was lower than theoretically calculated, which might be interpreted as an effect on the metabolic rate in general rather than as a result of a modified efficiency of ATP formation from fish oil oxidation. In any case there was no evidence of a facultative thermogenesis induced by the oils.

Table 2. Energy relationships between GE, ME, and RE in Experiment II.

Y	X	Animal P>F	$\beta_0 \pm s_{\beta 0}$	$\beta \pm s_\beta$	RSD	$t(\beta)$
Starch						
ME	GE	0.07	-6.62 ± 0.59	0.988 ± 0.018	0.39	54
REcnrq	ME	0.65	-18.76 ± 1.61	0.899 ± 0.064	1.35	14
Olive oil						
ME	GE	0.06	-6.71 ± 0.67	0.992 ± 0.020	0.53	49
REcnrq	ME	0.31	-20.84 ± 1.52	0.995 ± 0.058	1.50	17
Fish oil						
ME	GE	0.47	-7.32 ± 0.78	1.015 ± 0.024	0.61	43
REcnrq	ME	0.80	-21.04± 1.36	1.010 ± 0.052	1.37	19

Model Y = Animal-Effect + β_0 + β X.

Experiment II. The relationship between ME and GE intake shows that, in the case of excessive energy intake, all three supplements were completely digested as previously found in the first experiment with sows on maintenance (Table 2). Energy retention increased linearly with the level of supplementation. There was a difference in slope of 11-12% between starch and the rations with olive oil or marine oil. Between types of oils only a negligible difference could be observed which implies equal HP in both ration groups. The same conclusions can also be derived from the mean HP with supplementation level 1 and 2 respectively: olive oil 20.97 and 21.09 MJ, and fish oil 20.34 and 20.94 MJ, which gives a difference of only 1.9% on average (Kirchgessner & Müller, 1995). The high RE was associated with a considerable increase in the body weight of the sows, which rose linearly from 184 kg to the final average weight of 219 kg. However, none of the three types of nutrients had any influence on the body mass of the animals (205.5, 204.8, and 205.8 kg for starch, olive oil and fish oil, respectively).

Experiment III. The values in Table 3 represent the mean energy balance (EB) for the starch and fat diets. All of the balance criteria were almost identical between the two groups over a period of 21 weeks.

Table 3. Energy balance in Experiment III (kJ / kg $W^{0.75}$/day).

	Starch	Fat	Prob>F
Energy intake	563 ± 11	558 ± 14	-
Faeces	106 ± 6	108 ± 9	NS
Urine	18 ± 2	20 ± 1	NS
CH_4	6.2 ± 1.6	6.9 ± 1.3	NS
ME	432 ± 11	423 ± 14	NS
HPcn	404 ± 35	405 ± 33	NS
HPrq	422 ± 28	418 ± 39	NS
REcnrq	19 ± 33	12 ± 34	

The intake of GE, as well as the results on ME, HP and RE in the five experimental periods, are illustrated in Figure 1. When starch was fed, the HP fluctuated slightly more than with the fat diet, whereas the ME was very constant within both diets. The test of interaction between period and treatment showed for ME (P>F) = 0.99 and for HP (P>F) = 0.08. Nevertheless, the mean HP over the whole experiment did not differ in either group and RE was close to zero. The mean body weight of the sows was 178.3 kg with the starch, and 178.5 kg with the fat diet.

In conclusion, fish oil containing approximately 40% polyunsaturated fatty acids tended to reduce the HP in sows on maintenance. However, it was as effective as olive oil (mostly oleic acid) for energy retention. The observed difference in RE between starch and fats consisting of fatty acids with mainly one and two double bonds (olive oil, soybean oil) was close to zero at maintenance and about 11% at excessive energy intake. This agrees with anticipated biochemical outcomes in the thermogenesis of starch and fat oxidation and of body fat deposition. Thus, it is also concluded that a so-called non-obligatory thermogenesis does not arise from dietary fat addition.

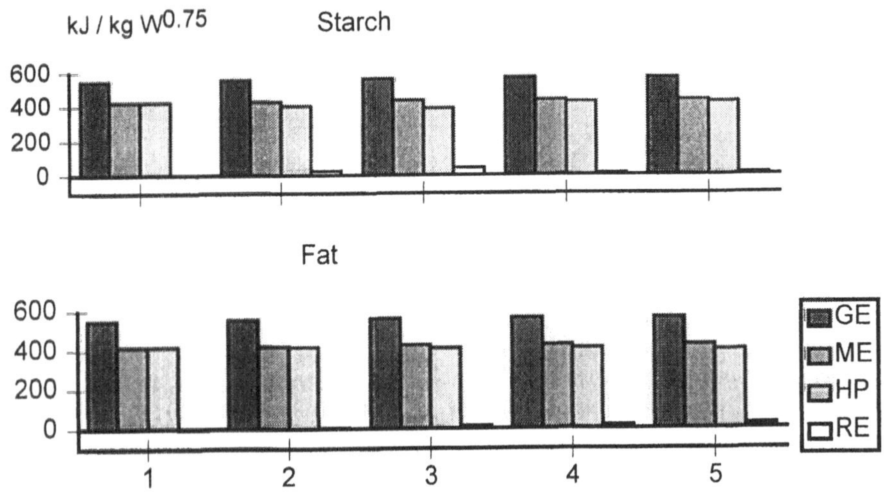

Fig. 1. GE, ME, HP and RE during the five periods of Experiment III.

References

Brouwer E. (1965). Report of Sub-committee on constants and factors. Proceedings of the 3rd Symposium on Energy Metabolism. *EAAP Publ. No.*11:441-443.

Flatt J.P. (1995). Body composition, respiratory quotient and weight maintenance. *Am. J. Clin. Nutr.* 62:1107S-1117S.

Hoffmann L., Beyer M. & Jentsch W. (1993). Abhängigkeit von Brutto-, verdaulicher und unsetzbarer Energiw des Futters sowie des Energgieansatzes vom Gehalt an verdaulichen Nährstoffen. *Arch. Anim. Nutr.* 44:123-137.

James W.P.T. (1992). Discussion of calorie balance versus nutrient balance. In:J.M. Kinney and H.N. Tucker (Eds) Energy Metabolism, pp. 134. Raven Press Ltd., New York.

Jéquier E. (1993). Body weight regulation in humans. The importance of nutrient balance. *News in Physiol. Sci.* 8:273-276.

Kasper H., Thiel H. & Ehl M. (1973). Response of body weight to a low carbohydrate, high fat diet in normal and obese subjects. *Am. J. Clin. Nutr.* 26:197-204.

Kirchgessner M. & Müller H.L. (1995). Thermogenese im Bereich der Überernährung bei Verabreichung von Olivenöl und Fischöl im Modellversuch an Sauen. *Z. Ernährungswiss.* 34:143-150.

Müller H.L. & Kirchgessner M. (1995). Thermogenese und Energieverwertung bei Verabreichung von Olivenöl und Fischöl im Modellversuch an Sauen. *Z. Ernährungswiss.* 34:206-213.

EFFECTS OF CELL-MEDIATED IMMUNE RESPONSE ON ENERGY METABOLISM IN WEANLING PIGLETS

H.K. Moon[1], I.K. Han[2], H.K. Parmentier[3] and J.W. Schrama[3]

[1]Nutrition and Physiology Division, National Livestock Research Institute, Suweon, Korea; [2]Department of Animal Science and Technology, College of Agriculture and Life Sciences, Seoul National University, Suweon, Korea; [3]Department of Animal Husbandry, Wageningen Agricultural University, Wageningen, The Netherlands

Summary

The effects of a cell-mediated immune response on energy metabolism were assessed in 24 weanling barrows with two immunization treatments : Control (CON : immunized with Incomplete Freund's Adjuvant (IFA)) or chronic inflammation (IMMU : immunized with Complete Freund's Adjuvant (CFA)). On d0, piglets were subcutaneously immunized at the medial side of the femur with 2 ml of IFA or CFA, respectively. Energy balances were measured for 13 days. Immunization with CFA containing *Mycobacterium butyricum* did not affect growth, feed intake, ME/GE, partitioning of energy, and rectal temperature ($P>0.10$). However, total antibody titers to *Mycobacterium butyricum* were significantly different ($P<0.01$) on d13 (1.8 vs. 2.5) and d20 (1.8 vs. 2.9) for CON and IMMU, respectively. IMMU group showed serious granuloma at injection sites. This study suggests that a chronic inflammation had no effect on energy metabolism. This might be due to encapsulation at the site of injection, which might induce an immune response locally rather than systemically.
Key words: Immune response, energy metabolism, pigs.

Introduction

Animals that are raised in an environment in which they are exposed to frequent microbial challenges grow at a slower rate, and have altered partitioning of nutrients, compared with animals raised in less stressful environments due to continuous (or chronic) immune stimulation (immunologic stress) (Klasing *et al.*, 1987). Little or no effect, however, was observed on energy metabolism in weanling piglets subjected to a humoral immune response to T-cell dependent antigens (Gentry *et al.*, 1996), and in five-week old pigs challenged with *Pasteurella multocida*-toxin (van Diemen *et al.*, 1995a, 1995b).

This study was designed to investigate the effect of a cell-mediated immune response (in the form of chronic inflammation) on the energy and nitrogen metabolism in weanling piglets.

Materials and Methods

The experiment consisted of 4 trials with 2 immunization treatments : Control group (CON: immunized with Incomplete Freund's Adjuvant (IFA)) or Immunization group (IMMU: immunized with Complete Freund's Adjuvant (CFA)). CFA (Cat No. 344289, CALBIOCHEM) used as immunogen consists of heat-killed *Mycobacterium butyricum* dry cells suspended in an emulsifying oil. IFA was identical with CFA except for the presence of *Mycobacterium butyricum* dry cells. In each trial three pairs of weanling barrows of approximately 4 weeks of age were assigned to one of two treatments. Each balance trial lasted 13 days, which consisted of two consecutive periods of 6 days and 7 days. On day 0 (the day of arrival), piglets were weaned, transported to the respiration chambers and subcutaneously immunized at the medial side of the femur with 2 ml of CFA or IFA, respectively.

During the balance period of 13 d, each group was housed in an open circuit, indirect climatic respiration chamber with inner dimensions of 1.0 m x 0.8 m x 0.97 m in length x width x height (Verstegen *et al.*, 1987). In the chambers the temperature was kept within the thermoneutral zone with the relative humidity of about 65-70%. Air velocity was <0.20 m/s. Artificial light was switched on from 0700 to 1900. Piglets were allowed *ad libitum* access to feed, a commercial pelleted starter diet, and water. Determined gross energy (GE) and crude protein (CP) contents were 17.6 kJ/g and 19.5%, respectively.

Rectal temperature was measured, blood samples were taken, and antibody titers to *Mycobacterium butyricum* were determined using ELISA (enzyme linked immunosorbent assay). The total heat production (HP_{tot}) per group was determined every 9 minutes from the measurement of exchanges of oxygen and carbon dioxide as described by Verstegen *et al.* (1987). HP_{tot} was calculated from these gaseous exchanges using the formula of Brouwer (1965). Total energy retention (ER), energy retention as protein (ER_p), energy retention as fat (ER_f) and maintenance ME (ME_m) were calculated according to Henken *et al.* (1991). Physical activity per chamber was continuously monitored by an ultrasonic burglar device using Doppler effect according to the method used by Wenk & van Es (1976) and recorded in the same intervals as HP_{tot}. The heat production not related to physical activity (HP_{cor}) was derived by subtracting HP_{act} from HP_{tot}.

All balance data were analysed with group as the experimental unit. The effects of immunization and balance period on the traits were tested by means of F-tests using a split-plot model (GLM procedure; SAS, 1989), with balance period values within groups taken as repeated measurements

Results and Discussion

Effects of immunization on the performance, feed intake, and availability and partitioning of energy are shown in Table 1. Only the means per treatment group are presented because the measured traits were not affected by the interaction between the immunization procedure and time (P>0.10). No significant differences were observed between treatments in average daily gain (ADG), feed intake, and availability and partitioning of energy (P>0.10). Average daily feed intake of both treatments was 0.33 ± .03 kg (Mean ± SE). In general, immunization or vaccination is expected to decrease feed intake, which is thought to be mediated by IL-1 (McCarthy *et al.*, 1985). However, feed intake was higher in the immunization group than in the control group in this study, as it was in the work of Gentry *et al.* (1996), which is contrary to the theory that interleukin-1 reduces feed intake. The energy losses in manure were not affected by the immunization of the pigs, as indicated by the similar ME/GE ratio of CON (83.5%) and IMMU (82.9%) group (P>0.10). This is consistent with the finding of Gentry *et al.* (1996) who found similar ME/GE ratio in the humoral immunization group (83.1%) and in the control group (82.4%).

Expressed as a percentage of HP_{tot}, HP_{act} was 16.18 and 17.08% for CON and IMMU group, respectively (P>0.10). In contrast to the present study, Gentry *et al.* (1997) found that piglets which mounted a humoral immune response had a lower HP_{act}/HP_{tot} ratio than control pigs (18.1 vs. 21.7%). The estimated ME_m was 427 and 419 kJ/kg$^{0.75}$/d for CON and IMMU group, respectively, which were a little smaller than the values of 438 and 455 kJ/kg$^{0.75}$/d estimated by Gentry *et al.* (1996) and van Diemen *et al.* (1995b), respectively.

Rectal temperature was not increased by immunization, and was maintained at about 39°C for both treatment groups throughout the experimental period (data not shown). However, total antibody titres to *Mycobacterium butyricum* were significantly different (P<0.01) on d13 (1.8 vs. 2.5) and d20 (1.8 vs. 2.9) for CON and IMMU group, even though overall titres were low because *Mycobacterium butyricum* mainly induces cell-mediated immunity rather than humoral immunity (Tizard, 1982).

The present study was designed to investigate the effect of a cell-mediated immune response on the energy metabolism in weanling piglets. Several researchers have reported that infection and/or a sustained immune response induced by multiple injections of a variety of immunogens might decrease growth rate and feed intake, and alter partitioning of nutrients, mostly from studies with chicks (Murray & Murray, 1979; Klasing *et al.*, 1991). However, no effect of immunization was observed on performance, feed intake or partitioning of nutrients in weanling piglets from the present study.

In the present study, we could observe serious granuloma formation at the injection site of piglets immunized with CFA (approximately 4 to 6 cm in length and 3 cm in width, data not shown). However, no increase of rectal temperature was found, though it is a well-characterized response to cell-mediated immune responses (chronic inflammation). The reason for the lack of effects of immunization, therefore, might be that the immune response was induced locally at the site of injection rather than systemically. It also might be possible that the site of injection was encapsulated by some material such as fibroblast, which prevented the immunogen from entering into the host body.

Table 1. Least squares means of performance, and the availability and partitioning of energy in weanling piglets as affected by immunization.

Trait	CON[1]	IMMU[2]	SEM[3]	P value
No. of groups	4	4	-	-
No. of animals	12	12	-	-
Initial wt, kg	7.7	7.8	0.27	0.916
Feed intake, kg	0.32	0.34	0.04	0.773
ADG, kg/d	0.26	0.28	0.03	0.656
Feed/gain	1.51	1.16	0.23	0.331
GE intake, kJ/kg$^{0.75}$/d	1,008	1,038	95.7	0.830
ME intake, kJ/kg$^{0.75}$/d	842	860	81.7	0.883
ME : GE, %	83.5	82.9	0.94	0.693
Heat production, kJ/kg$^{0.75}$/d				
Total (HP_{tot})	591	592	22.7	0.993
Activity-related (HP_{act})	96	100	5.3	0.548
Non-activity-related (HP_{cor})	496	491	19.2	0.874
Retention, kJ/kg$^{0.75}$/d				
Total energy (ER)	251	268	61.2	0.846
Protein (ER_p)	153	156	21.6	0.916
Fat (ER_f)	98	112	40.3	0.812
ME for maintenance (ME_m), kJ/kg$^{0.75}$/d	427	419	15.8	0.747

[1] Control group immunized with Incomplete Freund's Adjuvant

[2] Immunization group immunized with Complete Freund's Adjuvant

[3] Standard Error of the Mean.

Consequently, a cell-mediated immune response (chronic inflammation) in the present study had no effect on the performance, availability and partitioning of energy, despite granuloma formation and antibody responses. This might be due to encapsulation at the site of injection, which might induce an immune response locally rather than systemically.

References

Brouwer E. (1965). Report of sub-committee on consultants and factors. In: K.L. Blaxter (Ed) Academic Press, London Energy metabolism. *EAAP Publ. No.* 11:441-444.

Gentry J.L., Swinkels J.W.G.M., Lindemann M.D. & J.W. Schrama. (1997). Effect of haemoglobin and immunization status on energy metabolism of weanling pigs. *J. Anim. Sci.* 75:1032.

Henken A.M., van der Hel W., Brandsma H.A. & Verstegen M.W.A. (1991). Differences in energy metabolism and protein retention of limit-fed growing pigs of several breeds. *J. Anim. Sci.* 69:1443.

Klasing K.C., Johnstone B.J. & Benson B.N. (1991). Implications of an immune response on growth and nutrient requirements of chicks. In: W. Haresign and D.J.A. Cole (Eds) Recent advances in animal nutrition. Butterworth Heinemann, Oxford, U.K. pp. 135.

Klasing K.C., Laurin D.E., Peng R.K. & Fry D.M. (1987). Immunogolically mediated growth depression in chicks: Influence of feed intake, corticosterone, and interleukin-1. *J. Nutr.* 117:1629.

McCarthy D.O., Kluger M.J. & Vanderg A.J. (1985). Suppression of food intake during infection: Is interleukin-1 involved? *Am. J. Clin. Nutr.* 42:1179.

Murray J.J. & Murray A.B. (1979). Anorexia of infection as a mechanism of host defence. *Am. J. Clin. Nutr.* 32:593.

SAS (1989). *SAS User's Guide: Statistics*, Version 6 (4th Ed.). SAS Inst. Inc. Cary. NC.

Tizard I. (1982). Antigens and antigenicity. In: An introduction to veterinary immunology (2nd Ed) W. B. Saunders Company. pp. 29.

Van Diemen P.M., Henken A.M., Schrama J.W., Brandsma H.A. & Verstegen M.W.A. (1995a). Effects of atrophic rhinitis induced by Pasteurella multocida toxin on heat production and activity of pigs kept under different climatic conditions. *J. Anim. Sci.* 73:1658.

Van Diemen P.M., Schrama J.W., van der Hel W., Verstegen M.W.A. & Noordhuizen J.P.T.M. (1995b). Effects of atrophic rhinitis and climatic environment on the performance of pigs. *Livest. Prod. Sci.* 43:275.

Verstegen M.W.A., van der Hel W., Brandsma H.A., Henken A.M. & Bransen A.M. (1987). The Wageningen respiration unit for animal production research: A description of the equipment and its possibilities. In: M.W.A Verstegen and A.M. Henken (Eds) Energy metabolism of farm animals. Nijhoff, Dordrecht, The Netherlands, pp. 21.

Wenk C. & van Es A.J.H. (1976). Eine Methode zur Bestimmung des Energieauswandes fur die korperliche Aktivitat von wachsenden Kuken. *Schweiz. Landwirstsch. Monatsh.* 54:232.

EFFECT OF PLANT LECITHINS ON THE EFFICIENCY OF RUMINAL NUTRIENT DEGRADATION AND ON METHANE AND HYDROGEN LOSS *IN VITRO* (RUSITEC)

M. Kreuzer, H.-R. Wettstein and A. Machmüller

Institute of Animal Sciences, Animal Nutrition, Swiss Federal Institute of Technology (ETH), ETH centre/LFW, CH-8092 Zurich, Switzerland

Summary

Depending on their content of residual oil and their water-solubility, canola lecithins as dietary ingredients showed intermediate effects on methane release as compared with canola oil and seed, and a non-supplemented control. This indicates a less adverse effect on fibre degrading microbes than that of oil although nylon bag fibre disappearance did not respond. Furthermore, both canola and soy lecithins reduced disappearance of protein indicating that lecithins might cause a lower rumen-degradability of dietary protein.
Key words: Lecithin, Rusitec, ruminal nutrient degradation, methane.

Introduction

The use of plant oils in ruminant nutrition is restricted by adverse effects on rumen fermentation, particularly of fibre. Effects due to coating of fibre by lipids might be less pronounced with partially water-soluble lecithins (phospholipids) derived from the same plants. Investigations on effects of plant lecithins on rumen fermentation are limited to studies with soy lecithin (Yoon *et al.*, 1986; Jenkins *et al.*, 1989; Jenkins & Fotouhi, 1990) which might have additional inhibitory effects on cellulolytic microbes by its high concentration of polyenoic fatty acids. The objective of the present study was to investigate the effects of various canola lecithins on nutrient degradation and methane release using Rusitec (Czerkawski & Breckenridge, 1977).

Materials and Methods

The types of canola lecithins which were employed to model an increasing water-solubility were: raw (CL1), partially acetylated/hydrolysed (CL2), deoiled (CL3) and deoiled/ partially hydrolysed forms (CL4) (all lecithin forms distributed by Lucas Meyer GmbH & Co., Hamburg, Germany). The complete diets contained 5.2 to 6.4% of these lecithins (depending on their content of gross energy as well as on fatty acids). To be able to separate the effects of the fatty acids from the lecithin effects, these diets were compared with diets supplemented with whole crushed canola seed (CS), pure canola oil (CO) and deoiled soy lecithin (SL). An unsupplemented diet served as control. All diets consisted of maize silage, hay and concentrate (with the respective supplements), and were designed to contain overall 60 g total lipids/kg diet, except for the unsupplemented control, which resulted in a different lipid supply (Table 1). Diets were designed to provide equal amounts of metabolizable protein and energy and were isonitrogenous.

The Rusitec (rumen simulation technique) system consisted of 8 fermenters which allowed simultaneous evaluation of all diets. Each fermenter had a capacity of 1.1 l and was filled with 800 ml strained rumen fluid from cattle and 100 ml McDougall buffer (Czerkawski & Breckenridge, 1977). The buffer flow rate was 510 ml/d, and the nylon bags (70 mm x 140 mm) had a pore size of 100 µm. 8 repeat determinations were carried out each consisting of 10 days measurement with the first three days serving for adaptation to system and diet. The 8 diets were supplied in nylon bags at daily amounts of about 17 g dry matter and remained in rumen fluid for two days (Table 1). Rumen fluid pH, ammonia and redox potential (by respective electrodes), volatile fatty acids (VFA) concentrations (by gas chromatography) and bacterial and protozoal counts (counting chambers) were determined. Gas produced in each fermenter was collected in gas proof bags and contents of methane and hydrogen were analysed by gas chromatography. With the exception of bacterial counts (determined three times per period) daily measurements of all parameters were carried out for each fermenter. Nutrient degradation was determined by a pooled sample of the residue after 48h of fermentation collected from days 4 to 10. Statistical evaluation was carried out by analysis of variance and means statistically compared by the Tukey method.

Table 1. Mean daily supply of nutrients per Rusitec vessel.

Daily supply (g)		Oil-sources		Lecithins				
	Control	CS	CO	CL1	CL2	CL3	CL4	SL
Dry matter	16.8	16.6	16.6	16.9	16.9	17.1	17.1	17.1
Organic matter	16.0	15.7	15.8	16.0	16.0	16.2	16.2	16.2
Crude protein	2.4	2.6	2.5	2.5	2.5	2.5	2.6	2.5
Total lipids	0.4	1.0	0.9	1.0	1.0	1.0	0.9	1.0
Hemicellulose	2.6	2.6	2.7	2.7	2.7	2.7	2.7	2.7
Cellulose	2.9	3.1	3.2	3.2	3.2	3.2	3.2	3.2
Gross energy (kJ)	316	327	325	332	331	334	334	334

Results

Rumen fluid pH was significantly lower with control than raw canola lecithins (CL1) and with intermediate values for the other diets (Table 2). Ammonia concentration was significantly reduced by pure canola oil (CO) and each of the lecithins but not by canola seed (CS) when compared with control. The highest concentration of VFA occurred in the control with lower proportions of propionate and acetate and higher proportions of butyrate than in all other groups. Bacterial counts were reduced with CO and CS as well as with all lecithin containing diets with the highest effect from raw (CL1) and deoiled (CL3) canola lecithin. Protozoal counts tended to be higher in all lecithin groups. The release of methane (mmol/d) was significantly reduced by CO and CS (-30%) followed by acetylated/hydrolysed lecithin (CL2, -24%) and raw lecithin (CL1, -21%). Methane production was less reduced by deoiled lecithins (CL3, CL4 and SL, -13% to -16%). Hydrogen release from fermentation accounted for 0.6 to 1.0 mmol/d, with no significant differences between the diets.

Apparent ruminal degradation of organic matter (percentage disappearing from nylon bags during 48 h of incubation) was significantly reduced by each kind of supplement at an average of 2.3 percentage units for the oil sources and of 4.5 percentage units for the lecithins (Table 3). Similar variation occurred in the disappearance of gross energy and of crude protein from the nylon bags. The differences between the lecithin-supplemented diets were quite low. No significant effect on the degradation of cell wall carbohydrate fractions (hemicellulose, cellulose) was found, although lipid-supplemented diets tended to give a slightly higher fibre degradation rate.

Discussion

Lipid-supplemented diets usually reduce rumen degradation of organic matter (Machmüller *et al.*, 1997), energy (Jenkins & Fotouhi, 1990) and fibre. This was not found in the present study at, however, a quite low rate of degradation of hemicellulose and cellulose.

Table 2. Fermentation patterns obtained with various oils and lecithins.

	Control	Oil-sources				Lecithins		
		CS	CO	CL1	CL2	CL3	CL4	SL
Rumen fluid properties								
pH	6.56^{b}	6.66^{ab}	6.66^{ab}	6.66^{a}	6.66^{ab}	6.65^{ab}	6.62^{ab}	6.65^{ab}
Ammonia (mmol/l)	11.3^{a}	11.6^{a}	9.3^{b}	8.8^{b}	8.4^{b}	9.2^{b}	9.3^{b}	9.2^{b}
VFA (mmol/l)	154^{a}	148^{ab}	146^{b}	147^{b}	147^{ab}	148^{ab}	150^{ab}	148^{ab}
Acetate (molar %)	49.8	50.5	50.5	51.1	50.4	51.7	51.0	50.7
Propionate (molar %)	17.9^{b}	20.9^{a}	20.9^{a}	21.0^{a}	20.7^{a}	20.1^{a}	20.3^{a}	20.5^{a}
Butyrate (molar %)	25.1^{a}	21.1^{b}	21.4^{b}	20.9^{b}	21.8^{b}	21.5^{b}	21.9^{b}	21.9^{b}
Rumen fluid microbial counts								
Bacteria (10^9/ml)	4.04^{a}	3.86^{ab}	3.64^{abc}	3.33^{c}	3.63^{abc}	3.49^{bc}	3.66^{abc}	3.66^{abc}
Protozoa (10^3/ml)	4.26	4.02	3.99	4.33	5.22	5.83	5.22	4.81
Gaseous release								
Methane (mmol/d)	9.60^{a}	6.72^{c}	6.73^{c}	7.54^{bc}	7.34^{bc}	8.31^{b}	8.13^{b}	8.08^{b}
Hydrogen (mmol/d)	0.72	0.77	0.87	0.71	0.98	0.56	0.69	0.64

Table 3. Rates of disappearance of nutrients and energy from the nylon bags.

Disappearance		Oil-sources				Lecithins		
(% of supply)	Control	CS	CO	CL1	CL2	CL3	CL4	SL
Organic matter	56.4^{a}	53.5^{abc}	54.8^{ab}	52.1^{bc}	52.6^{bc}	51.5^{c}	51.4^{c}	51.7^{b}
Gross energy	54.1^{a}	51.1^{abc}	52.9^{ab}	49.1^{cd}	49.9^{bcd}	47.7^{d}	47.9^{cd}	48.1^{cd}
Crude protein	58.6^{a}	56.6^{a}	55.9^{ab}	50.6^{c}	49.1^{c}	49.6^{c}	49.4^{c}	51.6^{bc}
Hemicellulose	18.6	21.0	20.1	20.2	19.0	20.7	19.6	20.7
Cellulose	16.0	18.3	19.4	17.3	18.1	18.5	17.5	17.7

a, b, c, d - means within a row bearing the same superscript are not significantly different.

The slightly higher fibre degradation with CO and CS as compared with both control and lecithin diets, might be an artefact although Jenkins *et al.* (1989) also observed that deoiled soybean lecithin did not change fibre digestion *in vitro*.

Methane release from ruminal fermentation, as determined by Rusitec, has been demonstrated to be suitable for the quantitative measurement of lipid effects on fermentation (Machmüller *et al.*, 1997). Also, in the present study, all lipid supplements significantly decreased methane formation. This can be explained by direct toxic effects upon the methanogenic bacteria and indirect effects through the adverse influence on protozoa caused by the unsaturated fatty acids released by lipolysis. The present results also show that, in contrast to nylon bag fibre degradation, methane release is sensitive enough to significantly respond to different lipids. This is at least valid for the part of fibre degradation that is associated with the microbes which are involved in methane formation. The highest decrease in methane production was found with oil (pure canola oil and seed) and an intermediate reduction with the canola lecithins which contained a residual amount of oil. Deoiled lecithins obviously had weaker effects on methanogenic microbes. Nevertheless, if methane is considered as the most sensitive trait, even plant lecithins inhibit rumen fermentation but to a lesser extent than oils.

In the present study, all types of lecithin caused a significantly lower rate of disappearance of organic matter and gross energy compared with control and oil supplemented diets. This mainly resulted from a lower degradation of dietary crude protein (-8.5 and -6.2 percentage units respectively as compared with control for the lecithin and the oil-source groups). Degradation of crude protein was also lowered by the use of soy lecithins in the *in vitro* study of Jenkins *et al.* (1989). The difference between oil-source groups was significant for all lecithins although the reduction in rumen ammonia concentration as compared with control was similar. This difference in protein degradation might be due to the amphiphatic behaviour of the plant lecithins in contrast to triglycerides resulting in a high affinity for protein (Jenkins *et al.*, 1989). Lecithin-complexed protein might be less available for ruminal degradation. Jenkins *et al.* (1989) showed that there are two fractions within the phopspholipids, one rapidly-degradable and one slowly-degradable within the rumen. The latter ensures that at least part of the lecithins and, consequently, the associated proteins reach the small intestine.

Overall, the use of lecithins instead of oils could have two favourable effects: alleviation of the suppression of fibre degradation and a decrease in ruminal protein degradation. This might finally increase the supply of metabolizable protein if it is not compensated for by a lower microbial protein flow which cannot be totally excluded in view of the ruminal fluid microbial counts.

References

Czerkawski J.W. & Breckenridge G. (1977). Design and development of long-term rumen simulation technique (Rusitec). *Br. J. Nutr.* 38:371-384.

Jenkins T.C. & Fotouhi N. (1990). Effects of lecithin and corn oil on site of digestion, ruminal fermentation and microbial protein synthesis in sheep. *J. Anim. Sci.* 68:460-466.

Jenkins T.C., Gimenez T. & Cross D.L. (1989). Influence of phospholipids on ruminal fermentation *in vitro* and on nutrient digestion and serum lipids in sheep. *J. Anim. Sci.* 67:529-537.

Machmüller A., Ossowski D.A., Wanner M. & Kreuzer M. (1997). Potential of various fatty feeds to reduce methane release from rumen fermentation *in vitro* (Rusitec). *Anim. Feed Sci. Technol.* (in press).

Yoon C.-S., Jung K.-K. & Bae D.-H. (1986). Effect of crude soybean lecithin supplementation on the digestibility, rumen and blood fatty-acid composition in sheep. *Korean J. Anim. Sci.* 28:488-499.

EFFECT OF α_2 - ADRENERGIC STIMULATION AND FEEDING ON HEAT PRODUCTION OF GROWING BULLS

M. Derno[1], B. Löhrke[2], W. Jentsch[1] and H.-D. Matthes[1]

[1]Research Unit of Nutritional Physiology 'Oskar Kellner'; [2]Research Unit of Population Biology and Breeding Research, Research Institute for the Biology of Farm Animals, Dummerstorf-Rostock, Germany

Summary

The influence of α_2-adrenoreceptor stimulation in the fasting state and after resumption of feeding on heat production (HP) was investigated in growing bulls (live weight: 280-340 kg). The investigations were made at low, medium and high levels of the energy metabolism (EM_l, EM_m, EM_h) induced by different feeding levels and ambient temperatures. The α_2-adrenoreceptor agonist clonidine caused a significant reduction in HP at all levels with the strongest response at the high level (EM_l, EM_m, EM_h, 52%, 27%, 35%). Feeding induced a different rise in HP at different metabolic levels with the strongest at the high level (117% vs. 43% and 38%). Continuous measurement of the heart rate reflected the changes in HP. The involvement of the sympathetic nervous system (SNS) in the regulation of energy metabolism is discussed.

Key words: Cattle, heat production, α_2-adrenoreceptors, clonidine, heart rate.

Introduction

The SNS plays an important role in the regulation of energy metabolism (Landsberg & Young, 1983) as it is involved in the regulation of the cardiovascular system, supplying fuels to the tissues according to their metabolic demands. Thus, variation in SNS activity can be the reason for different levels of energy metabolism (Ravussin & Tataranni, 1996) leading to different energy expenditure for maintenance, growth and adaptation to a changed environment. The relatively limited number of investigations in this field on ruminants was reviewed by Christopherson *et al.* (1995), suggesting a role of the α-adrenergic system in the regulation of thermogenesis. According to Miaron *et al.* (1995) and Hunter *et al.* (1992) the α_2-adrenoreceptor agonist guanfancin reduced HP in cattle by 20%. In these studies, the α_2-adrenergic mediated reduction of the HP in the cold (- 9°C) and the hot (28°C) was different (23% and 8%, respectively). Moreover, the α_2-adrenergic effects on HP were dependent on the feeding level. In contrast, detailed data about the metabolic response to feeding under conditions of the α_2-adrenergic reduction of the sympathetic outflow seems to be lacking. Thus, the aim of the present study was to characterize the dynamics of α_2-adrenergic mediated changes in HP. To obtain information about the interaction between sympathetic outflow and HP, the metabolic rate was modified by feeding and altering ambient temperatures, and the sympathetic outflow was pharmacologically reduced in each case.

Materials and Methods

Black-Pied dairy cattle bulls (n=16), 280-350 kg in weight were used. They received a diet consisting of 20% barley and 80% roughage supplemented with vitamin-mineral mixture (crude protein: 14%, crude fibre: 24%, metabolizable energy (ME): 10.6 MJ/kgDM). The HP was measured by indirect calorimetry, using 4 climatized respiration chambers. Gas exchange was recorded continuously (every 10 min; one set of data for each chamber). HP was calculated according to Brouwer (1963). The animals were catheterized in the jugular vein. The blood levels of glucose, non-esterified fatty acids (NEFA), thyroid hormones (T_3, T_4) and cortisol were determined by commercial kits (Löhrke *et al.*, 1994). Standing time, position changes, rectal temperature and breathing frequency were monitored. The heart rate (HR) was measured continuously (1 record/min) by a biotachometer (Polar, Finland).

Three levels of HP were induced by altering feeding levels (FL) and ambient temperatures (T):

Low (I):	T = 18°C	FL = 1.0 fold ME_m	(n = 8)
Medium (II)	T = 18°C	FL = 1.6 fold ME_m	(n = 11)
High (III)	T = 4°C	FL = 1.6 fold ME_m	(n = 7)

ME_m = maintenance energy requirement [assumed to be 450 kJ/(kgLW$^{0.75}$d)]
(n = number of animals)

The adaptation time to the feeding level was 2 weeks, the low temperature was adjusted one day before the start of infusion measurements. Each trial lasted four days. On the first day the animals were familiarized with the chamber. On the second and third day the control and drug infusions were made and on the fourth day the long time reactions were measured. Blood samples were taken before (0 min) and at 15 min, 60 min, 5 h and 7 h after the saline and the drug infusion. The animals were fed 1.5 h and 6 h after an infusion.

The sympathetic outflow was reduced by the α_2-agonist clonidine (20 nmol/kg LW infused over 8 min). Clonidine mainly acts centrally. A control infusion of carrier (physiological saline) was also made. Statistical analysis was done by ANOVA reporting to a significance of $P<0.05$.

Results

Figure 1 shows the daily time-course of HP for the saline and clonidine infusions. The HP values are means over a 30 min period and are expressed in terms of HP per day for better comparison. The resting HP values (between 00:00 h and 06:00 h) of the control infusion show significant differences between the experimental variants. Table 1 demonstrates the mean values of the resting HP.

Table 1. Mean values of resting HP [kJ/(kgLW$^{0.75}$d)] for different treatments.

	FL=1.0 T=18°C	FL=1.6 T=18°C	FL=1.6 T=4°C
Saline	472[a]	533[b]	614[c]
Clonidine	479[a]	539[b]	631[d]

Values with different superscripts differ significantly ($P<0.05$) (n=14).

The time-course of HP after saline infusion shows the expected rise after the first and second feeding. The peak after the second feeding is clearly indicated. Clonidine lowered the HP response. The α_2-adrenergic agonist modified also the pattern of HP increase.
For clarity, the data are arranged in three periods (Table 2).

I	infusion - first feeding	(fasting state; 1.5 h)
II	first feeding - second feeding	(4.5 h)
III	second feeding and onwards for	4.5 h

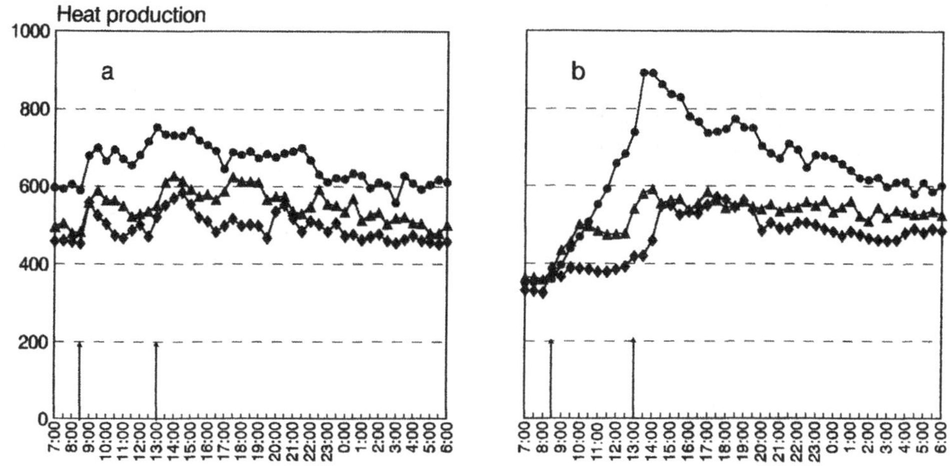

◆ - FL = 1.0 , T = 18 °C ▲ - FL = 1.6, T = 18 °C ● - FL = 1.6, T = 4 °C

Fig. 1. Daily time course of HP (kJ/(kg$^{0.75}$d)) for a) control infusion and b) clonidine infusion at different basic HP levels (arrows indicate feeding time).

Table 2. Heat production (kJ/(kgLW$^{0.75}$d)) in three periods: I - fasting, II - after first feeding, III - after second feeding.

	Saline			Clonidine		
	I	II	III	I	II	III
FL=1.0,T=18°C	479^{aA}	511^{abA}	540^{bA}	376^{aA} *	415^{abA} *	537^{cA}
FL=1.6,T=18°C	546^{aB}	588^{abB}	618^{bB}	419^{aB} *	525^{bB} *	582^{cA} *
FL=1.6,T=4°C	596^{aC}	704^{bC}	725^{bC}	391^{aA} *	608^{bC} *	847^{cB} *

Values with different superscripts differ significantly ($P<0.05$),
capital letters: between treatments (columns), lower cases: between periods,
* between saline and clonidine infusion (ANOVA repeated measurements).

The HP reactions are consistent with the HR response, except for the saline variant at FL=1.0, T=18 °C (Table 3).

Table 3. Heart frequency (beats per min) in three periods: I - fasting, II - after first feeding, III - after second feeding.

	Saline			Clonidine		
	I	II	III	I	II	III
FL=1.0,T=18°C	51^{aA}	62^{abA}	61^{bA}	38^{aA} *	55^{bA} *	76^{cA}*
FL=1.6,T=18°C	56^{aB}	69^{bB}	72^{bB}	43^{aB} *	61^{bB} *	76^{cA}
FL=1.6,T=4°C	68^{aC}	87^{bC}	94^{bC}	47^{aA} *	77^{bC} *	115^{cB} *

Values with different superscripts differ significantly ($P<0.05$),
capital letters: between treatments (columns), lower cases: between periods,
* between saline and clonidin infusion (ANOVA repeated measurements).

The linear regression between HP and the corresponding mean HR values in the three periods for all treatments gives the equation:

$$HP = 207\ (\pm 35) + 5.5\ (\pm 0.5)\ HF \qquad (n=18) \qquad R^2 = 0.89$$

The high R^2-value indicates the strong correlation between HP and HF.

Discussion

The differences in resting HP level, modified by changes in feeding level and ambient temperature, were 13% between low and medium level and 15% between medium and high level. The reduction of the sympathetic outflow by clonidine caused a strong reduction in HP in the fasting state (17 h after the last feeding). The α_2-adrenergic agonist reduced the HP at the low, medium and high levels by 27%, 35% and 52%, respectively, indicating a direct relationship between metabolic level and HP reduction. The results suggest that the clonidine action is dependent on the sensitivity of the sympathetic outflow to α_2-adrenergic signal transductions. Thus, sensitivity corresponds to the different resting HP. After the first feeding, HP rose in all cases, but the degree of this rise was dependent on the HP level. The second feeding induced a strong increase in HP of 117%. The values for the low and medium levels were 43% and 38%. The second feeding after saline infusion induced an increase in HP of 12%, 13% and 22% for the low, medium and high basic levels. The strong increase of HP after feeding at the high HP level indicates that, in comparison to the other variables, a higher amount of feed energy is converted to heat. The striking decrease in HP, induced by the α_2-adrenergic agonist in the fasting state, seems to be compensated for by a substantially larger amount of heat from the ingested feed energy. Like the HP, the HR responded to altered feeding level or ambient temperature as well as to the α_2-adrenergic agonist. The high demand for oxygen in the compensation of the excessive deficit, which arose from the reduction of the sympathetic outflow, corresponds to the striking increase in HP and HR. Taken together, the pharmacological approach provided evidence for different sensitivity of the α_2-adrenergic component of the sympathetic outflow induced by different feeding or ambient temperatures. The sensitivity seems to be related to the differences in the HP.

Acknowledgements

The investigations were supported by the Deutsche Forschungsgemeinschaft.

References

Brouwer E. (1965). Report of sub-committee on constants and factors. In: K. L. Blaxter (Ed) Energy metabolism of Farm Animals, *EAAP-Publ. No.* 11 pp. 441-443.

Christopherson R.J., Ekpe E.D., Moibi J.A. & Li B.T. (1995). Role of adrenergic receptors in regulating metabolism in ruminants. In: M. Ivan (Ed) Animal science research and development: moving toward a new century. Centre for Food and Animal Research, Agriculture and Agric-Food Canada, Ottawa pp. 205-223.

Hunter R.A., Sillence M.N., Gazzola G. & Spies W.G. (1993). Increasing annual growth rates of cattle by reducing maintenance energy requirements. *Aust. J. Agric. Res.* 44:579-595.

Landsberg L. & Young J.B. (1983). Autonomic regulation of thermogenesis. In: L. Girardier and M.J. Stock (Eds) *Mammalian Thermogenesis.* Chapman & Hall, London. pp. 99-140.

Löhrke B., Derno M., Jentsch W., Matthes H.-D. & Göbel D. (1994). Neurokrine und immunologische Gesichtspunkte zur Anpassung des Wärmehaushaltes an Umweltänderungen beim Rind. *Forschungsinstitut für die Biologie landwirtschaftlicher Nutztiere Dummerstorf-Rostock Schriftenr.* 7:123-157.

Miaron J.O.O., Christopherson R.J., Hardin R.T., Mosenthin R. & Cosgrove S. (1995). The effect of α_2-adrenoceptor stimulation with guanfancin on heat production of restricted fed steers kept at -9, 11 or 28°C, and on *ad libitum* fed steers acclimatised to -19°C and 22°C. *J. Therm. Biol.* 20: 291-298.

ENERGY AND PROTEIN METABOLISM AND ENDOCRINE RESPONSE IN GROWING RATS FED ON DIETS WITH DIFFERENT PROTEIN SOURCES - SOY PROTEIN OR CASEIN

M. Klein[1], R. Schadereit[1] and U. Küchenmeister[2]

[1]Research Unit of Nutritional Physiology 'Oskar Kellner'; [2]Department of Muscle Biology and Growth, Research Institute for the Biology of Farm Animals Dummerstorf-Rostock, Germany

Summary

Three experiments were conducted on male, growing rats, fed restrictively on isocaloric and isonitrogenous (10% crude protein) diets based on either casein, supplemented with methionine, or soy protein isolate (SPI). Feeding of SPI resulted in a lower efficiency of energy utilization as well as a lower protein utilization compared with casein-fed rats. Protein synthesis, expressed in g/kg $BW^{0.75}$, was unaffected by the protein source in experiments 2 and 3, but decreased in experiment 1 in the SPI-fed rats. Serum thyroid hormone levels were not enhanced, but rather tended to be lower in the SPI-fed rats compared with casein-fed rats. In experiment 3, a greater mitochondrial protein content of interscapular brown adipose tissue (BAT) following SPI feeding was evident. Observed differences in some variables of hepatic and cardiac mitochondrial respiration were not repeatable.
Key words: Protein quality, energetic efficiency, protein turnover, thyroid hormones.

Introduction

In growing animals, long-term feeding of diets which meet the energy requirement for growth but do not meet the requirement for essential amino acids, results in a retardation of growth and a decreased protein retention. The ingested surplus energy is retained as body fat or dissipated as heat. In the case of *ad libitum* feeding of low-protein diets, the first metabolic adaptation was demonstrated in chickens and the latter one in rats and pigs. In protein-deficient rats, an elevation in plasma T_3 levels was observed. Concerning the effect of deficiency of essential amino acids (caused by feeding proteins of low biological value) on energy metabolism there are only a few investigations on growing rats (Hoffmann & Schiemann, 1977; Aguilera & Prieto, 1987). These studies revealed no clear effects of protein quality on net energetic efficiency or energy requirements for maintenance (MEm). Under certain conditions, long-term feeding of vegetable proteins of somewhat lower biological value (soy protein, wheat gluten) resulted in higher plasma thyroid hormones compared with high-quality animal proteins (casein, whey and egg protein) in young rats and in minipigs (Barth *et al.*,1988; Scholz-Ahrens *et al.*,1990; and others). It is not clear whether the higher thyroid hormone levels represent a true change in biological activity and do play a role in dissipating excess energy. There are indications that thyroid hormones affect the thermogenic activity of BAT and other energy dissipating systems.

The objective of the present study was, in addition to our studies on growing pigs (Saggau *et al.*, 1997), to estimate the effect of protein quality on the efficiency of utilization of metabolizable energy (ME) in growing rats and to investigate whether there are relationships between serum thyroid hormone levels, protein turnover and the energetic efficiency of mitochondrial respiration.

Materials and Methods

Three separate experiments were conducted on male albino rats (Fischer 344 strain) with an initial weight of about 80 g at 34 days of age. Rats were kept singly in metabolic cages at 24°C and fed restrictively (9.5 g dry matter/day) on isocaloric (17.6 MJ/kg DM) and isonitrogenous (103 g crude protein/kg DM) semi-purified diets based on either soy protein isolate (SPI) or casein supplemented with D,L-methionine. The content of lysine was 8.0 and 6.1 g/kg and of methionine+cystine 4.4 and 2.7 g/kg for the casein and SPI diets, respectively. Prefeeding lasted 9 to 16 days for the casein diets and 17 to 30 days for the SPI diets. At the beginning of the energy balance periods, both dietary groups had similar body weights (BW). Energy balances were measured by indirect calorimetry through carbon and nitrogen balance on 9 rats per group during one (expts 1 and 2) or two (expt 3) 5 to 7 day periods. Heat production was calculated as the difference between ME intake and energy retention. The energy-balance measurements were followed by the estimation of the whole-body protein synthesis rates (end product method, 24 h) using a single oral dose of ^{15}N labelled amino acid mixture as tracer.

At the end of the experiments, blood samples were taken after decapitation of rats and analysed for serum total thyroxine (T_4), free T_4 (fT_4), total 3,5,3'-triiodothyronine (T_3) and free T_3 (fT_3) (expts 1 and 2) by radioimmunoassays. BAT was analysed for mitochondrial protein content (expts 2 and 3). Liver and heart were taken from 4 to 6 rats per group (in experiment 3 only after the first period from parallel animals) and pooled from two rats each before isolation of mitochondria. Mitochondrial respiration was measured polarographically using a Yellow Springs YSI model 5300 oxygen monitor (U.S.A.). The substrates were glutamate/malate (complex I activity) and succinate in the presence of rotenone (complex II activity). State 3 and state 4 respiration, respiratory control indices (RCI), ADP:O ratios and the quantity of ATP synthesized were estimated.

Results

The digestibility and metabolizability of energy averaged 93 to 95% and 92 to 94%, respectively, and were slightly, but significantly higher in the casein rats than in the SPI rats. At similar ME intakes (kJ/d), daily protein energy retention as well as fat energy retention were decreased ($P<0.05$) in the SPI rats compared with the casein rats (Table 1). When energy retention was corrected to equal ME intake per kg $BW^{0.75}$ (k_{pf}=0.72), the gross energetic efficiency (ERcorr./ME equalized) was lower in the SPI rats than in the casein rats ($P<0.05$) by 15, 4 , 9 and 18 percentage units in experiments 1, 2, 3.1 and 3.2, respectively. The Me_m values (kJ/kg $BW^{0.75}$/d), calculated by correction of ME intake to zero energy balance and assuming equal partial efficiency coefficients of protein and fat deposition for both dietary groups (k_p=0.61, k_f=0.83), were within the narrow range from 480 to 504 in the casein rats (426 in expt 3.2), but 520 to 635 kJ/kg $BW^{0.75}$*d in the SPI rats. ME_m was increased by 30, 8, 20 and 37% in the SPI fed rats compared with the casein fed rats.

Feeding of SPI resulted on each occasion, in decreased N balance in comparison to casein-fed rats. Protein turnover rates (g/kg $BW^{0.75}$d) showed no clear effect of protein sources. Whereas in experiment 1, synthesis and breakdown were significantly decreased in the SPI rats, in experiments 2 and 3 there were no differences in protein synthesis. Protein breakdown, calculated as the difference between synthesis and accretion, was consequently increased in the SPI rats, but statistical significance was reached only in one experiment. The synthesis:flux ratios were significantly elevated in the casein-fed rats in experiments 1 and 3.1 (0.84 vs. 0.71 and 0.79 vs. 0.68).

Estimates of various variables of mitochondrial respiration revealed no clear picture. Differences between the two dietary groups, observed for some variables, were not consistent between the three experiments. Overall, the magnitudes of the investigated parameters were lower for the SPI rats (with two exceptions in experiment 1 - RCI in complex I and state 3 respiratory rate in complex II of heart mitochondria), or equal to the values of the casein rats.

Mitochondrial protein content in BAT was not different in experiment 2 (3.0 and 3.2 mg per animal) but tended to be lower in experiment 3.1 in SPI rats (2.1 vs. 2.7 mg). BAT weight was also lower in experiment 3.1 (0.16 vs. 0.27 g, $P<0.05$). However, after the 2nd period of experiment 3, BAT weight and mitochondrial protein content were significantly increased in the SPI rats compared with casein rats (0.32 vs. 0.23 g and 4.7 vs. 2.7 mg).

Serum fT_4 levels were significantly lower (on average by 16 pmol/l) in the SPI-rats compared with the casein-rats in experiments 1 and 2. In experiment 1 a lower T_4 level and, in experiment 2, a lower fT_3 level, were observed in the SPI rats, whereas there were no differences in the T_3 levels. All levels were within the physiological range.

Table 1. Energy balance, protein turnover and thyroid hormones in rats fed on casein (cas) or soy protein (SPI) diets.

Parameters	Expt. 1		Expt. 2		Expt. 3.1		Expt. 3.2	
	Cas	SPI	Cas	SPI	Cas	SPI	Cas	SPI
Energy balance period, (n= 9)								
Age at start (d)	50	59	50	64	44	52	65	74
Average BW (g)	134	126 *	120	122	105	106	155	128 *
Gross energy,	165	165	167	167	167	157 *	166	164 *
Urine N (mg/d)	58.7	99.2 *	81.5	94.1 *	58.6	86.3 *	82.8	105 *
N balance (mg/d)	76.3	38.1 *	59.2	44.9 *	78.0	52.7 *	53.7	31.6 *
N balance/ digest. N (%)	56.5	27.7 *	42.1	32.3 *	57.1	38.0 *	39.4	23.1 *
Metab. energy (ME) intake (kJ/d)	155	154	157	156 *	157	154 *	155	150 *
ME intake kJ/kg $BW^{0.75}$* d	699	729 *	773	754 *	850	829 *	627	700 *
Heat production kJ/kg $BW^{0.75}$* d	542	660 *	544	569 *	585	658 *	472	612 *
Energy retention (ER) (kJ/d)	34.8	14.4 *	46.4	38.2 *	48.7	31.7 *	38.5	19.0 *
Fat ER (kJ/d)	23.4	8.7 *	37.5	31.5 *	30.0	23.8 *	30.5	14.2 *
ERcorr./Meeq. %[1]	23.6	8.2 *	29.2	25.2 *	30.8	21.5 *	28.5	10.8 *
MEm (kJ/d)	108	134 *	97.4	107 *	93.1	112 *	105	125 *
MEm (kJ/ kg) $BW^{0.75}$/d	487	635 *	479	518 *	504	605 *	426	584 *
Rates of protein turnover, g/kg $BW^{0.75}$/d (n= 5 or 6)								
Synthesis	7.54	5.87 *	7.74	7.71	7.67	7.36		
Breakdown	5.56	4.34 *	5.36	5.99	5.04	5.92 *		
Accretion	1.98	1.53 *	2.38	1.72 *	2.64	1.43 *		
Thyroid hormones (n=7 to 12)								
T4 (nM)	48.9	42.1 *	60.7	59.2	56.4	49.7 *	55.8	45.6 *
T3 (nM)	2.15	2.32	2.46	2.53	2.53	2.38	2.17	2.00
fT4 (pM)	34.3	18.2 *	59.1	42.1 *	38.4	22.7 *	45.1	28.5 *
fT3 (pM)	1.67	1.83	2.64	2.36 *	3.15	3.11	2.63	2.43

* Significant differences within an experiment ($P<0.05$, ANOVA and multiple range test)

[1] Energy retention corrected to equal ME intake per $kg^{0.75}$ within an experiment (k_{pf} 0.72).

Discussion

From the energy balance data, a lower energetic efficiency was evident in growing SPI-fed rats than in casein-fed rats. In contrast to our own results on growing pigs fed on higher energy levels (Saggau *et al.*, 1997), the rats on the SPI diet retained less protein energy and less fat energy per day as well as per kg $BW^{0.75}$ compared with the rats on the supplemented casein diet under the conditions of a limited energy supply and mild deficiency in protein. Thus, growing rats seem to respond to a low-quality protein diet in a similar way as demonstrated for rats on low protein-energy diets. The pronounced but different response to the SPI diet in the three experiments was possibly due to seasonal effects.

Estimates of protein turnover from two of three experiments suggest that lower protein accretion rates following SPI feeding were the result of increased protein breakdown rather than of decreased protein synthesis. This conclusion is supported by results from the literature, concerning the effect of low-quality proteins on protein turnover, but is not confirmed by our preliminary results on very young pigs (Saggau *et al.*, 1997). From an energetic point of view, it can be concluded that differences in protein turnover did not contribute markedly to the observed differences in total heat production between dietary groups.

SPI feeding was not connected with higher thyroid hormone levels, as reported by Barth *et al.* (1988) for young, Wistar rats under almost the same feeding and environmental conditions, despite a similar retardation in growth.

The differences in thermogenic response between the dietary groups were not accompanied by distinct differences in hepatic and cardiac mitochondrial respiration indicators which reflect the capacity or efficiency of energy utilization for ATP synthesis. Whereas a low protein level has frequently been shown to affect thermogenic activity of BAT, there seem to be no studies concerning the influence of protein quality on BAT. Our results suggest an increased amount of interscapular BAT and of mitochondrial protein after longer feeding of the SPI diet, which is consistent with the higher thermogenic response in this dietary group.

In conclusion, young rats undersupplied with essential amino acids by feeding of low-quality proteins, dispose of ingested surplus energy by increased heat production. Our results suggest that thyroid hormones play only a permissive role in the underlying mechanisms for the dissipation of energy in growing rats.

References

Aguilera J.F. & Prieto C. (1987). Effect of protein quality on energy requirements for maintenance in growing rats. *Arch. Anim. Nutr.* 37:881-894.

Barth C.A., Scholz-Ahrens K.E., Pfeuffer M. & de Vrese M. (1988). Endocrine response to animal and vegetable protein. In: C.A. Barth & E. Schlimme (Eds) Milk protein: nutritional, functional and technological aspects. Springerverlag, New York. pp. 62-67.

Hoffmann L. & Schiemann R. (1977). Energy requirements for the synthesis of body proteins during growth as determined in model experiments with rats. *Arch. Anim. Nutr.* 27:243-253.

Saggau E., Beyer M., Klein M., Schadereit R., Derno M., Jentsch W., Scholze H. & Hagemeister H. (1998). Influence of casein and soy protein isolate on thyroid hormone status and on energy and protein metabolism in growing pigs. In: K.J. McCracken, E.F. Unsworth & a.R.G. Wylie (Eds). Energy Metabolism of Farm animals. Proceedings of the 14th Symposium on Energy Metabolism, Sept. 14-20, Newcastle, Northern Ireland, CAB International, Wallingford, UK, pp. 119-122.

Scholz-Ahrens K.E., Hagemeister H., Unselm J., Agergaard N. & Barth Ch. A. (1990). Response of hormones modulating plasma cholesterol to dietary casein or soy protein in minipigs. *J. Nutr.* 120:1387-1392.

CARNITINE SUPPLEMENTATION DECREASES HEAT PRODUCTION IN RACING PIGEONS DURING FLIGHT SIMULATION

G.P.J. Janssens[1], J. Buyse[2], M. Seynaeve[1], E. Decuypere[2] and R. de Wilde[1]

[1]Laboratory for Animal Nutrition, University of Ghent, Merelbeke, Belgium; [2]Laboratory for Physiology and Immunology of Animals, Catholic University of Leuven, Heverlee, Belgium

Summary

Four groups (A,B,C,D) of nine, trained, male racing pigeons were starved for one day after two months of high energy ration intake and then subjected to a respiration chamber test. One week before, groups A and B were orally supplemented with 90 mg of L-carnitine daily, whereas C and D were given a placebo. A and C underwent flight simulation by electro-stimulation of the breast muscles. Flight simulation increased heat production, prevented a decrease in respiratory quotient due to feed deprivation, decreased T_4 plasma levels and increased weight loss. Feed deprivation caused lower plasma T_3 and T_4 levels. L-Carnitine decreased the rise in heat production during electro-stimulation but did not influence respiratory quotient, weight loss or thyroid hormones. L-Carnitine supplementation in pigeons improved efficiency of energy usage during heavy exercise, thus preventing energy losses due to anaerobic metabolism.
Key words: Pigeons, carnitine, exercise, heat production, thyroid hormones.

Introduction

L-Carnitine is a vitamin-like substance produced in all vertebrates. It transports acyl chains from the cytosol of the cell into the mitochondria. Therefore, it is an essential link in the provision of fatty acids as substrates to the citric acid cycle for energy production.

Supplementation with exogenous L-carnitine can support the generation of energy in cases where the level of L-carnitine in the body is sub-optimal, either through decreased availability or increased requirements (Janssens & de Wilde, 1994).

During races, pigeons need a considerable amount of energy to perform long-term intensive flight exercise. Previous research has shown that L-carnitine supplementation of pigeons subjected to flight simulation decreased plasma levels of fatty acids, lactate and creatine phosphokinase (Borghijs & de Wilde, 1992). These results indicated that muscle fatigue had been delayed by L-carnitine supplementation.

In the present experiment, respiratory parameters were monitored during flight simulation in order to determine the impact of L-carnitine supplementation on the energy metabolism of the exercising pigeon.

Materials and Methods

For two months, 36 trained male pigeons (*Columba livia domestica*) were fed a mixture of whole grains and seeds and a fat pellet supplement (vol/vol 5/1) (Versele-Laga, Deinze, Belgium). The apparent metabolizable energy content of the feed mixture was 3.34 MJ/kg. The given amounts were based on average intakes from earlier experiments conducted under similar conditions, being 27 ± 2 g (x ± SE). The pigeons were housed in 4 identical group cages of 9 individuals each. The cages were large enough for the pigeons to fly in and there was a passage to an outside wired cage which was freely accessible.

The pigeons were divided into nine groups of four animals. Each group began the experiment at a different time, but the relative time schedule was the same for each group.

The last seven days were used to give daily oral supplements of L-carnitine or placebo. Two pigeons per group received an intubation of 90 mg of L-carnitine (Lonza Ltd, Basel, Switzerland) dissolved in 0.5 ml distilled water. The other two animals were supplemented in the same way with a placebo solution of 0.5 ml distilled water. At the end of the two month period, the pigeons were starved and transported by basket to another laboratory where the basket with the pigeons was put in a dark, ventilated room. To enable them to acclimatize and the system to stabilize, the pigeons were put in four individual respiration chambers at about 18.00 h and respiration measurements were started. Each pigeon had been attached to a standard by means of crosswise bandages in order to enable head and wings to move freely. One carnitine supplemented pigeon and one placebo supplemented pigeon had been prepared for a flight simulation (CS,PS respectively). The other two pigeons in each trial - one carnitine supplemented, one placebo supplemented - were used as controls at rest (CR,PR respectively). For the flight simulation, a self-adhesive gel electrode was stuck on the skin of each of the two breast muscles (*M. pectoralis*), connected to an apparatus for electrical stimulation. The next morning at 08.30 h a flight simulation experiment was executed in the respiration chambers. The period between the start of the respiration measurements and the flight simulation test will be referred to as hour 0. The two pigeons, with electrodes attached, were electrically stimulated for 6 hours at a pulsating current of 20 V with a frequency of five pulses per second and 2 milliseconds per pulse, forcing them to flap their wings. Before and after the respiration chamber test, all pigeons were weighed individually and blood samples were taken.

Respiratory quotient (RQ) and heat production (HP) were measured in individual open circuit respiration chambers for indirect calorimetry, designed for small animals. Heat production (HP) was calculated according to the formula of Romijn & Lokhorst (1961):HP (MJ/h) = 3.88 O_2 consumption (l / h) + 1.20 CO_2 production (l / h).

Plasma T_3, and T_4 levels were measured by RIA using commercial antisera and ^{123}I-T_3 and -T_4.

Statistical analysis was performed with SPSS for Windows 6.0. Hourly HP and RQ data were analyzed with the Repeated Measures ANOVA procedure of SPSS. Body weight loss and thyroid hormone levels were analyzed by General ANOVA procedure of SPSS.

Results

Heat production. Mean HP at rest was 4.1 ± 0.2 W/kg$^{0.75}$ whereby no significant differences were seen between the four groups. During the respiration chamber measurements HP rose slowly in the pigeons at rest (Fig. 1). This rise in HP was significantly more pronounced in pigeons subjected to flight simulation. The supplementation of L-carnitine partly suppressed this exercise-generated increase in HP.

Respiratory quotient. The average RQ decreased slightly during the experiment. Linear regression gave the following equation : RQ = -0.0086h + 0.8357 (r^2 = 0.86) with h being the hours after the beginning of the flight simulation test. This lowering of the RQ was inhibited by flight simulation, so that RQ values remained rather stable at about 0.83. L-Carnitine did not show any effect on RQ.

Thyroid hormones. Both T_3 and T_4 decreased during the experiment (Table 1) due to feed deprivation ($P<0.001$). Plasma T4 levels did not alter due to L-carnitine supplementation but were lower in pigeons subjected to flight simulation ($P<0.001$).

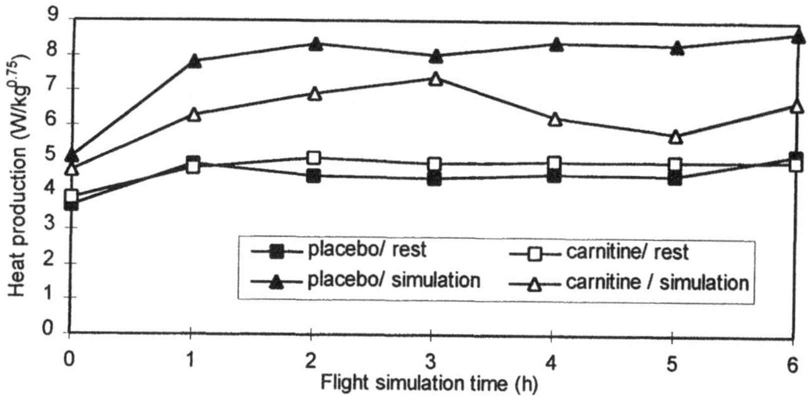

Fig. 1. Heat production during flight simulation test.

Table 1.

Decrease in:	Rest	Flight simulation
Plasma T_3 (ng/ml)	0.28	0.36
Plasma T_4 (ng/ml)	1.18[a]	6.2[b]
T_3/ T_4	0.031[a]	-0.045[b]
Body weight (g)	19[a]	30[b]

Different indices in a row indicate a significant difference at $P<0.05$.

Body weight. Both flight simulation and feed deprivation had a decreasing impact on the body weight of the pigeons (Table 1) ($P<0.001$). No evidence was found of L-carnitine supplementation influencing body weight change.

Discussion

Feed deprivation showed similar effects as in chickens, i.e. decreased body weight, decreased plasma thyroid hormones, especially T_4 (Kühn *et al.*, 1991) and lowered RQ due to more use of fat as an energy source.

Exercise in pigeons increased body weight loss, but also seemed to have an influence on the substrate use as the RQ tended to remain stable in contrast with the pigeons at rest. This can only mean that carbohydrates must have been generated and used for energy production, although the underlying mechanism is not clear. This carbohydrate use was also confirmed by the increased T_3/ T_4 ratio due to flight simulation, probably indicating a higher peripheral conversion rate of T_4 to T_3. This process is known to be very carbohydrate dependent.

The supplementation of L-carnitine limited the energy losses through heat production. This gives indications for a more efficient use of energy substrates. Different pathways can be pointed out for this purpose. Firstly, if the endogenous sources of L-carnitine are insufficient, supplementation can increase the flow of fatty acids into the mitochondria. To enable the citric acid cycle to cope with this increased substrate inflow, L-carnitine indirectly stimulates the production of the enzymes malic enzyme and fatty acid synthase (Roncero & Goodridge, 1992). These enzymes might speed up the citric acid cycle, enabling it to supply more energy per time unit. Secondly, it has been shown that L-carnitine supplementation stimulates pyruvate dehydrogenase, which in turn favours the aerobic pathway of pyruvate from glycolysis and inhibits the anaerobic conversion of pyruvate to lactate (Lopaschuk *et al.*, 1991). A third mechanism of action could be a shift in the substrates used for energy generation, but as RQ did not significantly react to L-carnitine supplementation, we can reject this hypothesis.

L-Carnitine supplementation might thus be a tool for improving flight performance in pigeons and can thus be proposed as an alternative to doping agents such as corticosteroids in the practice of pigeon racing.

References

Lopaschuk G.D., Broderick T.L., Quinney A.H. & Saddik M. (1991). Both carnitine and carnitine palmitoyl transferase I inhibitors stimulate glucose oxidation in isolated hearts perfused with high concentrations of fatty acids. In: A.L. Carter (Ed) *Current Concepts in Carnitine Research*, CRC Press, Boca Raton, pp. 231-244.

Borghijs H.K. & de Wilde R.O. (1992). The influence of two different dosages of L-carnitine on some blood parameters during exercise in trained pigeons. *J. Vet. Nutr.* 1:31-35.

Janssens G.P.J. & de Wilde R.O. (1994). Het gebruik van exogeen carnitine bij nutsdieren. *Vlaams Diergeneesk. Tijdschr.* 63:172-178.

Kühn E.R., Herremans M., Dewil E., Vanderpooten A., Rudas P., Bartha T., Verheyen G., Berghman L. & Decuypere E. (1991). Thyrotropin-releasing hormone (TRH) is not thyrotropic but somatotropic in fed and starved adult chickens. *Reprod. Nutr. Dev.* 31:431-439.

Romijn C. & Lokhorst W. (1961). Some Aspects of Energy Metabolism in Birds. In: E. Brouwer and A.J.H. Van Es (Eds) Energy Metabolism of Farm Animals. Proceedings of the 2nd Symposium of the European Association of Animal Production. Lunteren, The Netherlands, pp. 49-59.

Roncero C. & Goodridge A.G. (1992). Hexanoate and octanoate inhibit transcription of the malic enzyme and fatty acid synthase genes in chick embryo hepatocytes in culture. *J. Biol. Chem.* 267:14918-14927.

THE PARTITIONING OF ENERGY FROM DIGESTIBLE CARBOHYDRATE BY RAINBOW TROUT (*ONCORHYNCHUS MYKISS*)

D.P. Bureau[1], J.B. Kirkland[2] and C.Y. Cho[1,3]

[1] Fish Nutrition Research Laboratory, University of Guelph; [2] Department of Human Biology and Nutritional Sciences, University of Guelph, Guelph, Ontario, Canada, N1G 2W1; [3]Ontario Ministry of Natural Resources.

Summary

The partitioning of the energy from gelatinized potato starch by rainbow trout was examined in a series of feeding trials. The dietary protocol used in all of the trials consisted of feeding the same amount of a basal, low digestible carbohydrate diet, to the fish but with different quantities of supplemental (digestible) gelatinized starch. Feeding supplemental starch had very little effect on recovered energy and nitrogen retention, and did not result in any increase in oxygen consumption by the fish compared to that of fish fed the basal diet alone. An energy budget based on results from comparative carcass analysis and oxygen consumption accounted for about 90% of the energy intake of fish receiving only the basal diet, whereas this proportion decreased significantly for fish fed the basal diet with supplemental starch. Fish receiving supplemental carbohydrate excreted a significant amount of glucose in their urine. Urinary excretion of glucose represented a significant source of energy loss in fish fed high dietary levels of digestible carbohydrate. A large proportion of the supplemental carbohydrate energy fed to the animal could not be accounted for in this study based on comparative carcass analysis, respirometry and urinary glucose excretion. This study indicates that digestible carbohydrate (in excess of what was provided by the basal diet, i.e. 14%) contributes little in terms of net energy to rainbow trout. The energy from digestible starch is poorly retained and a significant proportion is lost via urine.

Key words: Fish nutrition, digestible carbohydrate, energy partitioning.

Introduction

Very few detailed studies on the partitioning of the energy from carbohydrate have been carried out with fish. As a result, the contribution of carbohydrate to the total energy requirement of fish remains unclear and the subject of debate. A study of data from available feeding trials shows that, at high levels of carbohydrate intake, a large proportion of the energy absorbed by fish seems unaccounted for when comparing intake of energy (IE) vs. recovered energy (RE) and faecal energy (FE). There have been very few attempts to directly or indirectly measure critical components of the energy budget such as heat increment of feeding (HiE) and urinary and branchial losses (UE+ZE) of fish fed high levels of carbohydrates.

Materials and Methods

A series of feeding trials was conducted using rainbow trout. In these trials, the fish were fed according to a dietary protocol and pair-feeding technique which consisted of feeding the same amount of a basal diet, low in digestible carbohydrate, to the fish with different quantities of supplemental (digestible) gelatinized starch. To do so, the basal diet was progressively diluted with gelatinized potato starch in the following fashion (ratio of basal diet to starch) : 65:0, 65:20 and 65:35 (Table 1). The three diets were pair-fed, block-wise, on the basis of the feed intake of the fish fed the diet containing the highest level of starch

(65:35). Those fish were fed to near-satiation three times daily and their intake represented 100%. The other diets were then fed at different levels in a manner allowing all the tanks of fish in the same block to receive the same amount of the basal diet. The basal diet (65:0) was, therefore, fed at 65% of the dietary intake of the 65:35 diet and served as control. The 65:20 diet was fed at 85% of the dietary intake of the 65:35 diet. The basal diet was also fed at the same level as the 65:35 diet to groups of fish that served as a satiation control (100:0).

Table 1. Formulation of the experimental diets.

	Diets			
Ingredients (g)	65:0	65:20	65:35	100:00
Fish meal, herring, 68% CP, 10% Lipid	45.5	45.5	45.5	70.0
Fish oil	6.5	6.5	6.5	10.0
Celite AW521	1.3	1.3	1.3	2.0
Vitamin+mineral premix	1.3	1.3	1.3	2.0
Starch, potato, gelatinized	10.4	30.4	45.4	16.0
Total	65 g	85 g	100 g	100 g

In the first (feeding) trial, 12 tanks of 100 rainbow trout each were fed according to the dietary protocol for 12 weeks. Comparative carcass analysis provided the basis for an energy budget. In the second (respirometry) trial, another 12 tanks of 100 fish each were fed according to the dietary protocol for a minimum of four weeks before the beginning of the respirometry measurements. Each tank was monitored successively for a seven-day period in one of two fish respirometers similar to the one described by Cho & Kaushik (1990). Heat production of the fish was estimated using an oxycalorific coefficient of 13.6 kJ per g of oxygen consumed. The heat increment of feeding (HiE) was calculated from the average oxygen consumption of the fed fish minus the oxygen consumption of these fish on the third day of fasting (estimate of basal metabolism, HeE). The estimate of HeE was adjusted for metabolic weight ($kg^{0.824}$) according to Cho & Kaushik (1990). HeE and HiE from the respirometry trial were used to calculate the heat production of the fish from the feeding trial.

A third trial was conducted to estimate the urinary glucose excretion of the fish. Twelve groups of four fish each (live weight = 269 g) were fed for 8 days according to the dietary protocol described earlier. Urinary glucose excretion was measured using the method of Curtis & Wood (1991). This method involved the determination of the urine flow rate of non-catheterized fish using a glomerular filtration marker, 1,2 ^{3}H polyethylene glycol, and spot sampling of the urine to measure glucose concentration.

Results

Feeding supplemental digestible starch to the fish resulted in a small, but significant, increase in weight gain and RE for the fish fed the 65:20 diet but not for the fish fed the 65:35 diet (Tables 2 and 3). The greater weight gain and RE of the fish fed the 65:20 diet compared to the 65:0 diet was due to an increase in viscera weight, since the dressed carcass weights of fish fed the 65:20 and 65:0 diets did not differ. There were no differences among diets 65:0, 65:20 and 65:35 in terms of nitrogen (N) gain which indicated that the supplemental starch had no sparing effect on dietary protein. Fish fed the 100:0 diet had much greater weight gain and RE than fish fed the 65:35 diet. The dilution of the diet with digestible starch may have limited the intake of net energy of the animal, which resulted in lower growth and RE. Increasing digestible carbohydrate intake did not lead to any significant increase in oxygen consumption of fish, indicating that the supplemental carbohydrate was not aerobically catabolized. The calculated heat production of the fish in the feeding trial showed a slight increase with an increase in starch intake (Table 3) but this could only account for a small portion of the ingested carbohydrate energy. At high carbohydrate intakes, a large portion of the energy absorbed by the fish was unaccounted for when comparing IE and RE, FE, UE+ZE (based on calculated N losses) and HeE and HiE estimated from oxygen consumption.

Table 2. Performance of the rainbow trout[1] fed the experimental diets for 12 weeks.

Parameters	Diets 65:0	65:20	65:35	100:0	SEM	LSD[2]
Feed (g/fish)	39.7	50.9	58.7	59.7	0.4	N/A[3]
Gain (g/fish)	52.7[c]	55.8[b]	51.6[c]	79.0[a]	0.3	1.6
Feed eff. (gain:feed)	1.33[a]	1.10[b]	0.90[c]	1.32[a]	0.01	0.05
Retained N (g/fish)	1.26[b]	1.35b	1.21[b]	1.98[a]	0.04	0.19

[1] Initial body weight = 8.1 g/fish; [2] least significant difference; [3] not applicable, independent variable. Values in a same row not sharing the same letter are significantly different (P<0.05).

Table 3. Energy budget of the fish fed the experimental diets for 12 weeks.

Components kJ/ fish per 84d	Diets 65:0	65:20	65:35	100:0	SEM	HSD
IE	823	1003	1125	1237	N/A	N/A
FE	54[c]	94[b]	142[a]	99[b]	0.4	2
UE+ZE[1]	37[b]	35[b]	37[b]	54[a]	1	6
RE	376[c]	410[b]	373[c]	635[a]	5	22
HeE	141[c]	176[b]	134[d]	220[a]	0.3	5
HiE	105[c]	79[d]	139[a]	113[b]	1	7
Deviation[2]	110[c]	205[b]	300[a]	115[c]	9	42

[1] Based on calculated N losses; [2] by difference: (IE-FE-(UE+ZE)-RE-HeE-HiE).

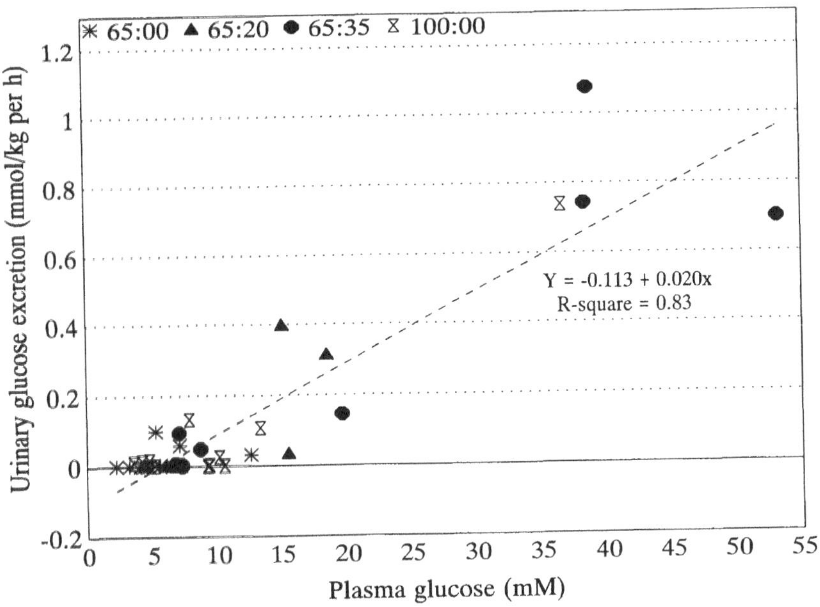

Fig. 1. Estimated urinary glucose excretion of rainbow trout.

Fish fed with high levels of digestible carbohydrate had high plasma glucose and excreted significant amounts of glucose in their urine (Fig. 1). The glucose excretion of the fish from the feeding trial over the 12-week experimental period was estimated at 0.2 g (3 kJ), 0.4 g (6 kJ), 3.3 g (52 kJ) and 1.4 g (22 kJ) for the fish fed the 65:0, 65:20, 65:35 and 100:0 diets respectively. Urinary glucose excretion represents a non-negligible proportion (ca. 25%) of the unaccounted carbohydrate energy for the fish fed the 65:35 diet. This could explain, at least in part, the apparent discrepancy in the energy budget of fish fed diets high in carbohydrate.

Conclusion

When fed in excess of a certain level, digestible starch contributes very little in terms of net energy to rainbow trout. The digestible carbohydrate did not appear to be metabolized aerobically and significant urinary glucose excretion was observed. Based on comparative carcass analysis, oxygen consumption and estimates of nitrogen and glucose excretion, a significant proportion of the energy from digestible starch fed to the animal remains unaccounted for in this study.

References

Cho C.Y. & Kaushik S.J. (1990). Nutritional energetics in fish - energy and protein utilization in rainbow trout (*Salmo gairdneri*). *World Rev. Nutr. Diet.* 61:132-172.

Curtis B.J. & Wood C.M. (1991). The function of the urinary bladder *in vivo* in the freshwater rainbow trout. *J. Exp. Biol.* 155:567-583.

EFFECTS OF BODY LIPIDS AND ENERGY BALANCE ON THE RESPONSE OF PLASMA NON-ESTERIFIED FATTY ACIDS TO A β-ADRENERGIC CHALLENGE IN THE LACTATING DAIRY EWE

F. Bocquier, A. Ferlay and Y. Chilliard

Laboratoire Ruminant Undernutrition, INRA-Theix, 63122 St-Genès-Champanelle, France

Summary

The lipolytic potential of adipose tissue can be estimated through the plasma NEFA response to an adrenergic challenge. In well-fed lactating ewes, the NEFA response to isoproterenol depends on body lipid mass, but not on energy balance. In underfed ewes however, the NEFA response depends on energy balance but not on body lipid mass. Most *in vitro* studies have revealed that stimulated lipolytic activity depends on adipose tissue cellularity, whatever the energy status of donor animals. These findings suggest that *in vitro* results may be incomplete in the case of underfed animals.
Key words: Ewe, undernutrition, energy balance, milk production, NEFA, adrenergic challenge.

Introduction

The capacity of an animal to support an energy shortage during lactation mainly depends on the amount of non-esterified fatty acids (NEFA) released from adipose tissue (AT). This release depends on numerous factors including the mass of AT (i.e. body lipids: Cowan *et al.*, 1982; Chilliard, 1992) and its responsiveness to lipolytic stimuli (Jaster & Wegner, 1981; McNamara, 1991). Among these lipolytic stimuli, catecholamines (epinephrine and norepinephrine) or synthetic β-adrenergic compounds (isoproterenol, etc.) are particularly efficient. The lipolytic potential of AT can be estimated by NEFA released by a lipolytic agent, either *in vitro* in the culture medium of AT explants or, *in vivo*, in the plasma. The aim of this work was to evaluate *in vivo* the effects of energy balance and body lipid mass (BLIP) on the AT lipolytic potential of the lactating ewe.

Materials and Methods

Twenty-four, lactating, Lacaune dairy ewes, on day 49 of lactation at the start of the trial, were placed in either a High group or a Low group according to the energy supply, for two successive periods (P1-P2) of 4 weeks. Ewes in the High group were fed so that an excess of energy (104 to 111%) compared to their total energy requirements (TER) was supplied, whereas ewes in the Low group received 83 to 86% of their TER. Their theoretical protein requirements were fully covered (>100%). Diets, given at 08.00, were individually adjusted (twice a week) to milk yield and consisted of rye-grass supplemented with concentrate (31% of total DMI). From the first to second period, the feeding treatments were crossed over (High-Low or Low-High). Milk production and composition were measured twice a week. Body composition was estimated *in vivo* by deuterium oxide dilution technique (Bocquier & Thériez, 1984) at the beginning and end of each period, allowing BLIP to be estimated.

The non-selective beta-adrenergic agonist (isoproterenol; ISO) was injected (20.5 nmol/kg $BW^{0.75}$) at 13.30 into the jugular vein on day 24 (P1) and day 50 (P2). Blood samples were taken for plasma NEFA determination, five minutes before (NEFA-5) and ten minutes after (NEFA+10) injections. This simplified procedure of blood sampling was adopted after observations that the NEFA level peaked at about ten minutes after an ISO injection (Chilliard *et al.*, 1994; Ferlay *et al.*, 1996). Morning pre-feeding blood samples were also taken 6h (NEFA-360) before ISO challenge.

Results and Discussion

Underfeeding of ewes (Low vs. High diets) induced a decline in milk yield, a mobilisation of BLIP, and an increase of NEFA that were all in good agreement with calculated energy balance (EBAL). Injection of ISO was followed by an increase of NEFA as has been previously reported in sheep (Basset, 1970) and the cow (Blum *et al.*, 1982; Gagliostro & Chilliard, 1991; Chilliard *et al.*, 1994; Ferlay *et al.*, 1996). The mean NEFA response to ISO (NEFA+10 *minus* NEFA-5) was significantly higher in underfed (+0.74 mM) than well-fed ewes (+0.45 mM). Part of this response can be explained by the pre-injection concentration (NEFA-5) which was already higher (+0.15 mM) in the Low group.

Table I. Mean characteristics of dairy ewes subjected to two feeding levels: consequences on plasma non-esterified fatty acids (NEFA) (n = 12 ewes per group).

Feeding level[1]	LOW		HIGH		SEM
Period[2]	P1	P2	P1	P2	
Body weight (kg)	65.4	66.9	72.6	65.4	1.4
Milk yield (l/d)	1.33[a]	1.09[b]	1.61[c]	0.93[b]	0.05
Net Energy intake (UFL/d)[3]	1.46[a]	1.29[c]	2.01[b]	1.49[a]	0.03
Energy balance (UFL/d)[3]	-0.29[a]	-0.25[a]	+0.07[b]	+0.08[b]	0.02
Body lipids (kg)[4]	15.4	17.6	17.3	12.8	0.8
Body lipid variation (g/d)	-95[a]	-93[a]	+8[b]	+16[b]	8
NEFA-360 (mM)[5]	0.533[a]	0.369[b]	0.180[c]	0.157[c]	0.021
NEFA-5 (mM)[5]	0.257[a]	0.229[a]	0.114[b]	0.090[b]	0.016
NEFA+10 (mM)[5]	1.138[a]	0.841[b]	0.653[bc]	0.465[c]	0.033

[1]: LOW : 85% and HIGH 111% energy requirements. [2]: two periods of 28 days each.
[3]: 1UFL = 7.11 MJ Net Energy. [4]: at the start of each period. [5]: see text.
[a, b, c]: figures with different letters differ significantly ($p < 0.05$).

There was a positive inter-individual correlation between basal values (NEFA-5) and post-ISO (NEFA+10) levels ($r=+0.65$; $n=48$; $P<0.001$), i.e. between spontaneous and stimulated lipolysis. The correlation with NEFA+10 was even higher ($r=+0.79$; $P<0.001$) when considering pre-feeding morning NEFA levels (NEFA-360). This suggests that basal NEFA concentration (NEFA-5 or NEFA-360) may partly be related to the beta-adrenergic component of the lipolytic activity of AT (Blum *et al.*, 1982; Ferlay *et al.*, 1996). NEFA+10 ($n=48$) was positively related to milk energy output ($r=+0.38$), BLIP ($r=+0.30$) and EBAL ($r=-0.72$). In fact, multiple regression revealed that NEFA+10 depends on animal nutritional status. For ewes in the High group, NEFA+10 was related to BLIP ($P<0.01$) and energy intake (EI ; $P<0.01$), but not (ns) to EBAL, whereas, for the Low group ewes, NEFA+10 mainly depended on EBAL ($P<0.01$), rather than BLIP (ns). This is apparent in Figure 1, where NEFA+10 was predicted from within-feeding treatment (High or Low) multiple regressions including three variables (EBAL, BLIP and EI).

Previous results obtained *in vitro* on lactating ewe AT (Vernon & Finley, 1985) showed a close relationship ($r=+0.96$) between stimulated lipolysis and adipose cell size, that is strongly correlated to BLIP (Robelin *et al.*, 1989). These *in vitro* results, with ewes presumably in positive EBAL, are in agreement with our *in vivo* results (High diets). Furthermore, for lactating dairy cows mostly in negative EBAL (Gagliostro & Chilliard, 1991), the *in vitro* adrenergic response depended on AT cellularity (P<0.01), but not on EBAL (ns). However, *in vivo* response in the same cows was related to EBAL (P<0.01) but not cellularity (ns). The latter result is in full agreement with what we have seen in underfed lactating ewes. Hence it seems that *in vitro* studies on AT lipolytic response illustrate principally the differences in AT cellularity (or BLIP) whatever the energy balance of the donor animal. Our results show that this may also be true *in vivo* for animals having a positive EBAL, but not for underfed animals. In underfed animals, the *in vivo* approach is more likely to take into account other factors that are known to modulate the expression of lipolytic potential, such as endocrine control (Vernon, 1992; Chilliard *et al.*, 1995).

Adrenergic challenge is also useful in explaining the differences in inter-individual adaptative strategies to underfeeding in the ewe. In the underfed ewes (Low diets), we observed that the relative variation in milk yield was negatively correlated to NEFA+10 (r= -0.51; P<0.01; n=23) : ability to sustain lactation was related to ability to mobilise BLIP. It would be interesting to study the effect of the milk genetic potential on the ability of AT to respond to adrenergic stimuli, as has been observed in dairy cows (McNamara, 1991).

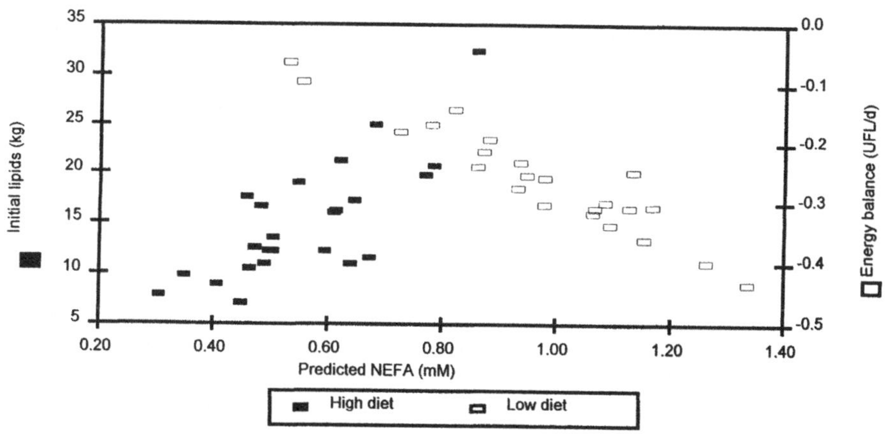

Fig. 1. Effect of initial body lipids (High diet) or energy balance (Low diet) on plasma NEFA concentration after isoproterenol challenge (predicted values).

Conclusions

In vivo estimation of AT lipolytic potential in the lactating ewe revealed that the roles of body lipid mass and energy balance depended on nutritional status. With undernutrition, when lipolytic activity was of particular importance for animal survival or for maintenance of lactation, this lipolytic potential is well related to the energy balance, as is quite logical. With well-fed animals, however, the AT lipolytic activity was basal and limited, and the NEFA response to adrenergic stimulation appeared to be proportional to the amount of adipose tissue present.

Acknowledgements

This work was supported by EEC project CAMAR n°8001-CT 91-0113.

References

Basset J.M. (1970). Metabolic effects of catecholamines in sheep. *Aust. J. Biol. Sci.* 23:903-914.

Bocquier F. & Thériez M. (1984). Prediction of ewe body composition at different physiological states. In: D. Lister (Ed) '*In vivo* measurements of Body Composition in Meat Animals', pp. 152-157.

Blum J.W., Froehli D. & Kunz P. (1982). Effects of catecholamines on plasma free fatty acids in fed and fasted cattle. *Endocrinol.* 110:452-456.

Chilliard Y. (1992). Physiological constraints to milk production : Factors which determine nutrient partitioning, lactation persistency and mobilisation of body reserves. *World Rev. Anim. Prod.* 27:20-26.

Chilliard Y., Després L., Ottou J.F. & Bocquier F.F. (1994). Response of plasma non-esterified fatty acids to beta-adrenergic challenge in underfed and overfed non-pregnant dry cows, or in lactating cows. VIII International Symposium on Ruminant Physiology. 25-30th Sept., Willingen, Germany. *Proc. Soc. Nutr. Phys.* 3:268.

Chilliard Y., Doreau M., Bocquier F. & Lobley G.E. (1995). Digestive and metabolic adaptations of ruminants to variations in food supply. In: M. Journet, E. Grenet, M-H. Farce, M. Thériez and C. Demarquilly (Eds) Recent developments in the nutrition of herbivores. Proceedings of the IVth International Symposium on the nutrition of Herbivores. INRA Editions, Paris, pp. 329-360.

Cowan R.T., Robinson J.J. & McDonald I. (1982). A note on the effects of body fatness and level of food intake on the rate of fat loss in lactating ewes. *Anim. Prod.* 34:355-357.

Ferlay A., Chilliard Y., Sala A.M., Durier C. & Bocquier F. (1996). Somatotropin treatment does not affect nonesterified fatty acid response to adrenergic injections in underfed or overfed nonlactating cows. *J. Nutr.* 126:945-954.

Gagliostro G. & Chilliard Y. (1991). Duodenal rapeseed oil infusion in early and midlactation cows. 4. *In vivo* and *in vitro* adipose tissue lipolytic response. *J. Dairy Sci.* 74:1830-1843.

Jaster E.H. & Wegner T.N. (1981). Beta-adrenergic receptor involvement in lipolysis of dairy cattle subcutaneous adipose tissue during dry and lactating state. *J. Dairy Sci.* 64:1655-1663.

McNamara J.P. (1991). Regulation of adipose tissue metabolism in support of lactation. *J. Dairy Sci.* 74:706-719.

Robelin J., Chilliard Y. & Agabriel J. (1989). Estimation of body lipids and proteins of Holstein Charolais and Limousin cows by dilution technique and adipose cell size. In Proceedings of the 11th Symposium On Energy Metabolism in Farm Animals. (Lunderen, Netherlands, 1988). *EAAP Publ. No.* 43:370-373.

Vernon R.G. & Finley E. (1985). Regulation of lipolysis during pregnancy and lactation in sheep. *Biochem. J.* 230:651-656.

Vernon R.G. (1992). Effects of diet on lipolysis and its regulation. *Proc. Nutr. Soc.* 51:397-408.

SUMMARY OF THE DISCUSSION

Scientific leader	P J Reeds
Scientific secretaries	J L Ramsey
	R L Baldwin

The papers in this session covered, as is probably appropriate for the subject, a wide variety of topics that ranged from studies of dietary composition, body composition and energy expenditure through specific effects of hormonal metabolic modifiers, to more mechanistic studies on the adrenergic and insulin mediated systems of metabolic regulation. The highpoint of the session was the paper by Romsos on the rapidly developing field of leptin and its multiple effects on physiological and metabolic systems. The extensive discussion on this humoral mediator of both food intake and metabolic rate was particularly apposite, as an attempt was made at the outset to focus the discussion on three phenomena.

(1) Organisms appear to have a genetically determined target for body size and composition.
(2) They are also able to sense the degree to which this programme has been achieved at any given stage of development or physiological state.
(3) At the same time they are capable of recognizing whether the targets are achievable given current environmental and nutritional circumstances.

In considering these points there appeared to be a consensus (a) that here may well be a central regulator and (b) that this regulator was capable of influencing the central-peripheral flow of neural information and interacting with the more 'classical'/humoral mechanisms, including the effects of local mediators and, moreover influencing feed intake.

At present, leptin is the closest approximation to such a central regulator. Its level is a good function of body fatness, it is quite clearly a major influence on food intake (at least in rodents) and it also has effects on energy expenditure and substrate metabolism. Romsos discussed each of these aspects of leptin physiology, highlighting the multiple points at which it can interact with food intake regulation and the important interactions between insulin (a stimulator of leptin secretion during feeding) and leptin (a potential regulator of tissue insulin sensitivity). However he also emphasized during the discussion that current 'leptin mania' should be tempered with the caution that it was probably not the answer to all questions related to obesity and the regulation of body fatness and that given the complexity of the regulatory problem it was highly likely that multiple and possibly redundant, systems existed. In answer to a question from the floor he also emphasized that, although attractive in view of the interactions between insulin and leptin, there was no compelling evidence to link derangements in leptin physiology to the development of type II (insulin-resistant) diabetes.

The discussion expanded these topics to consider leptin, reproduction and lactation; subjects that have been long term concerns of the meeting. Romsos re-emphasized that it was plausible that the role of leptin in reproduction was related to the necessity for there to be adequate energy stores, an illustration of point 3 above with regard to the ability to achieve developmental targets. Tyrrell asked whether leptin had a role in lactation in addition to its well established influence of reproduction.

MacNamara pointed out a parallelism between prolactin and leptin levels. However Rosmos pointed out that there was little direct information on a possible role for leptin in the regulation of body lipid stores during lactation. Reeds drew attention to work carried out by Butte that showed that leptin levels rose in late pregnancy in humans but that the changes were entirely commensurate with the changes in body fat mass.

The discussion was then broadened to other topics raised by the papers in the session. Overall metabolic regulation could be achieved by altering both the input (intake) and output (metabolic demand) sides of the equation. Reeds asked the meeting to what extent the participants believed that overall regulation (for example feed intake) was a response to changing demands or to what extent was intake the prime target of the regulatory systems. There appeared to be some consensus that demand was the primary stimulus and that changes in overall nutrient availability (i.e. intake) and local substrate availability (i.e. blood flow) were responses rather than initiators. Aharoni pointed to evidence that, in lactating dairy cows, frequent milking not only stimulates appetite but also apparently alters feed digestibility. There was also discussion of whether there is genetic regulation of voluntary appetite and whether maximum appetite is achieved before maturity. McCracken made the important point that in many cases where feed intake had approached an early asymptote the results were flawed by inappropriate feeding conditions and that, when properly conducted, most experiments indicated that voluntary appetite increased throughout the life span and in parallel with body mass.

The importance of the amount and quality of nutrient availability was highlighted by two papers (Saggau *et al.* and Klein *et al.*) concerned with responses of pigs and rats to variations in the protein intake (protein:energy ratio in the diet) and protein quality. In pigs 'low' protein diets or poor quality proteins were associated with an increase in the lipid:protein ratio in the carcass while in rats apparently similar diets appeared to be associated with lower overall energy storage. In this regard McCracken emphasized that the effect in rats was associated with improved efficiency of energy restricted animals associated with changes to a meal-feeding pattern. However, the key point emphasized by the authors was that the different responses were not a species difference but a reflection of dietary design. The pigs were fed *ad libitum* on relatively high protein diets, while the rats received restricted intakes of lower protein:energy rations. In other words the metabolic (regulatory) response was greatly dependent on nutrient availability. Reeds used this point to re-emphasize that sometimes in seeking tight statistical design the use of pair- or restricted feeding might alter the very response that the work was designed to understand and that there should always be groups of animals that were allowed free access to experimental diets. MacNamara re-emphasized the crucial point that feed intake is regulated by multiple and potentially redundant systems. Romsos also pointed out that the term feed intake is potentially misleading in as much as organisms consume organic nutrients and that the intake of each major class can be regulated separately. Leptin levels are, for example, more sensitive to carbohydrate intake and there are strain differences in leptin responses to high fat diets. In other words organisms have systems that sense the nutrient requirements of their metabolic demands at any stage of development and are quite capable of selecting the appropriate mix of nutrients, if they are available. Recent evidence also favours a specific effect of leptin on glucose oxidation.

The discussion then turned to peripheral regulation. Once again the existence of multiple systems was highlighted. It was emphasized that the autonomic nervous system, especially the sympathetic system played an important role in metabolic regulation. This had been studied most extensively in rodents where the role of brown fat in these species makes the adrenergic, especially the β3-receptor-mediated system of particular importance. However, although farm species do not have extensive brown fat stores, except immediately after birth, some β-agonists, notably clenbuterol, do exert a sub-acute thermogenic effect although others (e.g. salbutamol, Fernandez paper) have minimal effects. However, evidence from studies with the α2-agonist clonidine in cattle (Löhrke *et al.* and Derno *et al.*) showed that changes in energy expenditure and blood flow (heart rate) both in response to low environmental temperatures and the habitual level of energy intake were at least in part a function of adrenergic tone. Moreover, Löhrke *et al.* had presented evidence that these changes in overall energy expenditure could be associated with changes in the expression of plasma membrane ATP-sensitive K^+ channels. In addition the paper by Wylie and McGrattan provided an important illustration of the fact that species, strain, breed and gender could have substantial effects on the rates and sites of lipid storage mediated at the local level by insulin receptors. By altering the magnitude of the intracellular signal to a given level of effector, quite marked tissue differences could be generated.

In closing, Reeds re-emphasized that it is important to remember that a focus on energy tends to avoid the fact that organisms ingest and metabolize organic compounds and that the meeting should give consideration to the measurement of substrate metabolism at the whole body, tissue and cellular level.

SESSION 4

METHODOLOGY

DEVELOPMENT AND USE OF DEUTERIUM OXIDE DILUTION EQUATIONS TO PREDICT BODY COMPOSITION OF HOLSTEIN COWS

B.A. Crooker[1], W.J. Weber[1] and S.M. Andrew[2]

[1]Department of Animal Science, University of Minnesota, St Paul, USA; [2]Department of Animal Science, University of Connecticut, Storrs, USA

Summary

Individual cow D_2O dilution and chemical data from two studies (SMA, n=21; MTM, n=68) were used to develop equations to predict composition. Equations were tested by comparing actual values to values predicted by equations from each study. MTM cows received bST (0, 5 or 14 mg/d). A two-pool model provided reasonable estimates of empty body water (pool A) but estimates of pool B (gut and other water) did not agree with measurements. Each study used different one-pool models to estimate gut water. Results demonstrate equations can provide reasonable estimates of composition, but equations from reference animals may not provide accurate estimates for cows of different composition. Use of a single equation to predict composition may bias results and prevent detection of treatment effects.
Key words: Body composition, Holstein cows, D_2O.

Introduction

Prediction equations have been developed for dairy (Andrew *et al.*, 1995; Brown *et al.*, 1989) cattle but have not always provided accurate estimates of composition when applied to animals from other studies (Brown *et al.*, 1989; Maslanka *et al.*, 1996). Objectives of this study were to determine the ability of prediction equations developed from two independent D_2O dilution - slaughter studies to predict empty body composition of the cows used in each of the two studies and to determine the likelihood of developing a single set of prediction equations to accurately estimate empty body composition of multiparous Holstein cows.

Materials and Methods

Animals and management. Chemical composition and D_2O dilution measurements from individual, multiparous cows from two independent studies were used. Study MTM (Maslanka *et al.*, 1996; Weber *et al.*, 1996) cows and lactating cows from study SMA (Andrew *et al.*, 1994, 1995) were fed *ad libitum* quantities of total mixed rations (corn silage, alfalfa haylage, concentrate). Non-lactating SMA cows were fed chopped orchardgrass hay. Diets were designed to meet or exceed NRC (1989) requirements. Cows were housed in tie stall barns, milked twice daily, and fed once daily. MTM cows were slaughtered within a 7 d (" 3d) window in early lactation (EL, 10, 20, or 50 d postpartum (PP), n = 5, 2, 5), in late lactation (305 d PP) if non-pregnant (LLNP, n = 12), at 70 d prior to expected calving (pre-dry off (LLP), n = 22) or 20 d prepartum (PREP, n = 22). MTM cows (except EL) were treated with 0, 5, or 14 mg bST/d from 14 d PP until dry off or 305 d PP. SMA cows were slaughtered 5 to 7 d prepartum (PREP, n= 6), 38 to 110 d PP (EL, n=7), and 163 to 375 d PP (LLNP, n=6).
Dilution technique. Measurements were conducted 3-7 d prior to slaughter. D_2O (99.8% Cambridge Isotope Labs, Woburn, MA) was infused (0.22-0.30g D_2O/kg BW) via jugular catheter and quickly flushed with 30 ml physiological saline. Blood samples (10 ml) were obtained in heparinized tubes during the following 72 h. Water was extracted from blood by lyophilization and analysed for D_2O content by infrared spectroscopy (Byers, 1979).
Estimates of body composition. Cows were weighed, transported to the abattoir and slaughtered. Carcass and non-carcass (gravid uterus included) components were collected and analysed

(A.O.A.C., 1984) for dry matter, crude protein (macro-Kjeldahl N x 6.25), ash (600°C) and fat (ether extract). Neither study was able to obtain reasonable estimates of gut water (Pool B) from a two pool model of the D_2O dilution curve.

Study MTM used a one compartment model ($Y_1 = Ce^{-k3t}$) to estimate total body water (TBH) from the equation that described D_2O dilution after the inflection point (>4 hr post-infusion) of the curve (Shipley and Clark, 1972). Estimated TBH (D_2O Space, DSP, kg) was calculated as DSP = dose in g of D_2O / (C H 1000). Equations to predict empty body (EB) water (EBH), EB protein (EBP), and EB ash (EBA) were developed using a multiple step-wise regression (PROC REG) with BW, DSP, DMI, and milk yield as independent variables (SAS, 1994). Separate equations were developed for zero dose bST cows (ZDSOL, n = 32) at each stage of lactation (EL, n = 12; LLNP, n = 4; LLP, n = 8; PREP, n = 8).

Study SMA used the Marquardt option of the nonlinear procedure (PROC NLIN) of SAS (1994) to determine parameters for a two pool model ($Y= Ae^{-k1t} + Be^{-k2t}$) and used the estimate of pool A and BW to estimate EBH (EEBH) from multiple regression (SAS, 1994). TBH was estimated from a single pool model as in study MTM. Gut DM was estimated from the regression of actual gut DM (at slaughter) on mean daily DMI during the 7 d prior to slaughter. EBP and EBA were estimated from ratios (EBH/EBP, EBH/EBA).

Statistics. Effects were considered significant when $P<0.05$. Statistical analyses were conducted with procedures of SAS (1994). Effects of stage of lactation and bST on chemical composition were determined using PROC GLM. Differences between predicted and actual composition values were compared to zero using a pairwise t-test (PROC REG). Slopes and intercepts of linear equations describing relationships between predicted and actual measurements were compared to one and zero using a pairwise t-test (PROC REG).

Results

Although variation was evident, equations (Table 1) to predict EBH, EBP, EBA, and EBF from MTM and SMA provided estimates that generally did not differ from actual overall values (Table 2, appropriate ALL columns). Results suggest the prediction equations provide reasonable estimates of empty body composition of multiparous Holstein cows. Andrew *et al.* (1995) concluded that the compositions of EL, LLNP and PREP cows were similar and were predicted well with a single set of equations. Although results (Table 2) tend to support this evaluation, the equations under (EBF of LLNP) and over (EBH of LLNP, EBF of PREP) predict composition of some groups. Maslanka *et al.* (1996) suggested differences in empty body composition across stage of lactation warranted individual sets of equations and this need was more apparent when composition was altered by metabolic modifiers such as bST.

Good agreement between predicted and actual values were obtained when the overall MTM equations were used to estimate overall composition of SMA cows (Table 2). Extending these equations to subsets of SMA cows decreased this agreement, especially for EBH and EBF. Deviations were greater when SMA equations were used to predict composition of MTM cows. These deviations are due in part to the presence of bST-treated cows in the MTM data set. Multiparous cows treated with bST had increased EBH and decreased EBF (Weber *et al.*, 1996).

To eliminate the influence of bST-treated cows, they were removed from the MTM data set and a separate set of prediction equations developed (Table 1) that predicted the zero-dose cows (n=32) well and tended to improve the prediction of empty body composition of the individual slaughter groups (data not shown). Although differences between actual and predicted values were reduced when bST-treated cows were removed from the data set, considerable deviations were still apparent whether MTM or SMA equations were utilized to predict composition of the zero-dose cows.

Table 1. Equations to predict empty body composition of Holstein cows.

Study & Component[1]	Prediction Equation	r^2
MTM[2] - All cows		
EBH	(0.2953) BW + (0.4046) DSP - 31.06	0.82
EBP	(0.1124) BW + (0.04916) DSP - 1.787	0.85
EBA	(0.0188) BW + (0.02361) DSP + 4.303	0.49
EBF	(0.6264) BW - (0.5217) DSP - 92.79	0.71
MTM[2] - ZDSOL cows		
EBH	(0.3635) BW + (0.2230) DSP - 3.458	0.87
EBP	(0.1186) BW + (0.03560) DSP - 0.768	0.87
EBA	(0.01142) BW + (0.03902) DSP + 2.408	0.59
EBF	(0.5663) BW - (0.3879) DSP - 104.1	0.81
SMA - All cows		
EBH	(0.1844) BW + (0.4967) EEBH + 38.17	0.82
EBP	(0.0574) BW + ((0.4152)EEBH/3.51) + 14.57	0.81
EBA	(0.186) BW + ((0.5171)EEBH/12.6) + 0.202	0.66
EBF	(0.0933) BW + (0.865)FFEBW - 38.9	0.87
FFEBW	BW - (TBH + EBP + EBA + GIDM + FDM)	--

[1] EBH, empty body water; EEBH, estimated EBH; EBP, EB protein; EBA, EB ash; EBF, EB fat; FFEBW, fat-free EB weight; TBH, total body water; GIDM, GI tract DM; and FDM, foetus.

[2] All cows, cows treated with 0, 5 or 14 mg bST/d; ZDSOL cows, cows not treated with bST.

Table 2. Ability (predicted minus actual kg) of overall equations to predict empty body composition of Holstein cows.

		SMA Study Cows[1]			MTM Study Cows		
Study & Equation[3]	Mean SE	ALL	LLNP	PREP	ALL	LLNP	PREP
MTM							
EBH	Mean	-2.0[‡]	9.1[a,‡]	25.4[b]	0.0	10.8[d]	-6.5
	SE	5.0	3.2	6.2	0.0	4.4	3.4
EBP	Mean	-1.4[‡]	6.8[c]	5.3[a]	0.0	0.4[‡]	-2.0[b]
	SE	1.3	0.3	1.7	0.5	1.6	0.6
EBF	Mean	-1.8	-13.3	-12.6[*]	0.0	4.9	-7.4[b]
	SE	5.4	6.6	6.2	2.4	5.4	2.6
SMA							
EBH	Mean	9.0	20.0	4.0[*]	-33.0[c]	-15.7[a]	-42.3[c]
	SE	5.4	11.6	7.5	2.8	6.2	3.2
EBP	Mean	2.1	-1.4	0.8	-5.4[c]	-2.8	-8.5[c]
	SE	1.3	3.2	1.9	0.7	1.7	0.8
EBF	Mean	8.0	-14.7	-22.9[a]	5.0	3.1	2.7
	SE	5.8	7.1	6.9	5.1	6.4	3.3

[a,b,c] Means differ from zero; [a] P<0.05, [b] P<0.01, [c] P<0.001.

[*,†,‡] Intercept differs from 0.0, slope differs from 1.0, both differ (P<0.05).

[1] Andrew *et al.*, 1995. All, n=19; LLNP, n=6; PREP, n=6.

[2] Maslanka *et al.*, 1996. All, n=68; LLNP, n=12; PREP, n=22.

[3] EBH, empty body water; EBP, empty body protein; EBF, empty body fat.

Discussion

Results indicate that the overall prediction equations do not always provide accurate estimates of empty body composition of sub-groups within the overall data set. Inaccurate estimates occur when composition of the sub-group differs substantially from the overall group used to develop the equation or ratio. This helps explain why a set of equations derived from one study does not always provide accurate composition estimates of animals from another study (Andrew *et al.*, 1995; Brown *et al.*, 1989; Maslanka *et al.*, 1996). Over and under estimation will increase as the difference between composition of the animals used to develop the equations and that of the animals being predicted increases.

The dramatic alterations that can occur in body composition of lactating dairy cows makes it unlikely that a single set of equations will provide sufficiently accurate estimates of body composition for cows at different stages of lactation. In addition, results suggest treatment induced differences in empty body composition of animals may not be accurately estimated when the same prediction equations are applied to treated and untreated animals.

References

American Association of Analytical Chemistry. (1984). Official Methods of Analysis, 14th ed., AOAC, Washington, DC.

Andrew S.M., Erdman R.A. & Waldo D.R. (1995). Prediction of body composition of dairy cows at three physiological stages from deuterium oxide and urea dilution. *J. Dairy Sci.* 78:1083-1095.

Andrew S.M., Waldo D.R. & Erdman R.A. (1994). Direct analysis of body composition of dairy cows at three physiological stages. *J. Dairy Sci.* 77:3022-3033.

Brown D.L., Taylor S.J., De Peters E.J. & Baldwin R.L. (1989). Influence of Sometribove, USAN (Recombinant Methionyl Bovine Somatotropin) on body composition of lactating cattle. *J. Nutr.* 119:633-638.

Byers F.M. (1979). Measurement of protein and fat accretion in growing beef cattle though isotope dilution procedures. Ohio Beef Cattle Research Program Report, 79-1:36-47.

Maslanka M.T., Weber W.J., Crooker B.A., Johnson D.J. & Stanisiewski E.J. (1996). Body composition of multiparous lactating and non-lactating Holstein cows: development and use of prediction equations based on deuterium oxide. *J. Dairy Sci.* 79: 170.

National Research Council. (1989). Nutrient Requirements of Dairy Cattle. 6th rev. ed. Natl. Acad. Sci., Washington, DC.

SAS User's Guide: Statistics, Version 7 Edition. (1994). SAS Inst., Cary, NC.

Shipley R.A. & Clark R.E. (1972). Tracer methods for *in vivo* kinetics - theory and applications. Academic Press, New York, NY.

Weber W.J., Maslanka M.T., Crooker B.A., Johnson D.J., Otterby D.E. & Stanisiewski E.P. (1996). Composition of Holstein cows after receiving bovine somatotropin (Somavubove) during lactation. *J. Dairy Sci.* 79:169.

ESTIMATE OF THE POSTABSORPTIVE PROTEIN AND FAT METABOLISM IN GROWING PIGS AFTER LONG-TERM FEEDING ON DIETS WITH SOY PROTEIN ISOLATE OR CASEIN USING A ^{13}C,^{15}N-AMINO ACID APPROACH

P. Junghans, M. Beyer, E. Saggau, M. Derno, W. Jentsch, J. Voigt and U. Hennig

Research Institute for the Biology of Farm Animals, Department of Nutritional Physiology 'Oskar Kellner', Dummerstorf-Rostock, Germany

Summary

In order to investigate the postabsorptive carbon and nitrogen metabolism in growing pigs after 4-week feeding on diets with soy protein isolate (group A, 36±2 kg body mass, n=5) or casein (group B, 38±2 kg, n=5), [1-^{13}C]glycine and [^{15}N]glycine were simultaneously given. The stable isotope study which started after 16 h fasting was combined with indirect calorimetry. The dietary supply of both groups was isoenergetic and isonitrogenous. The protein intake was adjusted to 50% of the protein requirement. At the same dietary carbon and nitrogen intake rates of about 38 mol C/d and 1 mol N/d, respectively, for groups A and B, the corresponding excretion rates were found to be equal for breath carbon dioxide (about 19 mol C/d) in both groups, but different for urinary nitrogen (group A:0.42 mol N/d and group B: 0.20 mol N/d, $P<0.05$). The cumulative isotope excretions (% of tracer dose) of ^{13}C in breath CO_2 and ^{15}N in urinary total nitrogen were significantly increased ($P<0.01$) in group A (22.6 and 17.3, respectively) as compared with those of group B (11.9 and 4.7, respectively). Flux rates of carbon and nitrogen for group B (158.7 and 4.1 mol/d, respectively) were about twice those for Group A (83.5 ($P<0.01$) and 2.4 ($P<0.05$) mol/d, respectively). Heat production, measured by indirect calorimetry, was found to be equal for both groups (about 1.14 MJ/kg$^{0.62}$*d). Protein retention was significantly decreased in group A (0.49 mol N/d) in comparison to group B (0.77 mol N/d). Fat retention was not significantly different between group A (15.4 molC/d) and group B (16.1 molC/d). The results indicate that long-term feeding on different protein diets such as soy protein isolate or casein show an after-effect on the metabolism during the postabsorptive state. We propose to call the capability of the organism to keep its nutritional history in memory as 'metabolic hysteresis'.

Key words: Stable isotopes, ^{13}C, ^{15}N, indirect calorimetry, protein, fat, metabolism, soy protein, pig, metabolic hysteresis.

Introduction

Indirect calorimetry is a highly sophisticated method for quantifying the balance of nutrients and energy, but it is not able to assess dynamic processes such as protein and fat metabolism. For this purpose isotope techniques, in combination with indirect calorimetry, are increasingly applied (Schutz, 1997). A major number of publications refer to the acute dietary impact on metabolic parameters like flux and synthesis rates (Nurlan & Garlick, 1989). However, much less research has been undertaken to examine the presumed capability of the organism to keep its nutritional history in memory (Millward, 1995). Therefore, this study aimed to investigate the effect of long-term feeding on different diets with soy protein isolate or casein on nitrogen and carbon metabolism in growing pigs during the fasting state. The method of choice was a doubly labelled (^{15}N, ^{13}C) amino acid approach (Junghans *et al.*, 1992) to differentiate nitrogen and carbon metabolism which are closely related with protein and fat metabolism.

Materials and Methods

Stable isotope studies were performed in 10 growing male pigs. Five animals each were fed diets containing soy protein isolate (group A: 36±2 kg body mass) or casein (group B: 38±2 kg). The dietary supply of both groups was isoenergetic (about 2.0 MJ ME/($kg^{0.62}$ * d)) and isonitrogenous (about 1.5 g N/($kg^{0.62}$ * d)). The protein intake was adjusted to 50% of the protein requirement. The different diets were given during four weeks before the tracer experiment was started. After 16 h of fasting [1-^{13}C]glycine (45 µmol/kg, 99 atom%) and [^{15}N]glycine (23 µmol/kg, 99 atom%) were simultaneously administered as a single pulse into the vena jugularis. Thereafter, breath and fractionated urine samples were collected for 2 and 6 days, respectively. Breath $^{13}CO_2$ and total urinary ^{15}N were measured by means of isotope ratio mass spectrometry. Animals were fed for 8 h after tracer administration. During the tracer experiment the indirect calorimetry was carried out.

In order to quantify the metabolic processes we used the compartmental model (Fig. 1), proposed by San Pietro & Rittenberg (1953) for nitrogen metabolism:

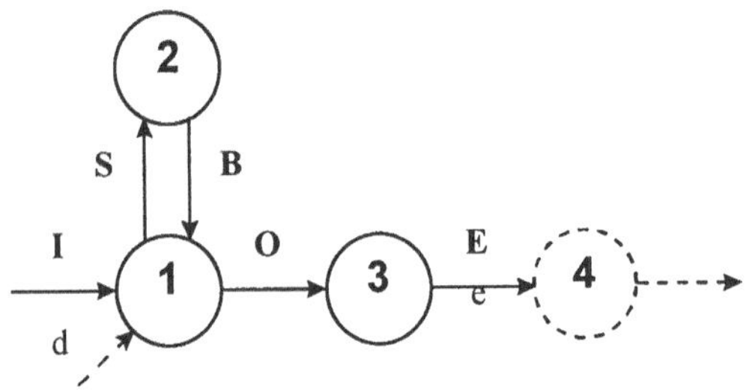

Fig. 1. Model of carbon and nitrogen metabolism. 1: rapid turn-over amino acid-C or N ('metabolic pool'), 2: slow turn-over C or N pool, 3: body bicarbonate-C or urea-N pool; 4: CO_2 of the respiration chamber (important for the ^{13}C kinetics), I, S, B, O and E represent intake, synthesis, breakdown, oxidation and elimination rates (mol/d) of C or N, respectively; flux Q of C or N is defined by Q = S + E = B + I, where E = O, e is the cumulative ^{13}C or ^{15}N elimination (mol) in breath CO_2 or urinary nitrogen in the isotope steady state, d is the ^{13}C or ^{15}N dose (mol).

The calculation of flux (Q) and synthesis (S) rates of carbon and nitrogen from the excretion of elements and the isotopes, ^{13}C and ^{15}N, respectively, was based on the following equations:

$$Q = E / (e/d) \quad (1)$$

$$S = Q - E \quad (2)$$

Results

At the same dietary carbon and nitrogen intake rates of about 38 mol C/d and 1 mol N/d, respectively, for the groups A and B, the corresponding excretion rates were found to be equal for breath carbon dioxide (about 19 mol C/d) in both groups, but different for urinary nitrogen (group A: 0.42 mol N/d and group B: 0.2 mol N/d), as presented in Table 1. The ^{13}C and ^{15}N kinetics, which have been selected from a representative experiment, are shown in Figure 2. The cumulative isotope excretions (% of tracer dose) of ^{13}C of breath CO_2 and ^{15}N of urinary total nitrogen were significantly increased in group A (22.6 and 17.3, respectively) as compared with those of group B (11.9 and 4.7, respectively), as seen in Table 1 and Figure 2. Flux and synthesis rates of carbon and nitrogen, calculated by means of equations (1) and (2), prove to be about twice for group B as compared with the respective values of group A, as listed in Table 1. Heat productions were found to be equal for both groups (about 1.14 MJ/$kg^{0.62}$*d)) which corresponds with equal carbon dioxide productions, reported above. Protein retention was significantly decreased in group A (0.49 molN/d) in comparison to group B (0.77 mol N/d). Fat retention was not significantly different between group A (15.4 molC/d) and group B (16.1 molC/d).

Table 1. Postabsorptive carbon and nitrogen metabolism in growing male pigs after four week feeding on diets with soy protein isolate or casein.

	Carbon metabolism				Nitrogen metabolism			
	Group A		Group B		Group A		Group B	
	Mean	SD	Mean	SD	Mean	SD	Mean	SD
Dietary intake I (mol/d)	37.8	1.0	39.5	0.9	0.96	0.04	0.99	0.03
Element excretion (E) in								
breath (mol/d)	18.7	2.7	18.8	2.0				
urine (mol/d)					0.42*	0.08	0.20	0.09
Isotope excretion (e/d) in								
breath (^{13}C) (% of dose)	22.6**	4.7	11.9	1.3				
urine (^{15}N) (% of dose)					17.3**	2.5	4.7	1.4
Flux rate Q (mol/d)	83.5**	5.8	158.7	19.2	2.4*	0.5	4.1	0.9
Synthesis rate S (mol/d)	64.8**	8.5	139.9	18.3	2.0*	0.5	3.9	0.9

Mean values for group A (soy protein isolate, n=5) were significantly different from those for group B (casein, n=5), * P<0.05, ** P<0.01 (unpaired t test).

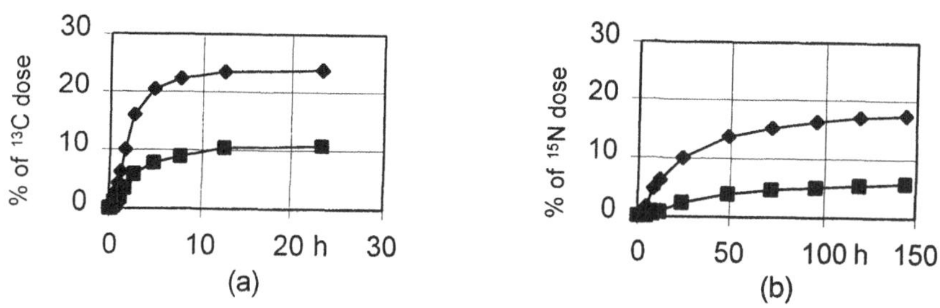

Fig. 2. (a) Cumulative ^{13}C amounts (in % of tracer dose) exhaled in breath CO_2 in dependence on the time after administration of [1-^{13}C]glycine as a single pulse into the vena jugularis; (b) Cumulative ^{15}N amounts (in % of tracer dose) excreted in urinary nitrogen in dependence on the time after administration of [^{15}N]glycine as a single pulse into the vena jugularis. Soy protein isolate (◆), casein (■).

Discussion

Major amounts of ^{13}C and ^{15}N from glycine (more than 75% of the tracer doses) were retained in the body. The high retention of ^{13}C and ^{15}N results probably from the low protein supply. Differences of the ^{13}C and ^{15}N elimination can be explained by their different pathways, which are illustrated by a three-compartmental model (Fig. 1). The ^{13}C or ^{15}N amino acid was administered into pool 1 ('metabolic pool'), which is turning over rapidly. The ^{13}C and ^{15}N label can leave pool 1 due to incorporation into slowly turning over body constituents or due to oxidation. Thus, ^{13}C from glycine was incorporated mainly into body protein, whereas a smaller part of ^{13}C was incorporated into fat and carbohydrates (pool 2). ^{15}N is only stored in body protein (pool 2). The other pathway for the elimination of ^{13}C and ^{15}N label from pool 1 is the oxidation of amino acids. The oxidation products enter pool 3 ('delay pool') before the irreversible loss of ^{13}C and ^{15}N in breath CO_2 or urine takes place. In the case of ^{13}C elimination there is an additional delay caused by the respiration chamber (pool 4), from which breath samples were collected. Results show a higher ^{13}C elimination as compared with ^{15}N for group B, whereas the difference for group A is not statistically significant, as shown in Table 1. The explanation for these findings could be, on the one hand, that ^{15}N, in comparison with ^{13}C, is more intensively transferred from glycine to other amino acids (transamination, reductive amination) and, thereafter, is incorporated into proteins (Krawielitzki *et al.*, 1996). However, glycine is reported to be an amino acid that transaminates poorly (Golden & Jackson, 1980). On the other hand, the ^{13}C label from the carboxyl position is known to be incorporated less into other metabolites like amino acids, fatty acids, glycerol etc. than that from other positions of the carbon skeleton (Elia, 1991).

The doubly labelled amino acid approach (Junghans *et al.*, 1992) has proved to be suitable to estimate the dynamics of nitrogen and carbon metabolism which are related to protein and fat metabolism. The approach needs to be validated by varying the tracer amino acid and the position of ^{13}C labelling.

An important question of our study was to examine diet-induced effects on the carbon and nitrogen metabolism during the fasting state. The results indicate clearly that long-term feeding on different protein diets such as soy protein isolate or casein affect the metabolism during the postabsorptive state. This finding corresponds with the observation, that the concentrations of plasma amino acids, especially the branched-chain amino acids, are increased in pigs fed soy protein diet as compared with casein-fed ones about 24 hours after the last meal (Hagemeister *et al.*, 1990). The mechanisms which enable the organism to keep its nutritional history in memory are still understood poorly. The diet-induced processes obviously go on although the acute dietary influence has already stopped. We propose to call this behaviour of the organism observed as 'metabolic hysteresis'.

References

Elia M. (1991). Estimation of short-term energy expenditure by the labelled bicarbonate method. In: R.G. Whitehead and A. Prentice (Eds) New techniques in nutritional research. Academic Press, New York, pp. 207-227.

Golden M.H. N. & Jackson A.A. (1980). Assumptions and errors in the use of ^{15}N excretion data to estimate whole body protein turnover. In: J.C. Waterlow and J.M.L. Stephen (Eds) Nitrogen metabolism in man. Applied Science Publishers, London. pp. 323-344.

Hagemeister H., Scholz-Ahrens K.E., Schulte-Coerne H. & Barth C. (1990). Plasma amino acids and cholesterol following consumption of dietary casein or soy protein in minipigs. *J. Nutr.* 120:1305-1311.

Junghans P., Hotz G., Strauch G., Hubald J., Faust H. & Häcker R. (1992). Assessment of the energy metabolism in man using a new stable ^{13}C,^{15}N isotope method. 14th Congress of the European Society of Parenteral and Enteral Nutrition, Vienna, August 7-9, 1992. In: *Clin. Nutr.* 11[Suppl.]:77-78.

Krawielitzki K., Kreienbring F., Schadereit R., Gabel M. & Schreurs V.V.A.M. (1996). Estimation of protein and amino acid metabolism using [^{15}N]- and [^{14}C] labelled tracer amino acids. *J. Anim. Feed Sci.* 5:187-199.

Millward D.J. (1995). A protein-stat mechanism for regulation of growth and maintenance of the lean body mass. *Nutr. Res. Rev.* 8:93-120.

Nurlan M.A. & Garlick P.J. (1989). Influence of nutrient intake on protein turnover. *Diabetes Metab. Rev.* 5:165-189.

San Pietro A. & Rittenberg D. (1953). Study of rate of protein synthesis in humans. Measurement of metabolic pool and rate of protein synthesis. *Biol. Chem.* 201:457-473.

Schutz Y. (1997). On problems of calculating energy expenditure and substrate utilization from respiratory exchange data. 1st International Rostock Workshop on Energy and Substrate Utilization, Rostock, September 5-7, 1996 (in press).

THE APPLICATION OF THE LABELLED ^{13}C-BICARBONATE METHOD FOR ESTIMATING ENERGY EXPENDITURE IN GOATS

The late C. Prieto, M. Lachica, F.F. Bermúdez and J.F. Aguilera

Animal Nutrition Department, Estación Experimental del Zaidín (CSIC), Profesor Albareda, 1. 18008 Granada, Spain

Summary

The suitability of the carbon dioxide entry rate (CER) technique using ^{13}C-bicarbonate to estimate CO_2 production and subsequently the energy expenditure of Granadina goats performing a variable physical activity (standing at rest or on an intermittent locomotion protocol) was investigated. The goats were prepared with catheters into the jugular vein and the parotid salivary duct. They were individually placed into an open-circuit respiration chamber and continuously infused for at least 15 h before starting saliva collection. Natural abundance of ^{13}C (AT^{13}C, %) measured in saliva samples decreased significantly ($P<0.05$) on increasing exercise load. Based on the CO_2 production: CER values, apparent mean recoveries of $^{13}CO_2$ between the resting and the intermittent locomotion conditions were not significantly different at 0.723±0.0437 and 0.779±0.0409, respectively. The absolute entry rate was slightly affected by the goats' physical activity. It is concluded that appropiate corrections are needed to allow accurate application of this stable isotope technique under different conditions.

Key words: Carbon dioxide entry rate, NaH^{13}CO$_3$, physical activity, goat, energy expenditure.

Introduction

The measurement of the apparent entry rate of CO_2 into the body pools using $^{13}CO_3H^-$ as a tracer has been proposed to estimate CO_2 production and subsequently to asses total energy expenditure of animals in free living conditions. The so-called carbon dioxide entry rate (CER) technique requires a constant infusion of labelled bicarbonate and frequent collection of some body fluids to determine the dilution of the isotope at equilibrium. There is comparatively little information on the influence of physical activity in the sequestration of labelled carbon. Available data only refer to human subjects and dogs (Wolfe, 1984). In the current study, the effect on goats of a change in exercise load on the apparent recovery of $^{13}CO_2$ has been addressed with the objective of contributing to validation of this technique for application to the unrestrained animal. Previous studies performed in our laboratory with Granadina goats fed at maintenance level and subjected to continuous infusion of NaH^{13}CO$_3$ were able to demonstrate that the time required to reach isotopic equilibrium was 8-10 h (Prieto *et al.*, 1991; Lachica, 1993).

Materials and Methods

Six adult female Granadina goats, of 50-62 kg initial body weight, were used. Throughout two experiments they were given, at about maintenance level, lucerne (*Medicago sativa*, 7.58 MJ·ME/kg DM) hay in a single meal 18 h before the start of measurements. Two catheters were surgically implanted in the goats: the first for saliva sampling was located into *Ductus parotideus* (3-4 cm) and exited the mouth via a perforation in the cheek; the second in the right jugular vein was used for infusion of NaH^{13}CO$_3$. Saliva was withdrawn by a peristaltic pump at a rate of 5 ml/h and poured into a sealed vacutainer provided with a valve containing soda lime to prevent contact with carbon dioxide in air. In Experiment 1 the animals were placed in metabolic cages. Several background samples of saliva were collected over 2 h intervals in the second and third day. Then they were moved to a treadmill and an intermittent locomotion programme was assayed. The goats walked continuously at the level for 30 min at 20 m/min and the treadmill was stationary during the preceding and subsequent 45 min. This schedule was repeated several times. Saliva samples were collected during each of these sets.

In Experiment 2 the goats were placed on the treadmill in a respiration chamber and were infused by a peristaltic pump at a constant rate of 15.7±0.29 ml/h with $NaH^{13}CO_3$ solution 15 h before starting saliva sampling. Several saliva samples were taken, each one being collected over 2 h intervals from each animal while standing at rest or performing the intermittent locomotion program described above. Expired CO_2 was recorded simultaneously. The schedule of measurements was that described for Experiment 1. The average concentration of $NaH^{13}CO_3$ in the infusates was 6.78±0.181 µM/ml.

The concentration of CO_2 in saliva samples was determined after precipitation as $BaCO_3$. The $Ba^{13}CO_3$ powder enrichment was determined by isotope ratio mass spectrometry (SIRA 12; VG Isogas, Middlewich, Cheshire) as described by McGaw *et al.* (1988). Fractional isotope enrichment over baseline was calculated according to Boutton (1991).

The experimental data were subjected to ANOVA. The Bonferroni test was used to ascertain the statistical significance of differences in natural abundance or in recovery of ^{13}C according to the level of physical activity.

Results and Discussion

Differences in natural abundance, expressed as $AT^{13}C$ (%), were significant ($P<0.05$), with average values of $1.09656 \pm 5.2*10^{-4}$ (n=25) and $1.09363 \pm 2.7*10^{-4}$ (n=24), for the standing at rest and the intermittent exercise conditions, respectively. The explanation of such a change in natural abundance remains unclear. Theoretically it may result from a change in the relative contribution of endogenous substrates to supply energy for an enhanced physical activity. Natural ^{13}C enrichment is known to be higher in glucose than in lipids. Wolfe (1984) reported that in fasted dogs a mild exercise led to an increase in the enrichment of the expired CO_2 from a mean of 0.0032±0.00004 APE (vs. standard) to 0.0058±0.00007 associated with a shift in RQ from 0.74±0.02 to 0.97±0.05. Exercise caused a marked increase in carbohydrate oxidation. However, it is doubtful that in our study the moderate increase in energy expenditure, produced by the mild exercise due to the intermittent locomotion imposed, could give rise to substantial changes in the mixture of energy yielding substrates being oxidized. In a previous study (Lachica *et al.*, 1997) we failed to find substantial differences in RQ values of goats standing still (0.89±0.016; n=52) or walking on a treadmill at 10 m/min (0.91±0.009; n=89), 20 m/min (0.85±0.010; n=86) or 30 m/min (0.84±0.012; n=87). Only a small tendency to increase O_2 consumption relative to CO_2 production on increasing exercise load was noticed, a fact which might indicate an increased contribution of lipid oxidation to support energy demand. The observed change in background enrichment with physical activity would demonstrate the need for determining natural abundance of ^{13}C in each physiological situation.

In Experiment 2, saliva sampling and measurements of CO_2 in expired air took place after at least 15 h of ^{13}C-bicarbonate infusion and therefore sufficient time was allowed for the label to recycle even through the slow-turnover pools (Elia *et al.*, 1988; Rocha *et al.*, 1994). Changes in total volume of CO_2 produced due to the exercise imposed involved opposite changes in ^{13}C enrichment (Table 1). Values of CO_2 entry rate obtained from enrichments of saliva exceeded the observed values for CO_2 production determined by respirometry (Table 1), indicating that sequestration of labelled carbon into body metabolites took place, therefore preventing quantitative excretion. Apparent mean recoveries of $^{13}CO_2$, calculated as values of the ratio CO_2 production: CER, were 0.723±0.0437 and 0.779±0.0409 for the standing at rest and the intermittent locomotion conditions, respectively. Therefore, due to ^{13}C sequestration, the CER technique overestimated CO_2 production by 38.3 and 28.4%, according to the physical activity of the goats. The recovery of labelled carbon has been reported to vary widely. Available data for sheep are in the range of 0.67-0.93 (White & Leng, 1968; Corbett *et al.*, 1971; Engels *et al.*, 1976; Prieto *et al.*, 1982; Rocha *et al.*, 1994) and the reasons for the differences observed between trials are unknown. In our study a lower recovery of ^{13}C was obtained when the goats were standing at rest than while they were performing the intermittent locomotion protocol. It is likely that the dilution of ^{13}C as a result of the moderate increase in CO_2 production due to the mild exercise tends to reduce the fixation of the labelled isotope to pools which turn over slowly, thus favouring its free exchange to sampling sites. Our results are in line with data reported by Wolfe (1984) who found, in human subjects, that light exercise (requiring only 30% of their maximum volume of O_2) increased ^{13}C recovery from 78 to 98% of the infusion rate.

The estimated CO_2 production (CO_{2r}) was related to energy expenditure (heat production, HP) by the regression equation:

$$\text{HP (J/kg BW/h)} = 24.07\ CO_{2r}\ \text{(ml/kg BW/h)}$$
$$(\pm 0.128)$$

obtained in our laboratory (Lachica, 1993) in goats standing at rest and walking on a treadmill belt at three speeds (10, 20 and 30 m/min) and five slopes (-10, -5, 0, 5 and 10%) in the same experimental conditions (type of diet, level of intake, laboratory conditions, etc.). In these trials the energy expenditure of the goats was calculated according to Brouwer (1965) except that no allowance was made for urinary excretion or for methane. The average RQ was 0.87±0.0006 (n=314). This equation is similar to those developed by Whitelaw *et al.* (1972), Oliveira (1987) and Sahlu *et al.* (1988). In the present study the average volume of CO_2 production estimated from CER values was 321.4 and 350.2 ml/kg BW/h for the resting and exercised animal, respectively (Table 1). Consequently, HP was estimated to be 7736 and 8429 J/kg BW/h, respectively. These values deviate 11.1 and 4.7%, respectively, from the energy expenditure which can be calculated from experimental data obtained by Lachica *et al.* (1997) in goats standing at rest (6960 J/kg BW/h) and walking on the level at 20 m/min (11329 J/kg BW/h).

Table 1. Natural abundance of ^{13}C ($AT^{13}C$, %), atoms per excess (APE), mean recovery of $^{13}CO_2$, CER values, expired CO_2 and heat production (HP) for goats standing at rest or on intermittent locomotion.

	Exercise level			
	Standing at rest	n	Intermittent locomotion	n
$AT^{13}C$ (%)	$1.09656 \pm 5.2 \cdot 10^{-4\,a}$	25	$1.09363 \pm 2.7 \cdot 10^{-4\,b}$	24
APE	0.0107±0.00049	13	0.0101±0.00049	14
Recovery	0.723±0.0437	13	0.779±0.0409	14
CER (CO_2, ml/kg BW/h)	444.6±22.40	13	449.6±27.67	14
$CO_{2resp.}$ (ml/kg BW/h)	314.9±15.05	13	338.8±12.83	14
HP (J·kg BW/h) *	7736		8429	

[a,b]Within the same row, values bearing different superscripts are significantly different ($P<0.05$).

*Calculated from: HP=24.07 CO_{2r} (Lachica, 1993); CO_{2r}= Apparent recovery of $^{13}CO_2$ CER value.

In conclusion, the data presented in this paper show that any change in the physical activity of the animal can alter the natural abundance of ^{13}C in sampling sites. Also sequestration of the labelled isotope may be quantitatively affected. These facts underline the need to determine appropriate corrections to allow accurate application of this stable isotope technique under different conditions.

References

Boutton T.W. (1991). Tracer studies with ^{13}C -enriched substrates: Humans and large animals. In: D.C. Coleman & B. Fry (Eds) Carbon isotope techniques. Academic Press, London. pp. 219-242.

Brouwer E. (1965). Report of Sub-committee on Constants and Factors. In: K.L. Blaxter (Ed) Energy metabolism of farm animals. Academic Press, London. pp. 441-443.

Corbett J.L., Farrell D.J., Leng R.A., McClymont G.L. & Young B.A. (1971). Determination of the energy expenditure of penned and grazing sheep from estimates of carbon dioxide entry rate. *Br. J. Nutr.* 26:277-291.

Elia M., Fuller N. & Murgatroyd P. (1988). The potential use of the labelled bicarbonate method for estimating energy expenditure in man. *Proc. Nutr. Soc.* 47:247-258.

Elia M., Fuller N.J. & Murgatroyd P.R. (1992). Measurement of bicarbonate turnover in humans: applicability to estimation of energy expenditure. *Am. J. Physiol.* 263:E676-E687.

Engels E.A.N., Inskip M.W. & Corbett J.L. (1976). Effect of change in respiratory quotient on the relationship between carbon dioxide entry rate in sheep and their energy expenditure. In: M. Vermorel (Ed) Energy metabolism of farm animals. INRA, Theix. pp. 339-342.

Lachica M. (1993). Estimación del gasto energético de la actividad física en ganado caprino. Comparación de técnicas calorimétricas y no calorimétricas. PhD Thesis. Univ. of Granada. 222 pp.

Lachica M., the late Prieto C. & Aguilera J.F. (1997). Estimation of the energy costs of walking on the level and on negative and positive slopes in the Granadina goat (Capra hircus). *Br. J. Nutr.* 77:73-81.

McGaw B.A., Milne E. & Duncan G.J. (1988). A rapid method for the preparation of combustion samples for stable carbon isotope analysis by isotope ratio mass spectrometry. *Biomed. Environ. Mass Spectrometry* 16:269-273.

Oliveira E.R. (1987). Maintenance energy requirement of free-ranging goats and sheep. Ph.D. Dissertation. Utah State University, Logan, UT. 115 pp.

Prieto C., MacRae J.C., Brockway J.M. & Lobley G.E. (1982). A comparison of different metabolic pools which can be sampled for bicarbonate activity in entry rate or substrate oxidation studies. In: A. Ekern and F. Sundstol (Eds) Energy metabolism of farm animals. *EAAP Publ. No.* 29:70-73.

Prieto C., Reyes M., Lachica M., Caballero E. & Aguilera J.F. (1991). Applications of the labelled ^{13}C -bicarbonate method for estimating energy expenditure of the goat under field conditions. In: International Symposium on Nuclear and Related Techniques in Animal Production and Health, International Atomic Energy Agency. FAO, Vienna. p. 150.

Rocha H.J.G., Connell A. & Lobley G.E. (1994). Whole body and trans-organ bicarbonate kinetics in sheep: effects of nutritional status. In: J.F. Aguilera (Ed) Energy metabolism of farm animals. *EAAP Publ. No.* 76:27-30.

Sahlu T., Jung H.G., Nienaber J.A. & Morris J.G. (1988). Development and validation of a prediction equation estimating heat production by carbon dioxide entry rate technique. *J. Anim. Sci.* 66:2036-2043.

White R.G. & Leng R.A. (1968). Carbon dioxide entry rate as an index of energy expenditure in lambs. *Proc. Aust. Soc.* Anim. Prod. 7: 335-341.

Whitelaw F.G., Brockway J.M. & Reid R.S. (1972). Measurement of carbon dioxide production in sheep by isotope dilution. *Q. J. Exp. Physiol.* 57:37-55.

Wolfe R.R. (1984). Tracers in metabolic research. Radioisotope and stable isotope/Mass spectrometry methods. Alan R. Liss, New York. 287 pp.

THE USE OF SULPHUR HEXAFLUORIDE FOR MEASURING METHANE PRODUCTION BY CATTLE

K.A. Johnson[1], H.H. Westberg[2], B.K. Lamb[2] and R.L. Kincaid[1]

[1]Department of Animal Sciences, [2]Department of Civil and Environmental Engineering, Washington State University, Pullman, WA, USA

Summary

Measurements of methane (CH_4) emissions were made from individual cows in a barn, groups of cows in a room and groups of cows on pasture. In all cases sulphur hexafluoride (SF_6) was used as a tracer gas. Emissions measurements from individual animals were consistent from day-to-day if the amount of feed offered and time of feeding were similar. When cattle had *ad libitum* access to feed (in a feeder or on pasture forage) within-day variation was reduced. The use of a gas tracer to measure CH_4 emissions from cattle housed in a room or barn resulted in accurate, repeatable estimates. Measurement of CH_4 emissions from groups of cows on pasture was possible if the environmental conditions were appropriate. Factors such as animal behaviour, site selection, and wind conditions should be considered. There is less variability associated with CH_4 emissions from individual animals than those from groups of animals in a room or grazing a pasture. Tracer techniques provide many options for the measurement of gaseous emissions from animals.

Key words: Cattle, methane, sulphur hexafluoride, pasture.

Introduction

Measurements of gaseous emissions from livestock have generally relied upon calorimetry techniques. While these techniques are both accurate and reliable, the expense and need for animal training limit the numbers of animals that can be measured. In addition, feed intake and animal behaviour in the chambers do not always reflect that seen in production situations which can impact on the measurements.

The use of atmospheric tracers such as SF_6 may allow CH_4 emissions to be measured from individual animals (Johnson *et al.*, 1994). However, the use of the SF_6 tracer technique in individual animals is limited in the U. S. by regulations imposed by the Food and Drug Administration. These limitations include a long withdrawal time after the permeation tube has been removed or has completed its release of SF_6. In addition, milk from treated dairy cows must be discarded. Therefore, there is a need for accurate measurement techniques to measure gaseous emissions that do not compromise the animal's value. The use of gas tracers allows the ventilation rate of a room or the dilution rate of gaseous emissions from animals outside to be measured. Tracer gases are commonly used to measure ventilation rates of houses and barns as well as gaseous emissions from landfills, termite mounds, and natural gas pipelines (Lamb *et al.*, 1985; Lamb *et al.*, 1995; Czepiel *et al.*, 1996; Marik & Levin, 1996). The objective of this project was to examine the use of SF_6 to measure CH_4 emissions from animals in production settings.

Materials and Methods

Individual animals. The repeatability among CH_4 emissions measurements from individual cows was determined from several studies. The data included limit-fed beef cows and heifers, *ad libitum* fed dairy cows, beef steers fed in individual pens and beef cows and heifers grazing pasture. In these studies feeding occurred at the same time each day. Measurements of CH_4 emissions from the cows were made as described by Johnson *et al.* (1994). Permeation tubes containing SF_6 were prepared, the release rate calibrated gravimetrically and the tubes placed in the rumen using a bolus gun. Animals wore horse halters modified with a 50 micron filter and a 54 cm piece of capillary tubing. The capillary tubing ran from the filter to a 2 l collection canister which was placed around the cow's neck. Sampling occurred for 5 consecutive days and 3 periods/d: 07:00-13:00; 13:00-19:00; and 19:00 to 07:00. The data were analysed using the GLM procedure of SAS. The main effects of the model included effect of cow, period, treatment, day and time of day as main effects. The interaction of cow x period x treatment was used as the whole plot error term and the interaction of treatment x day was used to test the effect of day.

Room measurements. A room tracer approach was used to measure CH_4 emissions from dairy cows (n= 108) fed three different levels of dietary fat from oilseeds, 2.3, 4.0, and 5.6% of the diet (n=4/trt). A 6.1m x 8.8m room was used. One side was a half wall so plastic was suspended to close the upper portion. Another side possessed 3 windows, which remained open. Two box fans were suspended from the ceiling in opposite corners of the room. A tank of 1024 ppm SF_6 fitted with a flow meter and controller was placed outside the room near one of the open windows. The release line entered the window and was dispersed in front of one of the fans. Five evacuated air sampling canisters fitted with 22.5 cm of capillary tubing were suspended in the room and two were placed outside to determine background concentrations of CH_4 and SF_6. A preliminary study was conducted to determine the release rate needed to reach a steady state. A release rate of 50.2 μg SF_6/min was determined to be adequate and used in all subsequent measurements. At least 4 h elapsed between initiation of the tracer release and measurements to insure SF_6 and CH_4 the room were at steady state. The duration of sample collection was 4 to 5 h. Cows had *ad libitum* access to food and water and were free to move around the room.

Pasture measurements. A 56 m^2 area of pasture was created using flexible electric fencing. The area contained a water source and enough pasture to feed the twelve cow-calf pairs. Four or five canisters of pure SF_6 fitted with calibrated flow meters were attached to fence posts on the upwind side of the pasture. Evacuated 2 l canisters fitted with 9-cm long restriction tubes were placed in two rows on the downwind side of the enclosure. A sampling canister used to quantify background CH_4 and SF_6 was placed well upwind of the tracer release canisters. One person dealt exclusively with the SF_6 release canisters and another person with the sampling canisters to prevent contamination of the air samples. Release of SF_6 (6 g/min) from the canisters began after the cows were moved into the sampling area. Following a 10-min equilibration period, collection of air by the evacuated canisters began. After a 2-h measurement period, the tracer release and sampling canisters were closed. To determine the repeatability of the measurements made with this method, collections taken from cows grazing in the same area at the same time were repeated 5 times. Wind direction was monitored for the duration of the measurement period. If the plume of CH_4 and SF_6 moved because of changes in wind direction, sampling canisters were relocated within the sampling plume.

Methane and SF_6 measurements and calculations. Methane concentrations were measured using a HP 5880 gas chromatograph fitted with a 1.0 ml gas sampling loop and flame ionisation detector. The 0.25cm x 1.22m Porapak N column was maintained at 50°C. Sulphur hexafluoride concentrations were determined using a HP 5880 gas chromatograph with a 1.0 cc gas sampling loop and electron capture detector. The molecular sieve 5A column (0.215 cm x 1.83 m) was maintained at 50°C. Methane emissions rate was calculated identically for each measurement method. The equation used was: $Q_{CH4} = Q_{SE6}$

x $[CH_4]$ / $[SF_6]$; where Q_{CH4} is the methane emission rate in g/h, Q_{SF6} the known release rate of the tracer gas source, and $[CH_4]$ and $[SF_6]$ are the measured concentrations of CH_4 and SF_6 in the sampling canister.

Results and Discussion

Individual animal measurements. We have not found a day of sampling effect associated with CH_4 emissions measurements using the SF_6 tracer technique. As expected, when cattle were fed once per day CH_4 emissions varied during the day, with the greatest emissions seen after feeding. When heifers were allowed to graze rapidly growing spring pasture and CH_4 emissions measured for 5 d, emissions were remarkably consistent among animals and across days. The 4 heifers produced 9.6, 9.1, 9.9, and 9.3 ± 0.39 g CH_4/h over the 5 d. On a daily basis emissions were 9.7, 9.2, 9.7, 9.3, and 9.6 ± 0.45 g CH_4/h. Emissions within a day tended to be different ($P<0.06$) with the highest emissions during the night, 9.0, 9.3, and 10.1 ± 0.33 g CH_4/h for morning, afternoon and night, respectively. The coefficient of variation (CV: SD/Mean) for daily methane production from these heifers was 12.5. Beef cows on pasture exhibited a similar response. Lactating dairy cows (n=4) also exhibited little variation between days although there was a greater difference between cows. In this case the CV was 9.3.

Room measurements. At two points during the lactation the same group was measured over several days to examine the repeatability of sampling. There were no differences associated with day of measurement. Initial measurements from the control group for two days were 21.1 g CH_4/cow/h and 20.5 g CH_4/cow/h. At the midpoint of the trial another measurement sequence took place and CH_4 emissions for the moderate fat group were 12.6, 13.6 and 15.8 g CH_4/cow/h. Marik & Levin (1996) recently published their results using SF_6 as a room tracer and found similar results in measurement repeatability and emissions of CH_4 from dairy cows.

In this study, there was no significant change in CH_4 production with the addition of dietary fat. The control (2.3% fat) cows emitted an average of 16.6 ± 1.2 g CH_4/cow/h (4.7% GEI) over the entire lactation and emissions declined from a high of 21.1 g CH_4 /cow/h to 12.6 g CH_4/cow/h as dry matter intake declined. Cows fed the moderate level of fat (4%) emitted 16.3 ± 1.5 g CH_4/cow/h and their emissions did not change over time. Dry matter intake of this group was greater ($P<0.05$) than the other treatments, 27.3 kg/cow/d versus 25.0 kg/cow/d and 26.9 ± 0.23 kg/cow/d for the control and high groups respectively. The high level of dietary fat (5.6%) resulted in the greatest CH_4 emissions, 20.5 ± 2.5 g CH_4/cow/h or 4.9% GEI. As with the control group, CH_4 production decreased linearly with increasing days in milk. These emission rates agree with those observed by Westberg *et al.* (1997) who found dairy cows fed a diet similar to the control group produced 17.6 g CH_4/cow/h. In addition, Westberg also noted no difference in hourly CH_4 production from dairy cows if they had *ad libitum* access to feed. Milk production was significantly greater for the moderate fat group than the others ($P<0.05$). The control, moderate and high group produced 32.4, 38.7, and 38.8 kg milk/d/cow, respectively. The CV for this measurement technique was 22.4.

Pasture measurements. Methane emissions measured from cattle grazing pasture using the tracer technique averaged 9.83 ± 1.7 g CH_4/cow/h. These values are well within the range of emissions observed by Westberg *et al.* (1997) for cows grazing similar pastures. During one period the wind ceased for the last hour of measurements. Under these conditions the tracer technique underestimated CH_4 production. Cows produced 4.1 g CH_4/cow/h which is 58% lower than other observations made with the same cows on the same pasture. Behavioural observations of the cow-calf pairs for this day indicated pairs behaved as they had before and during the other measurement periods. It is not surprising that the tracer technique did not work under these conditions because one of the requirements of the technique is mixing of the SF_6 and CH_4 plumes. For the tracer technique to be a good method of choice for grazing cattle several conditions must be met. The cows need to be tractable enough to be grouped in a relatively small area and should be accustomed to the presence of people so they will behave normally, graze, ruminate, etc. The area chosen should not have other large upwind sources of CH_4 or SF_6 emissions. The background air sample will allow for correction of ambient CH_4 concentrations from small sources. Wind direction should be monitored and the sampling cans moved if the plume switches directions. Finally the sampling cans should be far enough downwind to allow mixing of the SF_6 and CH_4 but not so far that the plume has dissipated.

References

Czepiel P., Mosher B., Shorter J., McManus B., Kolb C., Allwine E., Lamb B. & Hariss R. (1996). Landfill methane emissions measured by static enclosure and atmospheric tracer techniques. *J. Geophys. Res.* 101:16711.

Johnson K.A., Huyler M.T., Westberg H.H., Lamb B.K. & Zimmerman P. (1994). Measurement of methane emissions from ruminant livestock using a SF6 tracer technique. *Environ. Sci. Technol.* 28:359.

Lamb B., McManus J., Shorter J., Kolb C.C., Mosher B., Harriss R., Allwine E., Blaha D., Howard T., Guenther A., Lott R., Siverson R., Westberg H. & Zimmerman P. (1995). Development of atmospheric tracer methods to measure methane emissions from natural gas facilities and urban areas. *Environ. Sci. Technol.* 29:1468.

Marik T. & Levin I. (1996). A new tracer experiment to estimate methane emissions from a dairy cow shed using sulphur hexafluoride (SF_6). *Global Biogeochemical Cycles* 10:413.

Westberg H., Huyler M., Johnson K. & Lamb B. (1997). Methane emissions from U.S. ruminants (in preparation).

NUTRIENT METABOLISM AND CARCASS CHARACTERISTICS OF MALE THIN TAIL SHEEP RECEIVING CLENBUTEROL

I.K. Saka[1], D. Sastradipradja[2] and B. Kiranadi[2]

[1]Department of Animal Production, Faculty of Animal Husbandry, Udayana University, Denpasar, Indonesia; [2]Department of Physiology and Pharmacology, Faculty of Veterinary Medicine, Institut Pertanian Bogor (IPB-Bogor Agricultural University), Jalan Taman Kencana 1 no. 3, Bogor 16151, Indonesia

Summary

The effects of once every two days clenbuterol (CB) intramuscular injection at dose levels of 0, 5, 10 and 20 µg/kg live weight (W) on nutrient metabolism and carcass characteristics of growing male Javanese thin tail sheep (JTT) were studied for 64 days. Results showed no influence of CB on growth performance, metabolism of nutrients, carcass composition and characteristics, which may have been due to the relatively young age of weanling lambs used. There were no significant improvements in carcass weight (+4.7%) or carcass composition but a trend for increased protein content (+16.9%) and decreased fat (-18%) in the treated animals. By using subsampling of randomized block design and polynomial regression model, it was found that there were significant quadratic responses ($P<0.05$) to dose of CB (X) for average weights of 5 carcass cuts (I-Y) and 5 individual muscles (II-Y) of which the equations were respectively

(I) $Y = 1.36 + 0.05X - 0.002\,X^2$ ($P<0.001$; $R^2 = 0.68$; $S_{y.x} = 0.06$).
maximum 1.695 kg for CB 12.3 g/kg BW.

(II) $Y = 196.33 + 8.53X - 0.316X^2$ ($P<0.05$; $R^2 = 0.87$; $S_{y.x} = 10.45$).
maximum 253.8 g for CB 13.48 g/kg BW.

Key words: Javanese thin tail sheep, clenbuterol, growth, metabolism, carcass characteristics.

Introduction

β-adrenergic agonists (AA) alter nutrient metabolism, carcass characteristics and composition through their actions on accretion of protein and increase of lipolysis in many species of animals including sheep (Baker *et al.*, 1984; Kim *et al.*, 1989; Koohmaraie *et al.*, 1991). Clenbuterol (CB) increased daily fractional protein synthesis rates of the brachialis muscles by 3.3 percentage units and daily total protein synthesis in semitendinosus muscle by 0.62 percentage units in lambs fed with CB compared to control (Cleys *et al.*, 1989). Lambs fed 10 ppm CB increased protein percentage of the hindquarter by 2.1 percentage units and decreased fat by 5.7 percentage units (Baker *et al.*, 1984). An increase of 31.6% ($P<0.05$) in protein content of biceps femoris muscle was recorded in wether lambs fed with 4 ppm L644,969 compared to control animals (Koohmaraie *et al.*, 1991). JTT sheep is a lean, meat producing animal and the objective of this study was to evaluate the responses of extended multiple CB injection in growing JTT sheep.

Materials and Methods

Sixteen weaned JTT ram lambs, initial average weight 22.1 ± 4.82 kg, were fed on a pelleted diet in individual pens for 64 days. Throughout the study all lambs were given the same *ad libitum* access to a 18.0% CP and 17.2 MJ/kg pelleted diet. CB was intramuscularly injected every second day throughout the experimental period at dose levels of 0, 5, 10 and 20 µg/kg W for the treatment groups CB0, CB5, CB10 and CB20, respectively with 4 replicates for each group. For measurements of digestibility and balance of nutrients the lambs were kept in metabolism cages and subjected to total collection of excreta for 7 days. Nutrient contents of samples of feed and excreta were determined by standard proximate analysis. Balance trials were conducted at the end of the experiment. Numbers of replicates for these trials were 3, 4, 4 and 3 for CB0, CB5, CB10 and CB20, respectively.

All lambs were slaughtered after 64 days of the experimental period and data on carcass traits were collected. After ageing, the longissimus dorsi (LD), psoas (PS), semitendinosus (ST), biceps femoris (BF) and semimembranosus (SM) muscles were individually dissected from the right side of carcasses according to standard procedures. The left half of each carcass was cut into five wholesale cuts, each of which was weighed and afterwards they were dissected into muscle, fat and bone using the method proposed by Thompson *et al.* (1979). The dissected tissues were later mixed together and then finely ground and blended. The ground tissue was randomly sampled for chemical analysis. Non-carcass tissues were treated in a similar manner.

A one-way ANOVA and ANCOVA for a completely randomized design were used to analyse data according to procedures outlined by MINITAB release 6.1 (1988). Means comparisons were made within significant main effects using Duncan's multiple range test. Effects of CB dose levels on means of 5 cut weights and 5 individual muscles weights were partitioned using orthogonal polynomial procedures.

Results and Discussion

The final liveweight (FLW) of the lambs was 30.0 $\pm$ 4.33 kg and average daily gain (ADG) was 0.133 $\pm$ 0.017 kg. CB treatment had no effects on feed consumption, feed efficiency, FLW, ADG, carcass traits, retention of nutrients, carcass physical and chemical compositions, or body chemical composition (Table 1). CB-fed lambs however, tended to have higher hot carcass and dissected carcass lean weight, carcass length, fleshing index and ribeye area, and lower carcass fat weight and back fat thickness. Retained protein was maximal in treatment group CB10 in parallel with the average weight of 5 individual muscles. For average weight of carcass cuts, the values were respectively 29% ($P<0.05$) and 31.8% ($P<0.01$) above the control group. By using subsampling of randomized block design and polynomial regression model it was found (Figs 1 and 2) that there were significant quadratic responses to dose of CB (X) for average weight of 5 cuts (I-Y; $P<0.001$) and 5 individual muscles (II-Y; $P<0.05$) of which the equations were respectively:

(I) $Y = 1.36 + 0.05X - 0.002\,X^2$ ($P<0.001$; $R^2 = 0.68$; $S_{y.x} = 0.06$).
maximum 1.695 kg for CB 12.3 g/kg BW

(II) $Y = 196.33 + 8.53X - 0.316X^2$ ($P<0.05$; $R^2 = 0.87$; $S_{y.x} = 10.45$).
maximum 253.8 g for CB 13.48 g/kg BW.

The results of the present study were in agreement with those of Baker *et al.* (1984) and Kim *et al.* (1989) who found no effects of AA on ME, DM digestibility, FLW, ADG, feed consumption or feed efficiency. However, although there was no increase in hot carcass weight of CB-treated lambs, they reported a larger increase in ribeye area (33.1-33.8%). In contrast with the present study, Kim *et al.* (1989) observed a dramatic increment in N retention (+130%) which was associated with decrease in N loss (-30%) through urine with no effect on N digestibility in AA CIM-treated lambs. There were no improvements in carcass weight (+4.7%) or carcass composition but a trend for increased protein content (+16.9%) and decreased fat content (-18%) in treated animals. The results of the present study showed that CB injection to lambs resulted in no changes in weight of wholesale cuts. This is similar to the report of Shackelford *et al.* (1992). Lack of influence of CB on growth performance, metabolism of nutrients, carcass composition and traits in this study might have been due to the relatively young age of weaning lambs at the start of the experiment. Mersmann *et al.* (1987) proposed that this was due to differences in *in vivo* distribution, metabolism, excretion and pharmacodynamics of AA. It may also be due to low receptor number, low affinity of receptor to AA, poor receptor coupling to intracellular processes, down regulation in tissues or differences in metabolic regulation in responsive tissues between young and older animals. In addition, 64 days treatment may have been too long and the growth promoting effect might have disappeared. This is in agreement with the results of Kim *et al.* (1989) in chronic CIM-administered lambs. They suggested that the disappearance of the growth promoting effect of CIM at around the 6th week of the 13 week experimental period in CIM-fed lambs might be attributable to a decrease in the sensitivity of skeletal muscle to CIM due to a decrement in adrenoceptors density. Reduction of the number of adrenoceptor binding sites in the rat skeletal muscle of chronic

Table 1. Effects of clenbuterol on growth, carcass traits and nutrient balance (±SE) in JTT rams.

Trait*	Treatment group			
	CB0	CB5	CB10	CB20
Overall:				
Initial weight, kg	22.1 ± 2.9	22.1 ± 2.5	22.3 ± 2.5	22.5 ± 2.9
Final weight[a], kg	30.6 ± 2.0	29.8 ± 1.7	32.2 ± 1.7	30.8 ± 2.0
Gain[b], kg/d	0.14 ± 0.03	0.12 ± 0.03	0.14 ± 0.03	0.13 ± 0.03
DM intake[c], kg/d	0.76 ± 0.07	0.68 ± 0.06	0.74 ± 0.06	0.64 ± 0.07
Feed/gain, kg/kg	5.04 ± 1.39	7.97 ± 1.20	5.14 ± 1.20	6.52 ± 1.39
Carcass traits:				
Carcass weight, kg	14.1 ± 1.41	14.7 ± 1.22	16.4 ± 1.22	15.6 ± 1.41
Carcass length, cm	55.2 ± 1.7	55.6 ± 1.5	56.6 ± 1.5	56.7 ± 1.7
Ribeye area, cm^2	13.3 ± 1.8	13.5 ± 1.6	15.1 ± 1.6	15.1 ± 1.8
Fat depth (GR), mm	6.3 ± 1.1	5.6 ± 0.9	5.0 ± 0.9	5.0 ± 1.1
Carcass composition:				
Muscle, %	59.7 ± 1.7	61.5 ± 1.2	63.0 ± 1.5	62.9 ± 1.7
Fat, %	20.9 ± 1.7	21.3 ± 1.5	19.5 ± 1.5	14.8 ± 1.7
Bone, %	16.1 ± 0.9	15.4 ± 0.8	15.5 ± 0.8	14.8 ± 0.9
Digestibility & balance:				
GE, MJ/d	11.6 ± 1.34	10.2 ± 1.16	10.3 ± 1.16	11.6 ±1.34
DM digestibility, %	76.4 ± 0.58	75.0 ± 0.58	74.1 ± 0.58	75.5 ±0.70
DE[d], MJ/d	8.4 ± 1.05	7.7 ± 0.91	7.8 ± 0.91	9.0 ±1.05
ME , MJ/d	6.9 ± 0.95	6.5 ± 0.82	6.7 ± 0.82	7.7 ±1.47
FE[d], MJ/d	3.2 ±0.44	2.5 ± 0.38	2.5 ± 0.38	2.8 ±0.44
CP intake[d], g/d	120.2 ± 4.9	123.5 ± 4.9	116.1± 4.9	126.3 ± 5.9
CP digestibility, %	70.1 ± 1.74	72.1 ± 1.74	70.3 ± 1.74	70.0 ±2.13
Urinary N, g/d	7.14 ± 1.29	7.59 ± 1.29	7.28 ± 1.29	5.54 ±1.58
UE[d], MJ/d	0.25 ± 0.04	0.21 ± 0.04	0.22 ± 0.04	0.23 ±0.04
Methane, %GE	6.22 ±0.02	6.20 ± 0.02	6.21 ± 0.05	6.23± 0.02
HP[d], $MJ/kg^{0.75}/d$	0.68 ± 0.06	0.46 ± 0.05	0.44 ± 0.05	0.56± 0.05
Retention of nutrients[e]:				
RE, MJ/d	1.25 ± 0.43	1.25 ± 0.37	1.45 ± 0.37	1.24 ±0.43
Protein, g/d	20.0 ± 0.55	19.3 ± 0.55	20.8 ± 0.55	20.5 ±0.67
Fat, g/d	19.9 ± 1.56	21.9 ± 1.56	19.7 ± 1.56	18.1 ±1.91

* None of the traits was affected by treatment (P>0.05).

[a,b,c,d,e] Data adjusted with initial weight, length on pellet feeding, length of CB treatment, DM intake, and empty body weight gain, respectively as covariates.

β-AA isoproterenol-fed rats had been demonstrated in an earlier study by Vallieres *et al.* (1979) (cited by Kim *et al.*, 1989).

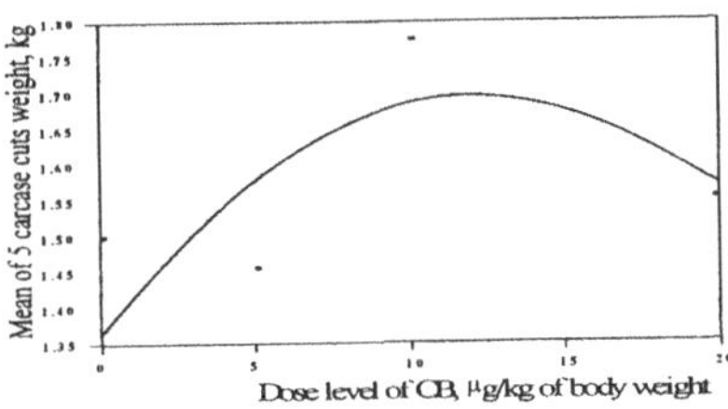

Fig. 1. CB dose-response curve for mean of 5 carcass cuts weights following the equation: $Y = 1.36 + 0.05X - 0.002X^2$ ($R^2 = 0.68$).

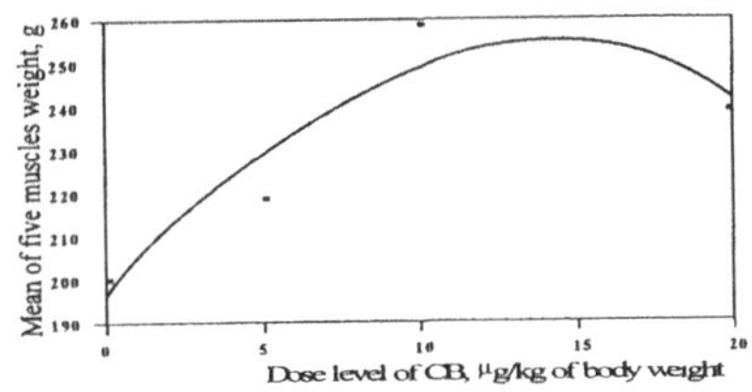

Fig. 2. CB dose-response curve for mean of five muscles weights following the equation: $Y = 196.33 + 8.53X - 0.316X^2$ ($R^2 = 0.87$).

Means of 5 wholesale cuts and 5 individual muscles however, indicated that CB significantly increased their weight with CB10 had the highest increases. CB10 significantly had heavier mean wholesale cut weight of 26.4% ($P<0.05$), 22.2% and 14.4% ($P>0.05$) than that of CB0, CB5 and B20, respectively. Mean of 5 individual muscle weight of CB10 was also the highest, i.e. 9%, 18.2% ($P<0.01$), and 8.4% ($P>0.05$) higher than that of CB0, CB5 and CB20, respectively. These increases were due to increases in protein deposition in skeletal muscles of the wholesale cuts or in the individual muscles as a result of reduction in protein degradation by μ-calpain which was inhibited by calpastatin. These increases in protein disposition led to muscle fiber hypertrophy (Koohmaraie *et al.*, 1991). In the present study however, there was no indication of muscle hypertrophy in the LD muscles of CB-injected lambs. The reasons for this disagreement were unknown.

References

Baker P.K., Dalrymple R.H., Ingle D.L. & Ricks C.A. (1984). Use of adrenergic agonist to alter muscle and fat deposition in lambs. *J. Anim. Sci.* 59:1256-1261.

Cleys M.C., Mulvaney D.R., McCarthy F.D., Gore M.T., Marple D.N. & Sartin J.L. (1989). Skeletal muscle protein synthesis and growth hormone secretion in young lambs treated with clenbuterol. *J. Anim. Sci.* 67:2245-2254.

Kim Y.S., Lee Y.B., Garrett W.N. & Dalrymple R.H. (1989). Effects of cimaterol on nitrogen retention and energy utilization in lambs. *J. Anim. Sci.* 67:674-681.

Koohmaraie M., Shackelford S.D., Muggli-Cockett N.E. & Stone R.T. (1991). Effect of the β-adrenergic agonist L644,969 on muscle growth, endogenous proteinase activities, and postmortem proteolysis in wether lambs. *J. Anim. Sci.* 69:4823-4835.

Mersmann, H.J., Hu C.Y., Pond W.G., Rule D.C., Novakofski J.E. & Smith S.B. (1989). Growth and adipose tissue metabolism in young pigs fed cimaterol with adequate or low dietary protein. *J. Anim. Sci.* 64:1384-1394.

Minitab Statistical Software. (1988). Reference Manual release 6.1. Minitab Inc. 3081 Enterprise Drive State College, PA 16801, USA.

Shackelford S.D., Edwards J.W., Smarr E.K. & Savell J.W. (1992). Retail cut yields of Rambouillet wether lambs fed the adrenergic agonist L644,969. *J. Anim. Sci.* 70:161-168.

Thompson J.M., Atkins K.D. & Gilmour A.R. (1979). Carcass characteristics of heavyweight crossbred lambs. III. Distribution of subcutaneous fat, intermuscular fat, muscle and bone in the carcass. *Aust. J. Agric. Res.* 30:1215-1221.

NON-INVASIVE OPTICAL DETERMINATION OF ANIMAL ENERGY METABOLISM USING NEAR-INFRARED LASER

J. Takahashi[1] and H. Eda[2]

[1]Department of Animal Science, Obihiro University of Agriculture and Veterinary Medicine, Obihiro, Hokkaido 080, Japan; [2]Technology Research Laboratory, Shimadzu Corporation, 380-1 Horiyamashita, Hatano, Kanagawa 259-13, Japan

Summary

Cerebral oxygen consumption and energy expenditure in sheep were determined non-invasively using near-infrared laser. Cerebral blood flow rates in animals were analyzed mathematically from optical density at 808 nm of two levels of indocyanine green (15 and 30 mg head) injected into jugular vein. Changes in oxyhaemoglobin and deoxyhaemoglobin contents were monitored by near-infrared spectroscopy at 775 nm, 808 nm, 825 nm and 800 nm. Pulmonary oxygen consumption and metabolic rate in the whole body were determined simultaneously with a fully-automated open circuit ventilated hood system. Cerebral blood flow rates were estimated 676 and 643 ml/min at 15 and 30 mg injection of indocyanine green, respectively. Cerebral oxygen consumption in the resting state was 13.6 ml/min. The postprandial increment of cerebral oxygen consumption was only 8.4% maximally, while summit oxygen consumption in whole body increased by 75.3%.

Key words: Non-invasive oxygen monitor, oxyhaemoglobin, deoxyhaemoglobin, energy metabolism, brain.

Introduction

Open-circuit respiratory procedures with ventilated chambers or hoods have been well established in the indirect calorimetry of farm animals (Blaxter, 1962). In particular, ventilated hoods have been used preferably in physiological studies (Takahashi & Young, 1991, 1992, 1994) due to relatively quick response in short-term pulmonary changes in animals. Further advances in the instrumentation are required to improve the understanding of local energy metabolism in tissues or organs (Tamura *et al.*, 1988; Hampson *et al.*, 1990; Elwell *et al.*, 1994; Homma *et al.*, 1996). A new procedure has been developed to determine quantitatively the energy expenditure from cerebral oxygen metabolism in sheep. The principles of this technology consist of non-invasive near-infrared (NIR) spectroscopy of oxyhaemoglobin (Hb-oxy) and deoxyhaemoglobin (Hb-deoxy) of tissues *in situ* according to Lambert-Beers Law. The aims of this study were 1) to derive equations to calculate changes in Hb-oxy and Hb-deoxy contents from the absorbance changes at the selected wave lengths 2) to estimate optically the cerebral blood flow as a parameter of oxygen consumption using indocyanine green and 3) to establish the non-invasive optical method for the quantitative determination of oxygen consumption and metabolic rate of animal tissues.

Materials and Methods

Four Cheviot ewes were used in *in vitro* and *in situ* NIR spectroscopy. For *in vitro* NIR spectroscopy Hb-oxy was prepared by bubbling pure oxygen into sheep blood. Hb-deoxy was obtained by the addition of sodium thiosulphate ($Na_2S_2O_4$) into the blood and then by bubbling of pure nitrogen. Each blood sample was added into 1/15

M phosphate buffer (pH 6.60) to make 20-, 50- or 100-fold diluted and haemolysed solutions. From the absorbance changes at 775 nm (λ1), 808 nm (λ2), 825 nm (λ3) and 800 nm (λ4) (Table 1), the equations to calculate changes in Hb-oxy, Hb-deoxy and total haemoglobin (Hb-total) contents were derived by the least squares methods for matrices as follows:

$$\Delta \text{Hb-oxy} = 3.119\,\Delta\lambda 1 - 8.106\,\Delta\lambda 2 + 7.195\,\Delta\lambda 3 - 2.443\,\Delta\lambda 4$$

$$\Delta \text{Hb-deoxy} = -0.312\,\Delta\lambda 1 + 2.852\,\Delta\lambda 2 - 3.256\,\Delta\lambda 3 + 0.948\,\Delta\lambda 4$$

$$\Delta \text{Hb-total} = 2.807\,\Delta\lambda 1 - 5.254\,\Delta\lambda 2 + 3.939\,\Delta\lambda 3 - 1.495\,\Delta\lambda 4$$

Table 1. Millimolar absorptivities (l/mmol/cm) and millimolar extinction coefficients of oxyhaemoglobin and deoxyhaemoglobin at selected wavelengths in near infrared.

	Oxyhaemoglobin			Deoxyhaemoglobin		
λ(nm)	OD	εHb-oxy	ε'Hb-oxy[1]	OD	εHb-deoxy	ε'Hb-deoxy
775	6.075	1.912	1.146	8.625	3.267	1.962
808	6.400	2.070	2.069	8.713	3.249	3.189
825	6.863	2.310	2.402	9.013	3.421	3.500
800	6.325	2.024	1.539	8.650	3.233	2.497

[1] Corrected by *in situ* ΔOD/*in vitro* ΔOD in different doses of indocyanine green at selected wavelengths.

In situ NIR signals were obtained using a non-invasive O_2 monitor (OM-100A, Shimadzu, Kyoto, Japan) (Fig.1). The light source consists of four high-power laser diodes, the light beams being emitted in turn by the time-sharing mode are coupled to one beam through the four-branched fibre optics and illuminate the tissue.

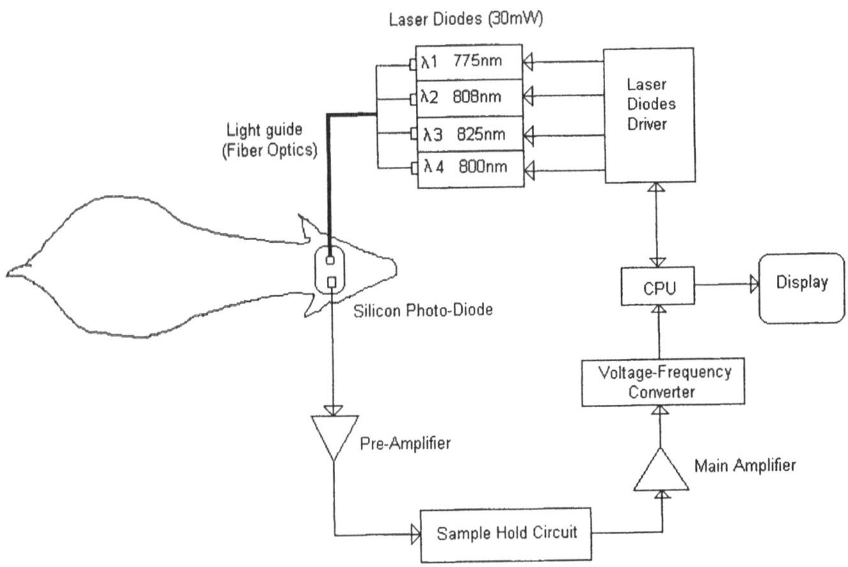

Fig. 1. Outline of *in situ* NIR spectroscopy.

The transmitted light is guided to photomultiplier through the receiving optical bundle fibre. Light guides were positioned over the forehead and spaced 3 cm apart in sheep. To evaluate the cerebral blood flow, 15 mg and 30 mg indocyanine green per head were administered into the jugular vein in physiological saline. This injection was executed in 20 minutes. The time-course changes in NIR signals at 808 nm were traced after intra-venous administration of indocyanine green solution. The synchronized determinations of pulmonary oxygen consumption and metabolic rate in combination with *in situ* NIR spectroscopy in subjects were conducted using a fully-automated ventilated hood system for open circuit indirect calorimetry.

Results

The estimated values of cerebral blood flow rates in ewes according to indocyanine green procedure are summarized in Table 2. There were no significant differences between 15 mg and 30 mg injection of indocyanine green ($P>0.05$). Cerebral blood flow rate of ewes in the resting state were presumed to be approximately 600 ml/min.

Table 2. Cerebral blood flow rates in ewe determined optically according to Indocinanine green procedure.

	Amount of Indocyanine green injected (per head)	
	15 mg	30 mg
ΔOD 808 nm	0.41±0.040[1]	0.94±0.047*
Blood flow rate (l/min)	0.68±0.079	0.64±0.175

[1] Mean ± SE, *P<0.01(15 mg vs.30 mg).

According to the Lambert-Beers Law, cerebral oxygen consumption in the resting state was estimated to be 13.56 ml/min from ΔHb-oxy in *in vitro* and *in situ* NIR spectroscopy (Table 3). The postprandial increment of cerebral oxygen consumption was only 8.4% maximally ($P<0.01$), while summit oxygen consumption in the whole body increased significantly by 75.3% ($P<0.01$).

Table 3. Resting and postprandial oxygen consumption and metabolic rate in the brain and whole body in ewes determined by non-invasive optical analyzer and ventilated hood.

	Whole body		Brain	
	Resting	Summit[2]	Resting	Summit
Oxygen consumption (ml/min)	221.67±6.49[1]	388.7±10.27[3]	13.6±0.17	14.7±0.35[4]
Metabolic rate (watts)	75.7±2.22	132.8±3.51[3]	4.7±0.06	5.0±0.12[4]

[1] Mean ± SE, [2] 18 min after feeding, [3, 4] P<0.01 (resting vs. summit).

Discussion

It is possible to estimate the cerebral blood flow using indocynanine green. Some improvements, however, are necessary for the practical use of this method. In this experiment total blood volume in brain was treated as a constant for convenience. A relatively lower postprandial increment was observed in cerebral oxygen consumption in comparison with that of whole oxygen consumption. This difference may originate from the fact that the cerebral blood flow rate was made a constant. To reflect the changes in cerebral blood flow to this method makes it possible to trace the rapid change in cerebral oxygen metabolism. According to the Kety method using N_2O, cerebral blood flow in intact humans is estimated to be approximately 750 ml/min (Kety, 1962). Thus, human cerebral oxygen consumption is calculated to be 49ml/min (Ganong, 1971). The present study demonstrates non-invasive determination of cerebral oxygen consumption and metabolic rate using *in situ* NIR signals as a variable according the Lambert-Beers Law. So far, reliability of cerebral oxygen consumption estimated by this technique is uncertain due to lack of information on the direct determination in sheep. Changes in *in situ* NIR signals reflect physiological activities in the subject clearly, although absolute values should be compared to the appropriate references.

References

Blaxter K.L. (1962). In: The Energy Metabolism of Ruminants. Hutchinson Scientific and Technical. London, pp. 53-61.

Elwell C.E., Cope M., Edwards A.D., Wyatt J.S., Delpy D.T., & Reynolds E.O.R. (1994). Quantification of adult cerebral hemodynamics by near-infrared spectroscopy. *J. Appl. Physiol.* 77:2753-2760.

Ganong W.F. (1971). In: Review of Medical Physiology, 5th ed. Lange Medical Publications. CA, pp. 443-447.

Kety S.S. (1962). Blood flow and metabolism of the human brain in health and disease. In: K.A.C. Elliot, I.H. Page and J.G. Quastel (Eds) Neurochemistry, the chemistry of brain and nerve, 2nd ed. Thomas. Springfield, IL.

Homma S., Eda H., Ogasawara S. & Kagaya A. (1996). Near-infrared estimation of O_2 supply and consumption in forearm muscles working at varying intensity. *J. Appl. Physiol.* 80:1279-1284.

Hampson N.B., Camporesi E.M., Stolp B.W., Moon R.E., Shook J.E., Griebel J.A. & Pintadosi C.A. (1990). Cerebral oxygen availability by NIR spectroscopy during transient hypoxia in humans. *J. Appl. Physiol.* 69:907-913.

Takahashi J. & Young B.A. (1991). Prophylactic effects of L-cysteine on nitrate-induced alterations in respiratory exchange and metabolic rate in sheep. *Anim. Feed Sci. Technol.* 35:105-113.

Takahashi J. & Young B.A. (1992). The modulation of nitrate-enchanced hypothermia by sulphur compounds in cold-exposed sheep. *Anim. Feed Sci. Technol.* 39:347- 355.

Takahashi J. & Young B.A. (1994). The regulation of energy metabolism on sheep by nitrate and L-cysteine. In: J. F. Auguilera (Ed) Energy metabolism of farm animals. *EAAP Publ. No.*76:387-390.

Tamura, T., Eda H., Takada M. & Kubodera T. (1988). New instrument for monitoring hemoglobin oxygenation. *Adv. Exp. Med. Biol.* 248:103-107.

THE INTEGRATION OF DEXA, ULTRASOUND SCANS AND GROWTH MODELLING TECHNOLOGY TO ESTIMATE BODY COMPOSITION IN GROWING PIGS

R. Geers[1], B. De Pauw[1], G. Spincemaille[2], H. Villé[1], J. Vits[1], G. Rombouts[2], W. Duchateau[1] and S. Perremans[1]

[1]Agricultural Buildings Research, K.U.Leuven, Belgium; [2]Seghers Hybrid, Buggenhout, Belgium

Summary

Thirty-six pigs of different lines were measured *in vivo* with DEXA and ultrasound scans during general anaesthesia. Body weight ranged from 20 to 75 kg. A sample of six pigs (10 to 25 kg) was killed humanely, cooked and homogenized for chemical analysis. Ash, fat and lean tissue were quantified by DEXA for all pigs. Backfat thickness, depth, circumference, cross-sectional area of the musculus longissimus dorsi were measured with a real-time ultrasound device. Body length and circumference of chest were also measured. Data were evaluated with analysis of variance within a generalized linear model.

DEXA measurements could be considered as representative for chemical analysis, at least for fat and lean tissue. Statistical significant differences were found between DEXA measurements and simulated results for fat and lean tissue. These were explained by a mismatch between model parameters related to maintenance requirements and efficiency for lean tissue growth, and real parameters. Body composition of growing pigs could also be estimated in the field by using measurements of body conformation and internal dimensions of tissues.

Key words: Growth, modelling, body composition, *in vivo* measurement.

Introduction

Growth modelling can be used as a tool to increase feed conversion ratio in animal production (Moughan *et al.*, 1995). Feeding the animals according to their needs will reduce intake and waste. Moreover, the quality of the end product will be targeted to the consumer's demand. However, taking into account animal variability, reference measurements of production should be collected in order to adjust the predicted feeding strategy.

Dual Energy X-ray Absorption (DEXA) has been developed to quantify osteoporosis and body composition of humans (Dequeker & Geussens, 1994). Precision of measurements with humans has been quantified by the calculation of a variation coefficient (VC) of repeated measurements on the same subject. Values varied from 2% (Dequeker and Geussens, 1994) to 5% (Verlooy *et al.*, 1991). For pigs weighing between 35 and 95 kg, VC was 2.5 to 4.7% for fat tissue and 0.6 to 2.3% for lean tissue (Mitchell *et al.*, 1996; Pintauro *et al.*, 1996; Svendson *et al.*, 1993). Mitchell and Conway (1994) found a determination coefficient of 0.97 between DEXA and chemical analyses for pigs weighing between 30 and 100 kg, while this coefficient was 0.99 for newborn piglets (Brunton, 1993). Recent studies indicated that ultrasound technology is also useful to estimate body composition (Cisneros *et al.*, 1996).

The aim of this study was to compare body composition of growing pigs as estimated by growth modelling with *in vivo* measurements by DEXA and ultrasound technology.

Materials and Methods

In a first experiment the comparison was made between DEXA and chemical analysis. A sample of six pigs (10 to 25 kg) was killed humanely, cooked (125°C, 1.2 atm, 18 h) and homogenized. Fat, protein, ash and dry matter were analyzed with standardized chemical methods, and compared with the DEXA-results. The DEXA device was a DPX-L from LUNAR (USA), and is maintained by LUNAR Europe (Brussels).

In a second experiment DEXA-measurements were predicted by the TMV-model (van der Peet-Schwering *et al.*, 1994). Thirty-six pigs of different lines (Seghers Hybrid, Belgium) were measured during general anaesthesia (intramuscular injection of azaperonum, intraperitoneal injection of methomidatum; Janssen Pharmaceutica, Beerse, Belgium). Body weight ranged from 20 to 75 kg. During sedation, backfat thickness, depth, circumference, cross-sectional area of the musculus logissimus dorsi were measured 5 cm from the midline after the last rib with a real-time ultrasound device based on the pulse echo principle. Accuracy on linear measurements was 3% (Toshiba, Tokyo). Also body length and circumference of chest were measured.

Data were evaluated with analysis of variance within a generalized linear model (SAS, 1994).

Results

Table 1 shows the comparison between results obtained by DEXA and chemical analyses. Correlation coefficients were respectively: fat = 0.98, lean = 0.99, ash = 0.93. The DEXA results have been corrected according to Hollis (1993) and Moughan *et al.* (1995). However, the difference with respect to ash % remained. Variation coefficients for chemical analyses were respectively: fat = 6.2%, lean = 2.2%, ash = 2.1%.

Table 1. Comparison of DEXA with chemical analysis (10 - 25 kg, n = 6) (mean ± SEM).

	Fat %	Lean %	Ash %
DEXA	10.3 ± 0.58	88.3 ± 0.73	2.74 ± 0.15
Chemical analysis	9.03 ± 0.52	86.0 ± 1.16	3.33 ± 0.12
P	> 0.05	> 0.05	< 0.01

Results of the second experiment are presented in Tables 2 and 3. Table 2 gives an idea about the repeatability of measurements of pigs within a weight class of 15 to 25 kg being measured two times at the same body weight. No differences were observed with respect to body weight between the DEXA method and the weighing scale. Mean daily growth

Table 2. Variation coefficient (%) of repeated measurements.

	Number	Mean	SEM
Fat	36	7.76	12.7
Lean	36	1.83	7.04
Ash	36	2.39	4.20

rate and feed conversion ratio were simulated correctly by the TMV-model, being respectively 760 g/day and 2.80. However, statistically significant differences were observed with respect to body composition, although initial body composition could be matched after correction of the DEXA measurements (Table 3).

Table 3. Predicted (TMV) and measured (DEXA) fat tissue and lean tissue (kg, MEAN ± LSMeans) at different body weights (n=36).

	20 kg	45 kg	75 kg
Fat - DEXA	2.67 ± 0.291	4.67 ± 0.291	9.37 ± 0.291
Fat - TMV	3.00 ± 0.291	7.00 ± 0.291	14.4 ± 0.291
P	> 0.05	< 0.001	< 0.001
Lean - DEXA	21.8 ± 0.384	39.3 ± 0.384	63.4 ± 0.384
Lean - TMV	21.1 ± 0.384	36.6 ± 0.384	58.3 ± 0.384
P	> 0.05	< 0.001	< 0.001

The following equations introducing external and internal body dimensions allowed the estimation of the DEXA measurements:

Fat tissue (g) = -3861 + 94 Circumference Chest (CC cm) - 123 Fat Thickness (FT mm) + 36 Depth M. LD (mm)
(n = 207; r^2 = 0.789; VC = 49%; root MSE = 739)

Lean tissue (g) = -4819 + 345 CC + 441 Body Length (cm) - 274 FT + 224 DMLD
(n = 207; r^2 = 0.959; VC = 14%; root MSE = 3309)

Discussion

In agreement with other data, corrected DEXA measurements could be considered as representative for chemical analysis of fat and lean tissue. An important advantage is that animals do not have to be slaughtered, so that growth curves of tissues can be studied within the same animal. However, due to large within animal variability of fat measured by DEXA, difficulties may arise when quantifying between animal differences. In the case of application of the slaughter method, Everts and Dekker (1994) drew attention to important potential errors related to the homogenization process of the animal before sampling, which cannot occur with DEXA. This may explain the difference with respect to the ash value between DEXA and chemical analysis.

The relative difference found between measured and simulated values of lean tissue at body weights of 45 and 75 kg fits within the accuracy of the model used (van der Peet-Schwering *et al.*, 1994). However, the absolute difference within lean tissue between measured and simulated values is reflected in the difference with respect to fat tissue, because of the applied calculation method of the model. As a result no differences have been found with respect to body weight. The model calculates firstly lean tissue deposition, and fat tissue growth equals metabolizable energy intake minus metabolizable energy for maintenance and lean tissue energy. The underestimation of lean tissue may be explained by the fact that model parameters used for either efficiency of lean tissue growth and/or maintenance requirements did not match the parameters of the experimental animals. Hence, these parameters should be available as inputs for the model, which will be investigated further on.

Body composition of growing pigs can be estimated in the field by using measurements from image analysis (external dimensions) and echography (internal dimensions of tissues) (Geers *et al.*, 1991). This information is necessary as feed-back to check the body composition as simulated by growth models. Hence, predicted carcass quality and mineral balance can be evaluated before slaughtering in view of the applied feed composition and feeding strategy. Further research will focus on the accuracy of estimation of body composition, feed intake and feed efficiency for those types of pigs which were not used within the procedure for developing and validating the TMV-model.

Acknowledgements

Funding was provided by the FWO, EC (AIR-262) and the Flemish Government. The group of Professor Dr R. de Wilde (Laboratory of Animal Nutrition, U. Gent) prepared the pigs for chemical analysis.

References

Brunton J.A. (1993). Evaluation of DEXA for body composition assessment in 1500 to 5500 g subjects. *Hologic Publications*, 1-8.

Cisneros F., Ellis M., Miller R.D., Novakofski J., Wilson E.R. & McKeith F.K. (1996). Comparison of tranverse and longitudinal real-time ultrasound scans for prediction of lean cut yields and fat-free lean content in live pigs. *J. Anim. Sci.* 74:2566-2576.

Dequeker J. & Geussens P. (1994). Measurements of bone mass. Spine: *State-of-the-art Reviews* 8:133-154.

Everts H. & Dekker R.A. (1994). Balance trials and comparative slaughter in breeding sows: description of techniques and observed accuracy. *Livest. Prod. Sci.* 37:339-352.

Geers R., Goedseels V., Parduyns G., van der Stuyft, E., Bosschaerts L., Deley J. & Neirinck W. (1991). Prediction of SKG-II grading of carcass lean content by body measurements and ultrasound *in vivo*. *Landbouwtijdschrift* 44:237-242.

Hollis G.R. (1993). Growth of the pig. CAB International, Wallingford.

Mitchell A.D. & Conway J.M. (1994). Measurement of energy deposition and body composition of pigs by DEXA. In: J.F. Aguilera (Ed) Energy metabolism of farm animals. CSIC, Granada, *EAAP Publ. No.*76:31-34.

Mitchell A.D., Conway J.M. & Scholtz A.M. (1996). Incremental changes in total and regional body composition of growing pigs measured by DEXA. *Growth, Development & Ageing* 60:113-123.

Moughan P.J., Verstegen M.W.A. & Visser-Reyneveld M.E. (1995). Modelling growth in the pig. Wageningen Pers, Wageningen.

Pintauro S.J., Nagy T.R., Duthie C.M. & Goran M.I. (1996). Cross-calibration of fat and lean measurements by DEXA to pig carcass analysis in the paediatric body weight range. *Am. J. Clin. Nutr.* 63:293-298.

SAS Institute Inc. (1994). SAS/STAT® User's Guide, Release 6.03, 4th Edition. SAS Institute Inc., Cary, NC.

Svendson O.L., Haarbo J., Hassager C. & Christiansen C. (1993). Accuracy of measurements of body composition by DEXA *in vivo*. *Am. J. Clin. Nutr.* 57:605-608.

van der Peet-Schwering C.M.C., Vos H.J.P.M., Verstegen M.W.A., Kanis E., Smits C.H.M., de Vries A.G. & Lenis N.P. (1994). Technical Model Pig Feeding. Research Institute for Pig Husbandry Rosmalen, Report P 1.117.

Verlooy H., Dequeker J., Geusens P., Goris M. & Nijs J. (1991). Body composition by intercomparison of underwater weighing, skinfold measurements and dual-photon absorptiometry. *Br. J. Radiology* 64:765-767.

EFFECTS OF SOLAR RADIATION AND FEEDING TIME IN A HOT ENVIRONMENT ON HEAT AND ENERGY BALANCE PARAMETERS IN CATTLE FED A HIGH ENERGY CONCENTRATION DIET

A. Brosh[1], G. Beneke[2], S. Fennell[2], D. Wright[3], Y. Aharoni[1] and B. Young[2]

[1]Department of Beef Cattle, Newe-Ya'ar, P.O.Box 1021, Ramat Yishay 3095, ARO, Israel; [2]Department of Animal Production, The University of Queensland, Gatton College, Australia; [3]Department of Companion Animal Medicine & Surgery, The University of Queensland, St Lucia, Australia

Summary

Effects of solar radiation and time of feeding of a diet of a metabolizable energy (ME) concentration of 10.8 MJ/kg DM, on heat and energy balance parameters were studied in growing female cattle during the summer. Feed intake was not affected by the higher heat load induced by solar radiation. The heifers coped successfully with the heat load by a higher respiration rate, and a larger fluctuation in rectal temperature, between morning and afternoon. It seems that the lowering of body temperature toward the morning is a physiological mechanism, used by the animals to prepare for the heat load that will develop during the day. Heat production (HP) and heart rate are affected mainly by the time of feeding and not by the heat load and cooling activation. Feeding in the afternoon will increase HP in the cooler hours of the day when conductivity and the radiation of heat to the environment are more efficient. Energy expenditure of the entire day was significantly reduced (4.7%) by solar radiation when the heifers were morning fed, but not when they were fed in the afternoon. The retained energy was not significantly affected by solar radiation.
Key words: Cattle, solar radiation, heart rate, oxygen consumption, respiration rate, rectal temperature.

Introduction

Summer heat load can cause a reduction in feed intake (Young & Hall, 1993), an increase in energy expenditure (EE) caused by the active cooling that can increase heat load and consequently reduce productivity. Protection from solar radiation has been advocated as a method to relieve the animals from heat load. Owing to the expense involved many beef feedlots are not shaded. The provision of new feed to the trough is the biggest stimulus to feeding (Fell & Clark, 1993). Eating and digestion activity induce an increase in heat production (Young & Webster, 1963; West, 1994). Most commonly, new feed is delivered in the morning. Eating in the morning may combine the HP resulting from alimentary activity and active cooling, with the burden imposed by the environment. This study was designed to determine the effects of solar radiation and time of feeding on heat balance parameters and energy balance in feedlot beef cattle during the summer in a subtropical region.

Materials and Methods

Effects of shade on feed intake, water consumption, rectal temperature (Tr), respiration rate (RR), heart rate (HR) and oxygen (VO_2) consumption, were measured in young growing female cattle during the summer (January-March) of 1993 in south-east Queensland. This region is subtropical with summer temperatures of over 30°C being common. The animals were kept individually in open feedlot pens each of 40 m^2. Shade was provided in half of the pens by galvanized iron sheet, 2.2 m high, over 11.5 m^2 plus 70% shade cloth, 4 m high over 12 m^2. A high energy diet of 80% concentrate and 20% roughage, 10.8 MJ/kg DM was offered *ad libitum* to the animals. Feed intake and water consumption were measured on a daily basis. Animal measurements and meteorological data: ambient temperature (Ta), black globe temperature (BG) and relative humidity (RH) were collected between 07:00 and 08:30 before feeding, and again between 14:00 and 15:30. Black Globe Humidity Indices (BGHI), Buffington *et al.* (1981) were calculated. Rectal temperature, HR and VO_2 consumption of the heifers in the pens were measured in two consecutive days on each treatment.

Ten 12-month-old trained Hereford heifers, 345±10.8 kg BW, were used in the cross-over designed study, in which heifers were subjected to either exposure (EXP) or protection (PRO) from solar radiation for 3-week periods. Within each period, the heifers were offered their feed once daily at 8:00 for 2 weeks, and then at 16:30 for another week. The first week was used for adaptation to the solar radiation treatment. Heart rate radio transmitters (Telonics, Mesa, Arizona) were implanted into six of the heifers approximately 1 month prior to the commencement of measurements. All the management variables were carried out in 10 heifers. Heart rate and VO_2 consumption, by face mask open system (Taylor *et al.*, 1982) were measured in 6 heifers. Heart rate was measured throughout the day, 5 min every half hour, for each animal. Energy expenditure of the heifers throughout the entire days of each trial combination was calculated by multiplication of the total daily heart beats by EE of one pulse (Brosh *et al.*, 1994).

Differences between treatment combinations were tested within heifers by paired t test, using significance level of $P < 0.05$. Data are presented as mean ± SE.

Results and Discussion

Ambient temperatures (^{O}C) were 23.5±0.8 in the morning and 30.2±0.7 in the afternoon, while relative humidity was 65±2% and 45±3%, respectively. BG temperatures were 25.2± 0.8 and 30.2±0.7 under the iron shade and 37.6±1.3 and 45.0±1.7 in the sun in the morning and in the afternoon, respectively. BGHI were 72.6±1.1 and 80.1±0.7 in the shade, 84.9± 1.3 and 92.1±1.6 in the sun in the morning and in the afternoon, respectively. Water consumption (ml/kg BW/day) was 108±10 and 127±12 in the shade and sun treatment respectively (NS).

The higher heat load induced by solar radiation and ambient temperature was expressed in the subjected heifers by a higher RR (Fig. 1. A) and a larger fluctuation in Tr (Fig. 1. B) between morning and afternoon. Feeding in the morning, compared with afternoon feeding, increased the fluctuation between the morning and afternoon of both parameters. The highest difference between morning and afternoon Tr (1.5^{O}C) occurred in morning fed, EXP heifers. According to the classic thermogenesis curve (Kleiber, 1975), when the environmental temperature is high, the animal increases its HP and is subjected, therefore, to an increased heat load. However, heat balance can be maintained by active evaporation. According to many studies, the heat load which increased heat production is involved in an increased HR which needed to propel an increased blood flow essential for increased oxygen consumption and for heat dissipation (Hales, 1974). In the presented study, in spite of the highly significant increase of the Tr and RR by solar radiation, whether the heifers were fed morning or afternoon, time of feeding was the main factor which affected HR and the VO_2 consumption (Fig. 2 A, B). Heat production during the day (Fig. 3) was increased in the hours following the introduction of fresh food to the trough, whether the food was offered in the morning or afternoon. Consequently, the morning-fed heifers had their peak heat production during the warmer hours of the day, whereas the afternoon-fed heifers had the lowest HP during these hours, and the highest HP at night.

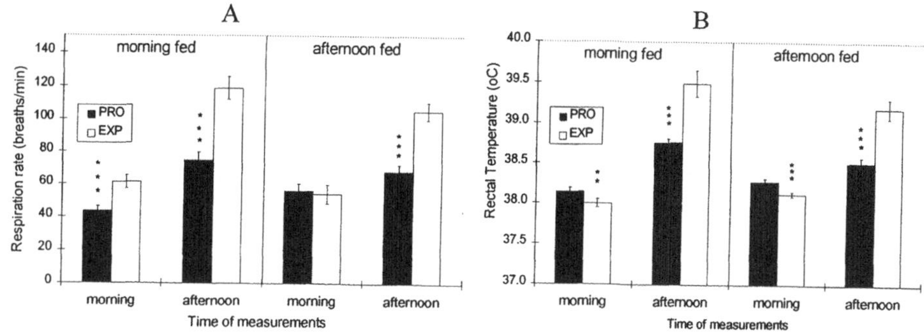

Fig. 1. Respiration rate (A) and rectal temperature (B) of the heifers (mean±SE). ** P<0.01, *** P<0.001.

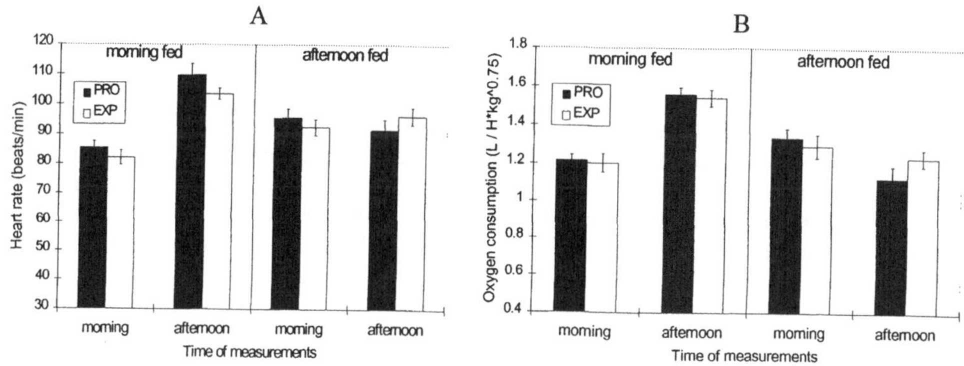

Fig. 2. Heart rate (A) and oxygen consumption (B) of the heifers (mean±SE).

The metabolizable energy intake (MEI) was not affected by solar radiation (Table 1). Energy expenditure of the heifers (Table 1) which were subjected to the highest heat load, the morning-fed exposed heifers, was significantly lower than that of the protected ones. Similar results were obtained by Brosh (1985) on Bedouin goats that were deprived of water for 4 days, in the summer, and exposed to solar radiation, compared with goats which were protected from solar radiation. In spite of the high effect of time of feeding on the distribution of heat production during the day, and the significant relief from the heat load enjoyed by the afternoon fed animals, total daily energy expenditure of the heifers was not affected by time of feeding. The retained energy (Table 1) also was not significantly affected by the heat load induced by solar radiation and feeding in the morning.

Table 1. Effects of solar radiation (exposed (E) and protected (Pr)) on the metabolizable energy intake (MEI), energy expenditure(EE) and retained energy (RE) , kJ/(kg$^{0.75}$*day), of six heifers (mean ± SE).

	Morning-fed		E-Pr	E-Pr	Afternoon-fed		E-Pr	E-Pr
	E	Pr	% of Pr	P<	E	Pr	% of Pr	P<
MEI	1190±37	1200±25	-0.6	NS	1169±40	1190±26	-1.7	NS
EE	624±15	655±11	-4.7	0.05	656±24	677±14	-2.8	NS
RE	567±29	545±27	5.0	NS	514±40	513±20	0.0	NS

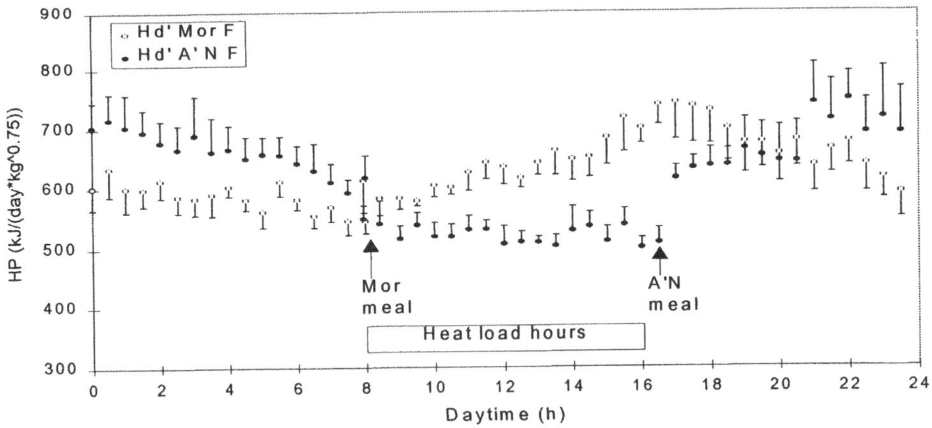

Fig. 3. Heat production during the day of 4 heifers (mean ± SE) that were fed morning (o), or were fed afternoon (•).

It may be concluded that solar radiation and alimentary activity are the main factors which induced heat load on the heifers. The main means implemented by the heifers to cope with the heat load were activated evaporative cooling, and increased tolerance for a temporary increase in body temperature during the day. Furthermore, the heifers adjusted themselves to the coming heat load day by lowering their body temperature at night, to have a lower start at day break. When fed in the morning the effect of both causes for heat load, environmental and alimentary, are accumulated. Changing the time of feeding to the late afternoon should relieve the situation. Feeding in the afternoon will increase HP in the cooler hours of the day when conductivity and radiation of heat from the animals' body is more efficient owing to higher temperature differences between the animals and their environment.

References

Brosh A., Beneke G., Fennell S., Wright D., Aharoni Y. & Young B.A. (1994). 13th Symposium on Energy Metabolism of Farm Animals, Mojacar, Spain. *EAAP Publ. No.* 76:35-38.

Brosh A. (1985). PhD Tel-Aviv University (in Hebrew). pp. 78.

Buffington D.E., Collazo-Arocho A., Canton G.H., Pitt D., Thatcher W.W. & Collier R.J. (1981). *Trans ASAE.* 24:711-4.

Fell L.R. & Clarke M.R. (1993). In: Farrell D.J. (Ed) Recent Advances in Animal Nutrition in Australia. University of New England, Armidale, N.S.W. pp. 107-16.

Hales J.R.S. (1974). Physiological responses to heat. In: Robertshaw D. (Ed) MTP international review of science, Physiology series 1 Vol 7. Butterworths, London, pp. 107-162.

Kleiber M. (1975). The fire of life, an introduction to animal energetics. Robert E Krieger Publishing Company, New York.

Taylor C., Heglund N.C. & Maloiy G.M.O. (1982). *J Exp. Biol.* 97:1-21.

Young B.A. & Hall A.B. (1993) Heat load in cattle in the Australian environment. In: B Coombs (Ed) Australian Beef. Morescope Pty Ltd, Melbourne, pp. 143-148.

Young B.A. & Webster M.E.D (1963). A technique for the estimation of energy expenditure in sheep. *Aust. J. Agric. Res.* 14:867-873.

West J.W. (1994). Interactions of energy and bovine somatotropin with heat stress. *J. Dairy Sci.* 77:2091-2102.

EVALUATION OF THE *IN VITRO* GAS PRODUCTION SYSTEM FOR ESTIMATING THE METHANE PRODUCING POTENTIAL OF RUMINANT FEEDS

A.R. Moss and D.I. Givens

Feed Evaluation and Nutritional Sciences, ADAS Drayton, Drayton Manor Drive, Alcester Road, Stratford-upon Avon, CV37 9RQ, UK

Summary

There is a need to be able to measure readily the methane producing potential of a range of feedstuffs under different rumen conditions in order to predict more accurately the amount of methane produced per animal. *In vitro* techniques using rumen fluid as an inoculum can be considered as models of *in vivo* rumen digestion and have been applied to estimate digestibility of feedstuffs. The *in vitro* gas production technique can be used successfully to estimate this but there have been limited attempts to use it to estimate the methane producing potential of feedstuffs. The objective was to assess the potential of the technique to do this. Two pure substrates (glucose and pectin) and seven mixed diets (grass silage based in varying forage to concentrate (F:C) ratios) were incubated in buffered rumen fluid for periods up to 48 h and gas production and composition were determined. Methane produced from the mixtures was compared with the volume produced when the same diet combinations were fed to mature wethers fed at maintenance. The methane production *in vitro* was lower and covered a narrower range than *in vivo* and the appearance of hydrogen for some diets may explain their lower values. The work indicates the potential of the gas production technique for ranking mixed diets for methane producing potential, but more work is required to optimize the technique for methanogenesis.
Key words: Methane, *in vitro, in vivo*, sheep, rumen stoichiometry.

Introduction

There is a need to be able to measure the methane producing potential of a range of feedstuffs under different rumen conditions in order to predict more accurately the amount of methane produced per animal. *In vitro* techniques using rumen fluid as an innoculum can be considered as models of *in vivo* rumen digestion and have been applied to estimate digestibility of feedstuffs. The *in vitro* gas production technique (Menke *et al.*, 1979) can be used successfully to estimate this but there have been limited attempts to use it to estimate the methane producing potential of feedstuffs. The *in vitro* gas production system as discussed earlier has primarily been used to describe the rate of degradation of feedstuffs and to study the formation of VFAs and stoichiometry of fermentation (Blümmel & Ørskov, 1993). The gas produced can be analysed for its methane content and with a limited number of straws at least, the methane production *in vitro* has been shown to correspond to stoichiometric predictions of *in vivo* production (Blümmel & Ørskov, 1993). The gas production technique described by Menke *et al.* (1979) has subsequently been developed into a system whereby the gas production is measured using a pressure transducer (Theodorou, 1993). The objective of this work was to assess the potential of the technique to measure methane producing potential of feeds.

Materials and Methods

Two pure substrates (glucose and pectin) and seven concentrates, barley (B), soya bean meal (SBM), molassed sugar beet feed (MSBF), fishmeal (FM), distillers' dark grain (DDG), distillers' light grain (DLG) and maize gluten (MG) were incubated in combination with freeze-dried grass silage in the proportions as fed *in vivo* (proportion of forage 1, 0.75, 0.5, 0.25 for all concentrates with the exception of FM which was 1, 0.95, 0.85, 0.75). One gram of each mixture was pre-wetted in 94 ml of medium D (Theodorou *et al.*, 1994), inoculated with strained rumen fluid (from wether sheep, 10 ml) and incubated without agitation for 48 h at 39°C. Gas production was measured at frequent time intervals with a manual pressure transducer and gas-tight syringe (Theodorou *et al.*, 1994) and replicate batch cultures were removed at 6, 24 and 48 h for VFA analysis (GC,FID detector) and organic matter (OM) degradation assessment. Gas was stored in Haemoguard vacutainers before analysis for H_2, CO_2 and CH_4 by GC (TCD detector). The above was repeated on three separate occasions for each feedstuff. The model of Ørskov & McDonald (1979) was used to fit the gas accumulation profiles recorded. Methane produced from the mixtures was compared with the volume produced when the same diet combinations were fed to mature wethers fed at maintenance (Moss & Givens, 1995).

Results

Cumulative gas production was 297 (SE 8.8) and 358 (SE 17.5) ml/g DM for glucose (G) and pectin (P) respectively. All substrates had reached maximum gas production by 72 h. The underlying rate of gas production was 0.073 (SE 0.0427) and 0.112 (SE 0.0142) per h for G and P respectively and P had a shorter lag. Total VFA concentrations were higher for P at 72 h compared with G (98, 80 mM respectively) and the molar proportions varied between substrates, 0.37, 0.69 acetate, 0.49, 0.21 propionate, 0.08, 0.09 butyrate for G and P respectively at 72 h. Proportions of gas generated were 0.645, 0.661 CO_2 from buffer, 0.300, 0.231 CO_2 and 0.053, 0.110 CH_4 from fermentation of G and P respectively at 72 h. Stoichiometric calculations estimated 19.2 and 79.6 ml CH_4 per g OM fermented (Demeyer, 1991) compared with 15.7 and 38.3 ml CH_4 per g DM measured for G and P respectively.

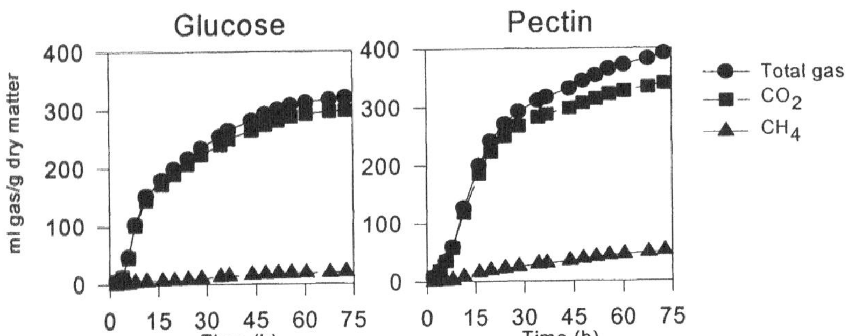

Fig. 1. Cumulative total gas production, carbon dioxide and methane from fermentation of glucose and pectin.

The work indicates the potential of the gas production technique for ranking substrates but highlights the limitations of simple stoichiometric models.

OM apparent digestibility determined *in vivo* was positively correlated with OM disappearance *in vitro* at both 24 and 48 h (R^2=73.3 and 56.9% respectively) for the diet combinations. Methane production expressed as l kg^{-1} DM determined *in vivo* was also positively correlated with methane produced *in vitro* at 24 and 48 h (R^2=26.4 and 19.8%

respectively). The low forage : concentrate ratio diets had a large influence on these relationships (Fig. 1). The 0.75 inclusion of B and MSBF produced the greatest amount of gas after 48 h (306 and 308 l/kg DM respectively) and the high protein feeds produced the least amount of gas. Hydrogen was negligible for all the feeds with the exception of the 0.75 inclusion of B and MSBF when 6.74 and 2.88 l/kg DM respectively was produced at 48 h. The concentration of propionic acid at 24 and 48 h was not well correlated with methane produced *in vitro* for the range of feeds studied (R^2=11.7 (SEP=3.86) and 2.5% (SEP=3.34) respectively), despite the good relationship between the two for glucose and pectin.

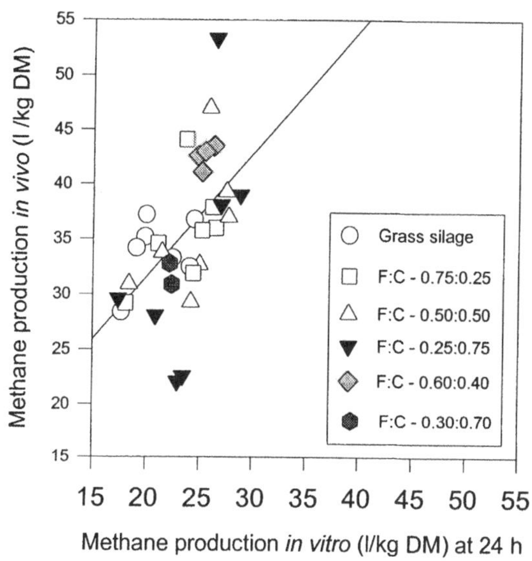

Fig. 2. Relationship between methane produced *in vivo* and *in vitro* from a range of diets.

Discussion

The methane production and fermentation established for glucose in the *in vitro* gas production system was comparable to standard stoichiometric equations, whereas for pectin the stoichiometric equations did not represent the fermentation *in vitro*. This raises doubt about either the stoichiometric models available or the *in vitro* system's ability to ferment a more complex carbohydrate like pectin. The methane production *in vitro* for the range of feedstuffs was lower and covered a narrower range than *in vivo*. The appearance of hydrogen for some diets may explain their lower values and suggests an inhibition of methanogenesis, possibly as a result of a low pH. The methane productions from the silages incubated alone were consistently lower for *in vitro* technique compared with the *in vivo* data. This may be a result of the presentation of the silage to the *in vitro* system (e.g. freeze-dried) compared with fresh *in vivo*. The fermentation acids associated with silage (particularly lactic acid) have a strong influence on the rumen fermentation pattern and these acids would essentially have disappeared during the drying process. The work indicates the potential of the gas production technique for ranking mixed diets for methane producing potential, but more work is required to optimize the technique for methanogenesis.

Acknowledgements

We wish to acknowledge the funding of this work from MAFF.

References

Blümmel M. & Ørskov E.R. (1993). Comparison of *in vitro* gas production and nylon bag degradability of roughages in predicting feed intake in cattle. *Anim. Feed Sci. Technol.* 40:109-120.

Demeyer D.I. (1991). Quantitative aspects of microbial metabolism in the rumen and hindgut. In: J. P. Jouany (Ed) Rumen microbial metabolism and ruminant digestion pp. 216-237.

Menke K.H., Raab L., Salewski A., Steingass H., Fritz D. & Schneider W. (1979). The estimation of the digestibility and metabolizable energy content of ruminant feedingstuffs from the gas production when they are incubated with rumen liquor *in vitro*. *J. Agric. Sci.* 93:217-222.

Moss A.R. & Givens D.I. (1995). Prediction of methane production by sheep from a range of grass silage based diets. *Anim. Sci.* 60:513-514 (Abs.).

Ørskov E.R. & McDonald I. (1979). The estimation of protein degradability in the rumen from incubation methods weighted according to rates of passage. *J. Agri. Sci.* 92:499-503.

Theodorou M.K. (1993). A new laboratory procedure for determining the fermentation kinetics of ruminant feeds. *Ciencia e Investigacion Agraria* 20:332-344.

Theodorou M.K., Williams B.A., Dhanoa M.S., McAllan A.B. & France J. (1994). A simple gas production method using a pressure transducer to determine the fermentation kinetics of ruminant feeds. *Anim. Feed Sci. Technol.* 48:185-197.

Van Soest P.J. (1994). Nutritional Ecology of The Ruminant. Cornell University Press, Ithaca, USA pp. 22-274.

OXIDATION OF NUTRIENTS IN GROWING CALVES

A. Chwalibog, A.-H. Tauson and G. Thorbek

Department of Animal Science and Animal Health, Research Department of Human Nutrition, The Royal Veterinary and Agricultural University, Copenhagen, Bülowsvej 13, DK-1870 Frederiksberg C , Denmark

Summary

Data from 127 balance experiments with 36 growing calves (100-400 kg LW) in which the energy metabolism was measured under different feeding regimes, were used to quantify some of the components in the model demonstrated. Oxidation-CO_2 (CO_2x) was separated from fermentation-CO_2 by assuming that CO_2x, l = total CO_2 - 3 x CH_4, l. A CH_4 production in the range from 13 to 36 l/kg ingested organic matter (IOM) was obtained by applying different types of roughages. With an ME intake between 0.7 and 1.2 MJ/kg$^{0.75}$ and digested protein between 10 and 12 g/kg$^{0.75}$ heat from oxidation of protein (OXP) was between 10 and 18% of heat production from nutrient oxidation (HEx), while gluconeogenesis was between 6 and 9% of ingested protein. Oxidation of carbohydrate (OXCHO) varied between 13 and 79% of HEx, while oxidation of fat (OXF) varied between 12 and 69% of HEx. Directly oxidized carbohydrate as OXCHO in relation to ingested carbohydrate varied between 9 and 58%, while carbohydrate oxidized as fat through SCFA or lipogenesis varied between 9 and 47%, related to the extent of the fermentation processes.
Key words: Oxidation of nutrients, heat production, methane production, calves.

Introduction

Previous data (Chwalibog *et al*, 1992; Chwalibog & Thorbek, 1995) demonstrated that quantitative oxidation of nutrients in growing pigs can be calculated from 24 h measurements of gas exchange and nitrogen excretion in urine, being in accordance with the theory of Lusk (1928) and the equations given by Brouwer (1958). As the fermentation processes in growing pigs are low, with a methane production generally less than 3 l/d, the measured CO_2 production was considered to be oxidation-CO_2. However, this is not the case in ruminants, where a certain part of the measured CO_2 is caused by fermentative processes and therefore, the total amount of CO_2 is partly composed of CO_2 produced from nutrient oxidation and partly derived from fermentation in the rumen. A separation between the two components is necessary in order to calculate substrate oxidation in ruminants.

The CH_4 production is followed by CO_2 production depending on the proportion between acetate, propionate and butyrate. It has been demonstrated by Fahey & Berger (1988) that a high grain diet will cause a proportion of 2:2:1 between the acids and a ratio of 3:1 between CH_4 production and fermentation-CO_2.

A general model of nutrient utilization in ruminants is shown in Fig. 1. The present investigation was an attempt to quantify the components in the model by using data from experiments with growing calves under different feeding regimes.

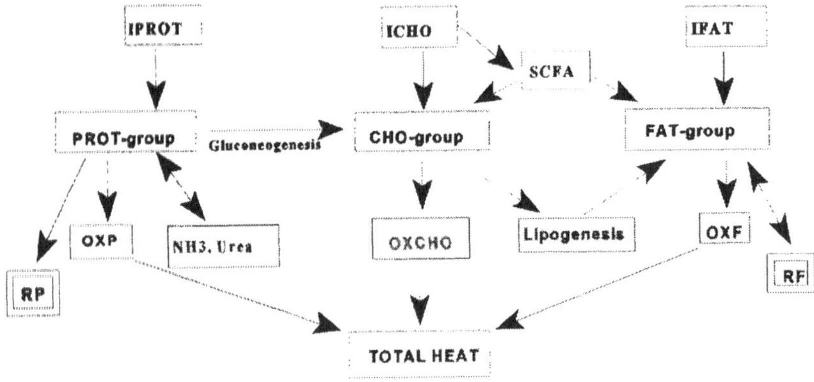

Fig. 1. Model of nutrient oxidation and retention in ruminants. IPROT: ingested protein, ICHO: ingested carbohydrate, IFAT: ingested fat, OXP: oxidized protein, OXCHO: oxidized carbohydrate, OXF: oxidized fat; SCFA: short chain fatty acids, RP: retained protein, RF: retained/mobilized fat.

Materials and Methods

The data (expt. I) were taken from measurements of the energy metabolism in 24 Holstein-Friesian bull calves fed an identical concentrate mixture supplied with clover grass hay (Ser. F), dried sugar beet pulp (Ser. G) or dried clover grass pellets (Ser. H), (Thorbek, 1980). The calves were fed near *ad libitum* and measured in individual balance and respiration experiments during the growth period from 100 to 250 kg live weight (LW). The data (expt. II) were taken from measurements with 12 Danish Red bull calves fed on a complete commercial feed compound supplied with barley straw (Chwalibog *et al*, 1996a). The calves were fed *ad libitum* and measured in individual balance and respiration experiments in two LW ranges from 150 to 250 kg (Period A) and from 300 to 400 kg (Period B).

The 24 h gas exchange was in all experiments measured in the middle of the collection periods by means of open-air-circuit respiration units. The calculations were carried out in accordance with the equations:

CO_2x, l = total CO_2, l - 3 x CH_4, l

RQnpx = CO_2x, l - (UN, g x 6.25 x 0.774) / O_2, l - (UN, g x 6.25 x 0.957)

HEx, kJ = 16.18 x O_2, l + 5.02 x CO_2x, l - 5.99 x UN, g

OXP, kJ = UN, g x 6.25 x 18.42

OXCHO, kJ = (-2.968 x O_2, l + 4.174 x CO_2x, l - 2.446 x UN, g) x 17.58

OXF, kJ = (1.719 x O_2, l - 1.719 x CO_2x, l - 1.963 x UN, g) x 39.76.

Results and Discussion

The mean intake of ME and digested protein varied depending on feed composition and LW of animals. The CH_4 production in relation to ingested organic matter reflected the differences in proportion between concentrates and roughages in the diets (Table 1).

Table 1. Mean live weight (LW), intake of ME and digested protein (DP), and CH_4 production in relation to ingested organic matter (IOM).

		LW kg	ME $kJ/kg^{0.75}$		DP $kJ/kg^{0.75}$		CH_4/IOM l/kg	
Expt.	n	Mean	Mean	SEM	Mean	SEM	Mean	SEM
I F	30	216	974	12.0	11.1	0.11	27.2	1.92
I G	24	206	879	11.2	11.6	0.22	22.4	1.09
I H	20	170	718	13.4	10	0.38	35.9	0.97
II A	23	193	1176	19.8	12	0.19	12.7	0.20
II B	30	342	1055	18.2	10.9	0.23	21.2	0.52

The intake of DP was between 10 and 12 $g/kg^{0.75}$ in all experiments and heat from OXP constituted about 16% of HEx in experiment I (Table 2), corresponding to results obtained in experiments with pigs (Chwalibog *et al*, 1992). In experiment II the values were lower, being between 10 and 12% of HEx, indicating that the norm applied was near the limit for obtaining maximum protein retention (RP) (Chwalibog *et al*, 1996b). Gluconeogenesis from protein calculated as a difference DP-OXP-RP was 9.4% of ingested protein in experiment I decreasing to 6.0% in experiment II-A, concurring with the low value of OXP.

Table 2. Mean heat production (HEx) and oxidation of protein (OXP), carbohydrate (OXCHO) and fat (OXF) in relation to HEx.

	HEx MJ/d		OXP/HEx %		OXCHO/HEx %		OXF/HEx %	
Expt.	Mean	SEM	Mean	SEM	Mean	SEM	Mean	SEM
I F	41.5	0.79	14.6	0.31	41.5	3.34	43.9	3.26
I G	37.9	0.97	16.2	0.37	36.5	1.86	47.3	1.65
I H	26.7	1.54	17.9	1.02	12.9	1.99	69.2	1.72
II A	43.9	1.10	9.5	0.28	78.5	1.17	12.0	1.26
II B	62.1	1.28	12.4	0.71	60.2	2.43	27.4	2.39

With CH_4 production increasing from 13 to 36 l/kg IOM, indicating increasing fermentation, OXF increased from 12 to 69% of HEx with reciprocal values of OXCHO. With increasing fermentation less glucose was absorbed directly and oxidized as OXCHO, while more carbohydrate was converted to SCFA and oxidized as OXF (cf. Fig. 1), being in accordance with the values of RQnpx decreasing from 0.96 to 0.75.

An attempt was made to calculate the partition of ingested carbohydrate (ICHO) between directly oxidized (OXCHO) or oxidized as fat (OXF(CHO)) assuming that the relatively small amount of digested fat was retained and that ICHO, kJ = Gross energy - Energy in IP and IF. The results are demonstrated in Fig. 2, indicating that in all experiments 65% of ICHO was oxidized. In Experiment II with calves fed a high level of concentrates, resulting in a low ruminal fermentation, only 58 and 44% was oxidized directly as OXCHO.

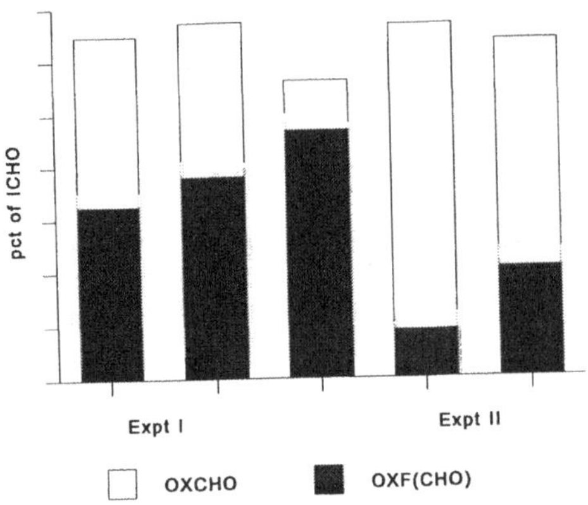

Fig. 2. Carbohydrate metabolism. Oxidation of CHO as carbohydrate (OXCHO) or as fat ((OXF(CHO)).

References

Brouwer E. (1958). On simple formulae for calculating the heat expenditure and the quantities of carbohydrate and fat metabolized in ruminants, from data on gaseous exchange and urine N. Proceedings 1st Symposium on Energy Metabolism, *EAAP Publ.No.* 8:182-194.

Chwalibog A., Jakobsen K., Henckel S. & Thorbek G. (1992). Estimation of quantitative oxidation and fat retention from carbohydrate, protein and fat in growing pigs. *J. Anim. Physiol. Anim. Nutr.* 68:123-135.

Chwalibog A. & Thorbek G. (1995). Quantitative partition of protein, carbohydrate and fat pools in growing pigs. *Arch. Anim. Nutr.* 48:53-61.

Chwalibog A., Jensen K. & Thorbek G. (1996a). Quantitative protein and fat metabolism in bull calves treated with beta-adrenergic agonist. *Arch. Anim. Nutr.* 49:159-167.

Chwalibog A., Jakobsen K. & Thorbek G. (1996b). The pattern of protein retention in pigs from 2 to 120 kg live weight. *Arch. Anim. Nutr.* 49:181-186.

Fahey Jr. C.C. & Berger L.L. (1988). Carbohydrate nutrition of ruminants. In: D.C. Church (Ed) The ruminant animal. Prentice-Hall, Englewood Cliffs, N.J. pp. 269-297.

Lusk G. (1928). The Elements of the Science of Nutrition. 4th Edn,Saunders, Philadelphia.

Thorbek G. (1980). Report from the National Institute of Animal Science, 498:104 pp.

SUMMARY OF THE DISCUSSION

Discussion leader: N.D. Scollan
Scientific secretaries J. van Milgen & A.D. Mitchell

The papers presented in this session were discussed in relation to several different categories, including: body composition measurement, methane measurement, oxygen consumption and carbon dioxide production, nutrient metabolism and heat production.

The use of deuterium oxide (D_2O) for measuring body composition of dairy cows was discussed. The use of a one pool model was thought to be appropriate due to lack of equilibrium with gut water. Factors that might influence the relationship between observed D_2O space and body composition were discussed. These included the contribution of mammary gland contents, the level of milk production, the use of bST, water turnover during the equilibrium period and the water content of visceral vs. subcutaneous fat depots. It was noted that the relative recovery of H_2O and D_2O during the analysis should be proportional as long as at least 2 to 3 ml are recovered during the lyophilization procedure. The recovery of D_2O was approximately 99%.

The other body composition measurement technique discussed was dual energy x-ray absorptiometry (DEXA). This technique was being used as a reference in developing a model for the prediction of body composition of pigs using non-invasive measurements. At the present, the model will predict growth, but not composition.

Some of the problems associated with the use of sulphur hexafluoride (SF_6) as a tracer for methane measurements were discussed. Recovery of SF_6 was slightly higher in chambers than with outdoor systems. Under pasture conditions, one of the main problems is changing wind conditions. The strategic placement of collection canisters around the pasture was offered as a possible solution. The environmental contribution to methane background levels by oil wells, bogs and swamps was also mentioned. An *in vitro* technique assessing methane production from feedstuffs was discussed. For comparing differences between forage and concentrate diets it was noted that there is no opportunity for adaptation with the *in vitro* technique. Carbon dioxide production might also be influenced by the type of buffer that is used.

Regarding the use of ^{13}C-bicarbonate (CO_2 entry rate) to measure CO_2 production, it was noted that both the level of activity and food intake could result in shifts between lipid and carbohydrate oxidation which could alter CO_2 production. The use of near-infrared laser technique was cited as a novel approach to measuring O_2 consumption and energy expenditure. Although the measurement was confined to O_2 consumption by the brain, it is hoped that correlations between whole body calorimetry and NIR values can be developed and that the technique can be applied to other body parts.

Difficulties were reported when using S-type transit-time ultrasound blood flow probes in the portal vein. These probes underestimate the blood flow by approximately 30% compared to the newer A-type probe and the dye-dilution technique (RAH), which is smaller and circular.

In discussing the use of ^{13}C, ^{15}N-amino acids for measuring post absorptive protein and fat metabolism it was noted that there were small flux rates that were not accounted for in the steady state model and also that metabolic differences between ^{13}C and ^{15}N can alter transfer kinetics.

When discussing the modest response of male thin tail sheep to clenbuterol treatment it was noted that this breed of sheep was fairly lean by nature and that, although not measured, there could also be a low density of receptors for the adrenergic agonist.

SESSION 5

GROWTH

MACRONUTRIENT UTILIZATION BY OFFSPRING FROM WILD AND SELECTED ATLANTIC SALMON

B. Grisdale-Helland and S.J. Helland

AKVAFORSK, Institute of Aquaculture Research AS, N-6600 Sunndalsøra, Norway

Summary

Macronutrient utilization by offspring from wild and farmed Atlantic salmon was compared in a 7-week growth and metabolism trial, six weeks after the fish were transferred to saltwater. The parents of the farmed fish had been selected for growth for 4 generations. When calculated in relation to body weight, feed intake and growth were similar in the two lines in this trial, but feed efficiency was significantly better in the selected line. This was related to the low fat digestibility in the wild line, resulting in greater losses of energy in faeces. Digestible energy was utilized similarly in the two lines, but the wild fish used more ($P<0.05$) absorbed, N-containing compounds for energy. It is hypothesized that low fat digestibility during the period of adjustment to saltwater may be a means of reducing osmotic stress associated with ketone body excretion.
Key words: Selection, feed efficiency, digestibility, energy metabolism, protein metabolism, salmon.

Introduction

In Norway, present-day farmed Atlantic salmon have been selected for growth rate (weight at 190 days after hatching and after 2 years in the sea (Gjedrem *et al.*, 1991)) in a combined family-individual selection programme for 5 generations. Together with improvements in nutrition and management, this has resulted in a reduction in production time over the last twenty years of approximately 1.5 years for 3.5 kg salmon. Because of the lack of an unselected control line, the effects that domestication and the selection for growth has had on feed utilization were investigated by comparing offspring from wild-caught salmon (Namsen river) and offspring from fish from the fourth generation of the selection programme.

Materials and Methods

A 7-week growth and metabolism trial was started six weeks after the fish were transferred to saltwater (9.4±0.1°C). Measurements of hypo-osmoregulatory ability made before the fish were transferred to saltwater indicated that the wild fish were better adapted for this transfer than the selected fish, although there was no mortality when the fish were transferred (unpublished results). Because differences in growth rate between the two lines were evident early and it was desired to start this experiment with fish of similar weight, the wild fish were raised in heated water for several months prior to the transfer to saltwater. At the start of the trial, the average weights were 90 g for the wild fish and 126 g for the selected fish. Initial slaughter groups of 15 fish from each line were taken at the start of the trial. Subsequently, six groups of fish from each line were placed in 150 l semi-open, semi-closed metabolism tanks (Grisdale-Helland & Helland, 1995) (2 fish kg/ tank) and were fed *ad libitum* an extruded, fishmeal, wheat and fish oil based diet (dry matter (DM) basis: protein, 50%; fat, 26%; starch, 14%; ash, 10%; energy, 23.8 MJ/kg). The diet contained 0.01% yttrium oxide as a marker for digestibility determinations. Daily feed intake of the fish in each tank was calculated by taking into account measurements of waste feed sieved from the effluent water (Helland *et al.*, 1996). Over a five day period during the second half of the trial, faeces were sieved from the effluent water from three tanks for each genetic line (designated the metabolism groups), using automatic collectors (Choubert *et al.*, 1982). Following a two day fast at the end of the feeding period, the other three tanks of fish from each line (designated the slaughter groups) were weighed and all of the fish slaughtered. After weighing the metabolism groups, they were fasted for a further 9 days and again weighed. In each of the metabolism groups, continuous measurements were done of ammonia and urea production during 6 days in the feeding period and of oxygen consumption during 26 days in the feeding period and during all of the fasting days.

The diet and faeces were analysed for starch content and the diet, faeces, carcass and viscera were analysed for DM, ash, nitrogen, fat, and gross energy (GE) as described by Grisdale-Helland & Helland (1997). Yttrium was analysed as described by Refstie *et al.* (1997). Complete energy and nitrogen budgets were determined for the metabolism groups. Heat increment was determined as the difference between fed and 4-6 days fasted oxygen consumption. Because of the need for fasting the fish, final body composition of these fish at the end of the feeding period was estimated from analysis of the slaughter groups. Energy used for maintenance and activity was estimated by subtracting heat increment from the total heat loss (13.6 kJ/g oxygen consumption; Elliot & Davison, 1975). The data were analysed by analysis of variance and are presented as means±SE. One outlier in the digestibility data (wild fish) and one outlier in the oxygen consumption data (selected fish) were removed from the statistical analyses.

Results

Relative feed intake and specific growth rate (Table 1) were not significantly different between the genetic lines. The feed efficiency ratio was significantly greater in the selected line than in the wild line. There were no significant differences between the genetic lines in carcass or viscera chemical composition.

Table 1. Performance of the wild and selected lines of Atlantic salmon (mean±SE, n=6).

	Start weight (g)	End weight (g)	Intake[1] (% BW/d)	SGR[2] (% BW/d)	FE[3]
Wild line	90±2*	155±5*	0.87±0.05	1.18±0.04	1.23±0.01*
Selected line	127±5	215±7	0.87±0.05	1.15±0.04	1.31±0.02

[1]Daily feed intake as a percentage of body weight (BW), dry matter basis.
[2]Specific growth rate [100*(ln(end weight) - ln(start weight))/days fed].
[3]Feed efficiency ratio (g wet gain/g dry feed intake).
*Indicates a significant difference between the genetic lines ($P<0.05$).

The apparent digestibility coefficients (Table 2) of protein and starch were not affected by genetic line, but fat and energy digestibility were significantly lower in the wild line than in the selected line.

The combined data from the slaughter and metabolism groups (Table 3) shows that, in comparison with the selected fish, the wild fish consumed more GE per kg fish produced, but lost more energy in faeces and deposited less energy than the selected fish. There were no significant differences between the lines in the amounts of energy lost in branchial and urinary excretions (ammonia and urea), heat increment, or the sum of maintenance and activity, per kg growth. A significant difference in retention of consumed energy (43.0±0.1% and 51.1±0.3%, respectively for the wild and selected lines) was found, but there were no differences between the lines in the efficiency of use of digestible energy for metabolizable energy (96%), net energy for production (66%) or total net energy (83%).

Significantly more N per kg growth was consumed by the wild line in comparison with the selected line (Table 3). A small but significant difference was found in the amount of N deposited per kg growth, and the percentage of consumed N which was retained in the body was greater ($P<0.05$) in the selected line (51.5±0.02) than in the wild line (47.0±0.12%). There were no significant differences between the lines in the estimates of N losses per kg gain in faeces, ammonia or urea. The wild and selected fish used 43.7±0.5 and 38.0±0.5% ($P<0.05$), respectively, of absorbed, N-containing compounds for energy.

Table 2. Apparent digestibility coefficients (%) of the macronutrients and energy in the diet fed to the wild (n=2) and selected lines (n=3) of Atlantic salmon (mean±SE).

	Protein	Fat	Starch	Energy
Wild line	83.5±0.5	46.7±1.4*	71.9±0.2	65.5±0.5*
Selected line	82.9±0.4	75.7±1.8	68.4±1.2	76.5±0.8

*Indicates a significant difference between the genetic lines ($P<0.05$).

Table 3. Partitioning of energy and nitrogen consumed by the fish, per kg growth (mean±SE, n=2).

	Intake	Faeces	AU[1]	Heat Increment	MA[2]	Growth
Energy (MJ						
Wild line	19.3±0.08*	6.7±0.07*	0.6±0.10	1.7±0.41	2.7±0.33	8.3±0.02*
Selected line	17.8±0.02	4.1±0.14	0.5±0.04	1.6±0.09	3.4±0.21	9.1±0.06
Nitrogen (g)						
Wild line	64.4±0.25*	10.6±0.29	25.1±4.02	-30.3±0.04*		
Selected line	59.5±0.08	10.1±0.44	21.4±1.45	-	-	30.7±0.05

[1]Sum of ammonia and urea excretions.
[2]Sum of maintenance and activity (difference between total heat loss and heat increment).
*Indicates a significant difference between the genetic lines ($P<0.05$).

Discussion

The performance and metabolism of offspring from wild fish and fish domesticated and selected for growth for four generations were studied in this trial. The experiment was done at a time in the life cycle of the salmon when they were recovering from their transfer from freshwater to saltwater. During the total three-year period in which the growth of these fish were compared, the fish from the selected line grew approximately twice as fast as the fish from the wild line. Therefore, it was not expected that the growth rates in these two lines would be equal in this trial. The better hypo-osmoregulatory ability of the wild line at saltwater transfer may have contributed to the similar growth of the two lines during the trial period, in comparison with the total period in which these groups were studied.

Selection for growth may influence both appetite and the efficiency of feed conversion. Better feed conversion may occur through a reduction in the maintenance requirement, an increase in the amount of energy available for gain in relation to that used for maintenance, or through a change in the composition of gain (McCarthy & Siegel, 1983). The difference in FER between the genetic lines seems to be mainly due to fat digestibility and consequently, the amount of metabolizable energy available for gain. In addition, the wild fish used more of the digested protein for energy, thus reducing the fraction available for muscle growth. Several reports have previously shown significant differences in protein digestibility between rainbow trout families (Austreng & Refstie, 1979) or strains (Medale, 1993) and in energy digestibility between families (Refstie & Austreng, 1981).

Assuming no lipogenesis from protein or starch, only 8% of the absorbed fat was used for energy in the wild line, compared with 31% in the selected line. Ketones are a source of energy for the brain of fasting Atlantic salmon (Soengas *et al.*, 1996). Excess production of ketones from fatty acids and their subsequent excretion, however, may be a source of osmotic stress for these fish during the period of adaptation to saltwater when feed intake is reduced. Thus, the reduced digestion and/or absorption of fat in the wild fish may be hypothesized to be a mechanism of defense against this situation. A study with these lines of fish when they weighed 500-1000 g showed normal fat digestibility coefficients (95%) in both lines, indicating that the low fat digestibility in this study was a transitory condition.

References

Austreng E. & Refstie T. (1979). Effect of varying dietary protein level in different families of rainbow trout. *Aquaculture* 18:145-156.

Choubert G., De la Noüe J. & Luquet P. (1982). Digestibility in fish: improved device for the automatic collection of faeces. *Aquaculture* 29:185-189.

Elliott, J.M. & Davison W. (1975). Energy equivalents of oxygen consumption in animal energetics. *Oecologia* (Berl.) 19:195-201.

Gjedrem, T., Gjøen H.M. & Gjerde B. (1991). Genetic origin of Norwegian farmed Atlantic salmon. *Aquaculture* 98:41-50.

Grisdale-Helland B. & Helland S.J. (1995). Methods to assess optimal nutrient composition of diets for Atlantic salmon. *J. Appl. Ichthyol.* 11:359-362.

Grisdale-Helland B. & Helland S.J. (1997). Replacement of protein by fat and starch in diets for Atlantic salmon (*Salmo salar*) at the end of the freshwater stage. *Aquaculture* (In Press).

Helland S.J., Grisdale-Helland B. & Nerland S. (1996). A simple method for the measurement of daily feed intake of groups of fish in tanks. *Aquaculture* 139:157-163.

McCarthy J.C. & Siegel P.B. (1983). A review of genetical and physiological effects of selection in meat-type poultry. *Animal Breeding Abstracts* 51(2):87-94.

Medale F. (1993). Relation between growth and utilization of energy substrates in three rainbow trout strains. In: S.J. Kaushik & P. Luquet (Eds): Fish nutrition in practice. INRA (Les Colloques, No 61), Paris, p. 37-48.

Refstie S., Helland S.J. & Storebakken T. (1997). Adaptation to soybean meal in diets for rainbow trout, *Oncorhynchus mykiss*. *Aquaculture* (In Press).

Refstie T. & Austreng E. (1981). Starch in rainbow trout diets. III. Growth and chemical composition of fish from different families fed four levels of starch in the diet. *Aquaculture* 25:35-49.

Soengas J.L., Strong E.F., Fuentes J., Aldegunde M. &. Andrés M.D. (1996). Post-feeding starch and ketone bodies metabolism in brain and liver of Atlantic salmon. *J. Physiol. Biochem.* 52:131-142.

EFFECT OF BREED AND BODY WEIGHT ON COMPONENTS OF HEAT PRODUCTION IN GROWING PIGS

J. Noblet[1], J. F. Bernier[2], S. Dubois[1], Y. Le Cozler[1] and J. van Milgen[1]

[1] Institut National de la Recherche Agronomique, Station de Recherches Porcines, 35590 Saint-Gilles, France; [2] Département des Sciences Animales, Université Laval, Ste-Foy, Québec, G1K 7P4, Canada

Summary

Heat production (HP) and its components (fasting heat production: FHP, activity HP: AHP and thermic effect of feed: TEF) were obtained in four groups of *ad libitum* fed growing pigs (Meishan and Large White castrates and Large White and Piétrain males) over the 20 to 60 kg live weight range. Animals were kept individually in the respiration chamber. Both daily FHP (700 vs. 990 kJ/kg $BW^{0.60}$) and the contribution of FHP to total HP (53 vs. 65%) were lower in Meishan compared to the other groups. On the other hand, TEF either as a proportion of total HP (42 vs. 29%) or relative to ME intake (21 vs. 16%) was higher in Meishan pigs. Energy loss due to physical activity contributed to about 5% of total HP. Overall, HP at zero physical activity and adjusted for similar ME intakes or similar amounts of protein and fat gains was lower (~12%) in Meishan pigs.

Key words: Growing pigs, breed, heat production.

Introduction

It has become usual to partition energy requirements of growing animals between requirements for maintenance and requirements for production. The maintenance fraction can vary widely with animal factors such as breed (Noblet *et al.*, 1991) and body composition or with behaviour and environmental factors. Under thermoneutral conditions, fasting heat production (FHP) is considered an accurate predictor of maintenance energy requirements. Requirements for growth are proportional to the amount of retained proteins and lipids which are mainly affected by breed, sex and body weight (BW). Alternatively, feed intake and composition will affect the amount of energy retained. Total heat production of a growing animal has therefore different origins: FHP, thermic effect of energy supplied by feed (TEF) when used for maintenance (at or below zero energy gain) and for energy deposition, heat due to physical activity (AHP) and energy generated specifically for thermoregulation when animals are kept below their lower critical temperature.

The present study was designed to study the effect of breed or growth potential on the three components of heat production (FHP, TEF and AHP) of growing pigs kept at thermoneutrality. The overall objective of the programme is to develop new approaches for predicting energy requirements of growing pigs. Methodologies for estimating FHP and the energy cost of physical activity for the present experiment have been described by van Milgen *et al.* (1997).

Material and Methods

Four groups of pigs (Meishan castrates: MC, Large White castrates: LWC, Large White males: LWM and Piétrain males: PM) were fed *ad libitum* a starter diet which contained 19.6% crude protein and 13.3 MJ ME per kg up to 25 kg and subsequently a cereal based growing diet containing 19.3% crude protein and 13.0 MJ ME per kg. MC can be considered as fat type pigs while LWC, LWM and PM correspond to lean type pigs, PM being the leanest. Animals were housed individually. At approximately 20, 35 and 55 kg

BW, animals were transferred to metabolism cages for measurement of feed intake and collection of faeces and urine over 8 to 10 consecutive days. Over the last 5 days of the collection period, the cage was introduced in a respiration chamber for continuous measurement of feeding behaviour and duration of standing of animals and oxygen (O_2) and carbon dioxide (CO_2) contents of the ingoing and outgoing air from the chamber. After these 5 days, feed was removed and measurements continued for 30 hours. Standing activity was measured with infra-red beams. Results on feed intake and faeces and urine losses and their chemical composition were used to estimate digestible (DE) and metabolisable (ME) energy values of feeds and the nitrogen balance of pigs, according to routine procedures. A total of 77 digestibility and energy balance measurements were carried out (Table 1).

Analysis of data recorded during the fasting period was carried out according to the model described by van Milgen *et al.* (1997). It provides for each measurement period 1) estimates of O_2 and CO_2 exchanges after prolonged fasting and at zero physical activity; and 2) the gas exchanges per unit of time of activity. Associated HP were calculated from gas exchanges with Brouwer equation without correction for urinary-N and CH_4 production.

Measurements on fed pigs provided estimates of their mean daily ME intakes, O_2 and CO_2 exchanges and duration of standing during this period. The gas exchanges and heat production associated with physical activity (AHP) while fed were estimated by multiplying the duration of standing by its unitary cost as obtained from measurements during the fasting period. The gas exchanges and heat production associated with the utilisation of feed (TEF) were then calculated as the difference between total gas exchanges or total HP and gas exchanges or heat production originating from both fasting and physical activity. The TEF component as calculated in this experiment corresponds to HP associated with feed energy utilisation for both maintenance and growth. The (partial) respiratory quotients for each component of HP were calculated as the ratios between CO_2 production and O_2 consumption. Total energy gain and its partition between protein and fat were also calculated.

Results were analysed according to variance or covariance models which took into account the main effects of group and stage of growth and their interaction.

Results

As indicated in Table 1, the effect of stage of growth was, as expected, significant for most criteria expressed on a daily basis with few significant interactions between stage of growth and group of animals. Therefore, in order to simplify the presentation of results, most attention will be paid to the effect of group which was the primary objective of the study.

Spontaneous energy intakes differed between groups with the lowest values in MC and PM and the highest in LWC; castration of LW males was associated with an increased energy intake (Table 1). Groups of pigs were also quite different with regard to the amount and the composition of energy gain: MC deposited the smallest quantity of protein and the highest amount of lipids and PM retained the smallest amount of lipids. According to these differences in ME intakes and composition of energy gain, RQ values were significantly different between groups with values ranging between 1.18 and 1.08.

As presented in Tables 1 and 2, FHP expressed either as MJ per day or per kg $BW^{0.60}$ was about 30% lower in MC than in the other groups; it was not affected by castration in animals from the LW breed and significantly lower in the lean type Piétrain pigs than in the conventional Large White animals. Despite visual perception, MC spent more time standing than the 3 other groups. In connection with a higher unitary energy cost of standing in PM (van Milgen *et al.*, 1997), their daily AHP was higher. The proportion of ME intake resulting in TEF was dependent on group with again a marked difference between MC (21% of ME intake) and animals of the 3 other groups (16% of ME intake). These differences between the fat MC and the lean LWC, LWM and PM are accompanied by a higher proportion of total HP associated with FHP in the lean animals. Even though groups differed in AHP, its contribution to total HP was small (5%). The partial RQ values were not highly affected by group. The high RQ for TEF (above 2) should be related to the method of calculation used in the present experiment which assumes that gas exchanges and RQ for FHP are the same in fasted and fed animals, which is not obviously the case.

Table 1. Effect of breed and sex on energy intake, energy gain, heat production (HP) and its components and RQ values in *ad libitum* growing pigs.

Breed	Meishan	Large White		Piétrain	RSD	Statistical
Sex[2]	C	C	M	M		signif.[1]
Measurements, n	17	17	21	22		
Body weight, kg	38.6	40.6	39.4	40.9	3.4	S**
ME intake, MJ/day	23.0	26.9	24.3	22.5	3.4	S** G**
Energy gain as						
Protein, MJ/day	2.56	4.74	4.69	4.36	0.75	S** G**
Fat, MJ/day	8.89	7.99	6.19	4.91	1.93	S** G** SG*
Total HP						
MJ/day	11.5	14.2	13.4	13.3	1.33	S** G**
% of ME intake	50.4	53.4	55.1	58.9	4.6	S** G**
Fasting HP						
MJ/day	6.02	9.09	9.01	8.61	0.77	S** G**
% of total HP	52.6	64.8	67.0	64.9	4.8	G**
Activity, min/day	214	153	148	159	37	G**S*
Activity HP						
MJ/day	0.57	0.64	0.66	0.90	0.17	S** G**
% of total HP	4.9	4.7	4.9	6.7	1.3	G**
Thermic effect of feed						
MJ/day	4.95	4.49	3.75	3.77	0.96	S** G**
% of total HP	42.5	30.6	28.1	28.3	4.9	G**
% of ME intake	21.3	16.2	15.5	16.6	2.5	S* G**
RQ values						
Total	1.18	1.14	1.10	1.08	.04	S* G**
Fasting HP	0.72	0.75	0.74	0.73	0.06	S*
Activity HP	0.91	0.78	0.86	0.86	0.15	
Thermic effect of feed	1.99	2.46	2.50	2.30	0.54	G*

[1] The analysis of variance included the effects of stage of growth (S; n=3), group (G; n=4) and the interaction between stage of growth and group (SG); RSD for residual standard deviation; levels of significance: *: P<0.05, **: P<0.01

[2] C for castrated males and M for males

In order to take into account differences in ME intake or differences in composition of gain between groups, covariance analyses were carried out (Table 2). The interactions between covariates and group were not significant. The higher TEF in MC attenuated the marked difference in FHP between MC and other groups; but the MC group still differed significantly from the 3 other groups with a 12% lower resting HP for the same level of "performance". Pigs from the 3 other groups had almost identical resting HP.

Table 2. Effect of breed and sex on heat production (HP) due to fasting, activity and at rest in *ad libitum* growing pigs (kJ/kg $BW^{0.60}$/day).

Breed	Meishan	Large White		Piétrain	RSD	Statistical
Sex[1]	C	C	M	M		signficance
ME intake	2704	2946	2747	2518	331	S** G**
Fasting HP	698	1005	1010	950	67	S* G**
Activity HP	65	72	74	99	18	G**
Resting HP[2]	1274	1425	1427	1421	78	G**
Resting HP[3]	1264	1444	1429	1411	103	G*

[1] See Table 1

[2] Resting HP calculated as total HP minus AHP; adjusted by covariance (G as main effect) for a similar ME intake (2710 kJ/kg $BW^{0.60}$); slope of adjustment: 0.27 ± 0.02

[3] Adjusted by covariance (G as main effect) for similar protein and fat gains (473 and 769 kJ/kg $BW^{0.60}$, respectively); slopes of adjustment: 0.39 ± 0.12 and 0.24 ± 0.06 respectively.

Discussion

An estimate of ME requirements for maintenance (ME_m) can be calculated by using FHP and the TEF obtained in the present experiment, even though TEF per unit of ME is lower below than above maintenance (Noblet *et al.*, 1993). Estimates of ME_m at zero physical activity are then 1.15 to 1.20 MJ ME per kg $BW^{0.60}$ per day in the lean breeds. These values are considerably higher than most literature estimates, especially those expressed per kg $BW^{0.75}$ (450 to 500 kJ ME). In addition, these latter estimates include the cost of physical activity. One explanation for this discrepancy might be related to the type of pigs which were fatter in the past with a subsequently lower FHP, as demonstrated by results obtained with MC in the present experiment. Another more likely explanation would be differences in methodologies. As explained by Noblet *et al.* (1991), most ME_m values have been obtained by regression methods with ME intakes much higher than ME_m and marginal efficiencies of ME for energy gain probably lower than the overall efficiency as measured in the present experiment. Indeed, the TEF estimate averaged 16% of ME intake in our conditions while values obtained by regression methods range between 25 and 40%. Moreover, the ME_m calculated here is based on the measured FHP in previously *ad libitum* growing animals which is probably physiologically different from an extrapolated value at zero energy retention of measurements conducted at variable energy levels.

The high rate of lipid deposition in MC pigs would have suggested a lower TEF value per unit of ME intake and a higher RQ for TEF. The contradiction between this assumption and the observed values challenges our understanding of energy metabolism and requires further investigations. For instance, the kinetics of adaptation of energy metabolism and nutrient utilization in pigs with different growth potential under nutritional challenges such as feed restriction has not been described adequately, especially according to more dynamic and mechanistic approaches.

References

Noblet J., Karege C. & Dubois S. (1991). Influence of growth potential on energy requirements for maintenance in growing pigs. In: C. Wenk & M. Boessinger (Eds) Energy Metabolism of Farm Animals. Juris Druck+Verlag, Zürich *EAAP Publ. No.* 58:107-110.

Noblet J., Shi X.S. & Dubois S. (1993). Metabolic utilization of dietary energy and nutrients for maintenance energy requirements in pigs: basis for a net energy system. *Br. J. Nutr.* 70:407-419.

ENERGY UTILIZATION IN REALIMENTED PIGS IN RELATION TO INITIAL BODY COMPOSITION

Skiba G. and Fandrejewski H.

The Kielanowski Institute of Animal Physiology and Nutrition, 05-110 Jabonna, Poland

Summary

Landrace gilts from 15 to 25 kg body weight (BW) were fed a control diet (C) (1.4 MJ/kg$^{0.75}$) or restricted by 40% for energy and protein (C60), or restricted for protein only (P60). From 25 to 70 kg body weight pigs were fed at standard level feeding or *ad libitum*. Animals were slaughtered at 15 (n=4), 25 (n=12) and 70 kg BW (34 pigs) and their bodies were analysed for physical and chemical composition. At 25 kg BW the body of P60 pigs contained more fat and less protein, and weights of internal organs were similar to those of the control group, whereas the C60 contained less fat in the carcass and organ weights were lower. From 25 to 70 kg BW the P60 and C60 pigs gained faster and deposited 9 g/d more protein than the C pigs. The C60 pigs consumed more food and deposited more lipid in the whole body and protein in viscera than the P60 group. The C60 pigs utilized ME most efficiently. This could be attributed to lower maintenance requirements as a result of smaller size of internal organs before realimentation. Increasing the initial store of energy in the body at 25 kg did not improve rate of protein deposition.

Key words: Energy, pigs, compensatory growth, body composition.

Introduction

It is well known that pigs can increase their growth rate and food utilization following nutritional restriction (Campbell *et al.*, 1983; de Greef, 1992; Bikker, 1994) but understanding of this phenomenon is still unclear. Physiologically, it seems that the most important factors which influence a compensatory response are connected with the intrinsic state of the animal at the end of the period of limitation, namely the size of metabolically active internal organs (Koong, 1982), which contribute a great part of fasting heat production (Fuller *et al.*, 1987). Moreover, the growth of realimented animals depend on the initial store of fat in their bodies (Kyriazakis *et al.*, 1991).

The aim of this study was to determine the energetic efficiency of gain in realimented pigs, which initially differed in body energy reserves, and in the size of internal organs.

Materials and Methods

Fifty Polish Landrace gilts from 15 to 70 kg BW were used in the study. Four animals were slaughtered at 15 kg BW as zero group. The remaining 46 animals were divided into three groups and fed from 25 to 70 kg BW on three feeding regimes: control C; group P60 containing 60% of protein in relation to group C; group C60 containing 60% of protein and energy in relation to group C. The control diet (based on cereals and soybean meal) contained 12.87 MJ ME/kg and 16.9% CP. The diet for P60 group was diluted by corn starch. At 25 kg four gilts from each group were randomly selected and slaughtered. The remaining pigs were fed the same diet from 25 to 70 kg on a standard feeding level (R) or *ad libitum* (A). The aim of this procedure was to eliminate the influence of higher voluntary intake during the realimentation period on the composition of body gain. The cereals-rapeseed meal mixture contained 12.44 MJ, 16.9% C.P. and 1.08% lysine. The ME content of the diets was estimated according to the Rostock system. Daily rations for pigs of R group were changed weekly according to their body weight. All animals were kept in individual pens. They were slaughtered after 16 hours of fasting. All internal organs and emptied small and large intestines were removed and weighed separately. Carcasses and offal were analysedd for chemical composition. Energy balance was calculated from comparative slaughter.

Results and Discussion

As was intended in the first period of growth (15-25 kg), the C and P60 pigs deposited daily similar amounts of fat (72 and 69 g), whereas the P60 and C60 pigs deposited similar amounts of protein (38 and 34 g, respectively). The ADG significantly differed among the C, P60 and C60 groups (513, 367 and 242 g, respectively, Table 1). At 25 kg the weights of offal and individual internal organs were lower in the C60 than P60 groups. This confirmed an earlier suggestion (Fandrejewski, 1994) that these organs are sensitive to nutritional stress resulting from limitation in feed intake, and are almost resistant to reduction of protein when energy intake is not limited.

Table 1. Performance and body composition of pigs in the restriction period (15-25 kg).

Item	Treatment			SEM
	C	P60	C60	
CP intake, g/d	180^{A}	114^{B}	110^{B}	1.19
ME intake, MJ/d	13.7^{A}	13.2^{A}	8.4^{B}	0.12
ADG, g	513^{A}	367^{B}	242^{C}	5.50
Protein deposition, g/d	66^{A}	38^{B}	34^{B}	1.38
Fat deposition, g/d	72^{a}	69^{a}	11^{b}	7.78
LW, kg	25.2	25.4	24.8	0.20
EBW, kg	23.0	23.1	22.7	0.24
Carcass, kg	17.95	18.45	18.64	0.15
Offal, kg	5.05^{A}	4.65^{A}	4.06^{B}	0.08
liver, g	623^{A}	612^{A}	411^{B}	13.3
small intestines, g	1007^{A}	868^{A}	711^{B}	27.4
large intestines, g	496^{A}	500^{A}	394^{B}	9.04
kidneys, g	113^{A}	99^{A}	83^{B}	2.60
Protein, g/kg of EBW	150^{A}	138^{B}	159^{A}	1.81
in the carcass, %	80.6^{a}	81.5^{a}	83.6^{b}	0.32
in the non-carcass part, %	19.4^{a}	18.5^{a}	16.4^{b}	0.32
Fat, g/kg of EBW	129^{A}	184^{B}	88^{C}	1.76
in the carcass, %	91.5	92.4	93.3	0.37
in the non-carcass part, %	8.5	7.6	6.7	0.37
Energy, MJ/kg of EBW	8.71^{A}	10.61^{B}	7.29^{C}	0.14

$^{a, b, c}$ - $P<0.05$ $^{A, B, C}$ - $P<0.01$

Initial content in the body (at 15 kg): 2.12 kg protein and 1.47 kg fat.

The P60 and C60 pigs differed in energy content and chemical composition of whole body and protein distribution in the carcass and offal, but not in distribution of fat. At the end of the restriction period the P60 and C60 pigs were 8 and 22 days, respectively, older than controls. During the realimentation period (25-70 kg) the group A pigs, having free access to food, consumed 9% more feed, grew 12% faster, deposited more protein (9%) and fat (23%) and were more fat at 70 kg than restricted-fed animals (R). The previously underfed P60 and C60 pigs grew faster and deposited more protein (8-9 g/d) than the C pigs. Gadeken *et al.* (1980), Kyriazakis *et al.* (1991) and de Greef (1992) also reported that animals had an increased preference for protein accretion during the realimentation period. It was estimated from rate of body gain that the animals tended to higher protein production for 3-4 weeks, and this compensatory response occurred in both the carcass and offal tissue with slightly higher enhancement in offal of the C60 than P60 pigs (difference 1.6g/d, P<0.01). Increasing initial storage of fat in the body (P group) did not improve the rate of protein deposition. However, it was quite possible that P60 group animals mobilized some part of body fat as an energy source for protein synthesis because at 70 kg they contained relatively less fat than at 25 kg. The rate of fat deposition was highest in C60 group (198 g), lowest in P60 group (154 g) and intermediate in the C group (182 g/d). These differences between treatments were independent of rations R and A, although there was a treatment x feeding level interaction for daily food intake (P<0.05), as the P60 pigs given free access to feed ate at a greater rate (2.45 kg) than the C60 (2.24 kg), whilst the C pigs consumed 2.31 kg per day. The C60 pigs did not have a higher appetite immediately after a change from restriction to realimentation, but 4-5 weeks later.

This was probably due to smaller organs at the end of restriction period, as found in a study with ruminants (Ryan, 1989). On the other hand the P60 pigs ate less food throughout the whole realimentation period, as they were initially fatter. The results of the study indicated that feed intake is not constant in the realimented pigs but depended on type of restriction, and that gastrointestinal and nervous mechanisms of regulation may be involved at different rates. The C, P60 and C60 pigs did not differ in chemical body composition or organ sizes at 70 kg BW. Thus, compensation was almost complete, although the P60 and C60 animals were still older than controls.

Table 2. Performance and body composition of pigs in the realimentation phase (25-70 kg).

Item	Treatment			Feeding regime		SEM
	C	P60	C60	R	A	
	12	1	11	18	16	
Food intake, kg*	2.23	2.18	2.28	2.13[A]	2.33[B]	0.20
ADG, g	810[a]	833[ab]	867[b]	766[A]	908[B]	7.81
FCR	2.80[a]	2.63[b]	2.66[ab]	2.8[A]	2.6[B]	0.03
EBW, kg	65.5	65.4	65.2	66.0	64.7	0.23
Protein, g/kg of EBW	155	157	159	160[a]	154[b]	1.1
Fat, g/kg of EBW	197	192	187	180	196	3.7
Offal, kg	12.23	12.20	12.62	11.89[A]	12.81[B]	0.13
Protein deposition, g/d						
in the whole body	122[a]	129[b]	130[b]	122[A]	133[B]	1.6
in the carcass	105	110	109	105[a]	112[b]	0.3
in the offal	17[A]	19[B]	21[C]	17[A]	21[B]	0.3
Fat deposition, g/d						
in the whole body	182[A]	154[B]	198[C]	160[A]	196[B]	4.7
in the carcass	169[A]	144[B]	183[C]	149[A]	181[B]	4.5
in the offal	13[A]	10[B]	15[C]	11[A]	15[B]	0.3

a, b, c - P<0.05 and A, B, C - P<0.01.
* treatment x feeding system interaction (P<0.05).

Finally, the gross efficiency of utilization of ME for energy gain was affected by the treatments (Table 3). The ER/ME coefficient was lower in the P60 pigs than in controls (33.8 vs. 36.2%) due to higher deposition of protein and lower deposition of fat. In contrast, the C60 pigs utilized energy more efficiently (38.5%) and deposited more energy as fat. Assuming the ME-utilization for growth as k_p=0.37 and k_f=0.72, i.e. obtained by comparative slaughter method for growing gilts (Fandrejewski, 1991), the amount of energy for maintenance requirement was calculated to be 506 (C60), 570 (P60) and 584 kJ/kg$^{0.75}$ (C). Thus the C60 pigs, which were previously restricted in food intake, had a 12% lower maintenance requirement than the P60 previously restricted in protein. This could be attributed to smaller size of internal organs before realimentation.

Table 3. Energy balance (kJ/kg$^{0.75}$) in pigs from 25 to 70 kg.

	Treatment			Feeding system		SEM
	C	P60	C60	R	A	
ME	1560	1507	1576	1462	1633	14
$RE_{protein}$	163^{a}	172^{b}	173^{b}	160^{A}	177^{B}	2
RE_{fat}	405^{A}	340^{B}	434^{A}	351^{A}	434^{B}	10
RE/ME	0.362^{A}	0.338^{B}	0.384^{C}	0.350^{A}	0.373^{B}	0.4

a, b, c - $P<0.05$ and A, B, C - $P<0.01$

Conclusions

During restricted feeding phase the growth of internal organs is more sensitive to energy than protein deficiency in the diet. The compensatory mechanism in pigs is associated with enhanced protein deposition in both the offal and carcass. Increased body energy reserves at the beginning of realimentation had no significant influence on the subsequent rate of protein deposition. It seems the variability in energy efficiency during compensatory growth could be partially explained by the size of internal organs and their associated maintenance requirements.

This study is funded by grant No 5PO6E00410 from the Committee for Scientific Research.

References

Bikker P. (1994). Protein and lipid accretion in body components of growing pigs: effects of body weight and nutrient intake. PhD thesis, Department of Animal Nutrition, Wageningen Agricultural University.

Campbell R.G. & Dunkin A.C. (1983). Influence of feeding level and protein level in early life on the cellurality of adipose tissue and body fat content of growing pigs. *Br. J. Nutr.* 49:109-118.

Fandrejewski H. (1991). Energy metabolism in growing pigs. Proceedings of the 12th Symposium Energy Metabolism of Farm Animals. C. Wenk and M. Boessinger (Eds). *EAAP Publ. No.* 58:103-106.

Fandrejewski H. (1994). Effect of protein intake on chemical body composition, size of internal organs and energy retention in growing pigs. Proceedings of the 13th Symposium. Energy Metabolism of Farm Animals. J.F. Aguilera (Ed). *EAAP Publ. No.*76:265-268.

Fuller M.F., Cadenhead A., Mollison G. & Seve B. (1987). Effects of the amount and quality of dietary protein on nitrogen metabolism and heat production in growing pigs. *Br. J. Nutr.* 58:277-285.

Gadeken D., Bohme H. & Oslag H.J. (1980). Restriction of feed intake and compensatory growth responses in pigs. In Energy Metabolism. L.Mount. (Ed) *EAAP Publ. No.* 26:407-410.

Greef K.H. (1992). Prediction of production: Nutrition induced tissue partitioning in growing pigs. PhD thesis, Department of Animal Nutrition, Wageningen Agriculture University.

Koong L.J., Nienaber J.A., Peaks J.C. & Yen J.T. (1982). Effects of plane of nutrition on organ size and fasting heat production in pigs. *J. Nutr.* 112:1638-1642.

Kyriazakis I. & Emmans G.C. (1991). Diet selection in pigs: dietary choices made by growing pigs following a period of underfeeding with protein. *Anim. Prod.* 52:337-346.

Kyriazakis I., Stamataris C., Emmans G.C. & Whittemore C.T. (1991). The effects of food protein content on the performance of pigs previously given foods with low or moderate protein contents. *Anim. Prod.* 52:165-173.

Ryan W.J. (1989). Compensatory growth in sheep and cattle. PhD Thesis, School of Agriculture. The University of Western Australia.

ENERGY EXPENDITURE OF ADOLESCENTS IN STANDARDIZED AND FREE-LIVING CONDITIONS; VARIATIONS WITH AGE, GENDER, BODY COMPOSITION AND PHYSICAL ACTIVITY

M. Vermorel[1], A. Bitar[1], J. Vernet[1], N. Fellmann[2], B. Beaune[2] and J. Coudert[2]

[1]INRA Theix, Energy and Lipid Metabolism Research Unit; [2]Faculty of Medicine, Physiology and Sports Biology Laboratory, F 63000 Clermont-Ferrand, France

Summary

Body composition (fat-free mass, FFM, and fat mass, FM), physical capacities as well as daily (DEE), sleeping (SEE) and circadian energy expenditure (EE) of 101 healthy volunteers from 10 to 28 years of age were determined by whole-body indirect calorimetry. FFM, maximal oxygen consumption (Peak VO_2) and gender were the main determinants of DEE and SEE. The latter were significantly lower in females than in males, even after adjustment for FFM and Peak VO_2. DEE was determined in free-living conditions over 5 days using the heart rate (HR) recording method. On a schoolday, DEE was 13 to 22% lower in girls than in boys. Two hours of physical training induced a mean 16.3% increase in DEE. Intra-group variability of DEE averaged 12.5% during a schoolday and ranged from 14 to 28% during the free days because of wide differences in physical activity.
Key words: Human, puberty, body composition, energy expenditure, physical activity.

Introduction

Adolescence is a period of deep and rapid anatomical, physiological and behavioural alterations. At peak growth rate, annual height gain is almost doubled, and annual weight gain is three to fourfold greater than during the prepubertal phase. Furthermore, muscle mass is strongly increased, especially in boys. These alterations are expected to induce major changes in energy expenditure (EE) and consequently in energy requirements of adolescents. The aim of the present study was, therefore, to investigate the changes in body composition, physical capacities and EE of adolescents and young adults both in standardized conditions in order to exclude the effect of variations in physical activity, and in free-living conditions. The protocol was approved by the local Ethical Committee.

Materials and Methods

The study was carried out with 101 healthy, sedentary or moderately active volunteers of boths genders of 10-11, 12-13, 14-16 and 20-28 years of age, during spring and autumn (Table 1). After a thorough medical examination, body composition (FFM and FM) was estimated using both the bioimpedance analysis and the skinfold thickness methods. Physical capacities were estimated through measurement of Peak VO_2 during a continuous exercice on a cyclo-ergometer against increasing braking forces.

Daily (DEE) and circadian EE was determined in standardized conditions by indirect calorimetry. Volunteers spent 36 h in two open-circuit 11 m^3 calorimeters: one evening and one night for adaptation and 24 h for measurements of EE and HR. The standardized activity programme included four 15 min periods of exercise on a cycloergometer. EE and HR were recorded continuously. EE was partitioned between 6 periods: effective sleep (8 h), BMR (1 h), rest (10 h), meals (2 h), exercises + recovery (2 h) and miscellaneous activities (1 h). In addition, EE and HR were measured by laboratory tests using a face mask during lying, cycling, sitting, stepping, walking at three speeds, and jogging on a treadmill, including recovery periods. Individual relationships were computed between EE and HR recorded in the calorimeters and during the laboratory tests. The best fit was obtained with polynomial relationships of the third order (Bitar *et al.*, 1996).

In other respects, HR was recorded every minute by telemetry using Sport Testers (PE 4000, Polar Electro KY, Kemple, Finland) and activity patterns were recorded by the volunteers in free-living conditions during 5 days (3 schooldays and one weekend). EE was computed from HR recordings using the computed individual relationships.

Values are expressed as means and SD. Data were analysed using SAS Software (SAS, 1987). Group comparisons were done using two-or-three-way analyses of variance. Stepwise multiple regression was used to determine the major determinants of EE. Single and multiple-linear regression analysis was performed using the GLM procedure.

Results and Discussion

Intra-group variability in height, BW, FFM and FM was high and the mean values were close to those recently recorded (Tichet & Vol, 1997, pers comm.) so volunteers could be considered as being representative of the French population. There were no gender differences in these parameters, except for FFM in 15-year-old adolescents. However, all these parameters were significantly different between men and women. Boys gained 27.0 cm, 22.6 kg BW, 20.9 kg FFM and only 1.6 kg FM on average between 10.5 and 15.0 years of age. The corresponding figures were 29.6 cm, 21.1 kg, 16.5 kg and 4.6 kg, respectively, in girls. The percentage of FM decreased from 21.3 to 15.3 in boys between 10.5 and 23 years of age (Table 1). Peak VO_2 was not measured satisfactorily in adults. In children and adolescents it increased significantly with age and was significantly higher in males than in females from 12 years of age onwards, even after adjustment for FFM.

For the same activity programme in the calorimeters, DEE was 5% (NS), 9%, 19% and 25% ($P<0.001$) lower in females than in males at 10.5, 12.6, 15.0 and 23 years of age, respectively. The results were similar for SEE with differences of 7, 9 (NS), 18 and 21% at the same ages ($P<0.001$). Furthermore, DEE and SEE increased with age but plateaued at 12.6 and 15.0 years of age in girls and boys, respectively, in spite of further increases in height, BW and FFM (Table 1). The main determinants of DEE and SEE in children and adolescents were FFM ($R^2 = 0.81$), Peak VO_2 ($R^2 = 0.03$), and gender ($R^2 = 0.02$, $P<0.001$), whereas FM was not a significant factor. DEE and SEE adjusted for FFM and Peak VO_2 were 6.1 and 3.8%, respectively, lower in girls than in boys ($P<0.001$).

Differences in adjusted DEE and SEE between genders or growth stages could be attributed to alterations of FFM composition, protein turnover and hormonal status during puberty. The proportion of less metabolically active tissues such as muscle increases at the expense of metabolically active organs such as brain, kidneys, heart, liver and gut (Elia, 1992). Protein turnover decreases since protein synthesis is similar but proteolysis is 12% lower in pubertal than in prepubertal children (Arslanian & Kalhan, 1996). Finally, hormonal changes during puberty may induce changes in energy metabolism with advancing age during adolescence (Brown *et al.*, 1996), which could explain the larger differences observed in girls than in boys between 12.6 and 15.0 years of age.

DEE of 12-13 and 14-16 year-old adolescents in free-living conditions were determined during 5 days from HR recordings: a schoolday, a schoolday including 2 h physical training, a Wednesday (school in the morning and free activities in the afternoon), a Saturday and a Sunday. DEEs of girls during the schoolday were 13 and 22% ($P<0.01$) lower than those of boys of 12.6 and 15.0 years of age, respectively. DEE of girls did not vary significantly with age, whereas DEE of boys increased by 10% ($P<0.02$). The intra-group coefficient of variation (cv) averaged 12.5%. Two hours of physical training induced a significant 16.3% increase in DEE on average. Mean DEEs were not significantly different between Wednesday, Saturday and Sunday on one hand, and the schoolday on the other hand. However, the intra-group cv (17% in girls and 28% in boys)

Table 1. Physical characteristics, maximal oxygen consumption (Peak VO_2) and energy expenditure of subjects either following a standardized activity programme in the calorimetric chambers or in free-living conditions (means and SD).

Group (years of age)	10 - 11		12 - 13		14 - 16		20 - 28	
Gender	M	F	M	F	M	F	M	F
Number	12	9	16	15	16	15	9	9
Age (years)	10.6 (0.5)	10.4 (0.4)	12.6 (0.3)	12.6 (0.4)	15.1 (0.8)	14.9 (0.7)	24.2 (4.4)	21.8 2.1)
Height (cm)	141.0^{a} (6.4)	136.0^{a} (5.8)	153.5^{b} (6.7)	153.8^{b} (7.5)	168.1^{c} (8.5)	165.6^{c} (5.7)	179.0^{d} (5.8)	164.0^{c} (4.3)
Weight (kg)	32.9^{a} (5.6)	30.2^{a} (3.1)	45.1^{b} (9.3)	43.2^{b} (5.8)	55.5^{c} (9.6)	51.3^{c} (5.1)	70.4^{d} (9.7)	58.3^{c} (7.1)
FFM (kg)	25.9^{a} (3.7)	23.2^{a} (1.8)	35.2^{b} (6.7)	33.8^{b} (4.0)	46.9^{d} (7.7)	39.7^{c} (3.7)	60.7^{e} (7.9)	45.3^{d} (5.0)
FM % BW	20.9^{a} (5.7)	22.9^{a} (4.7)	21.3^{a} (5.8)	21.5^{a} (5.0)	15.3^{b} (3.5)	22.6^{a} (2.5)	13.6^{b} (2.7)	19.4^{a} (2.3)
Peak VO_2 (ml/min)	1.56^{ab} (0.24)	1.40^{a} (0.12)	1.92^{c} (0.29)	1.67^{b} (0.28)	2.62^{d} (0.44)	1.93^{c} (0.26)	-	-
DEE (MJ/d)	7.48 (1.03)	7.13^{a} (0.65)	10.16^{d} (1.40)	9.27^{c} (1.12)	11.78^{e} (1.46)	9.52^{cd} (1.01)	11.40^{e} (1.00)	8.58^{b} 0.92)
SEE (kJ/min)	3.45^{a} (0.39)	3.21^{a} (0.26)	4.53^{c} (0.59)	4.10^{b} (0.43)	5.28^{d} (0.82)	4.32^{bc} (0.49)	4.90^{cd} (0.60)	3.88^{b} (0.33)
FREE LIVING CONDITIONS: Schoolday (MJ/d)			10.59^{b} (1.87)	9.20^{a} (1.04)	11.63^{c} (1.32)	9.08^{a} (0.89)		
Schoolday + 2 h PT (MJ/d)			11.99^{b} (1.90)	$10..55^{a}$ (1.18)	13.97^{c} (2.12)	10.64^{a} (1.09)		
Wednesday (MJ/d)			0.87^{ab} (2.55)	9.51^{a} (1.49)	11.27^{b} (3.58)	9.97^{a} (1.82)		
Saturday (MJ/d)			11.02^{b} (2.73)	9.30^{a} (1.39)	11.81^{b} (2.38)	8.98^{a} (1.75)		
Sunday (MJ/d)			10.68^{b} (2.61)	9.20^{a} (1.14)	11.34^{b} (2.15)	9.03^{a} (1.45)		
Extrapolated to one week (MJ/d)			11.02^{b} (2.00)	9.57^{a} (1.06)	11.94^{c} (1.49)	9.5^{a} (1.04)		
Mean DEE over a week (kJ/kg BW)			243.0^{c} (31.7)	223.4^{b} (27.0)	218.9^{b} (32.9)	185.2^{a} (15.6)		

BW: body weight. M: males; F: females; FFM: fat-free mass; FM: fat mass; DEE: daily energy expenditure; SEE = sleeping energy expenditure. PT: physical training; Wednesday morning, school; afternoon; free.
$^{a, b, c, d, e}$: on the same line, values with different superscripts are significantly different ($P<0.05$).

was higher than during the schoolday, due to wide interindividual differences in physical activities.

DEE was estimated for the two other schooldays in order to calculate mean DEE over a week (DEEw). The latter did not vary with age in girls, but increased by 8% ($P<0.05$) in boys. DEEw of girls were 13.2 and 20.5% lower than those of boys at 12.6 and 15.0 years of age, respectively ($P<0.01$). DEEw per kg BW decreased significantly by 10% in boys and 17% in girls ($P<0.04$ and 0.01, respectively) between 12.5 and 15.0 years of age. Furthermore, DEEw/kg BW and the cv obtained with these 4 groups of adolescents were very close to those observed with groups of 8 to 12 subjects of similar characteristics in free-living conditions using the doubly labelled water method (Davies *et al.*, 1991).

Conclusion

The present study showed the main determinants of SEE and DEE of adolescents determined in standardized conditions and the significantly lower DEE and SEE of girls compared to those of boys after adjustment of FFM, Peak VO_2 or age. The results obtained in free-living conditions underlined the high interindividual variability of DEE in adolescents of similar ages, due to differences in anthropometry, body composition and especially physical activity. Further calculations will allow estimation of EE for the main types of activities according to age and gender. In addition, regression equations will be computed to predict mean weekly DEE of individuals from age, gender, anthropometry, body composition and detailed activities with a view to improving current recommended dietary allowances.

References

Arslanian S.A. & Kalhan S.C. (1996). Protein turnover during puberty in normal children. *Am. J. Physiol.* 270:E79-E84.

Bitar A., Vermorel M., Fellmann N., Bedu M., Chamoux A. & Coudert J. (1996). Heart-rate recording method validated by whole-body indirect calorimetry in 10 year-old children. *J. Appl. Phys.* 81:1169-1173.

Brown D.C., Kelnar C.J. & Wu F.C. (1996). Energy metabolism during male human puberty. I. Changes in energy expenditure during the onset of puberty in boys. *Ann. Hum. Biol.* 23:273-279.

Davies P.S.W., Livingstone M.B.E., Prentice A.M., Coward W.A., Jagger S.E., Stewart C., Strain J.J. & Whitehead R.G. (1991). The energy expenditure during childhood and adolescence. *Proc. Nutr. Soc.* 50:14A.

Elia M. (1992). Organ and tissue contribution to metabolic rate. In: J.M. Kinney and M.N. Tucker (Eds) Energy metabolism: tissue determinants and cellular corollaries. Raven Press Ltd, New York, p. 71-79.

Statistical Analyses System Institute, Inc. (1987) SAS/STAT guide for personal computers. Version 6. SAS Institute Inc, Cary, NC.

EFFECTS OF DIET ENERGY DENSITY AND PROTEIN (AMINO ACID) CONTENT ON INTAKE, BODY COMPOSITION AND ENERGY METABOLISM OF BOARS AND GILTS FED *AD LIBITUM* FROM 22 TO 46 KG

K.J. McCracken[1], V.E. Beattie[2], R.N. Weatherup[2], S.G. McIlroy[3] and R.W. Henry[4]

[1] Department of Agricultural and Environmental Science, The Queen's University of Belfast and Department of Agriculture for Northern Ireland, Newforge Lane, Belfast BT9 5PX, Northern Ireland; [2] Agricultural Research Institute of Northern Ireland, Large Park, Hillsborough, Co Down, BT26 6DR, Northern Ireland; [3] John Thompson and Sons Ltd, 35-39 York Road, Belfast BT15 3GW, Northern Ireland; [4] Nutec Ltd, 14 Victoria Road, Bangor BT29 5EX, Northern Ireland

Summary

48 Large White x Landrace pigs, balanced for sex, were used in a 4 x 2 x 2 factorial experiment from 22 to 46 kg. The factors were: diet energy content (12.4, 13.4, 14.4, 15.4 MJ/kg of DE); digestible lysine (low, 9.5; high 12.0 g/kg diet); sex (entire males, females). They were killed at 46 kg for body composition analysis. Digestibility coefficients increased with diet energy content and were lower for the low lysine diets. Dietary protein did not affect body composition or gain of protein, fat or energy. ME intake, empty body (EB) gain, ME:gain and crude protein (CP) gain were similar across dietary energy levels. Fat and energy gain increased with diet energy level and heat production tended to fall. ME intake was 10% higher for gilts than for boars but EB gain was the same so that ME:gain was 10% poorer. Protein gain was the same for both sexes but fat gain and energy gain were higher for gilts.

Key words: Diet energy, lysine, sex, digestibility, metabolizable energy, body composition, energy retention, heat production, efficiency of energy utilization.

Introduction

There is a considerable volume of literature on the interactions of energy and protein intake with growing pigs from 40 to 100 kg but little with modern strains of pig in the early growth phase. Feed intake would appear to be a major constraint on growth in the period up to 40 kg and the studies of Rao & McCracken (1992) showed that the effects of energy restriction in the 40-90 kg phase are greater with pigs of high genetic potential. Bikker (1994) studied a wide range of lysine levels in combination with two levels of energy intake with gilts of high genetic potential for lean growth and concluded that the digestible lysine requirement was approximately 0.6 g/MJ of DE if a linear-plateau model was applied but close to 0.8 g/MJ using a quadratic model. There are environmental and economic reasons why protein (lysine) levels should not be in excess. The present study was part of a larger performance study on individually fed pigs and was specifically designed to examine the effects on body composition and energy metabolism of a wide range of dietary energy contents in conjunction with one of two (low or high) levels of protein (lysine) in diets of similar protein quality fed 'to appetite' to growing pigs.

Materials and Methods

The experimental design was a 4 (energy) x 2 (lysine) x 2 (sex) factorial with 3 pigs per treatment. The planned energy levels were 12.4, 13.4, 14.4, 15.4 MJ DE with either 9.5 g (low) or 12.0 g (high) digestible lysine per kg diet fed. The two low-energy diets were based on barley, wheat and wheat-feed while the high-energy diets incorporated wheat and maize with added fat. The main protein source was soya the inclusion of which was kept as constant as possible within each protein level. The analyses of the diets are given in Table 1.

Table 1. Analyses (g/kg DM) of diets.

Energy	12.4	13.4	14.4	15.4	12.4	13.4	14.4	15.4
Protein (lysine)		High				Low		
Crude protein	261	254	249	248	215	205	209	206
Ether extract	39	37	42	69	36	34	51	75
Ash	65	55	51	51	74	64	58	55
Gross energy (MJ/kg)	18.5	18.6	19.0	19.5	18.8	18.9	19.0	19.6

Forty-eight Large White x Landrace pigs (equal numbers of males and females), of mean initial weight 22 kg, were allocated at random to one of the eight dietary treatments and fed the diets to appetite in individual pens. Feed allowances were adjusted daily to ensure that feed intake was close to *ad libitum* but avoiding wastage. Pigs were weighed weekly and at the end of the experiment. At about 46 kg, pigs were slaughtered for body composition analysis. In addition 6 pigs were used to estimate initial body composition. The carcass and offal components (including head and feet) were analysed separately. Crude protein (CP) and fat deposition were determined and energy retention calculated using the values 23.8 and 39.6 MJ/kg respectively for protein and fat deposition.

A 7 d balance was conducted on a separate set of pigs (4 per treatment) to determine the digestibility of nutrients and ME contents of the diets using a separate latin-square design for each protein level.

Results

Digestibility increased (P<0.001) with increasing dietary energy content and the determined values for DE (Table 2) were in good agreement with the formulations. The digestibility coefficients for DM and energy were 2% lower (P<0.05) for the lower lysine diets and crude protein digestibility was 5% lower (P<0.001).

Table 2. Effects of dietary energy and protein (lysine) level on digestibility coefficients and on DE and ME contents of diets.

	Energy level						Protein level			
	12.4	13.4	14.4	15.4	P=	SEM	High	Low	P=	SEM
Dry matter	0.77	0.82	0.87	0.88	***	0.003	0.84	0.83	*	0.003
Crude protein	0.81	0.84	0.88	0.88	***	0.004	0.87	0.83	***	0.004
Energy	0.77	0.81	0.87	0.88	***	0.002	0.84	0.83	*	0.003
DE MJ/kg DM	14.3	15.2	16.5	17.3	***	0.06	16.0	15.7	**	0.05
ME MJ/kg DM	13.9	14.8	16.1	16.8	***	0.06	15.5	15.2	*	0.06

* P < 0.05; ** P < 0.01; *** P< 0.001

There were no significant effects of protein level on ME intake, empty body (EB) gain, ME:gain or any aspect of body composition or energy metabolism (Table 3).

There were no significant effects of dietary energy level on ME intake or EB gain but values tended to be lower for the 13.4 diet and higher for the 14.4 diet. Total body DM, fat and energy increased (P<0.01) with increasing dietary energy but total CP was unaffected. CP gain (g/d) was similar at all energy levels but fat gain increased (P<0.01) with energy level. When the energy data were related to metabolic body size ($W^{0.75}$), ME intake was not significantly different across treatments but energy gain increased linearly with energy level. Although not statistically significant, there was a 7% reduction in calculated heat production from the low to the high dietary energy level.

ME intake of females was 10% higher ($P<0.01$) than for males (23.5 vs. 21.3, SEM 0.45). Empty body gain was the same for both sexes with the result that ME:gain was higher ($P<0.01$) for gilts (29.7 vs. 26.9 MJ/kg, SEM 0.61). Associated with the higher ME intake, total body DM, fat and energy of gilts were higher ($P<0.001$) and total body water was reduced ($P<0.001$) but total CP was the same for gilts and boars (Table 4). CP gain (g/d) was the same for boars and gilts but fat gain was 40% higher ($P<0.001$) for gilts. ME intake and energy retention (ER) were 11 and 20% higher ($P<0.001$) for gilts and heat production (HP) was 6 % higher ($P<0.05$).

Table 3. Effects of energy and protein level on performance, body composition and energy metabolism of pigs from 22 to 46 kg (df 31).

	Energy level						Protein level		
	12.4	13.4	14.4	15.4	P=	SEM	High	Low	SEM
ME intake	22.3	21.5	23.6	22.4	NS	0.63	22.6	22.3	0.45
EB gain (g/d)	789	752	841	811	NS	25.0	799	797	18.0
ME:gain	28.6	28.6	28.3	27.7	NS	0.86	28.4	28.1	0.61
Total body									
DM (kg)	13.0	13.1	13.7	13.9	**	0.21	13.4	13.5	0.14
CP (kg)	7.4	7.6	7.4	7.4	NS	0.09	7.5	7.4	0.06
Fat (kg)	4.3	4.2	5.0	5.0	**	0.24	4.6	4.7	0.16
Energy(MJ)	344	346	371	374	**	8.7	359	359	5.6
CP:EBW	173	177	171	172	NS	1.8	174	172	1.3
Fat: EBW	99	98	117	118	**	5.0	107	109	3.5
CP gain(g/d)	151	150	154	152	NS	4.8	153	151	3.1
Fat gain(g/d)	96	89	124	126	**	9.7	108	110	6.3
Energy metabolism									
ME (kJ/kg$^{0.75}$)	1580	1532	1658	1593	NS	42.5	1597	1585	30.1
ER (kJ/kg$^{0.75}$)	518	501	606	605	**	26.3	558	557	18.6
HP (kJ/kg$^{0.75}$)	1062	1032	1052	988	NS	29.7	1039	1028	21.0

Table 4. Effects of sex on body composition and energy metabolism of pigs from 22 to 46 kg (n, 24; df 31).

	Total body				Gain g/d		Energy (kJ/kg$^{0.75}$)		
	DM	CP	Fat	Energy	CP	Fat	ME	ER	HP
Boars	13.0	7.4	4.1	339	150	90	1508	567	1001
Gilts	14.0	7.5	5.1	379	153	127	1673	608	1066
P=	***	NS	***	***	NS	***	***	***	*
SEM	0.14	0.06	0.16	5.6	3.1	6.3	30.1	18.6	21.0

Discussion

The lack of effect of protein (lysine), even with the high energy diet, suggests that the value of 0.6 g digestible lysine (Bikker, 1994) is adequate for these pigs at the levels of feed intake achieved. However, these were considerably greater than those normally observed with group-fed pigs and the results cannot be directly translated to commercial practice.

The 10% difference in intake between gilts and boars contrasts with the results of Cole & Chadd (1989) and McCracken & Stockdale (1989) with older pigs, fed to appetite, where no differences were observed. The results suggest that, under the *ad libitum* conditions prevailing, the gilts had attained their ceiling of genetic potential for lean growth and the additional energy was stored as fat. Moreover this fat storage was associated with reduced total body water with the result that EB gain was not increased and feed:gain was poorer.

Although ME intake was similar at the highest and lowest dietary energy levels there was a significant increase in fat and energy retention with no increase in protein retention and a concominant reduction in calculated heat production. Assuming the value for maintenance requirement reported by Rao & McCracken (1991), the calculated values for efficiency of energy utilization for gain were, respectively, 0.53, 0.54, 0.58, 0.62 for the 12.4, 13.4, 14.4 and 15.4 diets reflecting the lower expected efficiency of energy utilization with the diets of lower digestibility (Noblet *et al.*, 1993) and the relatively high protein: fat ratio in the gain.

References

Bikker P. (1994). Protein and lipid accretion in body components of growing pigs. PhD thesis, Wageningen University.

Cole D.J.A. & Chadd S.A. (1989). Voluntary feed intake of growing pigs. *Br. Soc. Anim. Prod.* Occasional Publication, No. 13:61-70.

McCracken K.J. & Stockdale R.I. (1989). Voluntary feed intake of pigs of high genetic potential fed pellets to appetite: effects of sex and dietary crude protein content. *Br. Soc. Anim. Prod.* Occasional Publication, No. 13:117-118.

Noblet J., Fortune H., Dupire C. & Dubois S. (1993). Digestible, metabolizable and net energy values of thirteen feedstuffs for growing pigs: effective energy system. *Animal Feed Sci. Technol.* 42:131-149.

Rao D.S. & McCracken K.J. (1991). Effect of energy intake on protein and energy metabolism of boars of high genetic potential for lean growth. *Anim. Prod.* 52:499-507.

Rao D.S. & McCracken K.J. (1992). Energy:protein interactions in growing boars of high genetic potential for lean growth. 2. Effects on chemical composition of gain and whole body protein-turnover. *Anim. Prod.* 54:83-93.

ENERGY METABOLISM OF THE HEAVY PIG: RESULTS OF DIFFERENT TRIALS

G.M. Crovetto, G. Galassi, A. Tamburini, A. Sandrucci, L. Rapetti and G. Succi

Istituto di Zootecnia Generale, Facoltà di Agraria, via Celoria 2, 20133 Milano, Italy

Summary

Three energy balance experiments were performed in barrows of 120, 150 and 170 kg on average. Intake energy decreased with increasing body weight and so did retained energy even if expressed as percentage of intake or metabolizable energy. Me_m, calculated by regression was 441 kJ/kg $W^{0.75}$. The efficiency of utilization of ME for growth (k_g) was 73.4%, a value in agreement with those reported by literature.
Key words: Pig, energy metabolism, energy utilization.

Introduction

In Italy more than 80% of pig production is destined to the Italian traditional ham ("prosciutto") which represents 27% of total pig weight, but 60% of the whole commercial value of the animal. In order to obtain a ham of standard quality the pig must be slaughtered at least at 160 kg of body weight (W). Many experiments have been conducted on growing pigs weighing up to 100-110 kg W (a.o.: Fernandez *et al.*, 1985; Bakker & Dekker, 1991; Laswai *et al.*, 1991; Susenbeth & Menke, 1991; Bikker, 1994; Quiniou *et al.*, 1996), but to our knowledge only one study has been carried out to study energy metabolism in the fattening heavy pig (above 120 kg) (Noblet *et al.*, 1994). With the present work we studied the energy metabolism of pigs between 120 and 170 kg to estimate maintenance requirements and the partition of intake and metabolizable energy in this fattening phase.

Materials and Methods

Seventy two 24 h energy balances were carried out in two different cross-over hierarchical designed experiments utilizing, for each experiment, 8 Landrace x Large White castrated male pigs allocated in individual metabolic cage. The first experiment was divided into two phases (120 and 150 kg, on average); the second experiment was done at an average weight of 170 kg. The data of two animals one of which had health problems were excluded from the analysis. Thus, data processing was done on 66 daily energy balances.

In the first experiment the four diets tested included mainly maize, wheat bran and soybean meal (Crovetto *et al.*, 1996) and differed for the base liquid (water or milk whey, 20% of diet DM) and the protein level: normal (15.5 and 13.3% as fed in the first and second phase, respectively) or low (11.5 and 10.2%, plus lysine, methionine, threonine and triptophane). Four pigs were always fed the water based-diet, the other four the whey-diet. Within each group of four animals, for each phase, each pair of pigs was fed the two diets (normal N and low N+AA) in succession (cross-over design).

In the second experiment the four diets differed in physical treatment (negative control or expanded) and the addition of enzymes (negative control or β-glucanase). The base diet included: 50% barley, 28.8% maize, 10.5% soybean meal, 4% wheat bran, 3% sugarcane molasse, 1% tallow plus minerals and vitamins. The analysis (on DM) was: 14.7% CP, 4.0% EE, 5.1% CF, 0.77% lysine.

In both trials digestibilities were determined individually, while respiratory exchanges were recorded by indirect calorimetry for each pair of pigs (three 24 h cycles during each 8 d period of digestibility). DM was offered at the level of 7.4% $BW^{0.75}$ at 120 kg BW and gradually reduced until 6.8% $BW^{0.75}$ above 180 kg BW. Total heat production (HP) was determined from the respiration exchanges using the Brouwer equation. Samples of feedstuffs, faeces and urine were taken daily. Gross energy of all samples was determined in an adiabatic calorimeter. DE, ME and RE, as well as fat deposition and protein accretion, were calculated for every treatment in each trial. By regression analysis on the overall data set the ME for maintenance (ME_m) and the efficiency of utilization of ME above maintenance (ME_p) for growth (k_g) were also computed.

Results and Discussion

The main results obtained in the experiments are summarized in Table 1. Energy intake per metabolic weight decreased with increasing body weight. Digestibility and metabolizability were lower in Experiment 3; this is probably due to a higher fibre content of diets used in trial 3 as compared to diets fed in trials 1 and 2. Heat production ranged from 47 to 51% of intake energy and, if expressed as a percentage of the ME, increased from trial 1 (120 kg BW) to trial 2 (150 kg) and trial 3 (170 kg). RE varied consequently with a negative trend between the first and the third experiment.

Table 1. Daily energy utilization and partition obtained with pigs of different body weight (BW) in three experiments.

Trial		1	2	3
BW, kg		120	150	170
IE	kJ/kg $W^{0.75}$	1337	1211	1126
	% IE	100	100	100
DE	kJ/kg $W^{0.75}$	1179	1071	952
	% IE	88.2	88.4	84.6
ME	kJ/kg $W^{0.75}$	1155	1051	912
	% IE	86.5	86.7	80.9
	% ME	100	100	100
HP	kJ/kg $W^{0.75}$	626	618	552
	% IE	46.9	51.2	49.2
	% ME	54.2	59.0	60.8
RE	kJ/kg $W^{0.75}$	529	433	360
	% IE	39.6	35.5	31.7
	% ME	45.8	41.0	39.2

Protein and fat deposition in trials 1, 2 and 3 were 4.75, 3.12, 1.18 and 11.16, 9.61, 7.97 g/kg $W^{0.75}$/d, respectively. Metabolizable energy for maintenance (ME_m) was computed with the following equation:

$$RE = 0.734\ MEI - 324\ (r^2=0.88;\ RSD=31.6;\ P<0.001) \qquad (1)$$

where retained energy (RE) and metabolizable energy intake (MEI) are expressed as kJ/kg $BW^{0.75}$ (see Fig. 1). Assuming RE=0, ME_m resulted to be 441 kJ/kg $BW^{0.75}$. This value is similar to that calculated by Bikker (1994) (456 kJ/kg $BW^{0.75}$). Equation (1) indicates an efficiency of utilization of metabolizable energy for growth (k_g) of 0.734.

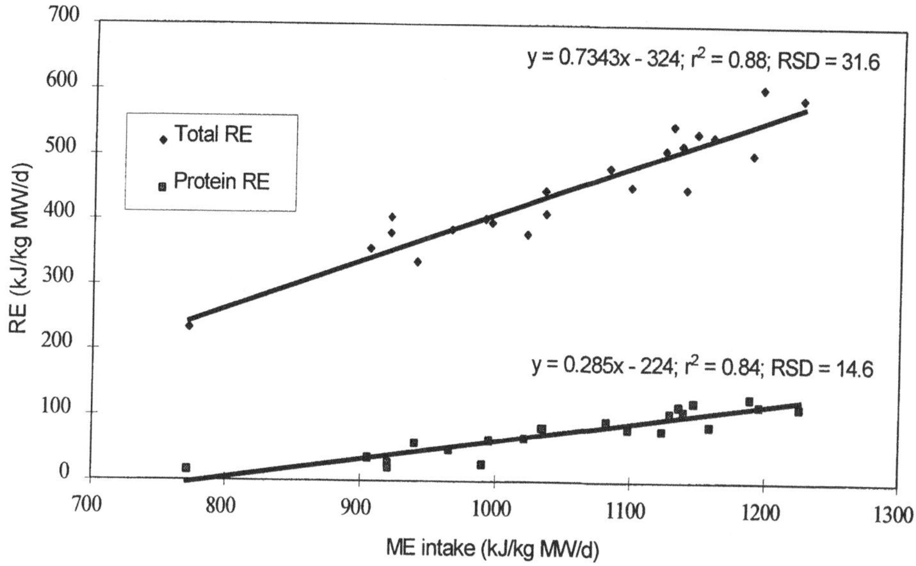

Fig. 1. Relationships between ME intake and total RE and RE as protein in pigs of 120-170 kg of body weight (BW). ME and RE are expressed per metabolic weight ($MW=BW^{0.75}$).

From Figure 1 it can be seen that the total RE and the RE as protein are well described by linear regression. Consequently, also RE as fat (calculated by difference) followed a linear trend. In our experimental conditions, the ratio between the increments in fat and protein energy deposition per unit of MEI ($\Delta FD/\Delta PD$) was constant and equal to 1.58.

The k_g value (0.734) agrees with the value found by Noblet *et al.* (1994) and with those reported - for pigs of lower weight - by Susenbeth and Menke (1991). In fact, the data cited in their paper, referring to pigs of different body mass and different "protein RE/RE" ratios, did not affect systematically the k_g values. This would confirm, as reported by Noblet *et al.* (1994), that k_g does not differ substantially between stages of growth. Laswai *et al.* (1991) and Bikker & Verstegen (1994) obtained k_g values slightly lower in comparison with our result.

From Figure 2 it is evident that the pigs of 170 kg liveweight ingested lower amounts of ME per metabolic weight and had a higher fat to protein deposition (FD/PD) ratio due to the very low nitrogen retention typical of mature animals. FD/PD ratios were similar between the 120 kg and the 150 kg pigs, while at lower weights the MEI per kg $W^{0.75}$ was higher.

Conclusions

The data obtained seem to indicate that the metabolizable energy requirement for maintenance of the heavy pig is similar, if expressed per metabolic weight, to the requirement of pigs of lower body mass. Moreover, the k_g value in the heavy pig was rather constant and similar to the k_g of younger animals, due to the linear relationships between the protein-RE and ME and between the fat-RE and ME.

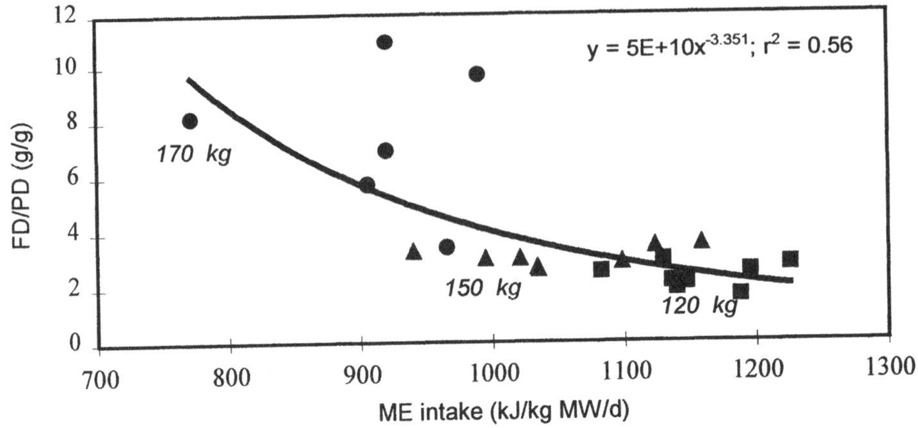

Fig. 2. Relationship between ME intake and the ratio between fat and protein deposition in pigs of 120-170 kg. ME is expressed per MW.

References

Bakker G.C.M. & Dekker R.A. (1991). Effect of source of carbohydrate and amount of fat supplemented on the energy balance of growing fattening pigs. In: C. Wenk and M. Boessinger (Eds) Energy metabolism of farm animals. Schriftenreihe aus dem Institut für Nutztierwissnschaften Gruppe Ernährung ETH-Zürich, *EAAP Publ. No.* 58:95-98.

Bikker P. (1994). Protein and lipid accretion in body components of growing pigs: effects of body weight and nutrient intake. PhD thesis, Wageningen Agricultural University, Wageningen.

Bikker P. & Verstegen M.W.A. (1994). Tissue deposition and energy metabolism in growing pigs: effects of energy intake and nutritional history. In: J.F. Aguilera (Ed) Proceedings of the 12th Symposium on energy metabolism of farm animals. *EAAP Publ. No.* 76:225-228.

Crovetto G.M., Galassi G., Rapetti L., Sandrucci A., Succi G. & Tamburini A. (1996). Energy and nitrogen balance of pigs fed diets with different protein content. Abstract In: J.A.M. van Arendonk (Ed) Book of Abstracts of the 47th Annual Meeting of the European Association for Animal Production. Wageningen Pers, The Netherlands, p. 101.

Fernández J.A., Jorgensen H. & Just A. (1991). The influence of sex and breed on energy utilization and body composition in growing pigs. In: P.W. Moe, H.F. Tyrrell and P.J. Reynolds (Eds) Energy metabolism of farm animals. Agricultural Research Service, United States Department of Agriculture, Beltsville, Maryland 20705, Rowman & Littlefield, pp. 114-117.

Laswai G.H., Close W.H., Sharpe C.E. & Keal H.D. (1991). Nutrient partition and energy metabolism in pigs of different body weights and growth rates. In: C. Wenk and M. Boessinger (Eds) Proceedings of 11th Symposium on energy metabolism of farm animals. *EAAP Publ. No.* 58:90-94.

Noblet J., Shi X.S., Fortune H. & Dubois S. (1994). Effect of diet composition on the efficiency of metabolizable energy for maintenance or growth in pigs. In: J.F. Aguilera (Ed) Proceedings of the 12th Symposium on energy metabolism of farm animals. *EAAP Publ. No.* 76:221-224.

Quiniou N., Dourmad J.Y. & Noblet J. (1996). Effect of energy intake on the performance of different types of pig from 45 to 100 kg body weight. 1. Protein and lipid deposition. *Anim. Sci.* 63:277-288.

Susenbeth A. & Menke K.H. (1991). The independence of the efficiency of energy utilization for growth (k_{pf}) of the composition of body mass gain in pigs. In: C. Wenk and M. Boessinger (Eds) Energy metabolism of farm animals. *EAAP Publ. No.* 58: 99-102.

EFFECT OF CHLORTETRACYCLINE MEDICATED FEED IN THERAPEUTIC LEVELS ON ENERGY AND PROTEIN METABOLISM IN GROWING PIGS

C. Lütolf[1], C. Wenk[2], G. Bee[2], A. Zabka[1], U. Huwyler[1] and M. Wanner[1]

[1]Institute of Animal Nutrition, University of Zurich, Switzerland; [2]Institute of Animal Production-Nutrition Group, ETH Zurich, Switzerland

Summary

The antibiotic effect of chlortetracycline in nutritional doses on the digestive processes as well as a possible anabolic effect on protein deposition in pigs is well documented. In the present study, the effect of therapeutic doses on energy and protein metabolism was evaluated. In 6 respiration experiments with 24 young castrate males from 14 to 24 kg body mass (BM) growth performance, protein and energy metabolism were evaluated. Pigs were divided into two groups with a normal commercial feed without antimicrobial additive (control group, CP) or with 500 mg/kg chlortetracycline (CTC).

Rate of mass gain increased with the CTC-supplementation by 23%, daily protein retention was increased by 28%, whereas daily fat gain was reduced by 23%. Of the retained energy 29 to 30% was deposited as protein and 70 to 71% as fat in the control group, whereas in the chlortetracycline fed group 36 to 44% and 56 to 64% were deposited respectively. The energy costs for maintenance amounted to 584 kJ ME/BM$^{3/4}$. These data demonstrated that chlortetracycline has a distinct growth-promoting and anabolic effect at therapeutic dose rates.
Key words: Energy metabolism, growth performance, pigs, chlortetracycline.

Introduction

Antimicrobial feed additives are commonly used in pig production as growth promoters. Such supplements have mainly selective effects on the microflora and on the strucure of the mucosa in the digestive tract. Improved nutrient utilization and reduced faecal losses are the consequence, although antimicrobial agents may also influence the intermediate nutrient utilization. A higher protein retention and a better feed conversion ratio in growing pigs are good indicators for such effects (Dierick *et al*., 1986; François & Michel, 1968; Greife & Berschauer, 1988; Klaus, 1956; Wenk & Schürch, 1976).

With the ban of antimicrobial feed additives as growth promoters in some countries the use of therapeutic dosages of antibiotics in healthy young pigs is of increasing importance. It is still not known whether similar modes of action on the nutrient utilization can be expected also at such high dosages. Therefore in a series of experiments with piglets, the influence of a therapeutic dose of chlortetracycline on energy utilization and protein retention was studied. With the same antimicrobial agent in therapeutic doses, Zabka (1997) evaluated the influence on nutrient absorption and morphology of the digestive tract of healthy weaned piglets. In her experiments CTC affected the ileal and faecal digestibility of organic matter significantly. Only ileal digestibility of nitrogen was increased by CTC supplementation significantly. Reduced volatile fatty acid and ammonia concentrations indicate furthermore reduced nutrient losses from the digesta.

Material and Methods

In a total of 6 respiration experiments with pigs from 14 to 24 kg body mass, 24 young castrate males were divided into two equal body mass groups. One group was fed with a normal commercial feed (control group). The other group received the same ration with supplementary 500 mg chlortetracycline per kg feed. The main ingredients of the ration were barley, maize, wheat, wheat bran, maize gluten, fish and meat meal, amino acids, minerals and vitamins. The mean nutrient content was 13 MJ digestible energy, 190 g crude protein and 48 g crude fat per kg. According to the recommendations of ARC (1980) the following amino acids were supplemented: lysine, methionine, threonine and tryptophan. Chlortetracycline was added to the second diet as a 10% premix as Aurofac[R]. For the determination of the digestibility of energy and nitrogen celite 545 was added as an indicator (HCl insoluble ash) (Prabucki *et al.*, 1975).

The method of indirect calorimetry was used for the measurement of oxygen consumption and carbon dioxide production in two separate respiration chambers. The piglets were kept in metabolic cages in groups of two individuals in order to reduce stress during the experiments. Each experiment of the duration of 3 weeks was divided into three collection periods, each of the length of 4 days. The first period was mainly used for adaptation of the piglets to the experimental conditions. Faeces and urine were collected in a mixed sample and body weight was recorded according to the three periods. Food intake was registered daily. The piglets were fed *ad libitum* at room temperatures declining from 26 to 21°C.

Results and Discussion

The results are shown in Table 1 and are the means in the different periods with the corresponding standard deviations (SD). Different letters indicate significant differences ($P < 0.05$).

Table 1. Growth rate, feed conversion and nutrient retention of piglets fed a diet medicated with and without chlortetracycline.

Treatment		Control			CTC		
Period		1	2	3	1	2	3
Body weight gain	g/d	-	683	657[a]	-	732	916[b]
	SD		190	159		121	97
Feed intake	g/d	995	1126	1240	887	1163	1372
	SD	269	279	251	342	331	371
Feed conversion ratio	g/g	1.48	1.68	1.98	1.23	1.60	1.51
	SD	0.17	0.19	0.59	0.24	0.15	0.21
Digestibility							
energy		0.781	0.775	0.769	0.781	0.775	0.774
	SD	0.012	0.009	0.010	0.021	0.031	0.036
nitrogen		0.763	0.759	0.776	0.796	0.790	0.799
	SD	0.021	0.010	0.030	0.020	0.033	0.039
Energy retention							
protein	kJ/d,$BM^{3/4}$	243[a]	241[a]	232[a]	281[b]	321[b]	316[b]
	SD	50	44	24	43	40	35
fat	kJ/d,$BM^{3/4}$	622	631	562	394	537	577
	SD	215	195	152	173	96	103
energy	kJ/d,$BM^{3/4}$	865	871	794	675	858	893
	SD	246	219	174	209	123	134

Despite the short periods of only one week the average daily gain was higher in the treatment with the CTC-supplementation compared to the control diet (significantly in period 3). That difference could only partly be explained by a higher feed intake. Also faecal energy digestibility was not affected by the treatment with CTC. This is in contrast to findings with low doses of CTC or other supplements added to the diets (Berschauer *et al.*, 1978;, Klaus, 1956; Wenk & Schürch, 1976).

Nitrogen digestibility was 4% higher with the CTC diet although not significant. Furthermore, the piglets in the treatment with CTC deposited significantly more protein than the control animals. Of the retained energy 29 to 30% was deposited as protein and 70 to 71% as fat in the control group, whereas in the chlortetracycline fed group 36 to 44% and 56 to 64% were deposited, respectively. The energy costs for maintenance were calculated to be about 529 kJ ME/BM$^{3/4}$. The rate of weight gain increased by 23%, daily protein deposition and gain of muscles were increased by 28% with the application of chlortetracycline, whereas daily fat gain was reduced by 23%. These data support that chlortetracycline has a distinct growth-promoting and anabolic effect also in therapeutic dosage.

In Figure 1 energy retention (RE) is plotted against the intake of metabolizable energy (ME) per kg metabolic body weight. Each point corresponds to a respiration experiment with two animals over the collection period of 4 days.

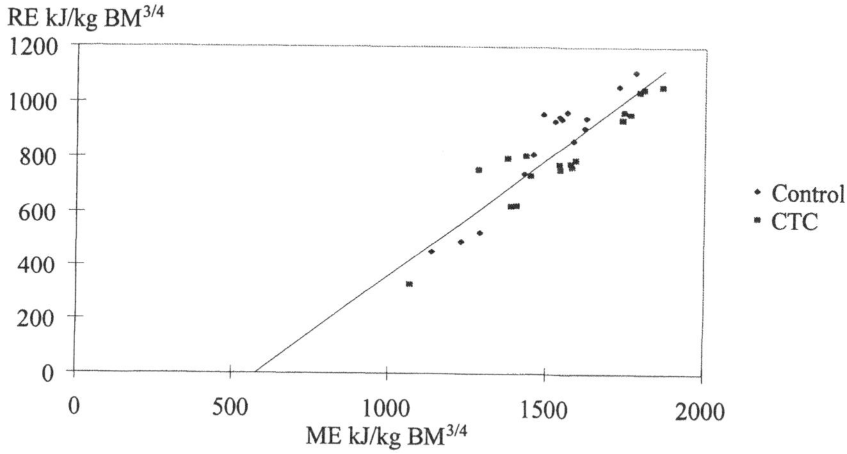

Fig. 1. Retained energy (RE) in relation to the intake of metabolizable energy (ME) in piglets that received a diet with and without chlortetracycline (CTC) in therapeutic doses.

Utilization of ME for growth followed a linear regression with a high regression coefficient (r = 0.90). Although some values for the control diet were below almost all corresponding values of the CTC - diet, no clear difference could be observed between the two treatments. This suggests that the CTC-supplementation did not affect ME utilization for growth. Therefore the following observations concern only the whole set of data.

From the regression line in Figure 1 energy requirements for maintenance (ME_M) of the piglets amounted to 584 kJ ME/BM$^{0.75}$. Partial efficiency for growth was 0.87. These values are rather high compared to other findings (Jentsch *et al.*, 1991; Thorbek *et al.*, 1984; Wenk, 1995). As discussed by Wenk & Kronauer (1987) ME_M and k_g are not independent. It must therefore be concluded that some of the variation of energy turnover for growth was charged to the maintenance component due to the high intercorrelation. If protein and fat deposition were considered separately ME_M amounted to 542 kJ ME/BM$^{0.75}$. For the partial efficiencies for protein and fat deposition k_p and k_f 0.45 and 1.39 were found respectively. Due to the low amount of fat deposited an overestimation of k_f may have occurred (Wenk & Kronauer, 1987).

References

ARC (1981). The nutrient requirements of pigs. Commonwealth Agricultural Bureaux, Slough.

Berschauer F., Ehrensvärd U. & Menke K.H. (1978). Zur Wirksamkeit von Olaquindox und Chlortetracyklin in der Ferkelaufzucht. *Z. Tierphysiol., Tierern. und Futtermittelk.* 41:280-292.

Dierick N.A., Vervaeke I.J., Decuypere J.A. & Henderickx H.K. (1986 b). Influence of the gut flora and of some growth-promoting feed additives on nitrogen metaolism in pigs. *Livest. Prod. Sci.* 14:177-193.

Drochner W. & Meyer H. (1991). Verdauung organischer Substanzen im Dickdarm verschiedener Haustierarten, Z. Tierphysiologie. *Beih. Fortschr. Tierphysiol. Tierern.* 22:18-40.

François A.C. & Michel M.C. (1968). Mode d'action des antibiotiques sur la croissance. In: J.C. Somogy and A.C. Francois (Eds.) Antibiotics in agriculture. Proceedings of the 5th Symposium of European Nutritionists, S. Karger, Basel.

Greife H.A. & Berschauer F. (1988). Leistungsförderer in der Tierproduktion: *Stand und Perspektiven, Übers. Tierern.* 16:27-78.

Jentsch W., Hoffmann L. & Beyer M. (1991). Untersuchungen zum Energieumsatz wachsender Schweine im Lebendmassebereich von 10-50 kg. *Arch. Anim. Nutr.* 41:623-635.

Klaus W. (1956). Untersuchungen über den Einfluss von Penicillin und Aureomycin auf die Verdaulichkeit der Nährstofe und die N-Retention bei Schweinen. *Z. Tierphysiol., Tierernährg. u. Futtermittelkde.* 11:1-20.

Prabucki A.L., Rennerova L., Vogtmann H., Wenk C. & Schürch A. (1975). Die Verwertung von 4N HCl-unlöslicher Asche als Indikator zur Bestimmung der Verdaulichkeit. Misc.-Papers II, Landbouwhogeschool Wageningen, NL, 113-114.

Thorbek G., Chwalibog A. & Henckel S. (1984). Nitrogen and energy metabolism in pigs of Danish Landrace from 20 to 120 kg live weight, Report from the National Institute of Animal Science No. 563 Copenhagen: Landhusholdningsselskabet.

Wenk C. & Schürch A. (1976). Untersuchungen über4 die Wirkungsweise wachstumsfördernder Futterzusätze, Schweiz. *Landwirtschaftl. Monatshefte* 54:293-301.

Wenk C.P.C. & Kronauer M. (1987). Effect of feeding level on maintenance requirements of growing pigs. In: Y van der Honing & W.H.Close (Eds) Energy Metabolism in Farm Animals, *EAAP Publ. No. 43*:425-439.

Wenk C. (1995). Effizienz: Aspekte der Nährstoffverwertung beim wachsenden landwirtschaftlichen Nutztier. Tagungsbericht ETH, Zürich, pp 14-64.

Zabka A. (1997). Die Wirkung von oral verabreichtem Chlortetracyclin in therapeutischer Dosierung auf den Magen-Darm-Trakt von Ferkeln, Vet. Med. Diss., Zürich, 105 pp.

ENERGY DEPOSITION AND BODY COMPOSITION MEASUREMENT OF PIGS OF DIFFERENT RYANODINE RECEPTOR GENOTYPES BY DUAL ENERGY X-RAY ABSORPTIOMETRY

A.D. Mitchell[1] and A.M. Scholz[2]

[1]United States Department of Agriculture, Agricultural Research Service, Beltsville Agricultural Research Center, Beltsville, Maryland 20705, USA; [2]Institute of Animal Sciences, Humboldt University, Berlin, Germany

Summary

Dual energy x-ray absorptiometry (DXA) was used for *in vivo* assessment of body composition and energy deposition in pigs that were either homozygous stress non-sensitive (*NN*), heterozygous (*Nn*) or homozygous stress sensitive (*nn*). DXA evaluation of pigs at 30 kg revealed no differences among genotypes with respect to total body fat content, however, it was estimated that the *nn* pigs contained more protein than either the *Nn* or *NN* pigs. During growth from 30 to 60 kg, the *nn* pigs deposited less fat than the *NN* pigs, while *Nn* were intermediate. This pattern continued during growth from 60 to 90 kg, resulting in the *nn* pigs having less fat and less total energy than the other two genotypes. There were no differences in protein deposition during these periods. Furthermore, based on the DXA results and measurements of energy intake during these periods, there were no differences among genotypes in the efficiency with which dietary energy was retained.
Key words: Pigs, ryanodine, body composition.

Introduction

Malignant hyperthermia, the condition associated with a mutation in the DNA sequence encoding the function of the ryanodine receptor 1 (Fujii *et al.*, 1991), results in changes in energy metabolism. The most dramatic changes have been shown to occur in pigs exposed to stress. *In vivo* ^{31}P-NMR studies have demonstrated halothane induced depletion in muscle ATP and phosphocreatine and accumulation of inorganic phosphate in homozygous stress sensitive (*nn*) pigs (Geers *et al.*, 1992; Scholz *et al.*, 1995) with moderate responses occurring in heterozygous (*Nn*) pigs. Evidence of alteration in energy metabolism under resting conditions is suggested by higher plasma glucose concentrations (Otten *et al.*, 1993) and higher intramuscular concentrations of phosphocreatine (Decanniere *et al.*, 1993) and phosphomonoesters (Geers *et al.*, 1996). In the long term these changes in energy metabolism may be associated with decreased lipid deposition and an altered efficiency of the utilization of dietary energy for tissue growth (k_g). The purpose of the present study was to use dual energy x-ray absorptiometry to monitor *in vivo* changes in body composition and energy deposition in pigs of different ryanodine receptor genotypes.

Methods and Materials

A total of 32 pigs were evaluated by live DXA scanning at 30 and 60 kg. The pigs were identified by the ryanodine receptor (RYR1) gene test described by Brenig & Brem (1992) as being either homozygous stress non-sensitive (*NN*, 14 pigs), heterozygous (*Nn*, 11 pigs) or homozygous stress sensitive (*nn*, 7 pigs). In addition, seventeen of these pigs (*NN*, 6 pigs; *Nn*, 6 pigs; *nn*, 5 pigs) were scanned again at 90 kg. Diet was withheld for 18 hours prior to scanning and, immediately prior to scanning each pig was weighed on an electronic scale (± 0.2 kg). All pigs were anaesthetized. The DXA scans were performed using a Lunar DPXL densitometer. Estimates of energy deposition were based on DXA measurements of total body fat and total body protein calculated from DXA lean body mass measurements. The use of DXA for measuring total body fat and protein of pigs within this range in body size has been validated through slaughter and chemical analysis of pigs from this and other studies (Mitchell *et al.*, 1996): FAT (%) = [493] - [349 DXA R value] and PROTEIN (g) = [-1020] + [0.25 DXA lean]. All pigs were fed a common diet (180 g crude protein and 14.1 MJ/kg) at 0.95 *ad libitum*.

Results

The DXA measurements at 30 kg live weight indicated that there was no difference among genotypes with respect to total body fat (Table 1). During the period of growth from 30 to 60 kg, there was an indication of less fat deposition in the *nn* pigs; however at 60 kg there was still no difference in total body fat among genotypes. During the 60 to 90 kg growth phase, the *nn* pigs deposited ($P<0.05$) less fat, resulting in less total body fat at 90 kg than either the *Nn* or *NN* genotype. The fat content of the *Nn* pigs was intermediate between the *nn* and *NN* genotype at both 60 and 90 kg. The pattern of energy deposition was such that there was a tendency toward increased fat deposition with increasing body size in the *NN* pigs, somewhat less in the *Nn* pigs and a slight decrease in the *nn* pigs.

Table 1. Energy status of pigs of different genotypes at 30, 60 and 90 kg.

Weight group	Genotype	(number)	Fat (MJ)	Protein (MJ)	Total (MJ)
30 kg	*nn*	(7)	198±13[a]	148±3[a]	346±15[a]
	Nn	(11)	197±11[a]	133±4[b]	330±13[a]
	NN	(14)	195±6[a]	129±3[b]	324±8[a]
60 kg	*nn*	(7)	485±18[a]	289±8[a]	773±18[a]
	Nn	(11)	508±20[a]	275±3[b]	783±20[a]
	NN	(14)	532±15[a]	271±3[b]	803±16[a]
90 kg	*nn*	(5)	670±20[a]	379±15[a]	1048±31[a]
	Nn	(6)	787±40[b]	380±7[a]	1167±35[b]
	NN	(6)	860±39[b]	367±3[a]	1227±39[b]

Using the DXA measurement of lean mass to estimate total body protein, it was observed that at 30 kg live weight the *nn* pigs contained more total protein than the *Nn* or *NN* genotype (Table 1). This difference persisted through 60 kg; however, at 90 kg there was no difference among genotypes. The pattern of energy deposition as total body protein appeared to be little different among genotypes and there was no significant difference in protein deposition among genotypes for either the 30 to 60 or 60 to 90 kg growth. The summation of total body fat and protein is included in Table 1 and is indicated as total energy (MJ). As a result of the reduced energy deposition as fat in the *nn* pigs, these pigs contained less total body energy at 90 kg than either the *Nn* or *NN* genotype.

Figure 1 compares the differences in the ratio of fat mass and lean mass as pigs of the three genotypes grew from 30 to 90 kg body weight. In the *NN* pigs, this ratio remained relatively constant. However, especially during growth from 60 to 90 kg, the ratio declined moderately in the *Nn* pigs and a dramatic decline occurred in the *nn* pigs.

The utilization of dietary energy for fat (k_f), protein (k_p)or total body growth during the 30-60-90 kg growth phases is shown in Table 2. There was a trend toward a decrease in the utilization of energy for fat deposition in favour of protein deposition by the *nn* and *Nn* genotype relative to the *NN* genotype. The only significant difference appeared during growth from 30 to 90 kg when the *nn* pigs deposited more protein relative to energy intake than did the *NN* pigs, with the *Nn* pigs being intermediate. There were no differences among genotypes in k_g. Likewise, Figure 2 reveals that there were no differences in the gross efficiency.

Table 2. Efficiency of energy utilization for fat, protein or total body growth in pigs of different genotypes at 30, 60 and 90 kg.

Weight group	Genotype	(number)	Fat (g/MJ)	Protein (g/MJ)	Total (g/MJ)
30-60 kg	*nn*	(7)	6.98±0.36[a]	5.71±0.27[a]	30.5±1.23[a]
	Nn	(11)	6.95±0.29[a]	5.32±0.22[a]	28.9±0.98[a]
	NN	(14)	7.55±0.25[a]	5.36±0.19[a]	29.7±0.87[a]
60-90 kg	*nn*	(5)	5.32±0.76[a]	4.01±0.31[a]	21.7±1.68[a]
	Nn	(6)	6.05±0.69[a]	3.75±0.29[a]	21.3±1.53[a]
	NN	(6)	6.54±0.69[a]	3.51±0.29[a]	20.9±1.53[a]
30-90 kg	*nn*	(5)	6.11±0.42[a]	4.98±0.25[a]	26.5±2.72[a]
	Nn	(6)	6.54±0.39[a]	4.46±0.23[ab]	24.8±2.49[a]
	NN	(6)	7.02±0.39[a]	4.17±0.23[b]	28.2±2.49[a]

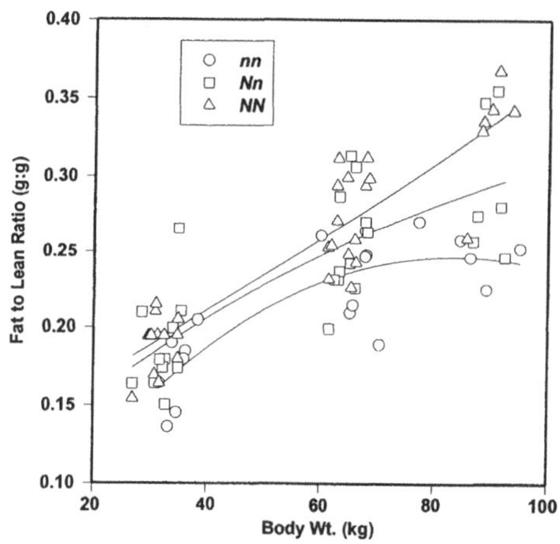

Fig. 1. Ratio of DXA measured fat and lean tissue mass in pigs of different ryanodine receptor genotypes during growth from 30 to 90 kg.

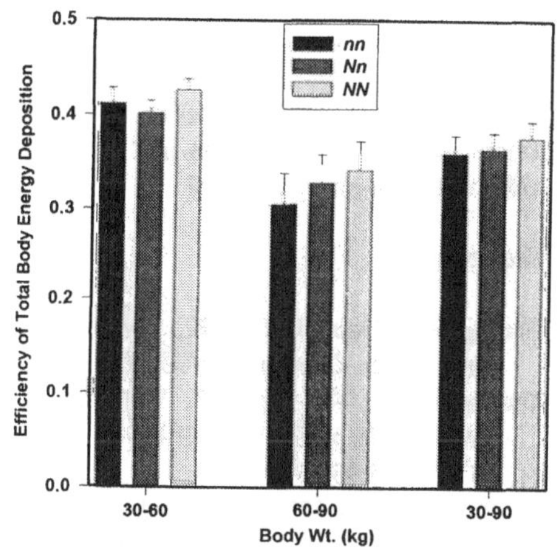

Fig. 2. Efficiency of energy deposition in pigs of different ryanodine receptor genotypes during the 30-60-90 kg growth periods.

Discussion

The use of DXA permitted the *in vivo* evaluation of energy utilization in pigs of different ryanodine receptor genotypes during growth from 30 to 90 kg body weight. The results indicated that during growth from 30 to 60 kg and from 60 to 90 kg, the *nn* pigs deposited less energy as body fat. At 30 kg the *nn* pigs already contained more total protein, deposited more protein during growth from 30 to 90 kg, but there was no difference in protein deposition between 60 and 90 kg.

Although metabolic events associated with the ryanodine receptor defect would suggest that the *nn* pigs might be less efficient with respect to energy utilization, DXA results revealed no differences in k_f or k_g. In fact, during growth from 30 to 90 kg, protein deposition per MJ of energy intake was more efficient in the *nn* pigs. However, k_g, k_f and k_p decreased ($P<0.05$) in all three genotypes in the 60-90 kg growth phase when compared with the 30-60 kg growth phase (Table 2).

From this study, inferences regarding the measurement of energetic efficiency and its association with the ryanodine receptor defect would be difficult considering the small number of pigs observed during the 60-90 kg growth phase. However, the results (Figure 2) suggest that energetic efficiency during the 60-90 kg growth phase decreased in the order *NN* >*Nn* > *nn*. One could speculate that the reduced energetic efficiency in pigs carrying the defective *n* allele may be due to higher maintenance energy requirements resulting from partial energy decoupling probably caused by latent increased intracellular Ca^{2+} concentration and a higher energetic cost for increased protein synthesis. Reduced lipid deposition could result from less available energy and/or indirect activation of the lipolytic system via catecholamine induced cAMP turnover.

The use of DXA body composition values of fat and lean mass for measurement of energy status depends on accurate validation of fat measurements and conversion of the lean mass value into protein and water components. It appears that the prediction equations applied to the DXA measurements gave reasonable values for both total body fat and protein. DXA can also quantify fat and lean masses into different regions of the body, but does not distinguish between subcutaneous, intramuscular or internal fat, nor does it partition lean mass into muscle, visceral organs or connective tissue, etc. Nevertheless, the results of this study indicate that DXA does give total body measurements that are useful for total body energy status assessment. Certainly, the ability to make sequential measurements on the same pig increases the overall reliability of observations regarding differences in energy status.

References

Brenig B. & Brem G. (1992). Molecular cloning and analysis of the porcine "halothane" gene. *Arch. Tierz., Dummerstorf* 35: 129-135.

Decanniere C., Van Hecke P., Vanstapel F., Villé H. & Geers R. (1993). Metabolic response to halothane in piglets susceptible to malignant hyperthermia: an *in vivo* ^{31}P-NMR study. *J. Appl. Physiol.* 75:955-962.

Fujii J., Otsu K., Zorzato F., De Leon S., Khanna V.K., Weiler J., O'Brien P.J. & MacLennan D.H. (1991). Identification of a mutation in the porcine ryanodine receptor that is associated with malignant hyperthermia. *Science* 253: 448-451.

Geers R., Decanniere C., Rosier A., Villé H., Van Hecke P., Vandesande F. & Jourquin J. (1996). Variability of energy metabolism and nuclear T_3-receptors within the skeletal muscle tissue of pigs different with respect to the halothene gene. *J. Anim. Sci.* 74:717-722.

Geers R., Decanniere C., Villé H., Van Hecke P., Goedseels V., Vanstapel F., Bosschaerts L., De Ley J., Zhang W. & Janssens S. (1992). *In vivo* muscle ^{31}P nuclear magnetic resonance spectroscopy during treatment of halothane-sensitive and halothane-nonsensitive pigs. *Am. J. Vet. Res.* 53:613-616.

Mitchell A.D., Scholz A.M. & Evock-Clover C.M. (1996) The application of dual energy X-ray absorptiometry for live animal and carcass evaluation of pigs. 47th Ann. Meeting of Europ. Assoc. Anim. Prod., Lillehammer. PS5.22, 4 pp.

Otten W., Eichinger H.M. & Goldberg M. (1993). Effects of different malignant hyperthermia genotypes on energy metabolism in swine. 44th Annual Meeting of the EAAP, August 16-19, Aarhus.

Scholz, A., Mitchell A.D., Wang P.C., Song H. & Yan Z. (1995). Muscle metabolism and body composition of pigs with different ryanodine receptor genotypes studied by means of ^{31}P nuclear magnetic resonance spectroscopy and ^{1}H magnetic resonance imaging. *Arch. Tierz., Dummerstorf* 38:539-552.

INFLUENCE OF AN IONOPHORE AND/OR A BUFFER ADDED TO A MOLASSES/CEREAL DIET IN GROWTH, CARCASS CHARACTERISTICS AND ENERGETIC EFFICIENCY OF BULLS

A. Araba[1] and F. M. Byers[2]

[1] Institut Agronomique et Vétérinaire Hassan II, Rabat, Morocco; [2] Texas A&M University, College Station, TX, USA

Summary

Effects of supplementing a sugar beet molasses/cereal diet with monensin and Trona buffer on feedlot performance, carcass merit, and energetic efficiency were determined in bulls (n = 23) of Holstein x Brown Atlas and Holstein x Friesian, individually fed one of the four diets (Control (C), Monensin (M), Buffer (B) or Monensin and Buffer (MB)). The diets consisted of molasses, barley, sunflower meal, and straw, and all diets were balanced for protein and minerals. Cattle were fed *ad libitum*. They were started on the experiment on an individual basis at 289 kg and were taken off the study at 500 kg. Bulls were slaughtered, and carcass information collected at slaughter. Animals on C, M, B, and MB diets were individually fed to an average of 196 days ($P = 0.33$), gained 1.12 kg/d ($P = 0.26$), consumed 12.65, 9.65, 10.84, and 10.17 kg DM/d ($P < 0.001$), required 11.01, 9.79, 9.25, and 9.15 kg DM/kg live weight gain ($P = 0.03$). Dressing percentage averaged 55.2, 53.2, 55.1, and 54.6% ($P = 0.03$) for cattle fed diets C, M, B, and MB, respectively. Diet did not affect ($P > 0.05$) other carcass characteristics. Diet did not affect ($P > 0.05$) final empty body composition, and consequently daily tissue deposition. Energy utilization was affected by diets. Compared to C diet, M, B, and MB diets decreased ($P < 0.001$) ME intake 30%, 22%, and 28%, respectively, did not influence RE (average of 44.4 kcal/kg$^{0.75}$), depressed NE_m ($P = 0.002$) 7%, 3%, and 6%, respectively, and increased NE_g 29%, 29%, and 39%, respectively. No significant difference ($P > 0.05$) was found between M, B, and MB diets, in energy utilization. It is beneficial to include either monensin or buffer in a molasses/cereal based diet for growing-finishing bulls fed molasses/cereal based diet; however the summative effect of these additives was small.

Key words: Cattle, monensin, buffer, molasses/cereal, growth, energy efficiency.

Introduction

Ionophores and buffers are examples of technologies that may allow meat to be produced with less energy and cost. Araba *et al.* (1997) and Araba and Byers (1997a,b) examined the use of sugar beet molasses for beef production and showed that this feed can be incorporated up to 40% of the diet (DM basis). However, they reported an increase in ruminal butyrate molar percent, and a decrease in ruminal propionate molar percent with increasing molasses in the diet. As well, high levels of sugar beet molasses in the diet were found to cause some bloat (Araba & Byers, 1997a). Because monensin is known to increase ruminal propionate, and decrease butyrate and was proven to decrease bloat, and Trona buffer was shown to decrease the production of butyrate (Boerner *et al.*, 1987), a trial was conducted to test the effect of monensin (M), buffer (B), and their combination (MB) on growth performance, carcass characteristics, and energy utilization of bulls fed sugar beet molasses/cereal based diet.

Materials and Methods

A total of 23 bulls of Holstein x Brown Atlas and Holstein x Friesian genotypes, that were maintained at the Gharb Experiment Station of the Hassan II Institute of Agronomy and Veterinary Medicine, Morocco, were randomly allocated to one of four diets in a 4 x 2 factorial arrangement of treatments. Animals were individually fed *ad libitum* in a tie-stall barn, twice daily, at 800 and 1600. They were started on the experiment on an individual basis as they reached 289 kg and they were taken off the study at 500 kg. Growth rate, daily dry matter intake, and feed efficiency were computed. At termination, the bulls were slaughtered and carcass information was collected.

Diet digestibility was determined on five animals per treatment, at an average of 360 kg live weight. Chromium oxide (Cr_2O_3) was used as an external marker. Metabolizable energy was calculated as DE times 0.8 (NRC, 1996). Empty body weight (EBWT) was estimated from live weight (LWT) using the following equation developed by Araba and Byers (1997b) in cattle of the same biological type and raised similarly to those of the present study : EBWT = -4.48 + (0.902 * LWT), $R^2 = 0.98$. Empty body composition of animals was estimated for all cattle using the equation relating empty body components to empty body weight and empty body weight gain (EBG) in bulls (Araba & Byers, 1997b): Empty body fat = 21.87 + 0.0007878 * EBWT * EBG; and Lean body mass = 12.62 + 0.8142 * EBWT - 0.000647 * EBG^2. These equations used to assess EBWT and empty body composition were developed through procedures including deuterium oxide (D_2O) isotope dilution measurements (Byers, 1979a,b). Daily protein gain, daily fat gain and daily retained energy (RE) were determined.

From estimates of retained energy/$W^{0.75}$, which was calculated as grams of protein x 5.539 + grams of fat x 9.385 (Garrett and Hinman, 1969) and ME intake, daily RE and ME/kg of metabolic size were determined. The semilog transformation of heat production (HP), measured as ME-RE with HP equal to 77 at zero ME intake (Garrett *et al.*, 1971), all expressed as daily kcal/$W^{0.75}$, was used to calculate the ME required for energy equilibrium or maintenance (ME_m).

The efficiency of ME use for maintenance was calculated as 77/ME_m. The slope of RE vs. ME available for gain (ME_g) (ME intake minus ME_m) provided an estimate of the efficiency of ME_g use for gain. Net energy for maintenance (NE_m) and net energy for gain (NE_g) were calculated from the diet ME (kcal/g) (calculated as ME_m/DM Maint) times the respective efficiency of ME use for each function. Feedlot performance, carcass characteristics, and body composition and energetic efficiency parameters were analysed using the SAS GLM procedure (SAS, 1986).

Results and Discussion

Average initial live weight and final live weight were not affected by diet ($P > 0.05$) (289 and 500 kg, respectively), since these weights were set constant by design. Cattle were fed to an average of 196 days ($P = 0.10$), and gained ($P = 0.26$) 1.12 kg/d. The lack of treatment effect on growth rate is in agreement with Zinn and Borques (1993). Daily DM intake was affected by diet ($P < 0.001$), and so was the DM fed per live weight gain ($P = 0.03$). DM consumption decreased (23.7%) with M in agreement with NRC (1996) resulting in a 11.1% decrease in feed required per unit of gain.

The decrease in feed consumption is related to the increase in the rumen propionic acid concentration usually observed with monensin supplementation. The 14% decrease in DM fed per day with Trona buffer diet may be explained by the increased ruminal molar proportion of propionate with buffer supplementation observed by Boerner *et al.* (1987a). This would be more favourable in such high concentrate diets due to the energetic efficiency of propionate.

Feed conversion improved by 16%. The effects of M and B were not additive for growth and intake, and small benefit (3.9%) was apparent for both additives in feed per gain. There were no treatment effects of diet on digestibility of dry matter ($P = 0.09$), organic matter ($P = 0.09$), and energy ($P = 0.08$).

Carcass weight of monensin-fed cattle was lower than controls as was dressing percentage (3.6% decrease). All other carcass traits were not ($P > 0.05$) affected by supplemental monensin. Diet did not modify composition of gain ($P = 0.68$ and $P = 0.69$ for percentage protein and percentage fat in gain, respectively), and, as a consequence, final empty body composition did not change with diet.

Daily tissue deposition did not differ ($P > 0.05$) between diets. It averaged 1011 g/d ($P = 0.27$) for empty body gain, 701 g/d ($P = 0.30$) for lean body mass, 154 g/d ($P = 0.30$) for protein, 291 g/d ($P = 0.33$) for fat, and 3598 kcal ($P = 0.30$) for retained energy.

Composition of gain was not altered by diet and averaged 15.2% protein in gain ($P = 0.68$) and 28.9% fat in gain ($P = 0.69$). Since diet did not affect growth rate, animals were expected to exhibit the same body composition. This supports the findings in other studies where diet did not alter growth rates sufficiently to significantly affect body composition at any given empty body weight (Araba and Byers, 1997b).

Daily ME intake was affected ($P < 0.001$) by supplementation. It decreased 30% with M (from 483.2 to 336.9 kcal/kg$^{0.75}$), 22% with B, and 28% with MB. No significant difference ($P > 0.05$) was found between M, B, and MB (average of 352.7 kcal/W$^{0.75}$). Daily RE was not influenced ($P = 0.30$) by diet and averaged 44.4 kcal /W$^{0.75}$. Daily ME for maintenance and DM for maintenance increased ($P = 0.002$) 8% by M, 4% by B, and 6% for MB treatment. The MB treatment effect was identical to the average effects of M and B separately.

Efficiency of ME use for maintenance (k_m) and NE_m were depressed ($P = 0.002$) 7% with M, 4% with B, and 6% with MB. Efficiency of ME use for gain (k_g) and NEg responded more to M (+29%), B (+29%), and MB (+39%). While no significant ($P > 0.05$) effect was shown between M, B, and MB treatments, some additive (+8%) effect was observed with the combination of these two additives when compared to the average effect of M and of B. Findings of Byers (1980) supported the same trend for NE_g, but not for NE_m, but the magnitude of improvement in NE_g for this diet is much greater than has been reported in the literature. This is probably related to the nature of the fermentation pattern of the diet, i.e. the preponderance of butyric acid, and limited propionic acid, as a consequence of incorporation of sugar beet molasses (Araba *et al.*, 1997), resulting in a much important effect of M in improving production of propionic acid.

References

Araba A., Byers F.M. & Guessous F. (1997). Patterns of rumen fermentation in cattle fed cereal/molasses diets. *J.Anim. Sci.* (In press).

Araba A. & Byers F.M. (1997a). Feedlot performance, carcass merit and energy efficiency of cattle fed cereal/molasses diets. *J. Anim. Sci.* (In press).

Araba A. & Byers F.M. (1997b). Patterns and energetic efficiency of growth and development of beef cattle fed cereal/molasses diets. *J. Anim. Sci.* (In press).

Boerner B.J., Byers F.M., Schelling G.T., Coppock C.E. & Greene L.W. (1987). Trona and sodium bicarbonate in beef cattle diets : Effects on pH and volatile fatty acid concentrations. *J. Anim Sci.* 65:309-316.

Byers F.M. (1980). Effects of limestone, monensin and feeding level on corn silage net energy value and composition of growth in cattle. *J. Anim. Sci.* 50:1127-1135.

Byers F.M. (1979a). Extraction and measurement of deuterium oxide at tracer levels in biological fluids. *Anal. Biochem.* 98:208-213.

Byers F.M. (1979b). Measurement of protein and fat accretion in growing beef cattle through isotope dilution procedures. OARDC Beef Cattle Res. *Rep. Anim. Sci. Ser.* 79-1:36-47.

Garrett W.N. & Hinman N. (1969). Reevaluation of the relationship between carcass density and body composition of beef steers. *J. Anim. Sci.* 28:1-5.

Garrett W.N., Lofgreen G.P. & Hull J.H. (1971). Influence of processing methods on the energy utilization of feed grains. *Hilgardia* 41:123-128.

NRC (1996). Nutrient requirements of beef cattle. National Academy Press, Washington, DC.

SAS. (1986). General linear models. SAS Institute , Cary, NC.

Zinn R.A., & Borques J.L. (1993). Influence of sodium bicarbonate and monensin on utilization of a fat-supplemented, high-energy growing-finishing diet by feedlot steers. *J. Anim. Sci.* 71: 18-25.

ANABOLIC IMPLANT EFFECTS ON BODY COMPOSITION AND ENERGETICS OF BEEF STEERS

J.P. Hutcheson[1], D.E. Johnson[2], C.L. Gerken[3], J.D. Tatum[2] and J.B. Morgan[4]

[1] Hoechst Roussel Vet, Amarillo, TX, USA; [2] Department of Animal Science, Colorado State University, Ft. Collins, CO, USA;[3] USDA, Washington, DC, USA;[4] Department of Animal Science, Oklahoma State University, Stillwater, OK, USA

Summary

Six sets of four genetically identical Brangus steers (avg wt 409 kg) were used to determine the effect of different anabolic implants on visceral organ mass, chemical body composition, estimated tissue deposition and energetic efficiency. Steers within a clone set were randomly assigned to one of four treatments: C - no implant; E - oestrogenic; A - androgenic or AE - androgenic plus oestrogenic. Mass of the empty gastrointestinal tract (GIT) was reduced ($P<0.10$) approximately 9% in steers implanted with oestrogen alone or in combination with an androgen. Liver mass was increased ($P<0.10$) from 6 to 14% by implants. Androgen-oestrogen implanted steers had greater ($P<0.10$) daily protein accretion (163 g/d) when compared to either E (129 g/d) or A (137 g/d) and these improved gain above C treated steers illustrating additive effects on protein accretion. The NE_g requirements per kg of gain were estimated to be reduced by 19% by E and AE implants.
Key words: Anabolic implants, visceral organs, body composition, clones, bovine.

Introduction

Anabolic implants used in cattle production increase skeletal muscle protein accretion in growing cattle (Bergen & Merkel, 1991). Use of anabolic implants may also affect the weights of visceral organs which represent some of the most metabolically active tissues in the body. A change in the proportion of these tissues could impact the animal's basal energy expenditures (Johnson *et al.*, 1990). The study was conducted to determine the effect of different anabolic implants on visceral organ mass, chemical composition of tissue deposition and energetic efficiency in genetically identical Brangus steers.

Materials and Methods

Six sets of purebred Brangus clones, four genetically identical steer calves per set, were used in this experiment. Steers were fed daily a high concentrate diet (CP, 13.5%; ME 12.7 MJ/kg). Calves were individually fed once daily and allowed *ad libitum* access to the diet. One-hundred and twelve days before the projected slaughter date, steers (average weight 409 kg) in each set were randomly assigned to one of four treatment groups: control (C) - no implant; oestrogenic (E) - 20 mg oestradiol benzoate and 200 mg progesterone; androgenic (A) - 140 mg trenbolone acetate; androgen-oestrogen (AE) combination - 120 mg trenbolone acetate and 24 mg oestradiol 17-β.

All animals were weighed and slaughtered after 18 h with access to water only. Blood was collected, the entire gastrointestinal tract (GIT) weighed and the contents were removed by flushing the stomachs with water and physically stripping the intestines. After removal of contents, the entire GIT was weighed and this along with the carcass weight was used to determine empty body weight. Individual weights were also obtained for visceral organs and hide. Final empty body composition was determined at slaughter by whole body grinding and chemical analysis.

Protein and fat accretion were estimated from final body chemical analysis minus estimates of initial body composition. Equations used to calculate initial EBW and the intercept of the equations to predict initial fat and protein contents were based on data of Simpfendorfer (1973). The log of component growth was assumed to relate to log EBW following ARC (1980). Energy was calculated as grams of protein x 23.18 (kJ/g) + grams of fat x 39.27 (kJ/g) (Garrett & Hinman, 1969). Data were analysed using the GLM procedure of SAS (1988). Analysis of variance was performed using steer as the experimental unit. In all analyses, each set of clones was treated as a block.

Results and Discussion

Mud depths reached approximately 30 cm periodically during the winter and thus environmental and pen conditions probably reduced performance and energetic efficiency.

Individual anabolic implants produced unique effects on visceral organ mass (Table 1). Empty gastrointestinal tract (GIT) weights were lower ($P<0.10$) in steers implanted with estrogen alone or in combination with androgens (9.5% of EBW) compared to control or androgen alone treated steers (10.5% of EBW). Steers that received any implant had larger ($P<0.10$) liver mass when compared to C steers (1.22 vs. 1.11%). AE implanted steers had the heaviest livers which were also larger ($P<0.10$) than E treated steers. These results indicate that the effect of lower energy expenditures due to a smaller GIT in AE implanted steers may be negated by a larger liver mass in these animals. Total organ mass was smaller ($P<0.10$) in AE and E steers when compared to C and A steers (12.6 vs. 13.6% of EBW). Overall, changes observed in visceral organ mass in implanted animals may not be large enough to impact basal energy expenditures.

Table 1. The effects of anabolic implants on visceral organ mass.

Item	C	E	A	AE	SEM
Final EBW, kg	467.9^{b}	493.6^{c}	492.5^{c}	509.7^{c}	9.1
			% of EBW		
GIT	10.37^{b}	9.38^{c}	10.57^{b}	9.60^{c}	0.18
Spleen	0.20^{b}	0.21^{b}	0.22^{bc}	0.24^{c}	0.01
Liver	1.11^{b}	1.18^{c}	1.21^{cd}	1.27^{d}	0.03
Heart, lungs and trachea	1.67	1.79	1.76	1.78	0.08
Kidney	0.09	0.10	0.10	0 .10	0.01
Total organ mass^{e}	13.47^{b}	12.51^{c}	13.72^{b}	12.73^{c}	0.25
Hide	7.86^{b}	9.06^{cd}	8.71^{c}	9.35^{d}	0.23

[b,c,d] Means in a row with a different superscripts differ ($P<0.10$).

[e] Total organ mass = GIT + spleen + liver + heart, lungs and trachea + kidney.

Implanted steers averaged 5 kg greater ($P<0.10$) weight of protein in the empty body when compared to C steers Table 2. Steers implanted with AE had 8 kg more ($P<0.10$) protein in the empty body than controls which was also more ($P<0.10$) than steers with individual implants.

The combination implant had an additive effect on increasing protein mass. Implanted steers also had a greater ($P<0.10$) amount of moisture than non-implanted steers which reflects an increase in lean mass as well as higher ($P<0.10$) amounts of ash reflecting greater skeletal growth. There were no differences ($P>0.10$) in the amount of fat in the empty body among treatments. These results indicate that anabolic implants increase growth by accelerating nutrient deposition as protein but not at the expense of fat. However, implanted steers tended to average a lower ($P = 0.16$) percentage fat in the empty body when compared to non-implanted steers (32.3 vs. 33.8%) which may reflect the dilution of fat by increased protein deposition.

Table 2. Anabolic implant effects on initial and final composition.

Item	C	E	A	AE	SEM
			kg		
Estimated initial weights					
Shrunk BW	393.0	401.2	394.0	401.2	7.4
EBW	362.3	370.2	363.3	370.2	7.2
Protein	53.5	54.5	53.6	54.5	0.2
Fat	107.4	111.45	107.8	111.4	6.8
Final weights					
Shrunk BW	496.5[d]	520.2[e]	518.5[de]	534.7[e]	9.1
EBW	467.9[d]	493.6[e]	492.5[e]	509.7[e]	9.1
Protein	64.8[d]	68.9[e]	69.0[e]	72.8[f]	1.2
Fat	157.7	157.3	162.8	162.6	5.9
Ash	15.0[d]	15.9[de]	15.3[d]	16.7[e]	0.5
Moisture	228.9[d]	249.5[e]	244.2[e]	255.8[e]	4.9

[d,e,f] Means in a row with different superscripts differ (P<0.10).

While the absolute amounts of tissue deposited per day must remain tentative because initial amounts were estimated, strikingly different implant effects are indicated (Table 3). Overall, implanted cattle were estimated to gain 27 to 64% more protein than control steers. AE implanted steers had greater (P<0.10) daily protein accretion (163.4 g/d) than E (128.8 g/d), A (137.1 g/d) and C (101.1 g/d) treated steers illustrating the previously mentioned additive effects. Daily fat gain was not different (P>0.10) among treatments and averaged 453 vs. 449 g/d for implanted and non-implanted steers, respectively. Viewed in concert with absolute amounts of tissue deposited, the lower fat percentage with implants may be best considered a dilution by increased protein deposition rather than decreased lipogenesis rates *per se*.

Table 3. Anabolic implants effect on rate and composition of tissue deposition.

Item	C	E	A	AE	SEM
Tissue deposition, g/d					
Empty body	829.4[a]	987.5[b]	1039.9[bc]	1131.8[c]	50.0
Protein	101.1[a]	128.8[b]	137.1[b]	163.4[c]	13.0
Fat	449.0	410.4	491.0	457.5	55.5
Composition of gain, %					
Protein	12.2	13.1	13.3	14.3	1.1
Fat	54.3	42.2	46.7	41.6	5.3

[a,b,c] Means in a row with different superscripts differ (P<0.10).

Limitations of experimental design and experimental conditions preclude an absolute but not a relative evaluation of implant effects of maintenance and growth energetics which is presented in Table 4. Implants shifted the growth curve so that SBW was 43 to 86 kg heavier when steers reached 28% fat in the empty body. Thus, increased weight with similar energy gains results in a marked implant reduction in NE_g requirement per unit of gain. Estrogen, with or without androgen reduced NE_g requirements by 19% and androgen alone reduced NE_g requirements by 10%.

Table 4. The effects of anabolic implants on estimated energy utilization.

Item[a]	C	E	A	AE	SEM
EBW, $kg^{0.75}$	93.0	95.8	95.1	97.1	1.3
			kJ/kg $BW^{0.75}$		
MEI	1025.1	1023.4	1069.4	1047.7	18.8
RE	216.7	200.8	237.6	225.1	21.4
MEM	519.7	537.6	511.3	516.7	15.5
DMM	171.1	177.0	168.2	169.9	5.0
Efficiency of ME use, %					
Maintenance	62.2	60.3	63.0	62.1	1.8
Gain	42.8	41.8	43.0	43.0	3.8
NEg requirements			MJ/kg EBG		
Predicted by NRC 84	24.3	25.5	25.5	25.9	0.42
Predicted by NRC 96	26.4	25.1	25.5	24.7	1.13
Observed	24.3	19.7	21.8	19.7	0.50
SBW adjusted to 28% fat	401.9[a]	470.0[b]	444.6[ab]	487.7[b]	24.0

[a] MEI = metabolizable energy intake, RE = retained energy, MEM = metabolizable energy for maintenance, DMM = dry matter for maintenance, NRC = NRC 1984 or NRC 1996.
[b,c,d] Means in a row with different superscripts differ ($P<0.10$).

References

Bergen W.G. & Merkel R.A. (1991). Protein Accretion. In: A.M. Pearson & T.R.Dutson (Eds). Growth regulation in farm animals. Advances in Meat Research Vol. 7. Elsevier Science, Essex, UK. pp. 169-180.

Garrett W.N & Hinman N. (1969). Reevaluation of the relationship between carcass density and body composition of beef steers. *J. Anim. Sci.* 28:1-5.

Johnson D.E., Johnson K.A. & Baldwin R.L. (1990). Changes in liver and gastrointestinal tract energy demands in response to physiological workload in ruminants. *J. Nutr.* 120:649.

NRC (1984). Nutrient requirements of beef cattle. (6th Ed) National Academy Press, Washington DC.

NRC (1996). Nutrient requirements of beef cattle. (7th Ed) National Academy Press, Washington DC.

Simpfendorfer S. (1973). Relationship of body type, sex and energy intake to body composition of cattle. PhD dissertation. Cornell Univ., Ithaca, NY.

ENERGY DENSITY OF GROWTH IN BEEF CATTLE WITH RESPECT TO GROWTH REGULATION STRATEGIES

F.M. Byers

Department of Animal Science, Texas A&M University, College Station Texas, 77843 USA

Summary

Regulation of protein, fat and energy accretion with zeranol and oestradiol/progesterone or oestradiol/testosterone combinations was assessed in beef cattle. Crossbred yearling steers (n=90) averaging 331 kg were assigned to an initial or 8 treatment groups with 10 cattle per group. Growth regulation strategies were: (N) none, (Z) zeranol 36, (ZZ) zeranol 72, (EP) oestrogen/progesterone, (ET) oestrogen/testosterone, (2-EP) 2-oestrogen/progesterone, (2-ET) 2-oestrogen/testosterone, or (ZEP) zeranol 36 plus oestrogen/progesterone. All cattle were fed a whole shelled corn 7% cottonseed hull finishing diet for 112 d. Body composition was determined from carcass specific gravity on the initial group and on all cattle as they were terminated. Protein, fat and energy accretion were derived from empty body components vs. time. Daily empty body gain (DEBG) over the 112 d study averaged 1181 g/d respectively. Daily protein and fat gain averaged 159 and 494 g/d. At a DEBG of 1500 g/d, RE was 24.3, 20.9, 17.5, 20.4, 17.3, 21.5, 15.9 and 18.7 MJ/kg for N, Z , ZZ, EP, ET, 2-EP, 2-ET, or ZEP respectively. This resulted in a 32 and 34% reduction in energy density of gain for 2-ET vs. N cattle growing at 700 or 1500 g/d. Repartitioning of energy in growth toward growth of lower energy density is more extensive in the composition of tissue accrual effected through greater levels and with combinations of currently available oestrogenic growth regulators.

Key words: Cattle, energy density, zeranol, oestradiol, protein accretion, fat accretion, energy accretion.

Introduction

More effective systems of growth regulation than currently exist must be developed to allow greater rates of lean tissue growth and less fat deposition in beef carcass production with positive and without deleterious impacts on acceptability of the beef products produced. Response with growth regulating implants is variable, with impacts from 5 to 30+% (Solis *et al.*, 1989), depending on cattle type, nutritional regimen, (Lemieux *et al.*, 1988) endogenous/exogenous regulator strategies (Ono *et al.*, 1996) and carcass endpoints. While recent research indicates that responses to levels (Byers *et al.*, 1990) and combinations of oestrogenic (zeranol, oestradiol 17 B) and androgenic growth regulators in enhancing growth and efficiency in steers are additive and efficacious (Rumsey *et al.*, 1996), impacts on energetic density of growth are less well known, especially with the many possible strategies for administration of combinations. Further, options available with the myriad of growth regulation strategies now possible offer new opportunities to more precisely regulate composition of growth and energy content of growth. Information on impacts of optional regulation strategies on energy density of growth is essential to allow development of targeted growth regulation systems and is needed to allow provisioning of required energy sources to meet animal growth needs.

Materials and Methods

Regulation of energy partitioning into protein and fat deposition was investigated in beef cattle with zeranol and oestradiol/progesterone or oestradiol/testosterone and combinations. Crossbred yearling steers (n=90) averaging 331 kg were assigned to an initial (n=10) or 8 treatment groups (n=80) with 10 cattle/group and were fed for 112 d.

Growth regulation strategies were: (N) none, (Z) zeranol 36 (Ralgro), (ZZ) zeranol 72, (EP) oestrogen/progesterone (Synovex-S), (ET) oestrogen/testosterone (Synovex-H), (2-EP) 2-oestrogen/progesterone, (2-ET) 2 oestrogen/testosterone, or (ZEP) zeranol 36 plus oestrogen/progesterone. All cattle were *ad libitum* fed a whole shelled corn (74%) finishing diet with cottonseed hulls (7%), cottonseed meal (13%), molasses (4%) and mineral/vitamin premix (2%). The diet was formulated to contain 13.8% protein and 12.6 MJ ME/kg.

Body composition was determined from carcass specific gravity (Garrett & Hinman, 1969) on the initial group and on all cattle as they were terminated. Initial empty body composition of all animals was derived from regressions of body composition on ultrasound fat and weight of the initial slaughter group. DEBG, protein (DPG), fat (DFG) and energy accretion (RE/d) were derived from final minus initial empty body components on each animal. Energy density of growth was fitted in exponential models as ln RE/d vs. ln DEBG*growth regulator strategy. Resulting models were used to derive growth functions illustrating repartitioning of energy into protein or fat and resulting implications for energy density of gain with respective growth regulator strategies.

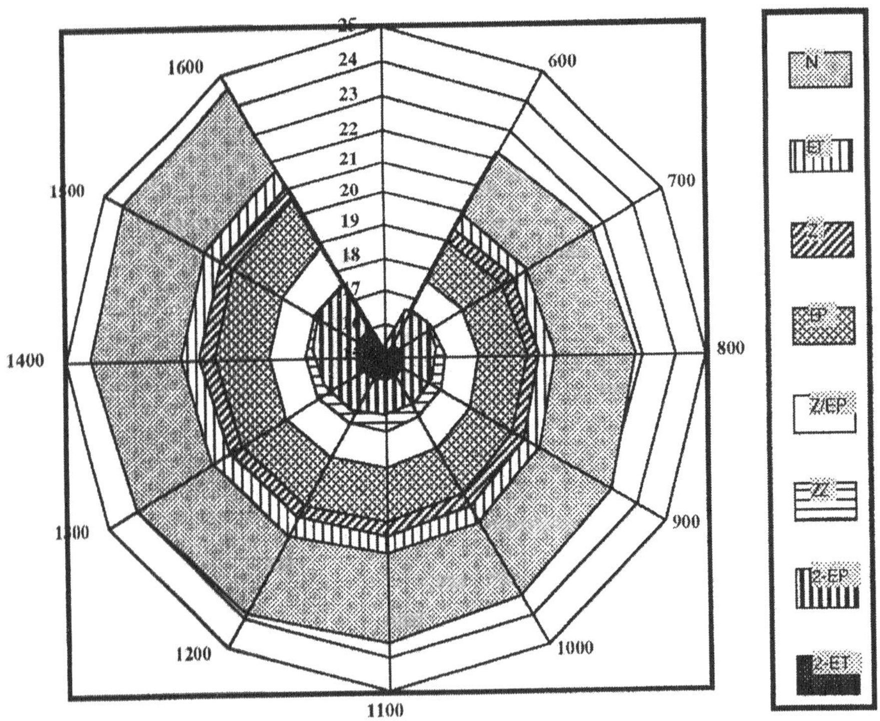

Fig. 1. Energy density of growth, MJ/kg for cattle vs. daily EB growth (g/d) for growth regulator.

Results and Discussion

Initial empty body weight averaged 331 kg, with 21.9% fat and 17.46% protein. Empty body weight at termination averaged 463.1 kg. All growth regulation strategies increased empty body protein accrued at termination, and multiple levels and the zeranol/oestradiol combination provided greater enhancement in total protein accrued than individual implants. Final EBP averaged 70.2, 71.3, 76.0, 76.3, 78.3, 75.4, 79.9, and 77.4 kg for N, Z, ZZ, EP, ET, 2-EP, 2 ET, or ZEP respectively. Maximum response was observed with the 2X level of ET, where EBP increased by 9.7 kg. Final liver weight reflected this growth regulation response and averaged 1.25, 1.34, 1.34, 1.32, 1.36, 1.28, 1.29 and 1.34% of EBWT for N, Z , ZZ, EP, ET, 2-EP, 2 ET, or ZEP respectively. To assess the repartitioning response to respective strategies, exponential growth models ($P < 0.001$, $R^2=0.46$, SÁ=1.0) were derived as in previous growth regulation investigations. Model coefficients were as follows: lnRE/d, MJ = 2.447 DEBG^(1.102, 1.081, 1.057, 1.078, 1.055, 1.085, 1.044, or 1.066) with respective DEBG exponents for N, Z , ZZ, EP, ET, 2-EP, 2-ET, or ZEP as indicated. Repartitioning in energy density of gain from these models is illustrated in Figure 1. As is evident, the DEBG exponent for non-implanted cattle was greater than for any of the groups receiving growth regulators, similar to and consistent with previous models from our laboratory (Byers *et al.*, 1989) for non-implanted cattle.

Exponents averaged 1.067 vs. 1.102 for cattle receiving or not receiving growth regulators respectively. Exponents for cattle receiving single growth regulators in the current study averaged 1.081 which compares favourably with values of 1.095 vs. 1.12 for individual oestrogenic growth regulators vs. no implants in previous research. In the range of growth rates included in this study, (700-1600 g/d), energy density of gain ranged from 15.4 to 24.5 MJ/kg. At a DEBG of 700 g/d, RE in MJ/kg was 22.5, 19.6, 16.8, 19.2, 16.6, 20.1, 15.4 and 17.8 for N, Z , ZZ, EP, ET, 2-EP, 2-ET, or ZEP respectively. At a DEBG of 1500 g/d, RE was 24.3, 20.9, 17.5, 20.4, 17.3, 21.5, 15.9 and 18.7 MJ/kg for N, Z , ZZ, EP, ET, 2-EP, 2 ET, or ZEP respectively. This resulted in a 32 and 34% reduction in energy density of gain for 2 ET vs. N cattle growing at 700 or 1500 g/d. The percentage of fat in gain ranged from 29 to 57%. At a DEBG of 1500 g/d, fat in gain was 56.1, 46.0, 36.1, 44.7, 35.4, 47.8, 31.5, and 39.6% for N, Z , ZZ, EP, ET, 2-EP, 2-ET, or ZEP respectively.

Conclusions

Clearly, repartitioning of energy in growth toward growth of lower energy density is more extensive with greater levels of and with combinations of currently available oestrogenic growth regulators than with single implant strategies.

References

Byers F.M., Schelling G.T. & Greene L.W. (1989). Development of growth functions to describe energy density of growth in beef cattle. 11th Symposium, Energy Metabolism of Farm Animals. *EAAP. Publ. No.* 43:195.

Byers F.M., Turner N.D., Cross H.R., Greene L.W. & Carstens G.E. (1990). Optimal growth regulator levels for enhancing carcass lean. *Proc. West. Sect. Am. Soc. Anim. Sci.* 41:422.

Garrett W.N. & Hinman N. (1969). Reevaluation of the relationship between carcass density and body composition of beef steers. *J. Anim. Sci.* 28:1.

Lemieux P.G., Byers F.M. & Schelling G.T. (1988). Anabolic effects of rate, composition and energetic efficiency of growth in cattle fed forage and grain diets. *J. Anim. Sci.* 66:1824-1836.

Rumsey T.S., Elsasser T.H., Kahl S., Moseley W.M. & Solomon M.B. (1996). Effects of Synovex-S and recombinant bovine growth hormone (Somavubove) on growth responses of steers: I. Performance and composition of gain. J. *Anim. Sci.* 74:2917.

Ono Y., Solomon M.B., Elsasser T.H., Rumsey T.S. & Moseley W.M. (1996). Effects of Synovex-S and recombinant bovine growth hormone (Somavubove) on growth responses of steers: II. Muscle morphology and proximate composition of muscles *J. Anim. Sci.* 74:2929.

Solis J.C., Byers F.M., Schelling G.T. & Greene L.W. (1989). Anabolic implant and frame size effects on growth regulation, nutrient repartitioning and energetic efficiency of feedlot steers. *J. Anim. Sci.* 67:2792-2801.

SUMMARY OF THE DISCUSSION

Discussion leader: C. Wenk
Scientific secretaries: G.E. Carstens & H. Jorgensen

Introduction

The utilization of energy in growing animals or in children is usually presented according to the scheme in Figure 1. The intake of metabolizable energy (ME) is plotted against the energy retention (RE) as protein, fat and ev. carbohydrates. In most cases a linear relation between ME intake and RE is accepted under the assumption, that ME and RE are corrected for metabolic body weight (BW^x), where the exponent x is, in most cases with adult animals equal to 0.75 (Kleiber, 1965), although many experiments with fast growing animals suggest that a value between 0.6 and 0.65 would be more accurate. For $ME/BW^{0.75}$ and $RE / BW^{0.75}$ the terms ME* and RE* are usually used.

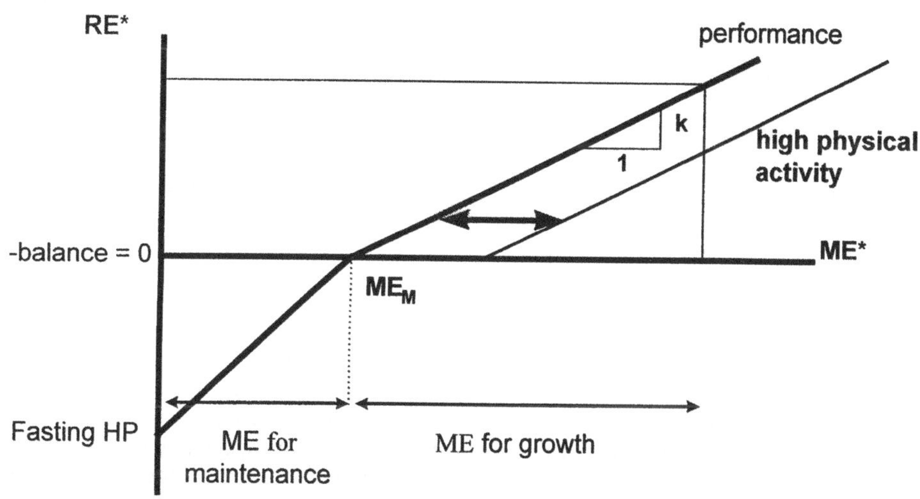

Fig. 1. Energy retention in relation to the intake of metabolizable energy.

Energy expenditure at ME intake = 0 is usually defined as fasting heat production (fasting HP), but this does not correspond to a stable physiological status for any animal. ME for maintenance (ME_M) is defined as the intake of metabolizable energy at zero energy balance which implies no energy deposition or mobilization. ME_M includes the energy costs attributable to feed intake at maintenance (thermic effect of feed) and moderate physical activity.

ME for growth is the amount of metabolizable energy that is transferred into nutrients deposited as growth, mainly protein and fat. From the slope of the linear equations between ME intake above ME_M and RE deposition the partial efficiency for growth (k_g) as well as protein (k_p) and fat deposition (k_f) can be described. In most experiments the linear approach to describe the energy utilization for growth gives satisfactory results for limited data sets. The extrapolation of such factors to general applications has to be carefully evaluated, because in this case, statistical values are interpreted in a physiological sense.

In growing animals the accuracy of estimation of the parameters ME_M, k_g, k_p and k_f usually depends on two factors. Often the experimental design does not allow wide variations of ME intake or, as a consequence, also of retained energy. Furthermore ME_M is not independent of ME for growth, e.g. protein deposition as growth is only a (small) part of total protein turnover. So most protein turnover is closely related to maintenance and only a part can be attributed to growth. Also energy deposition as protein and fat, in fast growing animals such as chickens or piglets, is due to physiological limitations which are not easily altered and give a narrow range of values.

Physical activity, in growing farm animals kept under practical conditions, amounts to about 20% of the maintenance requirements (ca. 100 kJ/$BW^{0.75}$). Under restricted room conditions, often the case in respiration chambers, lower values are found. So the ME_M values from respiration trials must be corrected for that difference.

The impact of activity on energy expenditure is important as this topic was addressed in a number of papers in both the growth and maintenance sessions. There is an obvious need for more research to address the impact on activity on energy expediture (EE) from several standpoints: (1) methodology to effectively partition, experimentally, EE associated with activity from EE associated with basal metabolism and heat increment, (2) methodology to estimate EE cost of activity of animals in "production" enviroments. Improvements in these methodologies would help us to better determine the effects of breed type, feeding level, diet, environment, stress, etc. on EE of farm animals.

General Discussion

The 11 papers covered a wide range of species, topics and methods of measurement. For simplicity they have been grouped by species.

Pigs

There are obviously only small differences in the efficiency of utilization of energy and nutrients for growth. Only in absolute terms such as protein or fat retention are there differences between breeds (e.g. Meishan versus Pietrain), sex or age.

In one study (Noblet *et al.*) fasting heat production was used as a predictor for maintenance. It was agreed that fasting heat production in growing animals should be interpreted with care. Some concern was expressed with regard to thermoneutrality. Usually experiments are carried out at temperatures below the alleged zone of thermoneutrality.

ME_M is mainly determined by body composition (chemical i.e. protein/fat ratio and anatomical i.e. size of viscera) and physical activity. Differences in ME_M in relation to growth are mainly caused by interdependence of the factors calculated. Differences in physical activity might, in some cases, explain some of the differences found between respiration measurements and the comparative slaughter technique.

It was pointed out that the estimation of body composition of live animals (e.g. by DEXA) is only reliable provided that regression equations which are valid for the data set in question are available.

Beef cattle

The use of ionophores, in a molasses-based diet, demonstrated that similar rates and composition of growth were maintained with a reduced feed intake through improvements in dietary NE_g. This ocurred even though ME_m was increased.

Two studies (Hutcheson *et al.* and Byers) demonstrated that anabolic implants had a distinct impact on nutrient partitioning. The use of these implants reduced retained energy as fat and reduced NE_g requirements by 19 to 30%.

Salmon

The better feed and energy utilization in selected salmon compared to wild salmon could, at least partly, be explained by a better nutrient digestibility. No differences in energy needs for maintenance were observed, but physical activity might be of some importance.

Human adolescents

Respiration experiments in combination with indirect measurements (heart rate monitors) showed, in male and female adolescents, a big variation in daily energy expenditure. That variation of daily energy expenditure could mainly be explained by differences in body composition and physical activity. The energy requirements which could be contributed to growth were only a minor part of total metabolizable energy requirements (max. 5-10%).

SESSION 6

ENVIRONMENTAL ASPECTS AND MAINTENANCE

EFFECT OF SHORT-TERM EXPOSURE TO HIGH AMBIENT TEMPERATURES ON HEAT PRODUCTION IN BOARS OF DIFFERENT BREEDS

A-H. Tauson[1], A. Chwalibog[1], J. Ludvigsen [3], K. Jakobsen[2] and G. Thorbek[3]

[1]Department of Animal Science and Animal Health; [3]Research Department of Human Nutrition, The Royal Veterinary and Agricultural University, Copenhagen; [2]Department of Nutrition, Danish Institute of Agricultural Science, Foulum, Tjele, Denmark

Summary

The effects of short-term exposure to high ambient temperatures on heat production (HE), respiration rate (RR) and rectal temperature (T_r) were evaluated individually in 28 boars of different breeds at about 100 kg live weight (LW). The animals included 4 Yorkshire (YS), 8 Duroc (DR), 8 Hampshire (HS) and 8 Danish Landrace (DL), out of which 3 were Halothane positive (DLH$^+$). The gas exchange was measured in 5 periods each of one hour, starting with a basal period (B) at 17°C, followed by two hours of heating to 35.0°C (Per I) and to 39.7°C (Per II), and then two hours cooling when temperature decreased to 21.8°C (Per III), and 18.2°C (Per IV). RR was recorded every 15 minutes and T_r after the end of Per IV.

HE increased in Per I to 18% above the basal level in DLH$^+$ but less than 10% in the other breeds. The increment in HE was pronounced for all boars in Per II with values around 34% for YS, DL and DR and about 56% for HS and DLH$^+$. The HE values in Per III were close to basal values for YS, DL and DR but remained 22% and 40% above B values for HS and DLH$^+$, respectively. In Per IV all breeds had HE values below their basal levels with a mean of 16%. Maximum RR was recorded in Per III with an average of 183 breaths/minute for all breeds, but RR increased earlier and more steeply in HS.

It was concluded that HS and DLH$^+$ were more severely affected by the heat stress than the other breeds, indicated both by a faster and steeper response to the increased temperature and a slower recovery during cooling, than in boars of YS, DL and DR.

Key words: Gas exchange, heat stress, respiration rate, rectal temperature, pig, breed.

Introduction

To tolerate extreme heat, such as ambient temperatures close to the body temperature, animals must be able to evaporate water from the skin and from the respiratory tract. Pigs have a comparatively poor heat tolerance since both these pathways are of limited importance. Estimates of the upper critical temperature (UCT) for the pig vary, but a level close to the normal rectal temperature has been indicated.

Pigs kept outdoors, and with access to a wet area solve the problem of heat dissipation by wallowing in water and mud, and pigs reared indoors may, if possible, use the dunging area for cooling the body. For pigs deprived of possibilities to wet the body, heat exposure leads to a gradual increase of rectal temperature and respiration rate (RR).

Different breeds of pigs may differ in some metabolic characteristics (Mersmann, 1991), and genotypic differences in susceptibility to the porcine stress syndrome, malignant hyperthermia, are well known (Webb, 1980). Furthermore, selective breeding of modern pigs may have influenced the animals' capacity to lose heat. The aim of the present study was to investigate the influence of brief periods of high ambient temperature on HE and RR in boars of different breeds, and to evaluate possible breed differences.

Materials and Methods

The animals used were 4 Yorkshire (YS), 8 Duroc (DR), 8 Hampshire (HS) and 8 Danish Landrace (DL) boars, of which 3 were stress susceptible (Halothane test) (DLH^+). All animals were fed the same commercial feed compound, containing 164 g crude protein, 29 g crude fat and 16.0 MJ gross energy per kg, and the intakes of metabolizable energy and digestible protein were 1.1-1.2 $MJ/kg^{0.75}$ and 11-13 $g/kg^{0.75}$, respectively. The mean LW of the boars before the heat stress experiment was 101 (SEM 1.7) kg.

The animals were fed in the pens at 07.00 h, then transferred gently to the respiration chambers at 08.00 h and exposed to the same pattern of temperature and relative humidity: the thermostats were set to 17°C. After one hour of rest basal gas exchange was measured from 09.00 h to 10.00 h (Per B). The thermostats were then set to maximum heating and no cooling, by which the temperature increased rapidly to 35.0°C, with a mean of 26.2°C (SEM 0.12) from 10.00 h to 11.00 h (Per I) and more slowly to 39.7°C, the mean being 38.2°C (SEM 0.19) from 11.00-12.00 (Per II). At 12.00 h the thermostats were set to no heating and maximum cooling, causing a rapid decline in temperature to to 21.8°C with a mean of 30.9°C (SEM 0.41) from 12.00 to 13.00 h (Per III) and slowly to 18.2°C with a mean value of 19.4°C (SEM 0.40) from 13.00 h to 14.00 h (Per IV). The animals were observed through the window of the chamber, their behaviour recorded, and RR was recorded each 15 minutes, and rectal temperature (T_r) was measured at the end of Per IV.

The temperature and relative humidity in the chamber was recorded every 15 minutes. The system was controlled by thermostats which worked with an accuracy of ±0.2°C. The gas exchange was measured according to the open-air-circuit principle for each hour from 09.00 to 14.00 and the heat production (HE(RQ)) was calculated according to the formula by Brouwer (1965). Chambers were as described by Thorbek (1969).

Statistical analyses were performed according to the GLM-procedure in SAS by a model comprising the fixed effects of breed and period, the interaction between breed and period and random animal within breed, and using animal within breed as error term when evaluating breed effects.

Results

During Per I the animals generally seemed comfortable, but as heating progressed during Per II all animals showed signs of distress. The respiration was audible, and several animals were breathing with open mouth and panting severely. Four animals, all HS, were salivating heavily. During cooling in Per III the animals returned to an apparently more comfortable behaviour, although panting occurred throughout this period, and RR was high. In Per IV most animals were at rest, lying down, some sleeping. RR (Table 1) showed the same general pattern for all animals, but there were profound breed differences, HS boars having the highest mean maximum value of 236 (SEM 9.2) breaths/minute by the end of Per II and with the highest individual observed value of 260 breaths/minute.

The highest T_r values (average 39.3°C) were recorded in HS animals, whereas rectal temperatures in other breeds ranged from 38.8°C (YS) to 39.0°C (DL, DLH^+). HS animals that were salivating during heat stress showed a tendency (P=0.15) for higher rectal temperatures than non-salivating animals (39.7°C vs. 39.0°C).

Table 1. Mean respiratory rate (breaths/minute), and mean heat production (kJ/h) in boars of different breeds exposed to high ambient temperatures.

	n	Period B 17°C	Period I 26°C	Period II 38°C	Period III 31°C	Period IV 19°C	SEM, breed
Mean respiratory rate, breaths/min							
YS	4	23	27	136	206	114	17.1
DL	5	24	33	144	176	69	13.6
DR	8	24	27	125	172	58	10.7
HS	8	21	43	182	192	98	12.2
DLH^+	3	18	33	125	169	83	18.4
SEM, period		0.8	2.7	7.3	6.5	7.7	
Heat production, kJ/h							
YS	4	838	902	1015	898	704	27.9
DL	5	878	913	1115	824	700	32.8
DR	8	780	832	1137	775	641	32.6
HS	8	917	1010	1438	1115	795	12.2
DLH^+	3	914	1080	1402	1280	842	68.5
SEM, period		17.7	28.0	46.5	45.4	21.4	

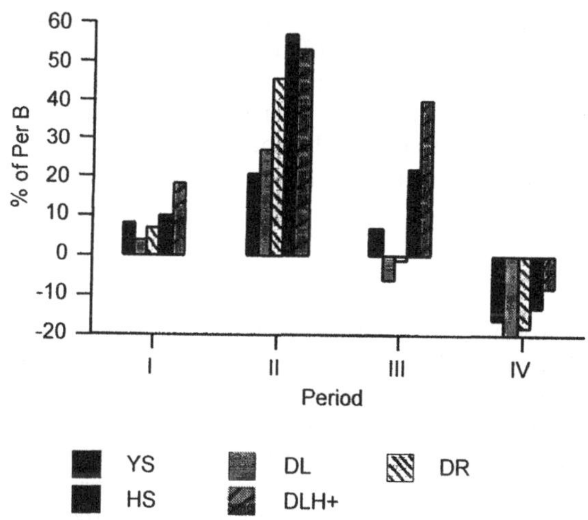

Fig. 1. Influence of heating (Per I and II) and cooling (Per III and IV) on heat production in boars of different breeds in relation to their basal levels (%).

The mean HE was about 860 kJ/h in Per B, with no significant differences between breeds (Table 1). In the first hour of heating (Per I) HE increased slightly to 18% above the level in Per B for DLH^+ but less than 10% for the other breeds (Figure 1), with no significant differences between breeds. The second hour of heating (Per II) caused high HE values for all breeds with a pronounced variation from 1015 to 1438 kJ/h in YS and HS, respectively, corresponding to an increment of 21% and 57% above basal level. The mean HE in Per II was 1102 kJ/h for YS, DL and DR, breed differences being non-significant, while it was 1428 kJ/h in HS and DLH^+, the differences between the two groups of breeds being highly significant. In the first hour of cooling (Per III) the HE values were around basal levels for

YS, DL and DR, but remained 22% and 40% above Per B for HS and DLH$^+$, respectively. In Per IV all breeds were below their basal levels with a mean of 16% and without significant differences between breeds.

Discussion

It was clearly evident that boars of YS, DL and DR differed from boars of HS and DLH$^+$ in their response to heat stress, the first group of breeds generally reacting less intensively both regarding heat production (YS, DL and DR) and respiratory rate (DL and DR) than the second group of breeds. The HS and DLH$^+$ boars could be considered as severely stressed, expressed by a more substantial increase in HE and an earlier and steeper increase in RR, which also reached higher maximum values (only HS).

The finding that DLH$^+$ responded more to heat stress than DL was not unexpected since these pigs had been found stress susceptible by the Halothane test. The HS pigs were not Halothane tested, so their stress susceptibility was not known. Therefore, their reaction to the heat stress could have been caused by the animals being stress sensitive, but may alternatively be explained by this breed simply being less able to withstand extreme heat.

In conclusion, this investigation has documented significant differences in HE and RR between pig breeds regarding their response to short-term exposure to high ambient temperatures. HS boars showed less tolerance to high temperature than did YS, DL and DR. Moreover, it was clearly evident that the stress susceptible pigs of DLH$^+$ reacted more negatively to the experimental treatment than did DL. The magnitude of the response was similar in HS and DLH$^+$, but the pattern somewhat different by HS, contrary to DLH$^+$, having a more rapid increase in RR, which also reached a higher peak value. Four out of eight HS animals salivated heavily during heat stress, and these animals had a tendency for higher rectal temperature when removed from the respiration chambers. These results point out that modern pig breeds may differ to a substantial extent as regards their tolerance to exposure to extreme heat, differences that may be of significance in practical pig breeding.

References

Brouwer E. (1965). Report of sub-committee on constants and factors. Proceedings 3rd Symposium on Energy Metabolism, *EAAP Publ. No.* 11:441-443.

Mersmann H.J. (1991). Characteristics of obese and lean swine. In: Miller, E.R., Ullrey, D.E. & Lewis, A.J. (Eds.) Swine Nutrition. Butterworth-Heinemann, Boston, pp. 75-89.

Thorbek G. (1969). Studies on energy metabolism in growing pigs. I. Construction and function of a new respiration plant for pigs. Rep. National Institute of Animal Science, Copenhagen, *373*, 46 pp.

Webb A.J. (1980). The incidence of halothane sensitivity in British pigs. *Anim. Prod.* 34:101-105.

NUTRITIONAL EFFECTS ON METABOLIC RATE AND PHYSICAL ACTIVITY IN GROUP HOUSED, 50 KG PIGS

J.W. Schrama[1], M.W. Bosch[1], M.J.W. Heetkamp[1], J. Haaksma[2] and M.W.A. Verstegen[1]

[1]Department of Animal Sciences, Wageningen Institute of Animal Sciences (WIAS), Wageningen Agricultural University, Wageningen, The Netherlands; [2]Institute of Sugar Beet Research (IRS), Nutrition Department, Bergen op Zoom, The Netherlands

Summary

Effects of non-starch polysaccharides (NSP) on energy metabolism were studied in group-housed pigs. Twelve groups of each 14 barrows were assigned to one of four diets differing in sugar beet pulp silage (SBPS) content: 0, 5, 10, and 15% (on DM basis). SBPS was exchanged for tapioca (≈ starch for NSP). Metabolizability decreased with increasing dietary SBPS content ($P<0.01$). Exchanging starch for NSP did not affect heat production ($P>0.1$). However, diet affected energy expenditure on physical activity ($P<0.10$). Pigs were more quiet when dietary SBPS content increased. From heat production, apparent digestibility of crude protein, crude fat, and NSP data, the estimated NSP value was 14.8 kJ/g of fermented NSP: being much higher than literature values. This energy value was partially associated with reduced activity. The estimated NSP energy value was 11.0 kJ/g when heat production data were corrected for activity.
Key words: Behaviour, pigs, energy evaluation, non-starch polysaccharides.

Introduction

Carbohydrates are the major dietary energy sources for growing pigs. Carbohydrates can be classified into a part (mainly starch) which is absorbed as glucose units after being digested by the pig's enzymes; and a part (non-starch polysaccharides, NSP) which is absorbed as volatile fatty acids due to fermentation by microbes. The NE value of digested starch is higher than that of fermented NSP (Just *et al.*, 1983; Hoffmann *et al.*, 1990; CVB, 1993; Noblet *et al.*, 1994). Physical activity of pigs varies. Factors causing between day variation are the animals' health status and level of exposure to stressors. Dietary factors can also induce differences in physical activity/behaviour. A previous study showed that fermentable NSP can reduce pigs' energy expenditure on activity compared with starch (Schrama *et al.*, 1996). In most energy evaluation systems (e.g. CVB, 1993; Noblet *et al.*, 1994), the maintenance energy requirements are assumed constant. Apart from this assumption of constancy, values used for maintenance vary largely between studies: the lowest value (CVB, 1993) being 70% of the highest value (Noblet *et al.*, 1994). Under practical conditions, pigs are kept in groups, whereby they have the ability to interact with each other and are relatively free in their movement. Most experimental conditions of research (e.g. Noblet *et al.*, 1994; Just *et al.*, 1983; Hoffmann *et al.*, 1990), which form the basis for various energy evaluation systems, differ largely from practical conditions: individual housing often coinciding with restriction in movement. Hypothetically, housing conditions may affect the energetic utilization of nutrients by influencing the pigs' physical activity. Therefore, the energetic utilization of fermented NSP was assessed in relation to physical activity in growing pigs being housed in groups.

Materials and Methods

The experiment consisted of a 2 week preliminary period followed by a 1 week experimental period. Twelve groups of 14 pigs each were assigned to one of four diets, which mainly differed in the starch and NSP content. Pigs were crossbred barrows [Great Yorkshire × (Dutch Landrace × Great Yorkshire)]. At the start of the preliminary period pigs were 15 weeks old and weighed 45.3 kg (SEM = 0.57). At the start of the preliminary period, each group (14 pigs) was placed in one of two identical, large, climatic respiration chambers (Verstegen *et al.*, 1987). Each group consisted of two sub-groups, which were housed in one of two pens of 8.3 m^2 each within each chamber (7 pigs per pen). Environmental temperature was kept at 20°C, relative humidity at approximately 65% and air velocity was < 0.2 m/s. Pigs were exposed to 12 h of light (from 07.00 to 19.00) and 12 h of dimness.

The four diets were formulated to be similar in composition, except for their contents of starch and NSP. The difference in the ratio between starch and NSP was created by exchanging sugar beet pulp silage (SBPS) for tapioca. The four diets, "0%SBPS", "5%SBPS", "10%SBPS", and "15%SBPS" contained on dry matter (DM) basis, respectively, 0, 5, 10 and 15% SBPS. Dietary nutrient compositions are given in Table 1. Pigs were fed according to their metabolic BW. During the experimental period, the feeding level was 2.5 times the maintenance requirements. The animals were fed per pen at 08.30 and 15.30 h. The daily amount of feed per pen was based on the average BW per pen, adjusted for an expected gain of 500 g/d. The feed was given as wet feed and water was added at 2.2 l per kg DM. Water was available *ad libitum* during the day.

Table 1. Analysed nutrient composition (g/kg DM) of the experimental diets with different amounts of sugar beet pulp silage (SBPS) fed to the growing pigs.

	Diet			
	"0%-SBPS"	"5%-SBPS"	"10%-SBPS"	"15%-SBPS"
Crude protein	185	190	196	200
Crude fat	37	37	38	38
Starch	415	382	341	314
Sugars	7	6	5	4
NSP[1]	275	305	340	365

[1]Non-starch polysaccharides (NSP) content was calculated as dry matter - crude protein - crude fat - starch - sugars - ash.

Individual BW was measured at the start and end of the 7-d experimental period. Apparent faecal digestibility of crude protein (CP), crude fat, and NSP were measured by using the acid insoluble ash marker method. On 5 days of the experimental period, grab samples of fresh faeces were collected during the morning feeding and subsampled within each group. Apparent faecal NSP digestibility was calculated assuming a 100% digestibility of starch and sugars. Energy and nitrogen balances were measured by quantitative collection of faeces with urine together per group. Heat production (H_{tot}) was measured in 9-min intervals by determining exchange of oxygen, carbon dioxide, and methane as described by Verstegen *et al.* (1987). Physical activity was recorded in the same time intervals as H_{tot} with a radar burglar device (two devices per chamber) (Wenk & van Es, 1976). One of two burglar devices per chamber was used in the analysis of the relationship between activity and H_{tot}. Calculation of the activity related heat production (H_{ac}) and heat production corrected for activity (H_{cor}) were done identical to Schrama *et al.* (1996). Average daily gain, energy, and N balance traits as well as H_{tot}, H_{ac}, and H_{cor} were analyzed for the effect of diet by linear regression on the dietary SBPS content (on DM basis). One group at the "0%SBPS" diet was excluded from the analyses due to health problems (diarrhoea and feed refusals).

Table 2. Mean daily intake (g/kg$^{0.75}$/d) of digestible nutrients (± SE) for diets differing in sugar beet pulp silage content (SBPS) fed to the growing pigs.

	Diet			
	"0%-SBPS"	"5%-SBPS"	"10%-SBPS"	"15%-SBPS"
Dry matter	70.2±0.31	70.7±0.18	71.0±0.49	70.4±0.17
Crude protein	9.7±0.11	10.2±0.06	10.3±0.06	10.1±0.07
Crude fat	1.7+0.01	1.7+0.02	1.7+0.02	1.6+0.03
Starch	29.1±0.13	27.0±0.07	24.3±0.17	22.1±0.05
Sugars	0.5±0.00	0.5±0.00	0.4±0.00	0.3±0.00
NSP[1]	12.7±0.22	15.1±0.12	17.2±0.19	18.5±0.09

[1]Non-starch polysaccharides (NSP) content was calculated as dry matter - crude protein - crude fat - starch - sugars - ash.

Results and Discussion

BW at the start of the experimental period was similar between diets (P>0.1) being on average 55.0 kg. During the experimental period, dietary SBPS content did not affect daily gain (P>0.1) being on average 670 g/d. Averaged over diets, apparent faecal digestibility of CP, crude fat and NSP were 74.3, 63.0, and 70.0%, respectively. Apparent faecal digestibility of CP and crude fat decreased by respectively 0.23 (P<0.01) and 0.22% (P<0.05) per percentage increase in dietary SBPS content, whereas apparent faecal NSP digestibility increased by 0.35% (P<0.01) per percentage increase in SBPS content. Table 2 shows intakes of digested nutrients for each of the experimental diets. According to the design, the main difference between treatments was in intake of digestible starch and fermentable NSP.

Table 3. Linear relationship between percentage of sugar beet pulp silage (SBPS) in the diet (on dry matter basis) and several energy balance parameters of group housed growing pigs.

Parameter	Diet	P-value of linear SBPS effect
Methane, kJ/kg$^{0.75}$/d	Y = 5.8 (±0.67) + 0.27 (±0.068) × SBPS	**
ME intake, kJ/kg$^{0.75}$/d	Y = 988 (±8.8) - 2.0 (±0.90) × SBPS	*
ME/GE, %	Y = 77.5 (±0.53) - 0.21 (±0.054) × SBPS	**
Heat production, kJ/kg$^{0.75}$/d:		
Total (H_{tot})	Y = 670 (±4.5) - 0.47 (±0.457) × SBPS	NS
Activity related (H_{act})	Y = 99 (±7.3) - 1.41 (±0.750) × SBPS	#
Corrected for activity (H_{cor})	Y = 571 (±4.6) + 0.94 (±0.475) × SBPS	#
Energy Retention (ER) kJ/kg$^{0.75}$/d	Y = 318 (±10.4) - 1.6 (±1.07) × SBPS	NS

[1]NS = not significant, P>0.1; #, P<0.10; *, P<0.05; **, P<0.01.

The linear SBPS effect on GE intake was not significant (P>0.1); being on average 1282 kJ/kg$^{0.75}$/d. With increasing dietary SBPS concentrations, ME intake decreased (P<0.05; Table 3). The decline in metabolizability (ME/GE) with increasing amount of dietary SBPS was mainly related to the decline in apparent faecal energy digestibility. Moreover, energy losses with methane increased with increasing dietary SBPS contents (P<0.01; Table 3). The increase in methane production equalled 4.1% of the energy content of the fermented NSP. Despite the lower ME intake, with increasing dietary SBPS content, both H_{tot} and total energy retention (ER) were unaffected by a linear SBPS effect (P>0.1; Table 3). However, increasing amounts of dietary SBPS resulted in a decline of physical activity, being indicated by the decline in H_{act} (P<0.10; Table 3). At 0, 5, 10, and 15% SBPS in the diet, H_{act} was 109, 87, 87 and 80 kJ/kg$^{0.75}$/d, respectively. In order to estimate the energetic utilization of fermented NSP (fNSP) and the effect of physical activity on it, both H_{tot} and H_{cor} were related to the daily intake of fNSP. However, to obtain unbiased relationships, both H_{tot} and H_{cor} were corrected ($CorH_{tot}$ and $CorH_{cor}$, respectively) for the heat increment induced by the intake of digestible CP, crude fat, starch and sugars using the Dutch NE (CVB, 1993) energetic efficiency of 46, 92, 77, and 77%, respectively. The estimated relationships of both $CorH_{tot}$ and $CorH_{cor}$ (in kJ/kg$^{0.75}$/d) with fNSP (g/kg$^{0.75}$/d) during the experimental period were as follows:

$$CorH_{tot} = 385(\pm 25.1) + 2.7 (\pm 1.54) \times fNSP \quad (n=11, R^2=0.25)$$

$$CorH_{cor} = 235(\pm 20.7) + 6.5 (\pm 1.27) \times fNSP \quad (n=11, R^2=0.75)$$

Based on the $CorH_{tot}$ data, the NE value of fNSP was 14.8 kJ/g, which is equal to an efficiency of 85%: being much higher than other literature values (57%, Hoffmann *et al.* 1990; 58%, Noblet *et al.*, 1994; 45%, Bakker, 1997). A possible explanation for the higher NE value of fNSP in the present study is that pigs were housed in groups and thus not restricted in their ability to perform physical activity/behaviour. Based on the $CorH_{cor}$ data, the NE value of fNSP was 11.0 kJ/g; being equal to an efficiency of 63%. Thus, this study demonstrated that variation in NE values of fNSP between studies might be related to differences in physical activity. In the present study, the energy saving by the reduced physical activity was 3.9 kJ/g fNSP intake. Another reason for the differences between studies is the differences in NE values used for maintenance in the calculations. This effect is quantified in Table 4. Varying the assumed NE requirements for maintenance from 293 to 425 kJ/kg$^{0.75}$/d resulted in a variation in the energetic value of fNSP from 5.2 to 13.2 kJ/g (Table 4).

Table 4. Effect of fixing the NE requirement for maintenance on the relationship between energy retention (ER; kJ/kg$^{0.75}$/d) and daily intake of fermentable non-starch polysaccharides (fNSP, in g/kg$^{0.75}$/d) in group housed growing pigs.

Assumed NE maintenance	Relation	R^2
Not fixed	ER = -425 (±34.7) + 13.2 (±2.13) × fNSP	0.81
293 kJ/kg$^{0.75}$/d	ER -293 = + 5.2 (±0.41) × fNSP	0.94
343 kJ/kg$^{0.75}$/d	ER -343 = + 8.2 (±0.33) × fNSP	0.98
393 kJ/kg$^{0.75}$/d	ER – 393 = + 11.3 (±0..27) × fNSP	0.99
425 kJ/kg$^{0.75}$/d	ER - 425 = + 13.2 (±0.41) × fNSP	1.00

References

Bakker G.C.M. (1996). Interaction between carbohydrates and fat in pigs: impact on energy evaluation of feeds. PhD dissertation, Agricultural University Wageningen. The Netherlands.

CVB (1993). Netto energie van voedermiddelen voor varkens; argumentatie en onderbouwing van de nieuwe Nev-formule. CVB-report nr 7, Lelystad, The Netherlands.

Hoffmann L., Schiemann R., Jentsch W. & Beyer M. (1990). *Arch. Anim. Nutr.* 40:209-223.

Just A., Fernandez J.A. & Jorgensen H. (1983). *Livest. Prod. Sci.* 10:171-186.

Noblet J., Fortune H., Shi X.S. & Dubois S. (1994). *J. Anim. Sci.* 72:344-352.

Schrama J.W., Verstegen M.W.A., Verboeket P.H.J., Schutte J.B. & Haaksma J. (1996). *J. Anim. Sci.* 74:2220-2225.

Verstegen M.W.A., van der Hel W., Brandsma H.A., Henken A.M. & Bransen A.M. (1987). In: M.W.A. Verstegen and A.M. Henken (Eds) Energy metabolism in farm animals: effects of housing, stress and disease. pp. 21-48. Martinus Nijhoff Publishers, Dordrecht.

Wenk C. & van Es. A.J.H. (1976). *Schweiz. Landwirtsch. Monatsh.* 54:232-236.

HEAT PRODUCTION ASSOCIATED WITH EATING IN CATTLE

A. Susenbeth[1], R. Mayer[2], B. Köhler[2] and O. Neumann[2]

[1]Institute of Animal Nutrition, Physiology and Metabolism, University of Kiel, Germany;
[2]Institute of Animal Nutrition, University of Hohenheim, Germany

Summary

Heat production associated with eating was determined in an experiment with 4 adult Hinterwälder steers of 625-675 kg body weight (BW) using indirect calorimetry in respiration chambers. Heat production was measured during 3 hours with and 3 hours without *ad libitum* access to wheat straw and the energy costs for eating calculated as the difference which amounted on average to 27 ± 13 J/min/kg BW (mean ± SD). Straw intake, time spent chewing and the number of jaw movements were also recorded. Values for the energy cost of eating reported in the literature were similar to those in this study and were around 27% of those for ruminating. The literature values varied greatly between feeds and feed treatments but were relatively constant when related to time spent eating and were similar for cattle, sheep and horses. The energy used for eating and ruminating was calculated to be 10% of the ME intake of high quality dried roughage and one third that of untreated straw. This may be the main reason for the lower efficiency of ME utilization in roughages.

Key words: Cattle, energy metabolism, eating, ruminating.

Introduction

The time cattle spend for eating and ruminating amounts to 13 to 17 hours per day, when animals have *ad libitum* access to diets containing a high proportion of roughage (Bosch *et al.*, 1992; Brouk & Belyea, 1993; Teller *et al.*, 1993; Susenbeth *et al.*, 1995). The capacity of ruminants for mechanically reducing feed particle size can be a limiting factor for feed intake. On the other hand, the energy required for eating accounts for a considerable proportion of the total energy requirement (Vermorel & Mormede, 1991). The energy needed for chewing thus reduces the amount of ME available for production, especially important when the production level is low. This fact might be mainly responsible for the lower efficiency of utilization of ME in roughages (Ørskov & MacLeod, 1990). The aim of this study was to determine the energy required for chewing in cattle and, using data from the literature, to quantify the factors which affect it.

Materials and Methods

Four adult Hinterwälder steers weighing between 625 and 675 kg were fed a diet consisting of 3 kg hay, 1.5 kg wheat straw and 2 kg concentrates per day which met the energy maintenance requirement. This diet was offered both before and after the experimental periods. On days when experiments were performed, the animals were put in open circuit respiration chambers at 07.00 h. From 08.00 to 11.00 h oxygen consumption and carbon dioxide production were measured, when animals had no access to feed. During a second period between 11.00 and 14.00 h, gas exchange was measured when animals had free access to unchopped wheat straw. According to the results of *in vitro* studies by Steingass & Menke (1986), the extent of the ruminal degradation of straw is negligible within the first 3 hours and therefore any additional heat production in the second period was unlikely to be caused by metabolization of absorbed nutrients. Heat production was calculated according to Brouwer (1965), the factor due to nitrogen excretion via urine was not considered. Each jaw movement was recorded by a computer so that the time spent for rumination and eating as well as the number of chews within the respective periods could be determined. Measurements were repeated four times per animal.

Results

The results of the experiments are given in Table 1. The average heat production during the three hours of the first period without feed was 8.0 MJ and it increased by 19% to 9.5 MJ during the second period due to eating activity. The average time spent eating was 97 min and the number of jaw movements for eating was on average 6500 during 3 hours. Although very variable, it correlated with the amount of straw ingested. In both periods, the time spent for ruminating was almost identical, therefore, additional heat production caused by chewing activity can be attributed solely to eating. Eating increased heat production by 0.28 kJ/s or 0.25 kJ/chew and, when related to BW, by 27 J/min/kg BW.

Table 1. Heat production and eating activity during 3 hours for four steers offered either no feed or wheat straw (mean ± SD[1]).

Body weight		kg	643 ± 24
Heat production	no feed offered	MJ	7.95 ± 0.52
	straw offered	MJ	9.46 ± 0.61
	relative change	%	19 ± 6
Eating activity	time	min	97 ± 36
	no. of chews		6500 ± 2540
	DM intake	kg	3.49 ± 1.28
Increase of heat production		kJ/s	0.28 ± 0.14
associated with eating		kJ/chew	0.25 ± 0.12
		J/min/kg BW	27 ± 13
		kJ/g DM	0.46 ± 0.16

[1] n = 15 observations for eating activity, n = 16 for heat production.

Discussion

The question, whether heat production associated with eating should be related to the time spent eating or to the number of chews, cannot be answered on the basis of the data of this experiment, because the chewing frequency was highly constant, which led to a similar variation for both references. Because the time spent for eating is chosen as the reference in the literature, this reference is used here as well.

Table 2 gives data on the energy requirement for eating and ruminating from the literature. The mean values are 30 and 9 kJ/min/kg BW for eating and ruminating, respectively. For sheep, similar mean values for eating (40 kJ) and ruminating (11 kJ) are reported in the literature reviewed by Susenbeth *et al.* (1997). Also similar values for eating were found in horses with 36 kJ (Zuntz & Hagemann, 1898) and 31 kJ/min/kg BW (Vermorel & Mormede, 1991). When these figures for eating are related to the metabolic BW, the values for cattle, sheep and horses differ to a smaller extent (124, 125 and 119 kJ/min/kg $BW^{0.75}$).

The energy requirement for eating is closely related to time spent eating. However it is not related to the amount of feed ingested and is therefore independent of the ingestion rate. This observation was made by Dahm (1910) and Adam *et al.* (1984) in cattle, by Ustjanzew (1991), Young (1966) and Osuji *et al.* (1975) in sheep and by Vermorel & Mormede (1991) in ponies.

Table 2. Energy required for eating and ruminating in cattle.

Authors	Body weight	Energy required for	
		Eating	Ruminating
	kg	J/min/kg BW	J/min/kg BW
Pächter, 1909	-	45	8.0
Dahm, 1910	240-275	32	9.3
Schiemann *et al.*, 1971	620-955	25	
Holmes *et al.*, 1976	55-120	26[1]	
Holmes *et al.*, 1978	115-405	42[1]	
Adam *et al.*, 1984	300-405	31	
Ku-Vera *et al.*, 1989	300	11[2]	9.3
	-	32	
Present study	625-675	27	
Mean ± SD		30 ± 10	9

[1] Time feed offered is taken as time spent eating. [2] Nutrition by intragastric infusion.

To estimate the energy requirement for eating and ruminating in cattle during 24 hours, results of an experiment reported by Susenbeth *et al.* (1995) were used for a first approximation. The animals were given *ad libitum* access to roughage of different qualities. Details of the calculation are given in Table 3. Feed intake decreased with decreasing roughage quality, whereas time spent ruminating and energy requirement for ruminating remained constant, and time spent eating increased only slightly. Energy requirement for eating and ruminating ranges from 8.3 to 11.0 MJ/d and from 0.8 to 2.4 MJ/kg DM ingested. This shows that the energy required for eating and ruminating can account for one third of the metabolizable energy (ME) in low quality roughages, which leads to a significant reduction in the proportion of ME available for maintenance and production. As mentioned by Ørskov & MacLeod (1990) this fact may be responsible for the lower efficiency of the utilization of ME, especially of low quality roughages. Therefore, energy requirement for eating and ruminating should be considered as a quantifiable component of the total energy budget instead of an indefinite consequence of feed composition or processing.

Table 3. Energy requirement for eating and ruminating in cattle fed different roughages[1].

Roughage		Grass hay high quality	Grass hay good quality	Wheat straw chopped	Wheat straw unchopped
Ad libitum intake of roughage DM	kg	10.3	8.1	5.1	4.5
Intake of ME from roughage	MJ	107	77	42	34
Energy requirement for					
Ruminating[2]	MJ	1.9	2.0	2.2	2.0
Eating[3]	MJ	6.4	7.0	7.4	9.0
Energy requirement for					
ruminating and eating	MJ/kg DM	0.8	1.1	1.9	2.4
	% of ME-intake	8	12	23	33

[1] Related to 24 h. Data for roughage intake and chewing behaviour of steers taken from Susenbeth *et al.* (1995). [2] 9 J/min/kg BW. [3] 30 J/min/kg BW.

References

Adam I., Young B.A., Nicol A.M. & Degen A.A. (1984). Energy cost of eating in cattle given diets of different form. *Anim. Prod.* 38:53-56.

Bosch M.W., Lammers-Wienhoven S.C.W., Bangma G.A., Boer H. & van Adrichem P.W.M. (1992). Influence of stage of maturity of grass silages on digestion processes in dairy cows. 2. Rumen contents, passage rates, distribution of rumen and faecal particles and mastication acitvity. *Livest. Prod. Sci.* 32:265-281.

Brouk M. & Belyea R. (1993). Chewing activity and digestive responses of cows fed alfalfa forages. *J. Dairy Sci.* 76:175-182.

Brouwer E. (1965). Report of sub-committee on constants and factors. In: K.L. Blaxter (Ed.) Energy metabolism. *EAAP Publ. No.* 11. 441.

Dahm K. (1910). Die Bedeutung des mechanischen Teils der Verdauungsarbeit für den Stoffwechsel des Rindes. *Biochemische Zeitschrift* 28:456-503.

Holmes C.W., Stephens D.B. & Toner J.N. (1976). Heart rate as a possible indicator of the energy metabolism of calves kept out-of-doors. *Livest. Prod. Sci.* 3:333-341.

Holmes C.W., McLean N.A. & Lockyer K.J. (1978). Changes in the rate of heat production of calves during grazing and eating. *N. Z. J. Agric. Res.* 21:107-112.

Ku-Vera J.C., MacLeod N.A. & Ørskov E.R. (1989). Energy exchanges of cattle nourished by intragastric infusion of nutrients. In: Y. Van der Honing, & W.H. Close (Eds) Energy metabolism of farm animals. *EAAP Publ. No.* 43.271-274.

Ørskov E.R. & MacLeod N.A. (1990). Dietary-induced thermogenesis and feed evaluation in ruminants. *Proc. Nutr. Soc.* 49:227-237.

Osuji P.O., Gordon J.G. & Webster A.J.F. (1975). Energy exchanges associated with eating and rumination in sheep given grass diets of different physical forms. *Br. J. Nutr.* 34:59-71.

Pächter. (1909). Respiratorische Stoffwechselforschung und ihre Bedeutung für Nutztierhaltung und Tierheilkunde. Dissertation, Berlin. Cit. in: Dahm, K. 1910. Die Bedeutung des mechanischen Teils der Verdauungsarbeit für den Stoffwechsel des Rindes. *Biochemische Zeitschrift* 28:456-503.

Schiemann R., Nehring K., Hoffmann L., Jentsch W. & Chudy A. (1971). Energetische Futterbewertung und Energienormen. p 37. VEB Deutscher Landwirtschaftsverlag, Berlin.

Steingass H. & Menke K.H. (1986). Schätzung des energetischen Futterwerts aus der in vitro mit Pansensaft bestimmten Gasbildung und der chemischen Analyse. 1. Untersuchungen zur Methode. *Uebers. Tierernahrang.* 14:251-270.

Susenbeth A., Aiple K.P., Lang G.U. & Mayer R. (1995). Effect of forage quality and particle size on voluntary feed intake and chewing activities of oxen. *Ann. Zootech. Suppl.* 44:251 (Abstr.).

Susenbeth A., Mayer R., Koehler B. & Neumann O. (1997). Energy requirement for chewing in cattle. (In press.)

Teller E., Vanbelle M. & Kamatali P. (1993). Chewing behaviour and voluntary grass silage-intake by cattle. *Livest. Prod. Sci.* 33:215-227.

Ustjanzew W. (1911). Die energetischen Äquivalente der Verdauungsarbeit bei den Wiederkäuern (Schafe). *Biochemische Zeitschrift* 37:457-476.

Vermorel M. & Mormede P. (1991). Energy cost of eating in ponies. In: C. Wenk, & M. Boessinger (Ed) Energy metabolism of farm animals. *EAAP Publ. No.* 58. 437-440.

Young B.A. (1966). Energy expenditure and respiratory activity of sheep during feeding. *Aust. J. Agric. Res.* 17:355-362.

Zuntz N. & Hagemann O. (1898). Cit. in: H. Bergner and L. Hoffmann. 1996. Bioenergetik und Stoffproduktion landwirtschaftlicher Nutztiere, p 202. Harwood Academic Publishers, Chur, Switzerland.

DAILY ENERGY EXPENDITURE AND PROTEIN REQUIREMENT OF WORKING FEMALE SWAMP BUFFALOES ESTIMATED FROM ENERGY BALANCE AND BODY COMPOSITION

I.G. Mahardika[1], D. Sastradipradja[2] and I.K. Sumadi[1]

[1]Department of Animal Nutrition, Faculty of Animal Husbandry, Udayana University, Denpasar-Bali, Indonesia; [2]Department of Physiology and Pharmacology, Faculty of Veterinary Medicine, Institut Pertanian Bogor (IPB-Bogor Agricultural University), Taman Kencana 1 no. 3, Bogor 16151, Indonesia

Summary

Four female swamp buffaloes (*Bos bubalis*) 280-380 kg liveweight were used in a study using a 4 x 4 latin square experimental design for the estimation of energy expenditure and protein requirements for draught, which consisted of 4 levels of work load: no work and loads amounting 450-500 Newton continuous traction for 1, 2 and 3 hours daily for 14 consecutive days. The buffaloes were fed King grass *ad libitum* and were subjected to energy and protein balance trials. The *in vivo* body density method was used to estimate body composition and energy expenditure (EE) was calculated from ME minus retained energy (RE). EE during rest was found to be 0.42 $W^{0.75}$ MJ/day and increased up to 1.99 times greater for the work of 3 hours. Daily energy expenditure for work was 9.56, 20.0 and 25.86 MJ for treatments 1 hour, 2 and 3 hours respectively. Fat retention was absent in all groups of working animals, while protein retention was negative for animals with a 3 hours work load. The requirement of protein for maintenance was 2.51 $W^{0.75}$ g/d, whereas for growth (P_{growth}) and work (P_{work}), they followed the equations: $P_{growth} = [(258 + 1.25W^{0.75})\Delta W_{kg}]$g; and $P_{work} = [12.59\ e^{0.95t}]$g, respectively.

Key words: Swamp buffalo, energy expenditure, work, protein requirement, *in vivo* body composition, mass densitometry.

Introduction

Large animal livestock husbandry in many third world countries generally functions as a source of draught as well as for meat and the production of offspring. Management and adequate feed are necessary for draught animals, but until now there are no reliable figures, especially for the energy and protein requirements in the humid tropical areas. Experimental data of local livestock, and of introduced livestock to the humid tropics, are needed on digestibility, metabolizability, and net-energy availability for gaining understanding of the animal's productive process. The mask calorimetry method involving respiratory gas exchange could be used (Clar *et al.,* 1992) to measure whole body heat production, but the obtained data are based on short period measurements only. Long-term daily energy expenditure measurement (EE) under field conditions is preferred over short-term determinations and an ideal method should be the double labelled water technique (Midwood *et al.,* 1989), but the method is expensive and provides only approximate estimates. The purpose of this research was to measure the energy balance of the working female swamp buffalo in order to determine energy and protein requirements. The approach was to estimate long-term EE of working buffaloes from an energy balance within a 14 d experimental period using the relationship that EE is equal to metabolizable energy (ME) minus retained energy (RE). RE was calculated from body composition estimated by the *in vivo* body density (water displacement) method (Kleiber, 1961) and daily gain of body weight.

Materials and Methods

The research used 4 female swamp buffaloes weighing 280-350 kg which had been trained to do daily work exercise for 3 hours. When the buffaloes were not doing field exercise, they were placed in individual pens with facilities required for balance trials. King grass (*Penisetum purpureum* x *P. tipoides*) aged 45-55 days as the sole feed was given at all times *ad libitum*. Water was given twice a day at noon and in the evening. A sledge which was the pulling load equipment furnished with a weight scale to measure draft force, a distance measuring device and a stop watch were used to quantify the treatment load. The experiment used a 4 x 4 latin square design. The field trial involved pulling loads of 450-500 N for 1, 2 and 3 hours continuously per day for 14 consecutive days starting at 07.00 h (4 treatments including non-working as control). Air temperature was 24-32°C and relative humidity 70-95%. Excreta and refused feed were collected, recorded and sampled for proximate analysis.

Body composition was determined *in vivo* by the mass densitometric method in water (Kleiber, 1961) using the relationship $W_f/W_b = [v_b - v_l]/[v_f - v_l]$, where W_b = body mass (kg), W_f = fat mass (kg), W_l = lean mass (kg), v_b , v_l and v_f were respectively specific volumes (l/kg) of body mass, lean and fat. Body protein was calculated using an equation relating the ratio of meat plus bone (R_d) -which are the components of lean minus blood- with body weight (W in kg): $R_d = 2.861 + 0.0109\ W$ ($r^2 = 0.99$; SD = 0.27) derived from data reported by Natasasmita (1978). Body protein was then calculated by multiplying protein percentage of swamp buffalo meat times the quantity of meat. From the changes in body-protein and -fat (and -energy) before and after the experimental period, RE was calculated assuming energy equivalents of 39.32 MJ/kg and 20.07 MJ/kg for body-fat and -protein respectively. The calculated heat production while resting ($EE_{resting}$), was obtained from the control group. $EE_{working}$ of a particular workload was calculated from EE of the working group concerned by subtracting $EE_{resting}$. The data obtained were subjected to analysis of variance and if significant differences were obtained between treatments, the data were further analysed by the Duncan Multiple Range Test (Steel & Torrie, 1986). Calculation of mathematical models to relate different measured variables employed multiple regression analysis using the Lotus 123 program.

Results and Discussion

Data on feed intakes and retained body constituents are presented in Table 1. DM intakes ranged from 2.02% BW (no workload) to 2.71% BW (3h workload). Considering *ad libitum* feeding, the level of energy output i.e. physical activity, is a factor regulating the level of energy intake. The $EE_{resting}$ value was found to be 0.42 MJ/W $kg^{0.75}$. Energy for work calculated as, $EE_{working}$ minus $EE_{resting}$, showed that the value for the 1 hour working group was 9.56 MJ/d, while those for the 2 hours and 3 hours working groups were 20.0 and 25.86 MJ/d, respectively. From the ME data it was calculated that ME intake (MJ/d) depended on length of work exercise t (hour) following the equation:

$$ME = \frac{59.5\, t^n}{0.53 + t^n}$$

Fat retention was absent in all working group animals, but protein retention was negative for animals with a 3 hours work load (Figure 1). The data indicated that intakes failed to match output resulting in a utilization of stores. Both fat and protein were utilized by the working animals for the energy demand of work.

Table 1. Category of energy intakes, protein intake and retained body constituents in working female swamp buffaloes.

Parameter	Work load			
	no work	1 hour	2 hours	3 hours
Energy and protein intakes:				
GE (MJ/d)	91.77^{a}	109.51^{b}	134.62^{c}	145.06^{d}
DE (MJ/d)	46.07^{a}	52.74^{b}	65.44^{c}	71.87^{d}
ME (MJ/d)	35.39^{a}	39.78^{b}	50.84^{c}	55.51^{d}
Protein intake (kg/d)	0.56^{a}	0.69^{b}	0.75^{c}	0.71^{bc}
Retained (R) body constituents:				
R fat (kg/d)	0.07^{a}	-0.04^{b}	-0.08^{bc}	-0.09^{c}
R protein (kg/d)	0.11^{a}	0.08^{a}	0.01^{b}	-0.10^{c}
RE (MJ/d)	4.96^{a}	0.13^{b}	-2.75^{c}	-4.92^{d}
ΔW (kg/d)	0.50^{a}	0.18^{b}	-0.11^{c}	-0.32^{d}
EE = ME - RE (MJ/d)	30.42^{a}	39.75^{b}	53.60^{c}	60.42^{d}

Values in a row with different superscript letters differ significantly ($P<0.05$)

Skeletal, muscular and organ development during growth are intricately associated with endocrine factors and proper nutrition. Epinephrine has a calorigenic effect in mammals and is intricately linked to activity. Epinephrine and nor-epinephrine stimulate the liberation of FFA from adipose tissue. Cortisol has an effect on gluconeogenesis from protein. The utilization of protein in our present study indicated that gluconeogenesis was permitted to proceed by provision of substances in the Kreb's cycle, thus supporting muscular activity.

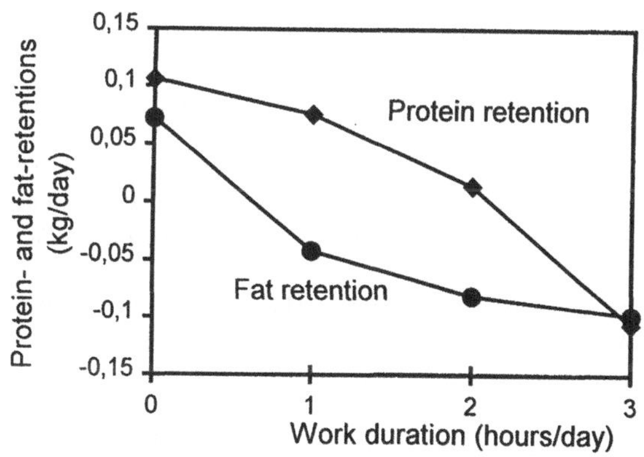

Fig. 1. Reduction in protein and fat retentions of working female swamp buffaloes due to length of daily work duration.

At maintenance when RE = 0, $ME_{maint.}$ equals $HP_{maint.}$ When ME intake increases with ΔME there will be an increase in RE of ΔRE. Mount (1970) postulated a ratio of 0.70 for $\Delta RE/\Delta ME$ which means that only 70% of the ME increment above maintenance would be converted into RE and the remaining will be lost as heat. So, $HP_{maint.} = ME - RE/0.70$ and consequently a value of 0.37 $W^{0.75}$ MJ/d was found for $ME_{maint.}$ for our swamp buffaloes. The same principle was followed for the calculation of ME requirement for growth for our working buffaloes. Calculations demonstrated that 14.6 MJ ME/d was required for each kg increase of body mass.

Similarly, protein requirement was estimated from the feeding trials and changes in body composition. The calculations required data on digestibility and the biological value of protein, arriving at a value of 2.51 $W^{0.75}$ g/d protein for maintenance, whereas the requirement for work depended on the length of carrying out work (for loads 450-500 Newton). The relationship between length of work and protein requirement followed the equation: $P_{work} = [12.59e^{0.95t}]$g. Protein requirement for growth of the working swamp buffalo was calculated from daily gain in body weight, body composition, the biological value of protein and nitrogen loss through urine. For each kg increase of body mass, 181 g of protein is needed (protein content of the body is 18.1%). Assuming a biological value of protein to be 70%, digestible protein with feed intake should be 181/0.70 = 258 g. In addition, assuming endogenous nitrogen excretion in urine = 0.2 $W^{0.75}$, then urinary protein would be 6.25 x 0.2 $W^{0.75}$ = 1.25 $W^{0.75}$. Based on these figures, protein requirement for the growing female buffalo would be: $P_{growth} = [(258 + 1.25\ W^{0.75})\Delta W_{kg}]$g.

In conclusion, the body density (or water displacement) method on swamp buffaloes proved to be very useful and appropriate because wading in water is a natural behaviour of swamp buffaloes. Combined with balance and exercise trials in the field, energy and nutrient requirements could be readily estimated.

Acknowledgements

Supported by research grant no. 90/P4M/DPPM/PHB II/1/93 from the Directorate General of Higher Education of the Department of Education and Culture of R.I. The authors wish to thank Prof. Dr. K. Becker of Hohenheim University for beneficial discussions.

References

Clar U., Becker K. & Susenbeth A. (1992). A mobile mask technique for measuring gas exchange in cattle. *J. Anim. Physiol. & Anim. Nutr.* 67:133-142.

Kleiber M. (1962). The fire of life. An introduction to animal energetics. John Wiley & Sons, Inc., New York- London. pp. 44-59.

Midwood A.J., Haggarty P., McGaw B.A., Robinson J.J. & Fuller M.F. (1989). Application of the heavy water technique for the measurement of energy expenditure to production animals. In: Y. van der Honing & W.H. Close (Eds) Energy metabolism of farm animals. *EAAP Publ No.* 43: 341-344.

Mount L.E. (1979). Adaptation to thermal environment. Man and his productive animals. (Contempory Biology). Edward Arnold (Publishers) Ltd., London. pp. 29-31.

Natasasmita A. (1978). Body composition of swamp buffalo. A study of developmental growth and of sex differences. PhD dissertation, University of Melbourne, Australia.

Steel R.G.D. & Torrie J.H. (1986). Principles and procedures of statistics. McGraw-Hill Book Co. Inc., New York.

ENERGY COST OF ACTIVITY (STANDING VERSUS LYING) IN GROUP-HOUSED, YOUNG PIGS

M.J.W. Heetkamp[1], J.W. Schrama[1], and J.W.G.M. Swinkels[2]

[1]Department of Animal Sciences, Wageningen Institute of Animal Sciences (WIAS), Wageningen Agricultural University, Wageningen, The Netherlands; [2]Research Institute for Pig Husbandry, Rosmalen, The Netherlands

Summary

The effect of regrouping young pigs directly before transport on metabolic rate in relation to behaviour was evaluated. Six groups of 20 pigs were allotted to one of two treatments. In three control groups, social structure of the two original litters was not altered. In the other three groups, the two litters were mixed just before transport. After transport, energy metabolism and activity were measured per group during two weeks. Additionally video recordings were used to quantify behaviour. Heat production, corrected for activity, was 450 and 464 kJ/kg$^{0.75}$/d for the control and mixed groups, respectively ($P<0.05$). On average 24.6% and 23.5% of animals were standing in the control and mixed groups, respectively ($P<0.05$). Energy expenditure on standing, on average 107 kJ/kg$^{0.75}$/d, was not affected by treatment. Most of the variation in heat production related to activity could be explained by variations in energy expenditure on standing. Consequently, the contribution of agonistic behaviour while standing as well as changes in posture were negligible. Regrouping before transport had no impact on energy partitioning, but reallocation of energy expenditure between activity/behaviour and other energy demanding processes seems to occur.

Key words: Pigs, energy metabolism, agonistic behaviour, transport, mixing.

Introduction

In common Dutch practice, 23 kg, 9-10 week old growing pigs are transported from multiplier unit/farm to finishing unit/farm. Coinciding with transport pigs usually are also regrouped. At this particular moment of life young pigs also have to cope with changes in housing system, feeding level and dietary composition (del Barrio *et al.*, 1993; Heetkamp *et al.*, 1995). This complex of stressors was absent in a previous study (Heetkamp *et al.*, 1995) where the effect of mixing pigs was evaluated. In the previous study it was demonstrated that regrouping of young, 20 kg pigs, without exposure to other stressors (optimal conditions), had no long-term effect on energy partitioning. It was hypothesized that the influence of regrouping of pigs is dependent upon the presence of other stressors. Heetkamp *et al.* (1995) did not find an effect of mixing, during 2 weeks after mixing, probably due to the optimal conditions experienced at the moment of mixing. In the present study, the effect of regrouping directly before transport, (intended to create sub-optimal conditions at the moment of mixing) on metabolic rate in relation to behaviour (standing vs. lying and agonistic behaviour) was evaluated during the first 2 weeks after transport.

Materials and Methods

Six groups of 20 pigs were allotted to one of two treatments. In three control groups, social structure of the two sub-groups (original litters) within each group was not altered. In the other three groups, the two original litters were mixed just before transport. During mixing one new group was formed with the five heaviest pigs of each litter and the other new group was formed with the five lightest pigs. After about 1 h (60 km) of transport, each sub-group within a group was placed in one of two pens (7.8 m^2 excluding the feeding trough) in one of two identical large open climate respiration chambers (Verstegen *et al.*, 1987). Room temperature was kept at 24°C and relative humidity at 65%. Pigs were fed a standard commercial diet, containing 16.3 kJ GE/gram and 18.0% crude protein, at a level of 1.5 times the metabolizable energy requirement for maintenance (440 kJ/kg$^{0.75}$/d). Pigs were fed twice daily at 08.00 and 15.30 h and were weighed weekly. Energy and nitrogen balances were measured per group during two 7 d balance periods. Therefore, heat production was measured in 9 min intervals during the last 6 d of each period by means of indirect calorimetry. Physical activity was measured continuously with a radar system. Additionally video recordings of the last day of each week were used to quantify behaviour. At intervals of 3 min, posture (number of animals standing and lying as a % of total number of animals) of the animals in one group was scored. In this way also agonistic behaving standing animals (as a % of total) were scored. The number of standing animals (% of total) not behaving agonistically was calculated as the difference of total number of standing animals and agonistically behaving standing animals. All these parameters, scored every 3 min, were averaged to 9 min data in line with the 9 min heat production data. Posture change (standing to lying position and lying to standing position) was calculated as the difference in number of animals lying and standing between the 3 min posture data. This parameter was added up to 9 min data (sum of posture changes as a % of total number of animals) in line with the 9 min heat production data. To calculate energy costs of total physical activity, standing, agonistic behaviour and posture change, continuous heat production data were related to the radar activity data and to the behaviour data (linear regression technique).

Results

As intended, gross energy intake per kg of metabolic weight, averaged over both experimental weeks, was exactly the same for control and mixed groups (761 kJ/kg$^{0.75}$/d; P>0.1).

Table 1. Daily gain, metabolizability (ME/GE), metabolizable energy (ME) intake, total heat production (HP), activity related heat production (HP_{act}), heat production corrected for activity (HP_{cor}) and energy retention (ER) in young pigs as affected by regrouping before transport.

	Treatment			
Trait	Control	Regrouped	SEM	P-value
Initial weight (kg)	19.6	19.1	0.35	0.298
Daily gain (g/d)	194	205	3.7	0.107
ME/GE (%)	81.7	82.0	0.34	0.600
ME (kJ/kg-$^{0.75}$/d)	622	624	5.8	0.829
HP (kJ/kg$^{0.75}$/d)	573	572	8.4	0.986
HP_{act} (kJ/kg$^{0.75}$/d)	123	108	5.6	0.146
HP_{cor} (kJ/kg$^{0.75}$/d)	450	464	3.1	0.032
ER (kJ/kg$^{0.75}$/d)	50	52	8.7	0.872

In Table 1, least square means over the total experimental period are presented for both the control and mixed groups. Metabolizability, total heat production (HP) and energy retention (ER) did not differ between treatments (P>0.1, Table 1). Activity related heat production (HP_{act}) was 123 and 108 kJ/kg$^{0.75}$/d for the control and mixed groups, respectively (P>0.1, Table 1). However, heat production corrected for activity (HP_{cor})was 450 and 464 kJ/kg$^{0.75}$/d for the control and mixed groups, respectively (P<0.05, Table 1). Due to the similar ME and HP, ER of the pigs was not affected by regrouping (P>0.1, Table 1). This is also reflected by the similar daily gain (200 g/d, P>0.1, Table 1) in both treatments.

Analysis of video recordings the last day of each experimental week are presented in Table 2.

Table 2. Results from analysis of behaviour in young pigs as affected by regrouping before transport using different models with different behavioural parameters.

Model with traits	Treatment: Control	Treatment: Regrouped	SEM	P-value
$Y = :\mu + \beta \times$ % standing [1]				
% standing	24.6	23.5	0.25	0.038
Energy cost of standing ($kJ/kg^{0.75}/d$)	456	434	38.2	0.699
Energy expenditure on standing ($kJ/kg^{0.75}/d$)	111	102	8.7	0.485
$Y = \mu + \beta_1 \times$ % agonistic standing $+ \beta_2 \times$ % standing not agonistic standing [1]				
% agonistic standing	3.4	3.0	0.28	0.462
Energy cost of agonistic standing ($kJ/kg^{0.75}/d$)	320	307	43.9	0.843
Energy expenditure on agonistic standing ($kJ/kg^{0.75}/d$)	10.7	8.9	1.06	0.286
% not agonistic standing	21.2	20.5	0.39	0.257
Energy cost of not agonistic standing ($kJ/kg^{0.75}/d$)	469	447	40.4	0.717
Energy expenditure on not agonistic standing ($kJ/kg^{0.75}/d$)	98.9	91.3	8.12	0.542
$Y = \mu + \beta_1 \times$ % standing $+ \beta_2 \times$ % posture change [1]				
% standing	24.6	23.5	0.25	0.038
Energy cost of standing ($kJ/kg^{0.75}/d$)	462	426	32.3	0.481
Energy expenditure on standing ($kJ/kg^{0.75}/d$)	113	100	7.5	0.303
% posture change	38.9	41.1	1.07	0.222
Energy cost of posture change ($kJ/kg^{0.75}/d$)	-4.5	8.5	13.56	0.535
Energy expenditure on posture change ($kJ/kg^{0.75}/d$)	-2.1	3.6	5.15	0.475

[1] β_1, β_2 calculated with fixed effect of day period (2 levels: light and dark) included in model

On average 24.6% and 23.5% of the total number of animals were standing in the control and mixed groups, respectively ($P<0.05$, Table 2). Energy cost of standing and energy expenditure on standing were not affected by treatment ($P>0.1$, Table 2), being on average 445 $kJ/kg^{0.75}/d$ and 107 $kJ/kg^{0.75}/d$, respectively.

Regrouping did not affect the number of animals behaving agonistically while standing being 3.4% and 3.0% for control and mixed groups, respectively ($P>0.1$, Table 2). Energy cost of agonistic behaviour during standing was not affected by mixing ($P>0.1$), being on average 313 $kJ/kg^{0.75}/d$. This estimation is lower than the energy cost of standing which was on average 445 $kJ/kg^{0.75}/d$. Theoretically, the difference (313-445) gives information about the extra energy cost of agonistic behaviour. In this case, estimation of energy cost of agonistic behaviour is probably underestimated because of low incidence of this type of behaviour.

Posture change was not affected by regrouping (P>0.1, Table 2). Despite the high average number of animals changing posture (40% every 9 min), the energetic cost and therefore energy expenditure of postural change was estimated to be very low.

Discussion

Regrouping of young pigs before transport has hardly any impact on energy partitioning, but reallocation of energy expenditure between different live processes seems to occur. When young pigs are mixed just before transport average number of standing animals is lowered significantly during 2 weeks after transport. Although not significant, the average number of mixed animals behaving agonistically while standing also decreased slightly. Relating total heat production to the radar activity counts shows the same trend: mixed pigs spent less energy on total physical activity. On the other hand, mixed pigs used more energy on non-activity related live processes. When mixed, animals seem to compensate extra energy requirements for non-activity related processes with a reduction in activity. Total energy partitioning between maintenance (including activity) and growth is not affected.

Most of the variation in heat production related to activity could be explained by variations in energy expenditure on standing. Consequently, the contribution of agonistic behaviour while standing and changes in posture were negligible. The low occurrence of agonistic behaviour may affect the estimation and therefore β may be underestimated. Energy costs of postural changes may be underestimated because it is difficult to distinguish from standing because both take place simultaneously. In this study agonistic behaviour and postural changes were scored only once every 3 minutes. When scored continuously agonistic behaviour and posture change could probably be estimated more accurately. Nevertheless energy expenditure on total physical activity (HP_{act}) and energy expenditure on standing is highly correlated (correlation coefficient: 0.85, P<0.001). Therefore it is doubtful whether other kinds of behaviour in a group of pigs, in addition to number of animals standing, will contribute significantly in explaining variation in heat production.

References

del Barrio A.S., Schrama J.W., van der Hel W., Beltman H.M. & Verstegen M.W.A. (1993). Energy metabolism of growing pigs after transportation, regrouping, and exposure to new housing conditions as affected by feeding level. *J. Anim. Sci.* 71:1754-1760.

Heetkamp M.J.W., Schrama J.W., de Jong L., Swinkels J.W.G.M., Schouten W.G.P. & Bosch M.W. (1995). Energy metabolism in young pigs as affected by mixing. *J. Anim. Sci.* 73:3562-3569.

Verstegen M.W.A., van der Hel W., Brandsma H.A., Henken A.M. & Bransen A.M. (1987). The Wageningen respiration unit for animal production research: a description of the equipment and its possibilities. In: M.W.A. Verstegen and A.M. Henken (Eds) Energy metabolism of farm animals. pp. 21-48. Martinus Nijhoff, Dordrecht, The Netherlands.

ENERGY EXPENDITURE IN GROWING PIGS INFECTED WITH PLEUROPNEUMONIA

H.J. Bray[1], L.R. Giles[2], J.M. Gooden[3] and J.L. Black[4]

[1]Pathobiology and Epidemiology Section, DLO Institute for Animal Science and Health, Postbus 65, 8200 AB Lelystad, The Netherlands; [2]Elizabeth Macarthur Agricultural Institute, NSW Agriculture, PMB 8, Camden NSW 2570, Australia; [3]Department of Animal Science, University of Sydney, PMB 3, Camden NSW 2570, Australia; [4]John L. Black Consulting, Locked Bag 21, Warrimoo NSW 2774, Australia

Summary

Pigs were intrabronchially inoculated with either saline (control group) or media containing *Actinobacillus pleuropneumoniae* (App), the cause of porcine pleuropneumonia. Oxygen consumption (VO_2), voluntary feed intake (VFI), body temperature (BT) and respiration rate (RR) were measured. Pigs inoculated with App had a reduction in VFI when compared to control pigs, however their VO_2 was maintained, indicating an increase in maintenance energy requirements.

Key words: Oxygen consumption, voluntary feed intake, pleuropneumonia, pigs.

Introduction

Irrespective of preventative measures, the prevalence of respiratory disease and its cost to the pig industry has not decreased. Although pneumonia is associated with poor growth rates in pigs the physiological mechanisms by which this occurs are unclear. Such knowledge is required if decision support computer software such as AUSPIG (Black *et al.*, 1986) which is used both in Australia and internationally as a management tool, is to accurately predict the performance of pigs over a range of conditions.

Little is known about changes in maintenance requirements in animals with respiratory disease. In addition, there is a poor understanding of the association between VFI, VO_2 and other physiological parameters, including BT and RR in pigs infected with App. The objective of this study was to examine the effect of challenge with App on VFI and VO_2 in growing pigs and the associations between these parameters and BT and RR.

Materials and Methods

Sixteen specific-pathogen-free hybrid gilts (Pig Improvement Company strain (Large White x Landrace); mean liveweight 65 kg, were fed *ad libitum* a commercial, pelleted, protein-adequate diet containing 12.6 MJ DE/kg. Pigs were housed in individual metabolism crates at 22°C and in constant light. Each pig was surgically prepared for the measurement of VO_2 using the pulmonary AV technique (Giles *et al.*, 1995) by which VO_2 is calculated from the product of cardiac output and the difference in blood oxygen content across the lungs. Cardiac output was recorded as blood flow through the pulmonary artery using transit-time ultrasound (Transonic Systems Inc., U.S.). Fibre optic catheters (Opticath, Abbott Laboratories, U.S.) were placed in the pulmonary and carotid arteries to measure mixed venous and arterial blood oxygen saturation respectively.

Electronic measurement and collection of VO_2 data began 2 weeks after surgery. VFI, BT and RR were recorded manually at intervals of 1 hour. After approximately 48 h the pigs were sedated with ketamine/xylazine and intra-bronchially inoculated with either 5 ml of media (Brain-Heart Infusion Broth: Oxoid, Unipath Ltd., U.K.) containing between 10^4 and 10^9 colony forming units of App (n=10) or 5 ml of phosphate-buffered saline (n=6).

Data collection continued for 96 to 144 h. Seven days post-inoculation the pigs were killed by pentabarbitone injection and the lungs were isolated and perfused with fixative. Right lung damage was calculated from image analysis of serial lung slices.

Results

All pigs challenged with saline had no right lung damage. All pigs inoculated with App showed superficial signs of lung damage except one. Lung lesions were characteristic of porcine pleuropneumonia. All pigs inoculated with saline had a BT less than 39.1°C post inoculation. BT for all pigs inoculated with App rose to a peak of 41.6°C or higher between 5 and 12 h post-inoculation, except for one pig which reached 41.8°C 31 h post-inoculation. Pigs with less than 10% lung damage had a BT near normal by day 2 post-inoculation. By day 5 BT had declined to pre-inoculation values in all pigs except for the pig with 87% lung damage, whose temperature fell below the pre-inoculation value.

Pigs with less than 3% right lung damage had little variation in their RR between the control period, day 2 and day 5. Pigs with 3% right lung damage or greater had an increased RR on day 2, with the pig with 87% lung damage having the highest relative RR (2.2 times that measured during the control period). The highest RR was 65 breaths/min 0-24 h post-inoculation in the pig with 46% lung damage measured at day 7. By day 5, the RR of all pigs except one had returned close to pre-inoculation values.

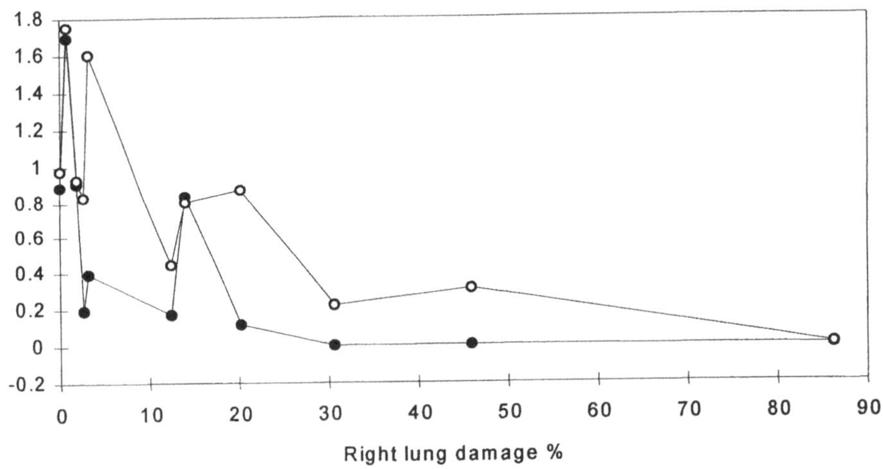

Fig. 1. Relative daily voluntary feed intake on day 2 (24-48 h) and day 5 (96-120 h) post-inoculation related to right lung damage measured 7 days post-inoculation for pigs inoculated with saline (open circles) or *Actinobacillus pleuropneumonia* (closed circles). Relative VFI was calculated by dividing the daily VFI by that pig's average daily VFI measured pre-inoculation.

All pigs with right lung damage at day 7 post-inoculation, including pigs with very small amounts of damage (less than 12%), had a period of inappetence following challenge. The overall daily intake of control pigs with no right lung damage was not affected. However, they did have a period of inappetence associated with sedation during inoculation but resumed eating after 12 h. Figure 1 shows the relative VFI on day 2 and day 5 related to lung damage. Pigs with greater than 30% right lung damage did not resume pre-inoculation VFI and had the lowest VFI at 96 h post-inoculation. The pig with 87% right lung damage did not resume feeding during the experimental period.

The average pre-inoculation VO_2 for the control pigs was 416 ± 9.9 ml/min while the average for App treated pigs was 423 ± 23.8 ml/min. The post-inoculation average for the control pigs was 409 ± 7.5 ml/min, while the average for the App treated pigs declined to 367 ± 7.0 ml/min. Both of these changes were found to be non-significant. The relative VO_2 on days 2 and 5 post inoculation is shown against lung damage in Figure 2. The pig with the highest right lung damage (87%), had the highest VO_2 on day 2 (24-48 h) post-inoculation with a relative VO_2 increase of 30% above pre-inoculation VO_2.

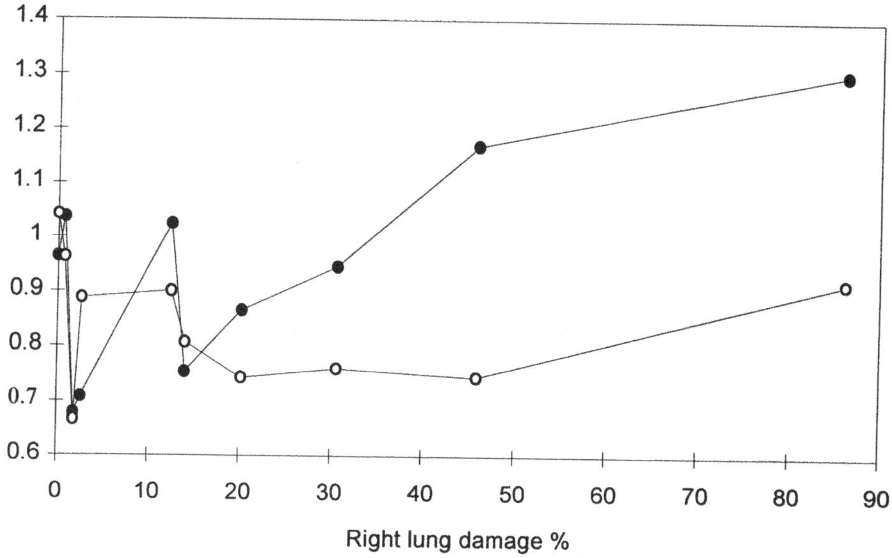

Fig. 2. Relative daily oxygen consumption on day 2 (24-48 h) and day 5 (96-120 h) post-inoculation related to right lung damage measured 7 days post-inoculation for pigs inoculated with saline (open circles) or *Actinobacillus pleuropneumonia* (closed circles). Relative VO_2 was calculated on same basis as for RVFI.

Discussion

It can be seen from Figure 1 that inappetence occurs with very low levels of lung damage and this has major implications for productivity in the pig industry. The results suggest that the reduction in VFI is related to disease severity rather than an elevation in BT. Previous studies with high ambient temperatures (Giles and Black, 1991) indicated a direct relationship between the reduction in VFI and the elevation of BT above 39.5°C. However, the current study with App infection indicated a poor relationship between VFI and BT. For instance, some pigs did eat following App infection despite elevated body temperature, whereas others ate little despite normal BT. This suggests that other mechanisms are at least partly responsible for the reduction in VFI. Possible factors include: a) the direct action of cytokines and other inflammatory mediators; b) changes in gut physiology; and c) reduced feeding activity. In mammals, interleukin 1, interleukin 6 and tumour necrosis factor are all anorexigenic (Klasing and Johnstone, 1991).

It can be seen from Figure 2 that, at day 2 post-inoculation, pigs with greater than 40% right lung damage had increased VO_2, compared to that measured during the control period. By day 5, VO_2 had declined to 0.75-0.9 of the pre-inoculation VO_2. Previous studies of VO_2 and heat production in pigs with respiratory disease by Intraraksa (1984) and Verhagen *et al.* (1987), predicted a reduction in VO_2 concomitantly with a fall in VFI.

However, on day 5 post-inoculation, VO_2 was maintained at an almost constant level in pigs inoculated with App regardless of VFI, illustrating an increase in maintenance requirements. This suggests that pigs with pleuropneumonia are similar to humans with sepsis, where increased VO_2 can increase heat production by 50% (Blaxter, 1989).

A maintenance of VO_2 relative to a decline in VFI indicates a decrease in the efficiency of energy utilization for liveweight gain. Increases in both RR and the oxygen cost of breathing will also increase maintenance requirements. However the RR of febrile pigs due to pleuropneumonia is much less than that for pigs that are febrile due to high ambient and the contribution is likely to be small. Other factors increasing maintenance requirements could be: a) an alteration in metabolic rate due to the action of cytokines, stress hormones and other endogenous mediators; b) the energy cost of protein turnover and tissue repair due to bacterial damage; and c) altered activity associated with feeding. The most likely of these during acute disease is the direct action of cytokines to induce fever by increasing oxygen consumption and conserving heat in order to raise body temperature (Moltz, 1993).

Results from previous studies have suggested that VO_2 will fall as feed intake falls in response to disease. Results from the current study suggest that this is not the case, and VO_2 is maintained while VFI is reduced due to acute disease. The direct effects of cytokines during acute disease are to reduce VFI, increase BT, and to generate and conserve heat by increasing VO_2 and limiting the increase in RR. Protein turnover, increased respiration rate, and altered activity are likely to play a role in the maintenance of VO_2 during chronic disease and this may in turn alter feeding activity and gut physiology to maintain the reduction of VFI in diseased pigs. The precise importance and contribution of each factor is an area for further research.

Acknowledgements

This study and the post-graduate scholarship of H.J. Bray was financed by the Australian Pig Research and Development Corporation.

References

Black J.L., Campbell R.G., Williams I.H., James K.J. & Davies G.T. (1986). Simulation of energy and amino acid utilization in the pig. *Res. Devel. Agr.* 3:121-145.

Blaxter K.L. (1989). Energy metabolism in animals and man. Cambridge University Press, U.K.

De Rijk R.H., Strijbos P.J.L.M., van Rooijen N., Rothwell N.J. & Berkenbosch F. (1993). Fever and thermogenesis in response to bacterial endotoxin involve macrophage dependant mechanisms in rats. *Am. J. Physiol.* 265:R1179-1183.

Giles L.R. & Black J.L. (1991). Voluntary food intake in growing pigs at ambient temperatures above the zone of thermal comfort. In: E.S. Batterham (Ed) Manipulating Pig Production III, *Proceedings of the Australasian Pig Science Association*, pp. 162-166.

Giles L.R., Annison E.F., Black J.L., Tucker R.G. & Gooden J.M. (1995). A new procedure for the continuous measurement of oxygen consumption in pigs. *J. agric. Sci. Camb.* 124:113-118.

Intraraksa Y., Engen R.L. & Switzer W.P. (1984). Pulmonary and hematologic changes in swine with Mycoplasma hyopneumoniae pneumonia. *Am. J. Vet. Res.* 45:474-477.

Klasing K.C. & Johnstone B.J. (1991). Monokines in growth and development. *Poult. Sci.* 70:1781-1789.

Leach R.M. & Treacher D.F. (1994). The relationship between oxygen delivery and consumption. *Disease-A-Month*, 40:301-368.

Moltz H. (1993). Fever: Causes and consequences. *Neurosci. and Behav. Rev.* 17:237-269.

Verhagen J.M.F., Groen A., Jacobs J. & Boon J.H. (1987). The effect of different climatic environment on metabolism and its relation to time of *Haemophilus pleuropneumoniae* infection in pigs. *Livest. Prod. Sci.*, 17:365-379.

METABOLIC ADJUSTMENTS IN THE CAMEL AND SHEEP DURING FOOD DEPRIVATION AND REFEEDING

A. Guerouali[1], N. Sbai[1] and M.F. Wardeh[2]

[1]Institut Agronomique et Veterinaire Hassan II, Rabat. Morocco; [2]The Arab Center for Studies of Arid Zones and Dry lands, Damascus, Syria

Summary

The present study was conducted with 5 female dromedaries and 6 ewes going through three successive periods. In the first period, camels and ewes were fed a ration composed of 2/3 barley grain and 1/3 lucerne hay corresponding to their maintenance energy requirements. During the second period, the two species were fasted for two days, but had free access to water. During the third period, they were offered the same amount of feed as in the control period. During the three successive periods, digestibility trials were conducted and animal heat production was determined by indirect calorimetry using calorimetric chamber for camels and a hood for ewes. Perturbations in feed intake were observed for both species in the refeeding period but were more pronounced in sheep. Dry matter digestibility averaged 61% and 67% in the control period. Camels and sheep showed comparable amounts of methane production in the control period, but the decrease during the fasting period was more important in sheep than camels. Lower respiratory quotient was observed in sheep indicating the nature of body metabolites oxidized during the fasting period. Heat production in camels was 20% lower than in sheep during the control period and was reduced by 27% and 20% after 2 days of fast for camels and sheep respectively. After 4 days of refeeding, heat production reached the level observed during the control period for ewes but was still 17% lower for camels.
Key words: Camel, sheep, fasting, methane production, respiratory quotient, heat production.

Introduction

Through evolution, camelids have specially adapted to arid environments and developed physiological mechanisms to face heat stress, dehydration (Guerouali *et al.*, 1994b) and shortage of nutrients. Consequently, camelids are known for their lower basal metabolism and their maintenance energy requirements are as low as 314 $kJ/kg^{0.75}$ (Guerouali *et al.*, 1994a) representing about two thirds of the requirements of cattle. The camelids feeding system in arid zones is based on grazing of large areas with little vegetation and animals go through periods of fasting for a few days in search of new grazing areas. The aim of the present study was to investigate the metabolic adjustments developed by the camel, compared to sheep, to face these periods of fasting.

Materials and Methods

Five adult female camels (average weight 300 kg, average age 10 years) and six adult ewes (average weight 35 kg, average age 5 years) were used in the present study. The experiment was divided into three successive periods (Control, Fasting and Refeeding). In the control period (P1) of 7 days, camels were fed 2 kg of barley grain and 1 kg of lucerne hay corresponding to the energy requirements for maintenance while ewes were fed 0.4 kg of barley grain and 0.2 kg of lucerne hay corresponding to maintenance energy requirements (Guerouali *et al.*, 1991) and both species had free access to water. During the second period, ewes and camels were fasted for 2 days but access to water was maintained.

Immediately after the fast, camels and ewes were fed the same amount of feed offered during the control period for 4 days (refeeding period). Digestive trials were conducted for 7 days in the control period and for 4 days in the refeeding period. Animal heat production (continuous 24h measurement) was determined at the last day of the control period, on the second day of fast and on the fourth day of the refeeding period using indirect calorimetry with chambers for camel and hoods for sheep (McLean, 1972).

Results and Discussion

In the refeeding period, perturbations in feed and water intakes were observed for both species, corresponding to the post-fast anorexia described in other species (Harris *et al.*, 1986), but were more pronounced in sheep compared to camel. Dry matter digestibility averaged 62% in camels and 68% in ewes during the control period (Table1). It seems likely that the diet offered was better digested by ewes than camels and the difference was due probably to lower efficiency of mastication and rumination in camels since a part of the barley grain offered in the ration was excreted in the faeces.

Table 1. Water intake, feed intake, digestibility and live body weight changes with respect to fasting period in camels and sheep.

CAMEL			SHEEP		
Feeding	Fasting	Refeeding	Feeding	Fasting	Refeeding
Water intake (litres/24h)					
5.2±1.22	2.4±0.44	3.5±0.41	1.78±0.42	1.18±0.54	1.63±0.41
Feed intake (kg/24h)					
3.0±0.0	------	2.81±0.21	0.6±0.0	------	0.47±0.11
Dry matter digestibility (%)					
61.6±3.8	------	------	67.5±6.3	------	------
Live body weight (kg)					
300±34	285±31	289±32	35.6±4.4	32.5±4.5	34.0±4.6

During the 2 days of fast, a 5% reduction in body weight was observed in camels which was lower than the 9% reduction observed in sheep (Table 1) and the value of 10% reported in cattle (Phillips *et al.*, 1991) indicating the superiority of camels in digesta retention and/or limitation in body reserves mobilization when fasted. Desert goats fasted for 4 days showed 11.6% reduction of their body weight (Ali *et al.*, 1984). Methane production, averaged 1.36 litres/kg$^{0.75}$/24h (Table 2) and was comparable in both species. It declined markedly during the fasting period (72% in camel and 93% in ewes), then increased progressively during the refeeding period without reaching the levels observed in the control period. Changes in methane production in dairy cattle with respect to 2 days fast were also reported by Bouvier (1977) showing that the pattern of changes was more comparable to that observed in sheep. The higher methane production observed in camel after 2 days of fast is in favour of more digesta retention in the digestive tract to allow fermentation. This result tends to confirm the lower rate of body weight loss found in camels compared to sheep. During the refeeding period, the observed perturbations in feed and water intakes may explain the limited methane production in both species.

The respiratory quotient of animals, representing the ratio between the O_2 consumption and the CO_2 production, averaged 0.99 and was comparable for both species in the control period (Table 2). During fasting it was reduced to 0.89 in camels and 0.79 in sheep which may indicate that camels were still deriving some of their nutrients requirements from fermentation products while sheep were using their body reserves to meet the required nutrients for maintenance. In another study with camels, sheep and cattle fasted for 5 days, Wensvoort *et al.* (1996) showed that camels had low rates of increase in serum NEFA and higher glucose levels compared to sheep and cattle and concluded that camels seem to do better than sheep and cattle in controlling their lipolytic and gluconeogenic rates in order to prevent the pathological state of ketosis after a period of fast. In the refeeding period, the respiratory quotient value returned to the level observed in the control period for camels but remained lower for sheep after 4 days of feeding which may show the rapid metabolic adjustments in camel compared to sheep.

Table 2. Methane production, respiratory quotient and total heat production during feeding, fasting and refeeding periods in camels and sheep.

CAMEL			SHEEP		
Feeding	Fasting	Refeeding	Feeding	Fasting	Refeeding
Methane production (litres/kg $^{0.75}$ /24 h					
1.35±0.13	0.39±0.08	0.99±0.09	1.37±0.25	0.1±0.01	0.98±0.18
Oxygen consumption (litres /kg $^{0.75}$ /24h)					
14.82±1.6	10.77±1.3	12.64±1.1	17.53±1.9	14.77±2.1	19.67±2.9
CO_2 production (litres/kg $^{0.75}$ /24h)					
14.78±1.2	9.64±1.1	12.81±1.0	17.23±1.8	11.68±1.8	16.69±2.8
Respiratory quotient					
0.99±.03	0.89±.07	1.01±.06	0.98±.05	0.79±.06	0.85±.11
Total heat production (kJ/kg $^{0.75}$ /24h)					
303 ± 33	220 ± 13	259 ± 29	380 ± 41	302 ± 42	403 ± 62

Heat production averaged 303 kJ/kg$^{0.75}$ in camels and 380 kJ/kg$^{0.75}$ in ewes when fed at maintenance energy requirements in the control period (Table 2) indicating that the metabolic rate in camelids is about 20% lower compared to sheep. After 2 days of fast camels showed more reduction in energy expenditure than sheep (27% vs. 20%) and after 4 days of refeeding, heat production reached the level observed during the control period for the ewe but was still 15% lower for camels. These data on energy expenditure demonstrate that, compared to sheep, fed camels use feeds with better efficiency and when fasted they mobilize less body reserves to meet the fasting metabolism.

Conclusion

Camels fasted for 2 days showed less body weight loss, more methane production and higher respiratory quotient when compared to sheep and therefore it appeared that they were able to maintain fermentation and to meet some of the nutrients required for fasting metabolism through fermentation products. Camels were able to develop digestive and metabolic adjustments in order to suffer less from the fast.

References

Ali B.H., Hassan T. & Musa N. (1984). The effect of feed restriction on certain haematological indices, enzymes and metabolites in Nubian goats. *Comp. Biochem. Physiol.* 79A (3):325-328.

Bouvier J.C. (1977). Etude descriptive des cinétiques des échanges gazeux chez le mouton. Thèse de doctorat en Sciences naturelles. Université de Clermont-ferrand. France.

Harris R.B.S., Kasser T.R. & Martin R.J. (1986). Dynamics of recovery of body composition after overfeeding, food restriction or starvation of mature female rats. *J. Nutr.* 116:2536-2546.

Guerouali A., Johson E.D. & Robertshow D. (1991). Energy balance in dry, pregnant and lactationg prolific sheep receiving two levels of feeding. Proceedings of the 12th Symposium on Energy Metabolism of Farm Animals. C. Wenk and M Boessinger (Eds) *EAAP, Publ. No.* 58:308-312.

Guerouali A., Zine Filali R., Vermorel M. & Wardeh M.F. (1994a). Maintenance energy requirements and energy utilization by camel at rest. Proceedings of the 13th Symposium on Energy Metabolism of Farm Animals. *EAAP Publ. No.* 76:367-370.

Guerouali A., Zine Filali R. & Wardeh M.F. (1994b). Metabolic adjustments of camel during heat stress and dehydration. Proceedings of the 13th Symposium on Energy Metabolism of Farm Animals. *EAAP Publ. No.*76 :363-366.

McLean J.A. (1972). On the calculation of heat production from open-circuit calorimetric measurements. *Br. J. Nutr.* 27:597-600.

Phillips W.A., Juniewicz P.E. & Vontungeln D.L. (1991). The effect of fasting on the source and amount of weight lost by steers of different ages. *J.Anim.Sci.* 69:2342-2348.

Wensvoort J, Kyle D.J, Orskov E.R. & Bourke D.A. (1996). Biochemical adaptation of camelids during fasting. The Arab Center for the Studies of Arid Zones and Dry Lands. *Camel Newsletter.* No 12:46-51.

BODY HEAT CONTENT OF DAIRY COWS

A.B. Hall and B.A. Young

Department of Animal Production, The University of Queensland, Gatton, Australia 4345

Summary

Body heat content is a balance of heat gained from metabolism and the environment, and heat lost to the environment. Seasonal variation, management systems and behaviour of the animal all influence heat balance. The body heat content of 4 high producing (35 to 38 l/d) Holstein Friesian dairy cows was monitored over 8 consecutive days during winter and during summer. The cows were housed under commercial dairy feedlot conditions with free access to shade. Two of the cows also had access to water spray cooling whenever the air temperature exceeded 26°C.

During winter, body heat content cycled diurnally by 5.68 kJ/kg body weight. On hot summer days (mean T_a > 27.7 °C), the body heat content of cows rose by 4.32 kJ/kg, but maintained a diurnal variation of 6.11 kJ/kg. Provision of spray cooling on hot summer days reduced the body heat content by 1.7 kJ/kg and increased diurnal variation to 7.40 kJ/kg. However, on cooler summer days (T_a < 25 °C), body heat content approached winter values.

This study showed that variation in the body heat content of dairy cows was dependent upon season and diurnal activity, and heat content was reduced by spray cooling during hot weather.

Key words: Dairy cow, body heat content, heat load, diurnal variation, season.

Introduction

Change over time in body heat content (kJ/kg body weight) (BHC) can be used as an indicator of the effects of environmental and nutritional factors on an animal's thermal state.

Typically, BHC has been measured under highly controlled situations using direct or indirect calorimetry (McLean & Tobin, 1987; Blaxter, 1989) and thermometry (Watts *et al.*, 1977; McLean *et al.*, 1982; McLean *et al.*, 1983a). While this original work provides a basis for understanding the individual physical factors affecting body heat, the validity of extrapolating results to an individual's natural situation have been questioned (Berman, 1968; Fuquay, 1981; Berman *et al.*, 1985; Maust *et al.*, 1972). Change in body heat needs to be investigated in commercial situations to determine the effects of normal management, exercise, radiant heat and wind and also allow for behavioural variations in the micro-environment by the animal.

A system for monitoring change in body heat content of cows under commercial management conditions was developed and then normal variations in BHC during winter and summer were monitored in high producing dairy cows.

Materials and Methods

The research was done at the Mutdapilly Research Station, Australia (latitude 27° 46' south, longitude 152° 40' east, elevation of 40 m) over 8 consecutive days each during winter and during summer. The climate was typically sub-tropical, with a winter mean daily temperature range of 4.3 ± 1.8°C to 22.0 ± 0.9°C and a summer mean daily range of 18.4 ± 1.4°C to 32.1 ± 1.5°C. Annual rainfall was 306.1 ± 58.4 mm.

Four second-lactation Holstein Friesian dairy cows were selected, with milk yields of 35 to 38 l/d, 480 to 690 kg body weight, and a coat colour between 50 and 70% black. The cows were housed under feedlot conditions with free access to corrugated iron roofed shade 3 m high over a concrete feed pad, with an adjacent unshaded dirt lot. The feedlot was situated approximately 100 m from the milking parlour. The cows remained in the feedlot except for concentrate feeding at 0700 h and 1400 h and milking at 0600 h and 1500 h daily. The concentrate feeding stalls, holding yards and milking parlour were shaded. Two of the cows also had access to water spray cooling under the shaded feedlot area (spacing = 1 m intervals; nossle delivery = 1.7 l/min; droplet size = 150 microns). The sprays were turned on automatically (cycle = 1.5 min on 15 min off) whenever the dry bulb temperature under the shade exceeded 26°C.

Thermometry was used to estimate change in body heat content (ΔHC) as described by McLean *et al.* (1983b), where:

$$\Delta HC = M\,c\,\Delta T_b \qquad (1)$$

with, M = mass of the animal (kg)
$\bar{c}$ = specific heat of the body (kJ/kg K)
$\Delta \bar{T}_b$ = change in mean body temperature (K).

The specific heat of tissue was assumed to be 3.47 kJ/kg °C (Mount, 1979; McLean and Tobin, 1987) and the body heat content was expressed in terms of kJ/kg body weight. Change in mean body temperature (ΔT_b) in eq. 1 was estimated from rectal temperature (T_r) and mean skin temperature (T_s) and using the relationship (McLean *et al.*, 1983b);

$$\Delta T_b = 0.86\ T_r + 0.14\ \bar{T}_s \qquad (2)$$

Rectal temperatures were recorded every 5 minutes 20 cm inside the rectum using a thermistor and microcomputer-based portable datalogger system (Mini Mitter, USA) (accuracy ± 0.05°C) held on each cow with a harness.

Skin temperatures were recorded every 2 hours between 0500 h and 0000 h (midnight) by a hand-held temperature probe, with a 0.33 mm copper/constantan fine wire sensor (Bailey Instruments Inc., Type MT-4), affixed to a 1.5 m plastic rod, with the sensor tip passing through the hair coat to touch the skin surface. The sensor was held in place until the reading on the digital recording instrument (Atkins Technical Inc) was steady (within 0.5°C) for 10 s. The plastic rod was of sufficient length to reach the various skin sites, with minimal interference to the animal, whether she was lying down, in the feeding stalls or in the milking parlour. Skin surface measurement sites were upper trunk (S_a), lower trunk (S_b), upper front leg (S_c), and lower front leg (S_d) with measurements repeated on both sides of the body (McLean *et al.*, 1983a; Young, *et al.*, 1984). Estimation of mean skin temperature (Ts) was calculated by averaging temperatures of the corresponding left and right sides for each site and then weighting them according to the surface area represented, (Nakamasu *et al.*, 1993; Purwanto *et al.*, 1994), i.e.

$$\bar{T}_s = 0.25\,(S_a + S_b) + 0.32\ S_c + 0.18\ S_d \qquad (3)$$

Results

Winter ambient temperatures did not exceed 26°C and no spray-cooling used. During summer water sprays were typically activated for about 4 to 6 hours each day. However, a 3 day heat wave (T_a > 27°C) was experienced at the start of the summer trial, where water sprays were on for more than 18 hours each day (Table 1).

During winter and summer the BHC in all cows typically cycled with two main peaks, the highest occurring at afternoon milking and the lower (by about 1 kJ/kg) around midnight. However, during the three day summer heat wave, BHC in cows under shade only had a single daily peak at the afternoon milking, while BHC in animals offered spray-cooling returned to the two peak cycle, although the afternoon peak was about 3.5 kJ/kg higher than the midnight peak.

Table 1. Average daily mean, minimum, maximum and range (± standard deviation) for ambient dry bulb temperatures.

	Average daily dry bulb temperature		
	Daily Mean	Minimum	Maximum
Winter	15.62 (±1.20)	7.03 (±2.22)	25.03 (±1.05)
Summer	21.47 (±0.62)	14.42 (±1.84)	29.08 (±0.94)
Summer heat wave*	27.71 (±1.14)	22.90 (±0.64)	36.37 (±0.60)

* 3 day period at the start of the summer trial.

Table 2 summarises the increase in daily mean, maximum and minimum level of body heat content above a base level of 0°C, per kg of body weight, for each treatment group. During winter, cows had a mean BHC of 130.48 (± 0.58) kJ/kg with a mean diurnal variation of 5.82 (± 2.53) kJ/kg body weight. During summer, the mean BHC of cows offered shade only was similar to winter values, while the diurnal variation was reduced. Animals offered water spray-cooling during summer had a mean BHC similar to winter values, but diurnal variation was greater.

For cows offered shade only, the excessive heat load conditions experienced during the three day heat wave produced an upward shift in the mean BHC by 4.4 kJ/kg and disrupted the biphasic diurnal cycle resulting in a monophasic rhythm which peaked at the afternoon milking. During the heat wave, cows that were offered spray-cooling reduced their BHC by 1.73 kJ/kg and returned to a biphasic rhythm.

Table 2. Average daily mean, minimum, maximum and diurnal variation (± standard deviation) body heat content (kJ above 0°C /kg body weight) of dairy cows during winter and during summer when offered shade or shade plus water spray cooling.

	Average body heat content (kJ/kg body weight)			
Treatment	Daily mean	Minimum	Maximum	Diurnal variation
Winter	130.48 (±0.58)	127.89 (±0.56)	133.71 (±1.52)	5.82 (±1.64)
Summer spray cooled	131.29 (±0.23)	128.54 (±0.32)	134.94 (±0.91)	6.40 (±1.06)
Summer shade only	131.72 (±1.13)	129.47 (±1.00)	134.57 (±1.44)	5.10 (±0.86)
Summer heat wave* spray cooled	134.43 (±0.47)	131.83 (±0.68)	139.23 (±0.81)	7.40 (±0.72)
Summer heat wave* shade only	136.16 (±0.73)	133.42 (±1.32)	139.52 (±0.95)	6.11 (±1.90)

* During the 3 day heat wave period at the start of the summer trial.

Discussion

Hahn *et al.* (1987) found that under thermoneutral conditions, restrained *ad libitum* fed steers had a monophasic diurnal tympanic temperature rhythm with the peak occurring around midnight. Furthermore, an ambient heat load produced a phase shift in the peak to an earlier time (1700 to 1900 h).

During our winter and normal summer periods, the midnight diurnal peak in BHC was observed. However, a second, higher peak was identified at the time of the afternoon milking. This second peak was possibly associated with milking activities and related exercise.

The present data indicates that shade alone was sufficient to maintain the daily mean value, amplitude and the biphasic structure of the cycle at similar levels to winter. However, during the heat wave period, cows offered shade only experienced an excessive heat load which increased the mean BHC and disrupted the 'normal' biphasic cycle. The provision of water spray-cooling during the heat wave did not significantly reduce the mean daily BHC, but did allow cows to return to a biphasic diurnal cycle.

References

Berman A. (1968). Nychthemeral and seasonal patterns of thermoregulation in cattle. *Australian J. Agric. Res.* 19:181-189.

Berman A., Folman Y., Kaim M., Mamen M., Herz Z., Wolfenson D., Arieli A. & Graber Y. (1985). Upper critical temperatures and forced ventilation effects for high-yielding dairy cows in a subtropical climate. *J. Dairy Sci.* 68:1488-1495.

Blaxter K. (1989). Energy metabolism in animals and man. Cambridge University Press, Cambridge.

Fuquay J.W. (1981). Heat stress as it affects animal production. *J. Anim. Sci.* 52:164-174.

Hahn G.L., Parkhurst A.M. & Nienaber J.A. (1987). Tympanic temperature rhythms in cattle in selected environments. Proceedings of the 8th Conference on Biometeorology and Aerobiology. American Meteorological Society, Boston, MA (Rept. B/A 4.4). 320-323.

Maust L.E., McDowell R.E. & Hooven N.W. (1972). Effect of summer weather on performance of Holstein cows in three stages of lactation. *J. Dairy Sci.* 55:1133-1139.

McLean J.A., Downie A.J., Watts P.R. & Glasbey C.A. (1982). Heat balance of ox steers (*Bos taurus*) in steady-temperature environments. *Am. Physiol. Soc.* 324-332.

McLean J.A., Downie A.J., Jones C.D.R., Stombaugh D.P. & Glasbey C.A. (1983a). Thermal adjustments of steers (*Bos taurus*) to abrupt changes in environmental temperature. *J. agric. Sci. Camb.* 100:305-314.

McLean J.A., Stombaugh D.P., Downie A.J. & Glasbey C.A. (1983b). *J. agric. Sci. Camb.* 100:315-322.

McLean J. & Tobin G. (1987). Animal and human calorimetry. Cambridge University Press, Cambridge.

Nakamasu F., Purwanto B.P. & Yamamoto S. (1993). *Anim. Sci. Technol.* (Japan). 64:629-636.

Purwanto B.P., Harada M. & Yamamoto S. (1994). *AJAS*, 7:179-182.

Watts P.R., McLean J.A. & Downie A.J. (1977). Heat balance in calves (*Bos taurus*) over 24-hr periods. *J. Thermal Biol.* 2:61-68.

Young B.A., McLean J. & Weingardt R. (1984). 63rd Annual Feeders' Day Report. Agriculture and Forestry Bulletin. Special Issue. Department of Animal Science, Faculty of Agriculture and Forestry. The University of Alberta pp. 51-52.

HEAT PRODUCTION AND BODY TEMPERATURE *OF BOS TAURUS* X *BOS INDICUS* CROSSBRED CATTLE IN THE HUMID TROPICS

J.B. Liang[1], J. Roch[2], M.A. Sharuddin[2] and S. Shanmugavelu[2]

[1]Department of Animal Science, Universiti Putra Malaysia, 43400 UPM Serdang, Malaysia; [2]Livestock Research Centre, MARDI, P.O. Box 12301, 50774 Kuala Lumpur, Malaysia

Summary

A study was conducted to examine how well *Bos taurus* x *Bos indicus* crossbred cattle adapt to the humid tropical conditions in Malaysia. The present paper reports the heat production (HP) and body temperatures (Tb) of the indigenous Kedah-Kelantan (KK) and Charolais x KK crossbred (CK) cattle kept in-door (ID) and out-door (OD). Heat production, rectal temperature (Tr), mean skin temperature (Ts) and mean body temperature (Tb) of the test cattle were measured at 10 regular intervals for 24 hours. The crossbred cattle exhibited a distinctly lower HP (ranged between 17.2 to 27.1 $kJ/kg^{0.75}/h$) than the indigenous KK cattle (ranged between 33.0 to 45.0 $kJ/kg^{0.75}/h$), irrespective of feeding levels and environmental treatments. Tb of cattle exhibited diurnal variation, high during the day and low during the late evening through to the early morning. Tb of ID cattle fluctuated in a narrow range between day and night (averaged 0.7°C and 0.9°C, respectively for CK and KK), while the fluctuation of Tb for OD cattle was more pronounced, averaging 1.7°C and 2.2°C respectively for CK and KK. The results of the present experiment indicated that CK crossbred cattle adapted to the hot and humid environments. They do so by maintaining a lower metabolic rate, and by their ability to quickly expel the excessive body heat, whenever their own micro-environment was favourable.
Key words: *Bos taurus* x *Bos indicus* crossbred, heat production, body temperature, humid tropics.

Introduction

Larger and faster growing temperate *Bos taurus* cattle are widely used to upgrade the indigenous *Bos indicus* cattle in the tropics. Crossbreeding of European breeds including Charolais, Limousine and Simmental with the indigenous Kedah-kelantan (KK) cattle had been reported to perform better than the KK under experimental conditions where the animals were kept in covered stalls (Ahmad *et al.*, 1995).

It is well documented that high temperature and humidity restrict animal productivity in the humid tropics. Heat arising from the metabolic process within the animal's body must be removed at the rate at which it is produced. It is, therefore, essential to know how well the *Bos taurus* x *Bos indicus* crossbreed can adapt to the humid tropical conditions. Understanding the degree of heat stress that might have been imposed on these cattle would provide information to improve their management.

Materials and Methods

Three Charolais x KK crossbred (CK) female yearlings (mean BW of 177.3 kg) and equal numbers and similar sex and age of KK cattle (mean BW of 123.0 kg) were used in the present study. The study consisted of 2 experiments. In the first, the cattle were fed at 3 intake levels, equivalent to 1% (L), 1.8% (M) and 2.5% (H) dry matter of their respective BW, in a double 3 x 3 Latin Square (one for each breed) under indoor (ID) conditions.

The same procedure was repeated in Experiment 2 with the cattle kept outdoors (OD).

The experimental diet consisted of 60% oil palm frond pellet and 40% concentrate pellet. The calculated energy content of the diet was 8.4 MJ ME/kg and containing 10.0% crude protein. During the experiment, the cattle were kept in individual stalls and were offered the allocated daily ration in 2 equal portions; at 10:30 and the balance at 22:00h.

Immediately prior to the start of the experiments, the relationships between heart rate (HR) and heat production (HP) for each animal were established. HPs were estimated based on oxygen consumption measured at 2 minutes intervals in an open circuit head hood and calculated using a procedure similar to that of Yamamoto *et al.* (1985). The corresponding HR of the animals were measured using a modified wireless mini-microphone attached to the chest-piece of a stethoscope to amplify the beats, which were picked up using a FM radio receiver. Measurements were carried out when the animals were at rest and during active feeding to obtain a wide range of the parameters. The linear relationships between HR (X in beat/min) and the HP (Y in $kJ/kg^{0.75}/h$) established (Table 1) were used to predict HP of the respective cattle based on the HR measured in the two experiments that followed. Rectal temperature (Tr) was measured using a digital clinical thermometer, while the skin temperatures were measured using a hand-held battery operated surface thermocouple probe.

Within each Latin Square experiment, each period consisted of 4 days adaptation and 1 day of test. On the test day, HR, Tr, skin temperatures at upper trunk, lower trunk, upper leg and lower leg were measured. Mean skin temperature (Ts) and mean body temperature (Tb) were calculated following equations of McLean *et al.* (1983). At the end of Experiment 1, the test cattle were gradually exposed to the OD conditions for 3 weeks before they were used in Experiment 2. Body weights (BW) of the cattle, measured at the beginning and at the end of each experiment were used for the calculation of feeding levels during the experiment.

Table 1. Mean and range of heart rate (HR) and heat production (HP) and linear regression analysis of HP on HR of Kedah-Kelantan (KK) and Charolais x KK crossbred (CK) heifers.

	No. of	Mean and (range)		Linear regression*		
Animals	Samples	HR	HP	(a)	(b)	(r)
KK184	108	105.8 (86–116)	35.0 (24.9-41.9)	3.131	0.301	0.64
KK374	217	92.4 (58-148)	30.0 (24.0-38.4)	19.353	0.115	0.88
KK048	187	82.5 (64-102)	22.3 (13.5-25.7)	3.391	0.228	0.64
KK Pooled	512	93.6 (58-148)	29.1 (13.5-41.9)	8.628	0.215	0.72
CK026	182	70.0 (46-120)	22.4 (14.2-31.4)	9.915	0.177	0.82
CK230	130	81.2 (56-130)	18.8 (15.1-26.0)	6.401	0.151	0.93
CK156	156	62.5 (44-104)	23.8 (16.9-33.5)	5.656	0.289	0.87
CK Pooled	468	71.2 (44-130)	21.7 (15.1-33.5)	7.324	0.206	0.87

* Y = a + bX, where Y= HP ($kJ/kg^{0.75}/h$) and, X= HR (beat/min).

Results and Discussion

Heat Production. Although HR has been shown to be a useful indicator to predict HP in unrestrained farm animals, the greatest disadvantage of this procedure is that HR varied among animals and days of measurement (Yamamoto *et al.*, 1979). The relationships of HR and HP for the test cattle were no exception (Table 1). Therefore, HP of the test animals in the present study was predicted based on their respective individual equations.

In almost all cases, the parameters measured (HP, Tr, Ts and Tb) for the medium (M) feeding level were between those for the low (L) and the high (H) levels. The parameters for M level were, thus, omitted in Figure 1 (for clarity of the graphs) and in the subsequent discussion of the results.

The HP of CK was distinctly lower that those of KK, irrespective of feeding level and environment (Fig. 1a). Within breed, HP was affected by feeding levels; being lower for cattle fed L level and higher for cattle fed H level. In all cases, HP increased rapidly during feeding (10:30h and 22:00h), reaching as high as 50.0 kJ/kg$^{0.75}$/h for OD-H KK, and 27.1 kJ/kg$^{0.75}$/h for OD-H CK. The rhythmic responses of HP during active feeding and rest are well documented (Yamamoto *et al.*, 1979). The distinctly higher HP (80%) recorded for KK as compared to CK in this study is interesting. The values recorded for CK (ranged between 17.2 to 27.1 kJ/k $^{0.75}$/h) were close to those reported for Holstein heifers (23.3 to 29.5 kJ/kg$^{0.75}$/h) kept under 30°C (Purwanto *et al.*, 1993). On the other hand HP recorded for KK (33 to 45 kJ/kg$^{0.75}$/h) resembled those reported for medium and high producing dairy cows which average 37.8 and 44.1 kJ/kg$^{0.75}$/h, respectively (Purwanto *et al.*, 1990). The high HP of KK, which is a low producing indigenous beef cattle with a potential growth of only 0.66 kg per day (Hogan & Liang, 1991) is unique, and requires further investigation. The relatively high HP of KK could explain their low energetic efficiency previously estimated in the same laboratory using other procedures (Liang *et al.*, 1988; Liang & Young, 1995).

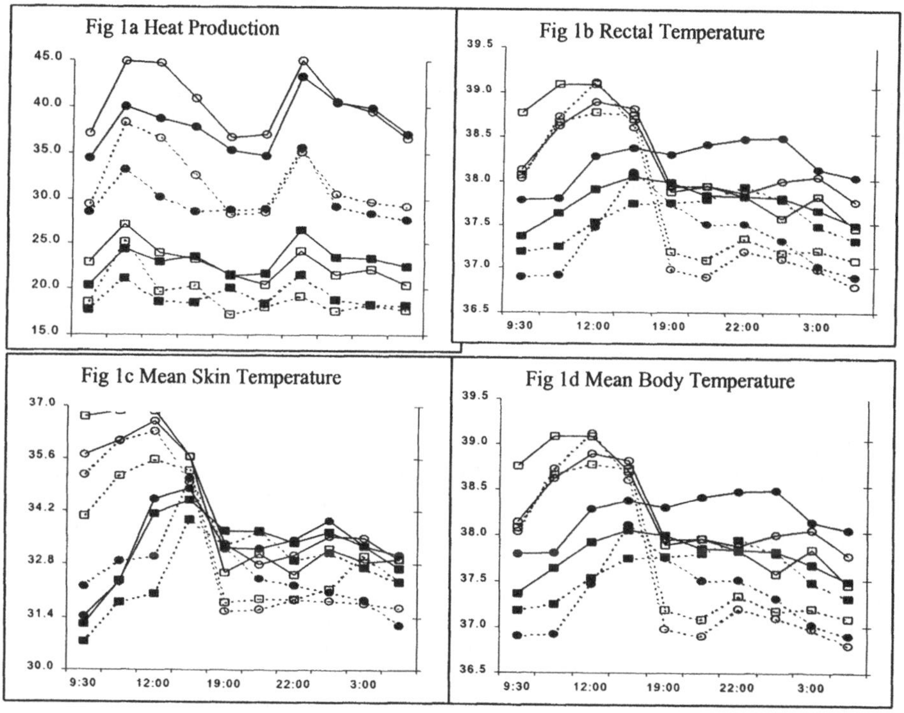

Fig. 1. Heat production and rectal, mean skin and mean body temperatures measured at various time of the day for CK kept indoor (■), outdoor (□) and KK kept indoor (●) and outdoor (O). ··········· indicating low and ——— high level of feeding, respectively.

<u>Body temperature</u>. In general, Tb of the cattle exhibited diurnal variation, high during the day and low during the night through to the early morning (Fig. 1d). Tb of ID cattle fluctuated within a narrow range between day and night (average 0.7°C and 0.9°C, respectively for CK and KK), while the fluctuation of Tb for OD cattle was greater

(average 1.7°C and 2.2°C, respectively, for CK and KK cattle). In all cases, the test cattle regulated their body temperature without showing any serious adverse effect, including loss of appetite. All cattle, including those of OD offered H feeding level, completely consumed the feed offered to them.

Data from the present study suggested that the cattle, particularly those kept OD, dissipated part of the excessive heat load by convective heat loss through the skin. Heat loss of the cattle started after mid-day and became most intense in the late afternoon after the 15:00h measurements. Mean skin temperature (Fig. 1c) of OD cattle at midday-high was 36.9°C and 36.6°C, respectively, for CK and KK fed H level intake, while the average ambient temperature at the same time was estimated to be 33°C. The 4°C gradient between the skin and the ambient temperatures, therefore, provided a favourable environment for heat transfer from the skin to its immediate environment. Since, the relative humidity (RH) at midday was between 55 to 65%, and the gradient between the skin and the environment was only 4°C, evaporative heat loss was probably the main mode of heat loss as indicated by the increased respiration rates of the cattle at that time. It is unfortunate that the respiration rates measured (by visual counting of the movement of the flank) for the test cattle were too incomplete and inconsistent to validate this speculation. Heat loss from the cattle during the late morning (from 09:30h to before 12:00h) was presumably restricted because of the high RH, estimated to range between 70 to 80% during that time.

The results of the present study suggested that the *Bos taurus* x *Bos indicus* crossbred cattle adapted well to the humid tropical conditions of Malaysia. They did so by means of maintaining a low metabolic rate (as seen in the lower HP of CK than that of KK) as well as their ability to dissipate the excess heat load quickly when the micro-environment immediate to them was favourable.

References

Ahmad A., Sukri I., Othman A., Ali M. & Shafit H. (1995). Performance and carcass characteristics of Limousine-KK, Charolais-KK and Simmental-KK in feedlot. Proceedings of the 17th MSAP *Conference*. 28-30 May 1995, Penang, Malaysia, pp. 61-62.

Hogan J.P. & Liang J.B. (1991). Balancing nutrients to improve production of tropical herbivores. *Proceedings of the 3rd International Symposium in the Nutrition of Herbivores*. 25-30 August 1991, Penang, Malaysia, pp. 237-246.

Liang J.B., Samiyah M.N. & Hirooka H. (1988). Energy requirements for growing Kedah-Kelantan (KK) and Brahman-KK heifers in Malaysia. *MARDI Res. J* 16:163-170.

Liang J.B. & Young B.A. (1995). Comparative energetic efficiencies of male Malaysian cattle and buffalo. *Livestock Prod. Sci.* 41:19-27.

McLean J.A., Stombaugh D.P., Downie A.J. & Clasbey C.A. (1983). Body heat storage in steers (*Bos taurus*) in fluctuating thermal environments. *J. agric. Sci. (Camb.)* 100:315-322.

Purwanto B.P., Nakamasu F. & Yamamoto S. (1993). Effect of environmental temperature on heat production in dairy heifers differing in feed intake level. *AJAS* 6:275-279.

Purwanto B.P., Abo Y., Sakamoto R., Furumoto F. & Yamamoto S. (1990). Diurnal patterns of heat production and heart rate under thermoneutral conditions in Holstein Friesian cows differing in milk production. *J. agric. Sci. (Camb.)* 114:139-142.

Yamamoto S., McLean J.A. & Downie A.J. (1979). Estimation of heat production from heart-rate measurements in cattle. *Br. J. Nutr.* 42:507-513.

Yamamoto S., Sumida M. & Kosako T. (1985). Evaluation of fast-response calorimetry, used for calibration of the relationship between heart rate and heat production in farm animals. *Japanese J. of Zootech. Sci.* 56:947-953.

DEPENDENCE OF HEAT PRODUCTION ON FEED LEVEL AND AMBIENT TEMPERATURE IN HEIFERS IN THE RANGE OF MALNUTRITION

A. Chudy

Forschungsinstitut für die Biologie landwirtschaftlicher Nutztiere, Forschungsbereich Ernährungspysiologie "Oskar Kellner", Justus-von-Liebig-Weg 2, D-18059 Rostock, Germany

Summary

During a 21 day experiment the 24h heat production of Holstein-Friesian heifers (n = 4; 225 kg liveweight, W) was estimated by indirect calorimetry at three environmental temperatures (5, 20 and 30°C) and three feed levels below maintenance (M) (0.90M, 0.75M and 0.50M) where maintenance was calculated as 550 kJ/kg $W^{0.75}$ and also, at a severe starvation level of 0.15 M. Animals were offered a diet of 80% hay : 20% barley (DMB) during the first three feeding levels and 1 kg straw on the 0.15M treatment. The results show that there was no significant correlation between ME intake and heat production. The efficiency of utilizationation of ME could, therefore, have been up to 100% with the heat production being determined by the requirement for heat by the animal for thermoregulation. There was an inverse relationship between environmental temperature and heat production; being 565 kJ/kg$W^{0.75}$ at 5°C compared with 515 and 525 kJ/kg $W^{0.75}$ at 20 and 30°C respectively. These data indicate that the thermoneutral zone for these animals was above 20°C. There were significant differences between animals in heat production.
Key words: Energy metabolism, heat requirement, maintenance, environmental temperature.

Introduction

The aim of this study was investigate the relationship between sub-maintenance feeding level and heat production and hence the efficiency of utilization of energy for maintenance and the identification of the thermoneutral zone for heifers maintained at these feeding levels.

Materials and Methods

During a 21 day experiment the 24h heat production of Holstein-Friesian heifers (n = 4; 225 kg liveweight) was estimated by indirect calorimetry at three environmental temperatures (5, 20 and 30°C) and three feed levels below maintenance (M) (0.90M, 0.75M and 0.50M) where maintenance was calculated as 550 kJ/kg $W^{0.75}$ and, also, at a severe starvation level of 0.15 M. Animal were offered a diet of 80% hay : 20% barley (DMB) during the first three feeding levels and 1 kg straw on the 0.15M treatment. The reason for offering straw during the lowest feeding level was to maintain rumen function and that of the remainder of the digestive tract, whilst determining heat production with the minimal influence of feed. The experimental treatments (below maintenance) were applied so that the animal's metabolism would be determined directly by the restricted dietary and environmental conditions. This approach precluded the use of long unnecessary adaptation periods.

However, before the experimental period, the animals were adapted to the type of ration offered and to one of four respiration chambers. The heat production of the animals was calculated from measurements every 10 min of respiratory exchange for two consecutive

23.5h periods at each feeding level and each environmental temperature giving a total of 84 measurements, as shown in Table 1.

Table 1. Experimental plan.

Day	Feed Level (FL)	Temperature (°C)
1-2	0.75	30
3-4	0.75	20
5-6	0.75	5
7-8	0.90	5
9-10	0.90	20
11-12	0.90	30
13-14	0.50	30
15-16	0.50	20
17-18	0.50	5
19	hunger	5
20	hunger	20
21	hunger	30
22	end	

The ratio of hay and barley offered and the mineral-vitamin supplement were constant throughout the experiment and water was offered *ad libitum*. Feed was offered twice daily at 07.00 and 14.30h. Total collections of faeces and urine were made and analysed. Standing and lying time and head and body movements were measured and the rectal temperature and respiration rate recorded throughout the experimental period.

Results

The results were statistically analysed using multiple range analysis of various procedures and those for heat production measurements are shown in Tables 2 and 3 and in Figure 1. The heat production was the highest at 5°C while the lowest heat production for this temperature treatment was found at feeding level 0.5 M. The heat production was not significantly different between the first and the second day of each temperature. The differences in activities of the animals between treatments were not significantly correlated.

Although there were big differences in ME intake, the only significant difference in the heat production between treatments was at feeding level 0.5 M which was 10% lower than for the other treatments. Heat production over all feeding levels at 5°C was 565 kJ/$W^{0.75}$ and was significantly higher ($P<0.10$) than at 20°C. There was no significant difference in

Table 2. Effects of feed level (FL) and ambient temperature (°C) on heat production (HP, kJ/kg $W^{0.75}$).

Feed level	Feed intake ME	Heat production 30°C	Heat production 20°C	Heat production 5°C
0.15	91	517±55	555±11	597±39
0.50	298	538±14	441± 3	498±37
0.50	298	468± 7	465±20	546±26
0.75	445	518±13	540±30	575±30
0.75	445	510±14	541±26	579±25
0.90	535	603±44	543±24	577±11
0.90	535	536±82	497±15	578±31

Table 3: Heat production (HP [kJ/kg$^{0.75}$]) (analysis of influence factors).

1. Feed level	0.15 FL	0.5 FL	0.75 FL	0.9 FL
ME-intake [kJ/ kg BW$^{0.75}$]	91	298	445	535
Heat production [kJ/ kg BW$^{0.75}$]	549 ± 12^{a}	492 ± 8^{b}	543 ± 8^{a}	555 ± 8^{a}
relative (FL 0,15 = 100 %)	100	90	99	101
2. Temperature	4° C	20° C	32 °C	
Heat production [kJ/ kg BW$^{0.75}$]	565 ± 8^{a}	513 ± 8^{b}	527 ± 8^{b}	
relative (5 ° C = 100 %)	100	91	93	
3. Animals	Animal 9	Animal 10	Animal 11	Animal 12
Heat production [kJ/ kg BW$^{0.75}$]	553 ± 9^{a}	548 ± 9^{a}	513 ± 8^{b}	526 ± 9^{a}
relative (Animal 9 = 100 %)	100	99	93	95

Significance: P<0.01 for unequal letter within the row.

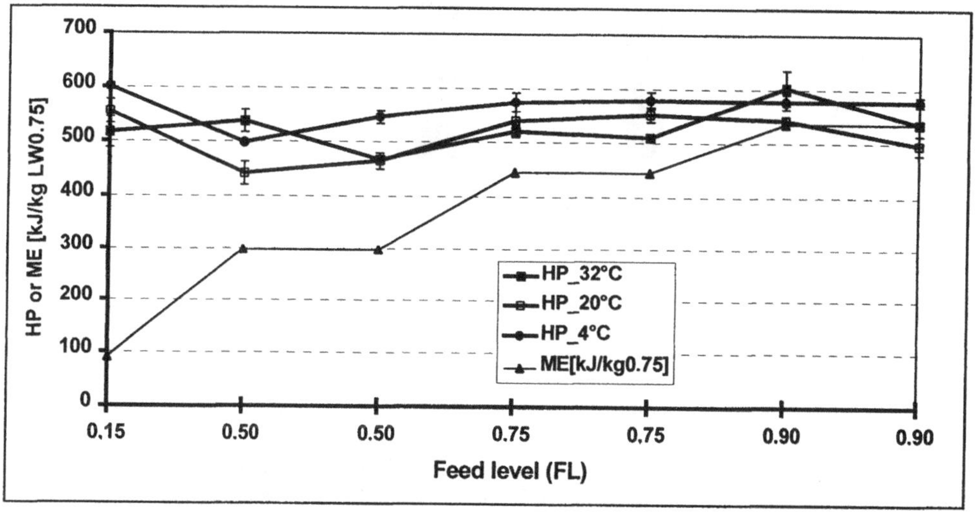

Fig.1. Effect of feed level and temperature on energy intake (ME) and heat production (HP).

heat production between 20°C and 32°C (Table 3). It should be noted that animals at 5°C, given the low feeding level, exhibited shivering but the heat production was about the same as for other feeding levels. This might imply that heat production was dependent on heat requirement. The difference in heat production between individual animals was about 10%.

The relationship between rectal body temperature (Y) and ambient temperature (X) (Y = 0.0625X + 37.3) was almost linear (Figure 2). The body temperature increased with ambient temperature. The increase of body temperature was about 0.0625°C per each degree ambient temperature.

Discussion

There was no significant correlation between ME intake and heat production (range 492 to 555 kJ/ kg^{0,75}). The efficiency of utilization of ME was approximately 100%; which implied that feed energy replaced body energy equivalently. At feeding level 0.15 M, the energy metabolism was mainly based on body nutrient reserves. In the range of undernutrition below the thermoneutral range the heat production was determined by the heat requirement. This fact questions the relevance of the "efficiency of utilization of ME for maintenance" under conditions below thermoneutrality. This agrees with earlier investigations with Chinchilla (Chudy, 1994). The significantly lower heat production at feeding level 0.5 M might be connected with the compensation of the additional energy requirement for gluconeogenesis by the starch in 330 g DM of barley. This agrees with the results of the experiments with bulls examining this problem (Chudy, 1997). These facts are important and need to be taken into consideration in further investigations on the effect of malnutrition in ruminants.

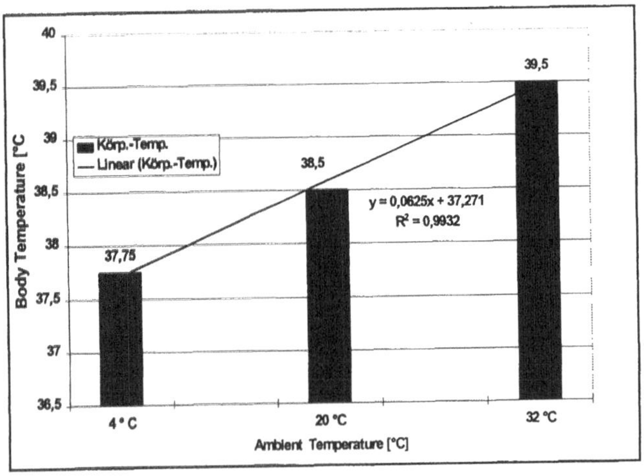

Fig. 2. Dependence of body temperature on ambient temperature.

The heat production (= heat requirement) is depended on environmental temperature. The heat requirement of black-white race heifers (225 kg BW) averaged 565 kJ/kg $^{0.75}$ at 5°C, 515 kJ/kg $^{0.75}$ at 20°C and 525 kJ/kg$^{0.75}$ at 32°C and was significantly higher at 5°C than at 20° or 32°C. The influence of this temperature difference is equal to daily losses of 110 g body mass with an energy content of 30 MJ/kg DM (54% fat and 46% protein). The thermoneutral zone for cattle is from 20°C and up to 30°C. The relationship between body and ambient temperature brings up interesting new questions in the field of thermogenesis, heat regulation and emission (delivery).

Significant differences of the heat (maintenance) requirement exist between animals and must be taken into consideration.

References

Chudy, A. (1994). Heat production of adult Blue Chinchilla in dependence on ambient temperature, feed level and sex. *EAAP Publ. No.* 76:379-382.

Chudy, A. (1997). Energy requirement for gluconeogenesis - measurements in bulls with duodenum fistulas. EAAP Symposium, Newcastle.

ENERGY REQUIREMENT FOR GLUCONEOGENESIS - MEASUREMENTS IN BULLS WITH DUODENAL FISTULAS

A. Chudy

Forschungsinstitut für die Biologie landwirtschaftlicher Nutztiere, Forschungsbereich Ernährungspysiologie "Oskar Kellner", Justus-von-Liebig-Weg 2, D-18059 Rostock, Deutschland

Summary

The investigation sought to estimate the energy requirement for gluconeogenesis by indirect respiration calorimetry. Experiments were performed using two duodenally-fistulated Black-Pied dairy bulls of 377 and 323 kg liveweight. The protein and energy metabolism of the bulls was measured in a thermoneutral environment during 6 two-day periods with and without the infusion of glucose (200, 300, 400 and 600 g/d) after the animals had become glucose deficient. Infusion of glucose decreased urinary nitrogen excretion and total heat production. The results indicated, in agreement with those of Chowdhury and Ørskov (1994), that the essential glucose requirement for bulls of 350 kg was 300-400 g/d (3.75-5.0 g/kg $W^{0.75}$/d) and the calculated energy requirement for gluconeogenesis amounted to between 140 and 180% (mean 158 ± 19) of the energy synthesized as glucose. The influence of the energy cost of gluconeogenesis on energy metabolism must be taken into account in the design and evaluation of experiments investigating the effects of undernutrition in ruminants.

Key words: Energy metabolism, gluconeogenesis, heat production, urinary nitrogen excretion, glucose, ambient temperature, thermoneutral conditions.

Introduction

Gluconeogenesis is a very important physiological process to meet substrate-specific requirements for glucose in ruminants and especially highly productive dairy cows. Gluconeogenesis occurs in the liver and is related to the high requirement for ATP. Chowdhury and Ørskov (1994) calculated that about 1/3 of the daily maintenance requirement (150 MJ/kg $W^{0.75}$ or 20 mmol glucose /kg $W^{0.75}$) is required to meet glycogenic demands. Stangassinger (1997) found that, in highly productive cattle only 16% of the total glucose requirement was absorbed from the digestive tract. Stansassinger (1997) also demonstrated in pygmy goats, using tracer techniques, a number of gluconeogenic pathways under different metabolic states, including glucose re-cycling via lactate/alanine. The energy requirement for these pathways is poorly understood and the current experiments sought to rectify this situation. The experimental animals, fitted with duodenal fistulas, were offered a restricted straw and urea diet rather than being starved to induce a glucose deficit. Heat production of the animals was measured whilst increasing levels of glucose were infused duodenally enabling the partial energy cost of gluconeogenesis to be estimated.

Materials and Methods

Two duodenally-fistulated Black Pied bulls (377 and 323 kg) were housed in climatically controlled respiration chambers which permitted the measurement of oxygen, carbon dioxide and methane exchange over 24h periods. Total glucose deficit was induced in the animals by feeding a basic daily ration of straw and urea (2.04 kg/d straw; 22 g/d urea and 60 g/d mineral/vitamin supplement) during a 7 d preliminary period when the animals were housed in the chamber and acclimatised to an environmental temperature of 28°C. During the following 12 days the energy metabolism of the animals was measured, under thermoneutral conditions (28°C), in 6 x 2 d periods. During periods 1 and 6 the animals received the basal diet only (control periods) but during periods 2, 3, 4 and 5 animals received 200, 300, 400 or 600 g/d glucose respectively by infusion in addition to the basal diet. The glucose dissolved in 3.5l water was administered continuously over 23 h by infusion pump into the duodenum *per fistula*. Blood samples were taken between 7.00 and 7.30 at the end of each period, cooled and analysed enzymatically for glucose concentration. Live weights of the animals were recorded at the beginning and end of each experiment.

Results

There were significant overall differences ($P<0.05$) between animals in respect of total heat production and urinary nitrogen excretion although there were no significant differences between first or second days within a period (Table 1). Compared to the values obtained during the control periods infusion of 300, 400 and 600 g/d glucose significantly ($P<0.05$) decreased heat production, whereas all levels of glucose infusion significantly ($P<0.05$) decreased urinary nitrogen excretion (Table 1). This observation demonstrates the compensation in gluconeogenesis as a consequence of glucose infusion. The decline in heat production was virtually directly proportional to the reduction in body fat mobilization as the amount of glucose infused increased.

Blood glucose concentrations were very stable (2.09 ± 0.23 mmol/l) and virtually independent of nutritional stress (Table 2), although they declined as the amount of glucose infused increased. There were no significant differences between animals in blood glucose concentration, whilst the level observed during the infusion of 400g/d glucose was significantly ($P<0.05$) lower than the value during the control periods.

Table 3 shows the relative changes in urinary nitrogen excretion and heat production as the rate of glucose infusion increased together with the apparent efficiency of glucose energy utilisation.

Discussion

The infusion of glucose compensated for the production of energy by gluconeogenesis and decreased the catabolism of glycogenic amino acids as indicated by the decline in urinary nitrogen excretion (Table 3). The apparent efficiency of glucose utilization of >100% was achieved by the fact that the heat production declined during glucose infusion. The drop in glucose efficiency with the infusion of 600 g/d glucose indicates that the essential requirement for glucose in the experimental animals was of the order of 300-400 g/d (or 3.75-5.00 g/kg $W^{0.75}$) which is in agreement with the results of Chowdhury & Orskov (1994). It is also possible to calculate that the energy requirement for gluconeogenesis was between 140 and 180% of the synthesized glucose energy. It is therefore important to take into account the influence of gluconeogenesis on the energy metabolism and glucose requirement of animals in the planning and evaluation of experiments with ruminants maintained on sub-optimal planes of nutrition. Further experiments are in progress in which glucose is being administered intravenously.

Table 3. Energy intake (ME). Urine-N. Body-energy gain and Heat Production (mean±S).

Animal	Glucose	Metabolizable Energy	Urine-N	Protein-retention	Fat-retention	Energy-retention	Heat Production
	g/d	MJ	g/day	MJ	MJ	MJ	MJ
1	0	17.6±0.13	34.7±0.95	-4.2±0.14	-16.0±0.39	-20.2±0.25	37.8±0.38
	200	21.0±0.31	26.6±0.74	-3.5±0.11	-10.5±0.48	-14.0±0.37	35.1±0.68
	300	22.2±0.08	28.7±1.73	-3.8±0.25	-7.4±0.02	-11.2±0.27	33.4±0.19
	400	23.6±0.07	26.8±0.78	-3.6±0.11	-5.8±0.37	-9.4±0.47	32.9±0.54
	600	27.0±0.40	28.7±1.73	-3.8±0.25	-2.6±0.34	-6.4±0.09	33.4±0.31
	0	17.6±0.31	33.9±3.26	-4.1±0.48	-15.1±3.11	-19.1±2.62	36.7±2.32
2	0	16.0±0.61	25.8±4.97	-3.6±0.75	-15.7±0.47	-19.2±0.28	35.2±0.33
	200	19.1±0.42	24.9±3.57	-3.0±0.53	-11.1±0.37	-14.0±0.91	33.1±0.48
	300	21.1±0.07	18.1±2.91	-1.9±0.43	-9.7±2.26	-11.7±1.82	32.8±1.90
	400	22.3±0.04	20.7±0.28	-2.3±0.04	-7.9±0.14	-10.3±0.18	32.6±0.04
	600	25.4±0.38	22.4±0.64	-2.6±0.10	-5.8±0.02	-8.4±0.12	33.8±0.50
	0	16.0±0.11	23.1±1.30	-3.2±0.19	-13.4±2.52	-16.5±2.33	32.5 ±2.22
1 and 2	0	16.8 ± 0.9	29.4 ± 5.8[a]	-3.75±0.6	-15.0 ± 1.9	-18.8 ± 2.0	35.6 ± 2.4[a]
	200	20.1 ± 1.15	25.7 ± 2.3[b]	-3.2 ± 0.45	-10.8 ± 0.5	-14.0 ± 0.6	34.1 ± 1.2
	300	21.7 ± 0.6	23.4 ± 6.4[c]	-2.9 ± 1.1	-8.6 ± 1.9	-11.45 ± 1.1	33.1 ± 1.2[b]
	400	22.9 ± 0.7	23.7 ± 3.6[c]	-2.9 ± 0.7	-6.9 ± 1.3	-9.8 ± 0.6	32.7 ± 0.4[c]
	600	26.2 ± 1.0	25.5 ± 3.8[b]	-3.2 ± 0.7	-4.2 ± 1.9	-7.4 ± 1.2	33.6 ± 0.4[b]
total	0-600	20.7 ± 3.5	26.2 ± 5.1	-3.3 ± 0.7	-10.1 ± 4.4	-13.4 ± 4.6	34.1 ± 1.9

Significant differences: [a, b] $P<0.05$; [a, c] $P<0.01$

Table 4. Glucose contents in the blood.

After Period	Glucose infusion g/day	Animal 1 mmol/l	Animal 2 mmol/l	Mean (± S) mmol/l
1	0	2.37	2.34	2.35±0.02
2	200	2.22	2.30	2.26±0.06
3	300	2.10	1.73	1.91±0.26
4	400	1.86	1.87	1.87±0.01[b]
5	600	1.89	2.06	1.97±0.12
6	0	2.37	1.98	2.17±0.27
1+6	0	2.37±0.00	2.16±0.25	2.26±0.19[a]
Total		2.13±0.23	2.04±0.24	2.09±0.23

[a, b] Significant differences (P<0.05).

The urinary N excretion decreased significantly as a result of glucose infusion: from 29.4 ± 5.8 g without glucose infusion to 25.7 ± 2.3 (P<0,05), 23.4 ± 6.4 (P<0.01), 23.7 ± 3.6 (P<0.01) and 25.5 ± 3.8 (P<0.05) by 200, 300, 400 and 600 g glucose infusion/day.

The results of the evaluation of the glucose infusion periods 2-5 in difference to the periods without glucose application (1 and 6) are shown in Table 5.

Acknowledgements

The author wishes to thank Dr. habil. W. Jentsch and the technical staff for help and advice and Dr. habil. J. Voigt for the analysis of blood glucose.

References

Stangassinger M. (1997). Zum Glucosestoffwechsel der Wiederkäuer unter Belastungszuständen (Glucose metabolism in ruminants under metabolic stress).

Chowdhury S.A. & Ørskov E.R. (1994). Implications of fasting on the energy metabolism and feed evaluation in ruminants. *J. Anim. Feed Sci.* 3:161-169.

POSTPRANDIAL ENERGY TURNOVER AND METABOLISM OF PHYSICALLY ACTIVE OR INACTIVE YOUNG MALE VOLUNTEERS

C. Wenk, P. Colombani, P. Frey-Rindova and W. Langhans

Institute of Animal Sciences, Nutrition Biology and Physiology, Swiss Federal Institute of Technology, Zurich, Switzerland

Summary

Postprandial energy turnover was measured in 24 healthy men (age 18 to 40) after a high-fat meal. The subjects were non-smokers, not overweight (body mass index < 27), and assigned to two groups according to their self-reported fitness level. In the group of physically active volunteers (A), the subjects had to perform at least four endurance training units with a duration of one hour or more weekly. In the non-active group (NA), the persons did not have a previous history in endurance sports and exercised less than one hour per week during the experimental period. Postprandial substrate oxidation was significantly different in the two groups. Before the test meal and at the end of the experiment (five hours after the meal), A had significantly lower RQ values than NA. This occurred despite the ingestion of similar amounts of macro-nutrients. A oxidized obviously more fat than the physically inactive volunteers. No significant group differences could be found for plasma levels of carbohydrate and fat metabolites. A significantly higher blood urea level in A can be used as an indication for a more active protein turnover. With an almost identical plasma glucose concentration in the two groups, A showed lower plasma insulin and higher glucagon levels than NA. It is concluded that a higher postprandial fat oxidation and more economic utilization of glucose can be expected as a consequence of physical activity.
Key words: Energy metabolism, high fat meal, exercise, male human volunteers.

Introduction

Postprandial substrate oxidation depends on several factors like meal composition (Acheson *et al.*, 1987; Flatt *et al.*, 1985; Surina *et al.*, 1993), body weight and body composition (Lennon *et al.*, 1985; Schutz & Jéquier, 1989). One possibility to influence the metabolic system is intensive physical activity (Tremblay *et al.*, 1986; Schulz *et al.*, 1991). Endurance trained muscles oxidize more fat, especially under work load (Åstrand & Rodhal, 1988, Saltin & Åstrand, 1993). In the present experiment the influence of a regular endurance exercise on postprandial energy metabolism at rest was evaluated.

Materials and Methods

Postprandial energy turnover was measured in 24 healthy men (age 18 to 40). The subjects were non-smokers, not overweight (body mass index < 27), and assigned to two groups according to their self-reported fitness level. In the group of physically active volunteers (A), the subjects had to perform weekly at least four endurance training units with a duration of one hour or more. In the non active group (NA), the persons did not have participation in endurance sports and exercised less than one hour per week during the experimental period. Table 1 shows characteristics and endurance capacity of the volunteers. Endurance capacity was evaluated with an incremental exercise test on a cycle ergometer. Body fat was calculated from skinfold fat thickness measured at four different body sites.

Table 1. Characteristics and endurance capacity of the volunteers (Mean ± SD).

		Non-active	Active	Student unp. t-test
n		11	13	ns
Age	y	27 ± 6	27 ± 4	ns
Height	cm	179 ± 6	180 ± 6	ns
Body mass	kg	70 ± 9	72 ± 6	ns
Body mass index	kg/m	22 ± 2	22 ± 2	ns
Body fat	%	14 ± 4	11 ± 3	$p < 0.05$
Lean body mass	kg	60 ± 6	65 ± 5	$p < 0.05$
Exercise per week	h	< 1.0	11.6 ± 3.5	$p < 0.05$
Performance at				
- 4 mmol lactate/l	W	165 ± 37	278 ± 30	$p < 0.05$
- RQ = 1	W	118 ± 25	186 ± 50	$p < 0.05$

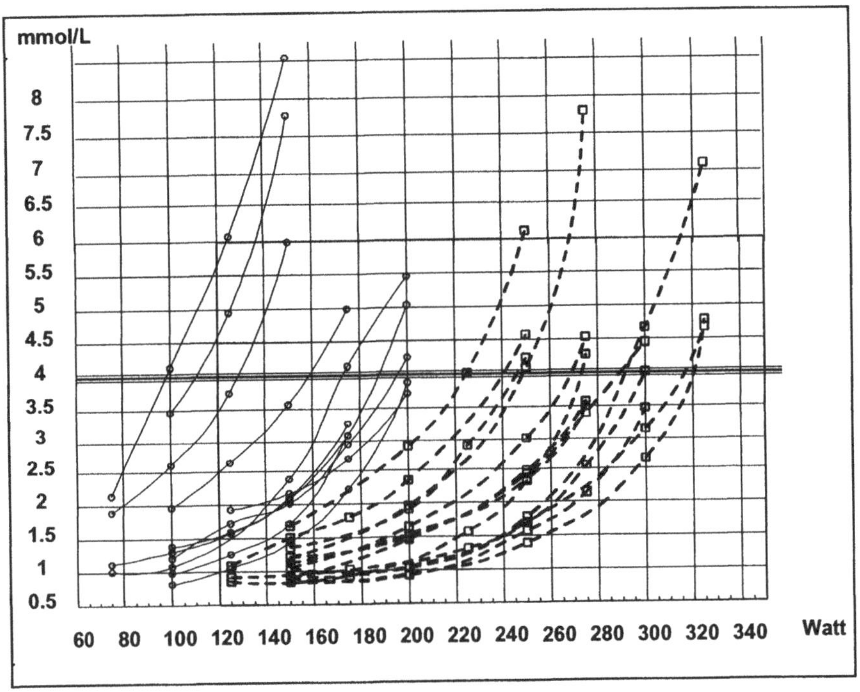

Fig. 1. Blood lactate level during the incremental cycle Ergometer Test (Dashed lines = Active volunteers; Straight lines = Inactive volunteers).

Nutrition was standardized the day before the main experiment. At a fixed time a vegetarian, carbohydrate-rich dinner had to be ingested. Furthermore, no intensive physical exercise, alcoholic beverages, or smoking were permitted. The following morning, subjects arrived at the laboratory after an overnight fast. Energy metabolism was measured with an open ventilated hood respiration system and blood samples were drawn from an antecubital arm vein by a catheter immediately before and after the ingestion of the test meal and for the following five hours in intervals of one hour. Blood samples were analysed for selected metabolites and hormones.

The experimental fat-rich meal consisted of 2 eggs, 40 g butter, 60 g dairy cream, 150 g bread, 22 g cheese, 220 g raspberries, and 11 g sugar (5.2 MJ ME: 13% derived from protein, 54% from fat, and 33% from carbohydrates).

Differences between the groups were assessed by an ANOVA for repeated measurements with grouping factor. Tukey's *post hoc* test was used to detect differences between time points within a group, and Student's t-test for unpaired data for differences at the same time point between the groups. The level of significance was set at $P < 0.05$.

Results

There were clear differences in body composition and endurance capacity between the two experimental groups (Table 1). Even the best trained non active person had higher plasma lactate values at corresponding performance levels than the least trained active person (Figure 1). Results of RQ and blood parameters are given in Table 2. No significant treatment differences were observed in plasma levels of carbohydrate and fat metabolites. In contrast, significant differences occurred in plasma insulin, glucagon, and urea concentration. The contribution of the macro nutrients to energy expenditure is plotted separately for the two groups in Figure 2. The contribution of protein was estimated to be 10% in both groups.

Table 2. RQ and Blood parameters before (0 min) and after (25 min and more) the high fat meal (Mean ± SE). An asterisk denotes a significant group difference.

Parameter	Group	0 min	25 min	85 min	145 min	205 min
RQ *	A	0.84 ± 0.03	0.93 ± 0.01	0.89 ± 0.02	0.87 ± 0.02	0.84 ± 0.01
	NA	0.93 ± 0.02	0.94 ± 0.02	0.92 ± 0.01	0.91 ± 0.01	0.88 ± 0.01
Glucose	A	5.21 ± 0.11	5.70 ± 0.16	4.52 ± 0.23	4.93 ± 0.18	5.13 ± 0.15
mmol/l	NA	5.46 ± 0.10	6.05 ± 0.24	4.73 ± 0.21	4.77 ± 0.25	5.29 ± 0.17
Lactate	A	0.94 ± 0.06	0.88 ± 0.05	1.23 ± 0.05	1.16 ± 0.05	1.08 ± 0.05
mmol/l	NA	0.89 ± 0.06	0.93 ± 0.07	1.38 ± 0.12	1.15 ± 0.06	1.01 ± 0.04
Free fatty acids	A	0.44 ± 0.07	0.26 ± 0.03	0.55 ± 0.09	0.37 ± 0.04	0.39 ± 0.05
mmol/l	NA	0.42 ± 0.05	0.33 ± 0.04	0.41 ± 0.05	0.32 ± 0.05	0.46 ± 0.09
Triglycerides	A	0.98 ± 0.06	0.94 ± 0.06	1.00 ± 0.10	1.83 ± 0.22	2.06 ± 0.24
mmol/l	NA	1.12 ± 0.13	1.09 ± 0.12	1.25 ± 0.16	2.23 ± 0.19	2.46 ± 0.23
HDL	A	1.20 ± 0.06	1.20 ± 0.06	1.24 ± 0.06	1.19 ± 0.07	1.18 ± 0.06
mmol/l	NA	1.11 ± 0.10	1.12 ± 0.10	1.09 ± 0.10	1.15 ± 0.08	1.08 ± 0.09
BHB	A	71 ± 23	68 ± 12	33 ± 4	15 ± 5	17 ± 6
µmol/l	NA	50 ± 5	60 ± 5	28 ± 4	11 ± 5	9 ± 3
Glycerol	A	78 ± 11	57 ± 8	323 ± 50	74 ± 15	73 ± 21
µmol/l	NA	76 ± 13	81 ± 17	270 ± 41	67 ± 20	82 ± 27
Urea *	A	5.36 ± 0.24	5.35 ± 0.25	5.36 ± 0.24	5.49 ± 0.21	5.66 ± 0.21
mmol/l	NA	4.23 ± 0.24	4.12 ± 0.22	4.20 ± 0.24	4.25 ± 0.25	4.38 ± 0.27
Insulin *	A	4.8 ± 0.4	20.1 ± 2.6	16.1 ± 1.8	16.1 ± 2.4	13.3 ± 2.8
µU/ml	NA	7.2 ± 0.5	29.9 ± 6.0	26.6 ± 2.6	16.5 ± 3.5	25.6 ± 5.3
Glucagon *	A	0.86 ± 0.09	0.27 ± 0.05	0.33 ± 0.04	0.35 ± 0.06	0.43 ± 0.07
pg/ml	NA	0.50 ± 0.06	0.16 ± 0.03	0.16 ± 0.02	0.16 ± 0.03	0.23 ± 0.05

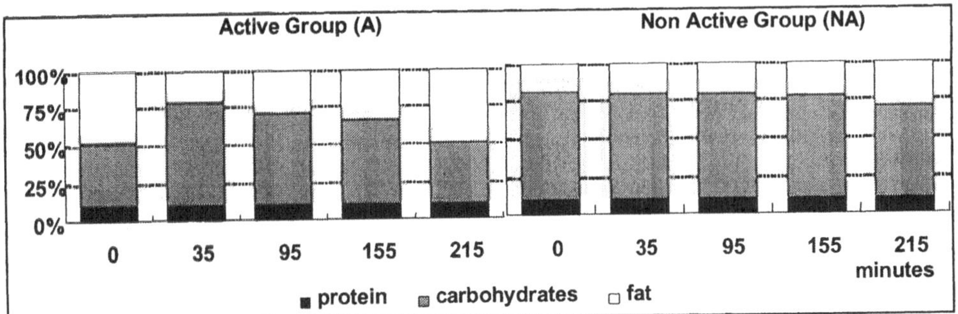

Fig. 2. Contribution of the nutrients to energy expenditure.

Discussion

Postprandial substrate oxidation was significantly different in the two groups. Before the the meal and at the end of the experiment (five hours after the meal), A had significantly lower RQ values than NA. This occurred despite the ingestion of similar amounts of macro nutrients. Group A oxidized obviously more fat than the physically inactive volunteers. No significant group differences could be found for plasma levels of carbohydrate and fat metabolites. A significantly higher blood urea level in A can be used as an indication for a more active protein turnover. With an almost identical plasma glucose concentration in the two groups, A showed lower plasma insulin and higher glucagon levels than NA. It is concluded that a higher fat oxidation and more economic utilization of glucose can be expected as a consequence of physical activity.

References

Acheson K.J., Schutz Y., Bessard T., Anantharaman K., Flatt J.-P. & Jéquier E. (1987). Glycogen storage capacity and *de novo* lipogenesis during massive carbohydrate overfeeding in man. *Am. J. Clin. Nutr.* 45:78-85.

Åstrand P.-O. & Rodhal K. (1988). Textbook of Work Physiology, 3rd edition, McGraw Hill, New York.

Flatt J.-P., Ravussin E., Acheson K. & Jéquier E. (1985). Effects of dietairy fat on postprandial substrate oxidation and on carbohydrate and fat balances. *J. Clin. Invest.* 76:1019-1024.

Lennon D., Nagle F., Stratman F., Shrago E. & Devines S. (1985). Diet and exercise training effects on resting metabolic rate. *Int. J. Obes.* 9:39-47.

Saltin B. & Åstrand P.-O. (1993). Free fatty acids and exercise. *Am. J. Clin. Nutr.* 57 (suppl):752-758.

Schulz L.O., Nyomba B., Alger S., Anderson T. & Ravussin E. (1991). Effect of endurance training on sedentary energy expenditure measured in a respiratory chamber. *Am. J. Physiol.* 260:E257-E261.

Schutz Y., Flatt J.-P. & Jéquier E. (1989). Failure of dietary fat to promote fat oxidation: a factor favoring the development of obesity. *Am. J. Clin. Nutr.* 50:307-314.

Surina D.M., Langhans W., Pauli R. & Wenk C. (1993). Meal composition affects postprandial fatty acid oxidation. *Am. J. Physiol.* 264:R1065-1070.

Tremblay A., Fontaine E. & Nadeau A. (1986). The effects of exercise-training on resting metabolic rate in lean and moderately obese individuals. *Int. J. Obes.* 10:511-517.

SUMMARY OF THE DISCUSSION

Discussion leader: K.A. Johnson
Scientific secretaries: A.B. Hall & E.M. Larsen

There was much variation in the papers discussed in this session and much of the discussion mirrored that of the earlier session on growth. The papers were grouped into three major categories for discussion. These were: 1) the energy cost of activity; 2) the impact of ambient temperature on heat production; and 3) the impact of malnutrition on heat production and metabolism.

Several papers examined the energy cost of activities such as chewing, working, standing or lying and the impact diet components have on activity. The major conclusion reached was that it is difficult to partition energy costs to specific activities because there are physiological and metabolic changes associated with measured behaviours or activities. For example, Susenbeth presented work examining the energy cost of chewing by cattle. Many questions were raised about other possible physiological activities that may be associated with the presentation of food to the animal, i.e. increased ruminal contractions or increased blood flow to the intestinal tract in response to the sight or ingestion of food. The degree to which these other events affect measurements of heat production was discussed. McCracken noted that a major part of the discrepancies seen between slaughter balance and respiration calorimetry could be attributable to activity. This had been the case with individually caged pigs. However lower feed intakes, under *ad libitum* feeding conditions, were also a possible aspect. As shown in the Figure in the previous discussion, (p 271), the slope of the line above maintenance tends to decrease when activity is quantified. Training will also reduce the slope of the line above maintenance as shown in the paper presented by Wenk.

The role of environmental temperature was also discussed in relation to several papers in the session. It was concluded that animals respond differently to heat loads and this ability to respond is associated with genotype, previous exposure and feed intake. The point was also made that the variation within breeds is as great as that seen between breeds and care should be taken in drawing conclusions about breed effects.

Several papers associated with metabolic adjustments due to training, food deprivation or malnutrition and infection were presented. Camels were determined to respond to a fast in a manner different than a sheep indicating different metabolic adjustments. Infection reduced energy available for growth by increasing maintenance requirements.

SESSION 7

MODELLING

DIETARY ENERGY UTILIZATION: SENSITIVITY OF A NET ENERGY MODEL TO VARYING ESTIMATES OF FOOD INTAKE, MAINTENANCE REQUIREMENT AND PROTEIN DEPOSITION

M.G. MacLeod

Roslin Institute (Edinburgh), Roslin, Midlothian EH25 9PS, UK

Summary

This paper analyses the effects of food intake, energy maintenance requirement and rate of protein accretion on Net Energy yield and other predictions of a simulation model of avian energy metabolism. Sensitivity was tested over the range ± 25% of means of food intake, maintenance and protein accretion, with simulation at 5 equidistant values (-25%, -12.5%, mean, +12.5%, +25%). The simulation program did not permit protein retention to reach a specified target at the two lowest food intakes and the two highest maintenance inputs. The range of variation in k_g (net availability of metabolizable energy for growth) and NE (net energy, kJ/g of food) from a test diet was only 1.2-2.0% when any one of food intake, maintenance or protein retention was varied by ± 25%. Variation in k_g and NE was related to the proportion of dry matter retained as protein, which varied from 0.33 to 0.49 in the simulations described.

Keywords: Model, net energy, poultry, protein deposition.

Introduction

A biochemically based simulation model of energy metabolism was outlined at a previous EAAP Energy Metabolism Symposium (MacLeod, 1994). Predictions from the model were validated against results from an experiment (MacLeod, 1990) in which diets of widely different composition were fed to growing broilers. The main objective of the model is to predict the Net Energy yield to animals of different mixtures of ingredients.

Since the Net Energy value of a diet or feedstuff is dependent on the biological use to which it is put by the animal, it is influenced by such factors as food intake, maintenance requirement and the composition of gain. The magnitude of the effects of such factors is important in considering the way the model might be used in practice. Assessing the influence of these key factors on the outputs of the model can be looked upon as a type of “sensitivity analysis” (Brown & Rothery, 1993). In this analysis, the parameter of interest is adjusted over a suitable range and the effect on the model output is quantified. If the model is particularly sensitive to adjustment of a parameter, it then becomes particularly important that the parameter is accurately known. The present paper analyses the effects of food intake, energy maintenance requirement and rate of protein accretion on Net Energy yield and other aspects of predicted energy metabolism. Preliminary analysis (MacLeod, 1994) indicated that the model gives a good initial prediction of the effects of diet composition on energy retention and expenditure, allowing the preliminary assessment of hypotheses which can then be validated by experiment.

Materials and Methods

The model uses files of chemical compositions (e.g. individual amino acids) of ingredients. Measured digestibility coefficients are used when available. Data files of the chemical compositions of the bird's synthesis products are also accessed by the model. Metabolism is simulated, using the equations of the major biochemical pathways; energy "accounting" within the model is done in terms of ATP generation and utilization, but the final outputs are in conventional nutrition units. The simulation was originally developed in PASCAL on a DEC VAX computer and has now been edited to run on a PC under Windows 95.

Table 1. Simulation of k_g (net efficiency of energy utilisation for growth) and NE (net energy per g of food). Food intake, maintenance energy requirement (MER) and rate of protein accretion (RPA) were varied in turn, while the 2 other parameters were held constant.

	Food intake (g)	TME intake (kJ/d)	Main-tenance (kJ/d)	Protein retention (g/d)	Propn. of dry matter gain	k_g	NE (kJ/g)
Food	79	987	525	5.88	0.49	0.732	9.14
intake	92	1149	525	7.39	0.55	0.735	9.18
varied	105	1311	525	8.74	0.44	0.737	9.20
	118	1474	525	8.74	0.37	0.743	9.28
	131	1636	525	8.74	0.33	0.747	9.33
	105	1311	394	8.74	0.38	0.741	9.26
MER	105	1311	460	8.74	0.41	0.739	9.23
varied	105	1311	525	8.74	0.44	0.737	9.20
	105	1311	591	8.62	0.46	0.734	9.18
	105	1311	656	8.34	0.49	0.733	9.15
	105	1311	525	6.68	0.35	0.745	9.31
RPA	105	1311	525	7.71	0.39	0.741	9.26
varied	105	1311	525	8.74	0.44	0.737	9.20
	105	1311	525	8.89	0.44	0.736	9.20
	105	1311	525	8.89	0.44	0.736	9.20

The midrange values for food intake, protein accretion and maintenance requirement were the means of these variates measured in MacLeod (1990). Sensitivity was tested over the range ± 25% of these means, with simulation at 5 equidistant values (-25%, -12.5%, mean, +12.5%, +25%). Each of the remaining 2 variates was held constant at its mean value while the variate being tested was adjusted over ± 25%. In a further test (Table 2) food intake was held constant while maintenance energy requirement was varied over ± 25% and rate of protein accretion was allowed to be determined by the model. The diet used in the simulation contained 12.5 kJ TME (true metabolizable energy) and 22.6 g crude protein per kg. The birds had a mean weight of 710 g and an age of 30 d (these also were the means from MacLeod (1990,1994)).

Results

A feature of the main sensitivity test (Table 1) was that the programme did not permit protein retention to reach the specified target at the two lowest food intakes and the two highest maintenance inputs. The range of variation in k_g and NE was only 1.2 to 2.0% when any one of food intake, maintenance or protein retention was varied by ± 25%. Variation in k_g and NE was correlated with the proportion of dry matter retained as protein, which varied from 0.33 to 0.49 under the conditions of the 15 simulations described. In the second test (food intake constant, maintenance varied by ± 25% and protein accretion determined by model) NE varied by 0.9%, while the proportion of dry matter retained as protein ranged from 0.41 to 0.49 (Table 2). In the latter case, protein retention as a proportion of weight gain increased as maintenance energy requirement decreased; however, absolute protein retention decreased, simulating increased catabolism of amino acids to satisfy the maintenance energy demand.

Table 2. Simulation of k_g (net efficiency of energy utilisation for growth) and NE (net energy per g of food). Food intake was held constant, while maintenance energy requirement was varied over ± 25% and rate of protein accretion was allowed to be determined by the model.

Food intake (g)	TME intake (kJ/d)	Maintenance (kJ/d)	Protein retention (g/d)	Propn. of dry matter gain	k_g	NE (kJ/g)
105	1311	394	9.42	0.41	0.739	9.23
105	1311	460	9.16	0.42	0.738	9.21
105	1311	525	8.89	0.44	0.736	9.20
105	1311	591	8.62	0.46	0.735	9.18
105	1311	656	8.34	0.49	0.733	9.15

Discussion

The proportion of energy retained as protein is within the zone where rate of change of this proportion with energy intake is very low (Boekholt *et al.*, 1994). This may account for the low sensitivity of k_g to food intake and maintenance requirement. By comparison, the sensitivity to diet composition is quite high (MacLeod, 1994). It is clear that, as in empirical experiments, the net availability of energy is strongly related to the proportion of energy supplied as protein and, particularly, the proportion used for protein synthesis (MacLeod, 1997). The maximum rate of protein accretion permitted by the simulation was similar to the rate recorded in MacLeod (1990) on a similar food intake. The relatively low sensitivity of the simulation model to the selected key parameters, combined with a higher sensitivity to diet composition, suggest that the simulation may be quite robust in its application to feed evaluation.

Acknowledgements

This work was funded by the Ministry of Agriculture, Fisheries and Food.

References

Boekholt H.A., van der Grinten P., Schreurs V.V.A.M., Los, M.J.N. & Leffering, C.P. (1994). Effect of dietary energy restriction on retention of protein, fat and energy in broiler chickens. *Br. Poult. Sci.* 35:603-614.

Brown D. & Rothery P. (1993). Models in Biology: Mathematics, Statistics and Computing. Wiley, New York.

MacLeod M.G. (1990). Energy and nitrogen intake, expenditure and retention at 20° in growing fowl given diets with a wide range of energy and protein contents. *Br. J. Nutr.* 64:625-637.

MacLeod M.G. (1994). Dietary energy utilisation: prediction by computer simulation of energy metabolism. In. J.F. Aguilera (Ed) Proceedings of the 13th Symposium on Energy Metabolism of Farm Animals. *EAAP Publ. No.* 76:237-240.

MacLeod M.G. (1997). Effects of amino acid balance and energy:protein ratio on energy and nitrogen metabolism in male broiler chickens. *Br. Poult. Sci.* 38 (in press).

PARTITIONING OF ENERGY IN PROTEIN AND FAT RETENTION AT INCREASING ENERGY INTAKES IN GROWING ANIMALS

H.A. Boekholt, V.V.A.M. Schreurs and R.A.H.M. ten Doeschate

Wageningen Institute of Animal Science (WIAS), Department of Human and Animal Physiology, Wageningen Agricultural University, Haarweg 10, NL-6709PJ Wageningen, The Netherlands

Summary

At increasing intakes, body fat content increases and protein content decreases. It is most often assumed that this is brought about because each increment in retention contains more fat and less protein. Experimental results, however, showed that this explanation is not true. Male broiler chickens were fed at levels of 60, 70, 80 and 90% of *ad libitum* intake. Body composition at 1500 g showed that at increasing intakes, body fat content increased, protein content decreased and more is retained in inedible tissues. Both fat and protein retention per day were linearly related to total energy retention (ER). This means that each increment in retention had the same fat and protein content (55% energy as fat, 38% as protein). At zero fat retention only protein was retained, about 50% of maximal retention. At zero ER protein was retained and fat was mobilized. The results show that total ER consists of two components: a constant basic daily protein retention and a variable additional ER with a constant ratio of protein and fat.

Key words: Growth, partitioning of energy, protein retention, fat retention, energy intake.

Introduction

Growth is largely determined by genetic, environmental and nutritional factors. Energy intake and diet composition can greatly influence carcass composition of meat producing animals. At increasing intakes, body fat content increases and protein content decreases. It is generally assumed that this is brought about because each increment in retention contains more fat and less protein. Experimental results, however, showed that this explanation is not true.

Materials and Methods

An experiment was conducted with male broiler chickens. Forty-eight one week old animals were divided in 4 groups and fed 60, 70, 80 or 90% of *ad libitum* intake individually. Animals were weighed weekly and fed according to body weight, increasing the feed intake per day. When the animals passed a body weight of 1500 g they were killed and body composition of 8 animals per group was determined.

Results

In earlier experiments it was found that daily retention of protein and fat was linearly related to energy retention ($ER/W^{0.5}$), whereas fat content in total retention increased as expected (Boekholt *et al.*, 1994a, 1994b). This linear relationship means that each additional increase in energy supply resulted in an additional retention of protein and fat in a constant ratio. Zero protein and fat retention were not reached at the same level of energy intake. At zero ER fat was mobilized whereas protein was retained. The results of the balance experiments were confirmed by this experiment.

Table 1 shows the results of the experiment. Average final weight of all groups was similar. Total feed consumption at the 60% level was 20 per cent higher than at the 90% level and the growth period increased from 31 to 57 days. Water content of the body decreased with level of feed intake. Protein content of the body decreased, whereas fat content markedly increased. This is more clearly demonstrated in the dry matter composition of the body as shown in Fig. 1. Protein content in dry matter decreased gradually from about 700 to 600 g/kg when intake increased from 60 to 90%, whereas fat content increased from 150 to 250 g/kg. The higher protein retention and lower fat retention at lower intake was accompanied by a shift to a considerably higher carcass weight.

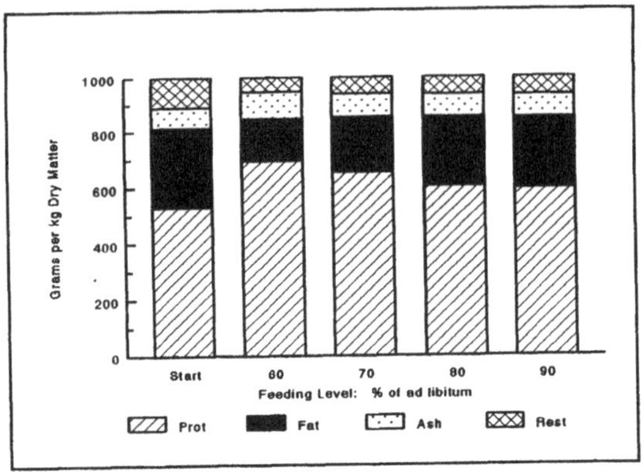

Fig. 1. Average body composition in dry matter at start and at the end of the experiment. Rest = DM - Protein - Fat - Ash.

Average fat retention and protein retention per animal per day were linearly related to total ER (Figs 2 and 3). At increasing intakes 55% of extra ER was retained in fat and 38 % in protein, independent of intake level. Fat retention was calculated to be zero at about 1.5 times maintenance level. Protein retention at that level was about 50% of maximal retention, as was found in the earlier experiments.

Discussion

The results of the experiments show that total protein retention consists of two components (Figs 2, 3 and 4). At zero fat retention (ER=105 kJ/day) only protein is retained. This basic protein retention is about half of maximal retention and is not dependent on energy intake. The earlier experiments revealed that daily basic protein retention is well related to metabolic body weight (Fig. 4). With increasing energy intakes the basic protein retention is combined with an additional amount of protein and fat in a constant ratio : 38/55 kJ.

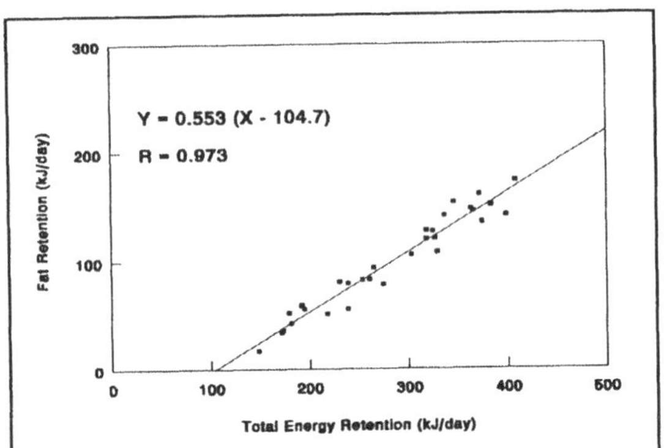

Fig. 2. Average retention of fat during the growth period from 160 to 1500 g body weight. Data per animal.

This ratio markedly differs from the ratio in the earlier experiments with chickens (16/84 kJ). Maybe this can be attributed to a higher activity of individually housed animals. The high contribution of fat in the additional variable protein plus fat retention means that at increasing intake levels relatively more energy will be retained in fat and less in protein and so fat content of total daily retention and also of the body increases.

Table 1. Summary of results.

% of *ad libitum* intake		60	70	80	90
Number of animals		7	7	8	8
Starting weight		159	158	156	155
Final weight		1534	1526	1532	1529
Feed intake		3005	2740	2672	2493
Experimental days		57	44	37	31
Body composition at end					
Water	g/kg	695	687	673	679
Dry matter, DM	g/kg	305	313	327	321
Protein	g/kg	211	204	197	191
Fat	g/kg	47	63	81	82
Ash	g/kg	30	27	27	26
Rest	g/kg	17	19	21	22
Energy	MJ/kg	7.26	7.77	8.40	8.33
Body composition in DM					
Protein	g/kg DM	694	654	604	596
Fat	g/kg DM	153	199	248	255
Ash	g/kg DM	97	85	84	82
Rest	g/kg DM	56	61	64	68
Energy	MJ/kg DM	23.8	24.8	25.7	25.9
Results per day					
Feed intake	g/day	53.0	62.6	72.2	81.0
Growth	g/day	24.3	31.3	37.2	44.7
Water	g/day	16.9	21.3	24.9	29.9
Dry matter	g/day	7.52	9.87	12.40	14.45
Protein retention	g/day	5.34	6.57	7.56	8.69
Fat retention	g/day	1.06	1.89	3.06	3.65
Ash	g/day	0.75	0.85	1.05	1.18
Rest	g/day	0.37	0.56	0.74	0.92
Energy retention	kJ/day	177.4	244.0	318.2	374.0
Protein retention	kJ/day	127.0	156.4	179.9	206.9
Fat retention	kJ/day	42.1	75.3	121.6	145.3
Rest	kJ/day	8.3	12.3	16.8	21.9
Feed conversion		2.19	2.00	1.94	1.81
N efficiency	%	46.3	48.1	48.1	49.0
Organ weight as % of body weight at 1500 g.					
Feathers, skin, head, legs	%	21.9	22.8	24.5	24.7
Intestines minus stomach	%	4.15	4.38	5.00	5.02
Heart	%	0.68	0.69	0.75	0.77
Liver	%	2.42	2.63	2.97	3.05
Abdominal fat pad	%	0.08	0.64	0.62	0.87
Carcass	%	66.6	64.5	60.7	60.6

It is worthwhile to realize that at higher intakes more is retained in rather inedible parts of the body, such as skin and intestines and less in the carcass.

The distinction between basic protein retention and additional daily deposition of protein and fat is important when using models for estimating growth from intake and metabolism.

These findings are in agreement with results of experiments with lambs, pigs, broilers, rats and bulls, from which could be derived that in young growing animals each additional unit of gain is composed of constant amounts of protein and fat (Boekholt *et al.*, 1994a). However the ratio between protein and fat will differ with species, age, sex, housing system and other factors. It remains interesting to know in what organs or tissues the extra retention is retained.

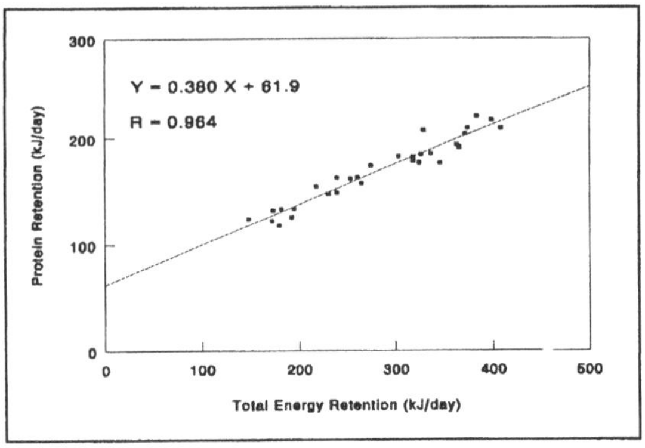

Fig. 3. Average retention of protein during the growth period from 160 - 1500 g body weight. Data per animal.

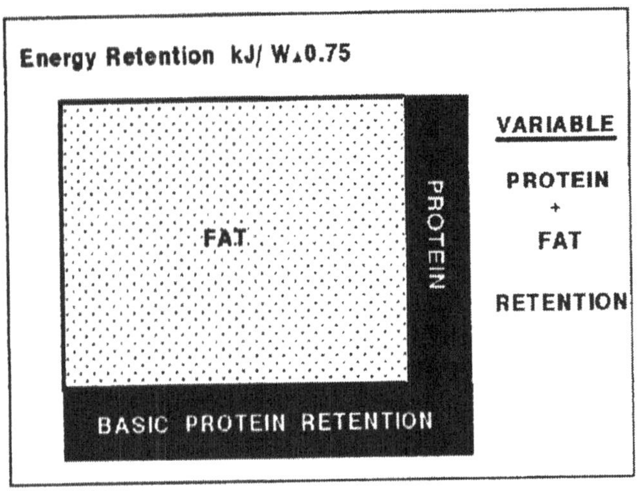

Fig. 4. Daily total energy retention consists of a constant Basic Protein Retention plus a variable ER with a constant ratio of protein and fat.

References

Boekholt H.A., van der Grinten Ph., Schreurs V.V.A.M., Los M.J.N. & Leffering C.P. (1994). Effect of dietary energy restriction on retention of protein, fat and energy in broiler chickens. *Brit. Poultry Sci.* 35:603-614.

Boekholt H.A., Schreurs V.V.A.M. & Verstegen M.W.A. (1994). Composition of daily growth in young growing animals receiving increasing amounts of energy. Proc. 13th Symposium on Energy Metabolism of Farm Animals, *EAAP Publ. No.* 76:229-232.

MODELLING DYNAMIC ASPECTS OF HEAT PRODUCTION IN PIGS

J. van Milgen[1], J.-F. Bernier[2], S. Dubois[1] Y. Le Cozler[1] and J. Noblet[1]

[1]Institut National de la Recherche Agronomique, Station de Recherches Porcines, 35590 Saint-Gilles, France; [2]Département de Sciences Animales, Université Laval, Ste-Foy, Québec, G1K 7P4 Canada

Summary

Heat production during fasting and physical activity were obtained in 20 to 60 kg growing pigs of 4 genotypes (Meishan and Large White castrates and Large White and Piétrain males) that were fed *ad libitum* prior to fasting. Fasting heat production (FHP) was proportional to the body weight raised to the power 0.55, but with genotype-specific proportionality parameters. Genotype effects could be removed by expressing the FHP as a function of muscle, viscera and fat: FHP (kJ/d) = $457(\text{muscle})^{0.81} + 1969(\text{viscera})^{0.81} - 644(\text{fat})^{0.81}$. For lean genotypes, FHP could be expressed as a function of muscle and viscera alone: FHP (kJ/d) = $508(\text{muscle})^{0.66} + 2011(\text{viscera})^{0.66}$. Both genotype and weight affected the duration of physical activity (i.e. standing or sitting), and the total cost of activity. Energetic cost of activity (kJ/h of activity) was proportional to the muscle mass raised to the power 0.91.

Key words: Fasting heat production, physical activity, pigs, mathematical model.

Introduction

The cost of feed energy represents a large proportion of the total cost of pig production. Therefore, it is important to match supply and requirement as closely as possible. Most, if not all feed evaluation systems are based on the concept of partitioning nutrients between maintenance and production. At maintenance, nutrient intake just equals the obligatory requirements and takes into account normal physical activity of the animal. Because it is impossible to directly measure maintenance requirements in producing animals, several predictors for maintenance have been proposed. If the efficiency of energy utilization below maintenance is known, fasting heat production (FHP) can be used to predict maintenance.

The present study is part of a larger experiment, which has the objective to study the effect of the animal on growth performance variables. The study presented here was designed to study the effect of body weight and genotype on the heat production due to fasting and physical activity in growing pigs. A mathematical model is used that allows studying rapid changes in O_2 consumption and CO_2 production (e.g. due to physical activity) in large, open-circuit respiration chambers.

Materials and Methods

Four genotypes (6 Meishan castrates (MC), 8 Large White castrates (LWC), 8 Large White males (LWM) and 8 Piétrain males (PM)) were *ad libitum* fed a grower diet containing 220 g protein/kg and 15.5 MJ ME. The experiment consisted of three measurement periods (corresponding to approximately 25, 40 and 60 kg live weight) during which feed intake, feeding behaviour, digestibility and heat production were measured. The animals were housed in individual metabolism cages.

At 25, 40 and 60 kg live weight, heat production was measured in an open-circuit respiration chamber. After 4 to 5 days of being fed in the respiration chamber, the animals were subjected to a 30 h (24 h at 25 kg) fasting period. Physical activity (sitting or standing) was recorded using infra-red beams.

Oxygen (O_2) and carbon dioxide (CO_2) concentrations were measured continuously. O_2 consumption and CO_2 production were estimated as described in detail by van Milgen *et al.* (1997). Briefly, this method considers the respiration chamber as a compartmental system of 3 gases (O_2, CO_2 and other gases). The changes in gas concentration are described by a model that accounts for the physical aspects of gas exchange (e.g. changes in temperature and pressure) as well as for the O_2 consumption and CO_2 production by the animal. The latter are further compartmented into O_2 consumption (and CO_2 production) during the resting state, physical activity, and the thermic effect of feeding. It was assumed that the resting O_2 consumption in the fed animal is higher than that after prolonged (24 or 30 h) fasting. This change between the fed and fasting state was described as a first-order process. Dependent variables in the model were $[O_2]$ and $[CO_2]$ in the respiration chamber whereas independent variables included time, time of physical activity, and (for fed animals) the time and quantity of feed intake. Model parameters were adjusted statistically to ensure maximum fit for both $[O_2]$ and $[CO_2]$. In the present study, only O_2 consumption and CO_2 production during the fasting period were considered and model parameters included the (asymptotic) resting O_2 consumption and CO_2 production after prolonged fasting (l/h), and the O_2 consumption and CO_2 production during physical activity (l/h of activity). Heat production during fasting (kJ/d) and physical activity (kJ/h of activity) were calculated from the respective O_2 consumption and CO_2 production as described by Brouwer (1965) excluding the correction for urinary-N and CH_4 production. The (partial) respiratory quotients were calculated as the ratio between CO_2 production and O_2 consumption.

Tissue compositions (muscle, fat, and viscera) of the animals were estimated from a data set of similar animals (published in part by Quiniou & Noblet, 1995) through allometric relations between empty body weight and body tissues. Parameter estimates for each animal obtained from the model were subjected to analysis of variance or non-linear regression using body weight or body tissues as dependent variables.

Table 1. Body weight, body composition, heat production during fasting and physical activity and duration of physical activity (sitting or standing) in growing pigs.

Group	MC*	LWC*	LWM*	PM*
		Body weight (kg)		
1	22.1	26.7	25.4	24.7
2	38.9	39.4	38.8	40.5
3	56.6	59.2	58.9	60.4
		Resting FHP (MJ/d)		
1	4.90	7.34	7.06	6.24
2	5.76	8.85	8.98	8.56
3	7.26	10.99	11.34	10.60
		HP activity (KJ/h of activity)		
1	129	195	206	225
2	169	246	253	318
3	211	342	360	545
		Duration of activity (h/d)		
1	3.52	2.14	1.71	3.24
2	3.17	1.90	1.82	1.96
3	2.76	1.40	1.69	1.94

* MC: Meishan castrate. LWC: Large White castrate; LWM: Large White male; PM: Piétrain male

Results

Table 1 lists the body weight, the heat production during fasting, physical activity and the duration of physical activity. The resting FHP expressed per $kg^{0.75}$ decreased with weight while a significant genotype effect was apparent. Parameter estimates of the model $FHP=a_iX^{b_i}$ (for genotype = i and X = body weight, muscle or viscera) indicated that FHP was not proportional to the metabolic body weight ($kg^{0.75}$). With weight as predictor, the exponent b (0.55 ± 0.03) did not differ between genotypes and differed significantly from 0.75 but not from 0.60. Nevertheless, FHP across genotypes could not be described sufficiently by a single predictor (i.e. without genotype-specific parameters). The regression coefficients (a_i) were 810, 1200, 1220, and 1120 kJ/$kg^{0.55}$/d for MC, LWC, LWM, and PM, respectively.

Alternatively, FHP can be expressed by multiple body components. Muscle and viscera are generally assumed to be key contributors to energy metabolism. Hypotheses concerning parameter estimates were tested against the full model (RSD = 710 kJ/d):

$$FHP = a_1(\text{muscle})^{b_1} + a_2(\text{viscera})^{b_2} + a_3(\text{fat})^{b_3} \quad (1)$$

The hypothesis of a single exponent across tissues was not rejected and the coefficients were: b = 0.81 (0.05), a_1 = 457 (93) kJ/(kg $\text{muscle})^{0.81}$/d, a_2 = 1969 (151) kJ/(kg $\text{viscera})^{0.81}$/d, a_3 = -644 (88) kJ/(kg $\text{fat})^{0.81}$/d, and RSD = 717 kJ/d. A large part of the variation in genotypes is due to the presence of Meishan in the dataset. Therefore, separate equations for the lean genotypes and Meishan were established. FHP in lean genotypes could be described by muscle and viscera whereas for Meishan a single predictor (weight) was sufficient:

Lean genotypes : FHP = a_1 (muscle) b_L + a_2 (viscera) b_L
Meisham : FHP = a_3 (BW) b_M

with b_L = 0.66 (0.04), a_1 = 508 (108) kJ/(kg $\text{muscle})^{0.66}$/d, a_2 = 2011 (188) kJ/(kg $\text{viscera})^{0.66}$/d, b_M = 0.44 (0.09), a_3 = 1194 (387) kJ/(kg $\text{BW})^{0.44}$/d and RSD = 673 kJ/d. Based on this equation, it can be shown that the contribution of viscera to the FHP exceeds that of muscle in the lean genotypes.

A general allometric model was used to describe the energetic cost of activity during fasting (kJ/h of activity; $HP_{activity}=a_iX^{b_i}$). With body weight as the independent variable, a common exponent b could be used for the 4 genotypes, but the regression coefficients were genotype-specific (a_i = 189, 283, 299, and 406 kJ/(kg $\text{BW})^{0.83}$/h of activity for MC, LWC, LWM, and P, respectively). With muscle as predictor, the four genotypes not only shared a common exponent, but also a common regression coefficient with a = 21.0 (5.2) kJ/h of activity/(kg $\text{muscle})^{0.91}$, b = 0.91 (0.08), and RSD = 72.0 kJ/h of activity. The exponent b neither differed from 0.75 nor from 1.00.

Neither genotype nor weight affected the $RQ_{fasting}$ and $RQ_{activity}$. However, $RQ_{fasting}$ was significantly lower than $RQ_{activity}$ (0.73 and 0.85, respectively). Both $RQ_{fasting}$ and $RQ_{activity}$ were significantly smaller than unity, indicating oxidation of fatty acids and/or proteins.

Discussion

In contrast to the customary mode of expression, FHP was not proportional to the metabolic body weight ($kg^{0.75}$). In growing pigs, it has been suggested that the exponent is lower than 0.75 (Brown & Mount, 1982; Noblet *et al.*, 1991) and this study further confirms this.

Literature values for the resting FHP in *ad libitum* fed, growing pigs range from 700 to close to 1000 kJ/(kg $\text{BW})^{0.60}$/d. The FHP in the lean genotypes used here are situated in the upper half of this range. However, a large part of this variation may be due to differences in body composition. For example, the difference in body fat alone between the current study (15.5%) and that of Tess *et al.* (1984; 18.9%) represents close to 100 kJ/(kg BW)$^{0.60}$/d in a 40 kg pig based on the equation established above. Differences in muscle and visceral mass can further accentuate the difference. Viscera contributed the most to the FHP. A very lean genotype such as the Piétrain has a FHP similar to the other lean genotypes due to the relatively small weight of its viscera.

The negative regression coefficient for fat in predicting FHP (equation 1) may, at first sight, be difficult to interpret and may be an artefact of the dataset. An alternative explanation can be found by reversing the logic of the regression equation. Maybe it is not the fact that fatter pigs have a lower FHP, but rather that pigs with lower FHP become fatter. This logic implies that animals with the same muscle and visceral mass can have a different FHP. Support for this can be found in an experiment by Kyriazakis and Emmans (1995) who fed Large White x Landrace and Meishan pigs the same feed allowance (formulated to be protein-limiting). Although both breeds had an equal protein gain, Meishan pigs gained 40% more lipid than the crossbreed. In view of the results presented here, a large part of this difference can be explained by the lower resting FHP in Meishan pigs. The reason why animals with identical muscle and visceral mass have a different FHP remains nevertheless unexplained.

Sitting or standing is reported to be a costly exercise in pigs compared to other species (Noblet *et al.*, 1993). During physical activity, the heat production increased between 65 and 125% compared to the FHP. Jakobsen *et al.* (1994) found that walking pigs weighing 80-90 kg spent two to three times more energy than resting pigs. The respiratory quotient declined during walking from 1.07 to 0.97, indicating that, although carbohydrates were the primary source of energy, some fatty acids are used as well. In the present study, the RQ increased from 0.73 to 0.85 during activity indicating that, in fasting animals, the contribution of carbohydrates to substrate oxidation increases during activity. It should be noted that the RQ during physical activity in the present study is still lower than that reported by Jakobsen *et al.* (1994), suggesting that the sources of energy during activity may be different between the fed and fasting state.

References

Brouwer E. (1965). Report of sub-committee on constants and factors. In: K.L. Blaxter (Ed.): Proceedings of the 3rd Symposium Energy Metabolism of Farm Animals. Academic Press, London. pp. 441-443.

Brown D. & Mount L.E. (1982). The metabolic body size of the growing pig. *Livest. Prod. Sci.* 9:389-398.

Jakobsen K., Chwalibog A., Henckel S. & Thorbek, G. (1994). Heat production and quantitative oxidation of nutrients by physical activity in pigs. *Ann. Nutr. Metab.* 38:1-7.

Kyriazakis I. & Emmans G.C. (1995). Do breeds of pig differ in the efficiency with which they use a limiting protein supply? *Br. J. Nutr.* 74:183-195.

Noblet J., Karege C. & Dubois S. (1991). Influence of growth potential on energy requirements for maintenance in growing pigs. In: C. Wenk & M. Boessinger (Ed.): Proceedings of the 12th Symposium Energy Metabolism of Farm Animals. *EAAP Publ* No 58: 107-110.

Noblet J., Shi X.S. & Dubois S. (1993). Energy cost of standing activity in sows. *Livest. Prod. Sci.* 34:127-136.

Quiniou N. & Noblet J. (1995). Prediction of tissular body composition from protein and lipid deposition in growing pigs. *J. Anim. Sci.* 73:1567-1575.

Tess M.W., Dickerson G.E., Nienaber J.A. & Ferrell C.L. (1984). The effects of body composition on fasting heat production in pigs. *J. Anim. Sci.* 58:99-110.

van Milgen J., Noblet J., Dubois S. & Bernier J.F. (1997). Dynamic aspects of oxygen consumption and carbon dioxide production in swine. *Br. J. Nutr.* 78:397-410.

INTERACTION OF GLUCOSE AND AMINO ACID METABOLISM IN LACTATING SOWS: ESTIMATING INTERNAL PARAMETERS OF A MODEL OF METABOLISM

John P. McNamara

Department of Animal Sciences, 233 Clark Hall, Washington State University, Pullman, WA 99164-6351, USA

Summary

The purposes were to determine milk production, body fat and body protein in sows consuming a range of energy and amino acid intakes, and to estimate model parameter values in lactating sows. Sows consumed 6.1 kg/d DM for a 20 d lactation. Protein intake was 863 (H), 767 (M) and 678 (L) g/d with 59, 53, and 47 g/d lysine at two levels of fat intake, 117 (L) and 410 (H) g/d. Litter weaning weights were 71, 74, 69, 69, 72 and 70 (SE 0.9) for sows fed HL, ML, LL, HH, MH, and LH, respectively. Body protein change from d 1 to d 21 of lactation was -1.4, -1.3, -2.1, +.29, -1.5 and -2.5 kg (SE 0.34). Body fat change was +1.4, -2.8, -5.0, +2.8, -3.51, and -2.2 kg (SE 1.2). The model was adequate in predicting milk component response to treatments. Body protein and fat loss were underestimated on all rations. Model behaviour in response to protein intake was adequate for body protein; response to fat intake was adequate for body fat. Interactions between fat and protein intake were inadequate. Maximal rates of proteolysis should be increased, and a better representation of the interactions between amino acids, fatty acids and glucose should be represented.

Key words: Lactation, amino acid, energy, model, metabolism, sows.

Introduction

Ability to predict performance and direct effective research in lactating sows depends upon a knowledge of the chemical interconversions of amino acids, fatty acids and glucose in body tissues during lactation (Baldwin, 1995; Black *et al.*, 1986; Pettigrew *et al.*, 1992). This knowledge is limited for many biochemical pathways (McNamara, 1995). Body fat and body protein are mobilized to meet the needs of lactation, especially when intake of these components is limited (Parmley *et al.*, 1996). Information on the partitioning of amino acids between milk production and body protein is limited.

Integrative biology requires testing complex hypotheses concerning interactions in the use of glucose, fatty acids and amino acids. Mechanism-based models (Baldwin, 1995; Pettigrew *et al.*, 1992) require identification and estimation of the parameters describing critical sub-cellular pathways and integration of these pathways to describe nutrient use in the whole body. To obtain valid independent estimation of parameters requires experiments to measure a range of responses and to measure several parameters simultaneously (Pettigrew *et al.*, 1991).

The hypotheses were: lactating sows producing large amounts of milk and consuming adequate diets would not lose significant amounts of body protein or fat; body protein will be mobilized as protein intake decreases; if protein intake decreases severely, milk protein output will eventually decline; increased fat intake will increase milk fat, spare maternal fat, and interact with dietary protein to affect milk production and maternal body composition. The null hypothesis was that parameter values in an existing mathematical model of metabolism of the lactating sow (Pettigrew *et al.*, 1992) are sufficiently adequate to predict responses in milk component production, body fat and protein use to changes in intake of protein and fat.

Materials and Methods

Sows were Landrace x Yorkshire crosses, parity 2 to 4. From day 1 post-farrowing, they were fed twice daily the following diets (g/kg): Low Energy, Low Protein: 20 added fat, 116 CP, 8 lysine; Low Energy, Medium Protein: 20 added fat, 13 CP, 9 lysine; Low Energy, High Protein: 20 added fat, 147 CP, 10 lysine; High Energy, Low Protein: 70 added fat, 116 CP, 8 lysine; High Energy, Medium Protein: 70 added fat, 13 CP, 9 lysine; High Energy, High Protein: 70 added fat, 147 CP, 10 lysine. Diets were fed up to 5 kg/d for week 1 and 6.8 kg/d for weeks 2 and 3.

Animals started diets on day 1 of lactation and lactation length was 20 days. Litters were adjusted to 10 to 12 pigs within 3 days of parturition. Sow body weights were taken 5 d prepartum and weekly postpartum. Litters were weighed twice weekly. Milk samples were taken on days 7 and 21 of lactation. Milk production was calculated using the equations of Noblet & Ettiene (1986). Fifteen sows were killed at day 1 of lactation for a comparative slaughter group; body fat and protein were determined. Six animals from each treatment were killed at day 21 for body composition.

The modelling simulations were conducted using an existing model of metabolism in lactating sows (Pettigrew *et al.*, 1992) running on Advanced Continuous Simulation Language Software (Baldwin, 1995). Experimental observations were compared with simulations. Dietary intake of protein, fatty acids and starch was simulated explicitly from experimental observations. Weight and composition of body were as determined on initial slaughter animals. The control group for the simulations was the Medium Protein, Low Energy treatment and simulated milk component production was set to this group. Simulations for milk production and body composition were compared with experimental group means.

Results

Protein intake below 900 g/d did not decrease litter growth (Table 1), but below 800 g/d, litter growth, milk, and milk protein production was diminished ($P < 0.05$) in week 3. Fat intake between 100 and 450 g/d did not affect litter performance. Sows consuming adequate rations maintained body fat and protein even when weaning 11.4 pigs and producing over 180 kg of milk, 9 kg of lactose and 6 kg each of fat and protein in 20 days. Animals consuming less than 900 g protein/d lost more ($P< 0.05$) protein than those consuming over 900 g/d (Table 2). Animals consuming low fat rations lost more fat than controls, and those consuming the LL ration lost the most body fat ($P < 0.05$), primarily that fat associated with body muscle.

The mechanistic model (Pettigrew *et al.*, 1992) aggregates pathways within the mammary gland, adipose tissue, lean body and viscera. Rates of milk component synthesis are functions of genotype (maximal) as well as nutrient supply. Synthesis of body fat from glucose and fatty acids, lipolysis, and lean and visceral protein synthesis and breakdown are stated. Equation forms are well-grounded in experimental observations or biochemical kinetics. Sensitivity constants and maximal velocities are set based on historical knowledge of situation (animals, genetics) being simulated (Pettigrew *et al.*, 1992). Most parameters are variables which can be initialized for each run, and may change during a simulation as a function of: stage of lactation, hormonal regulation, availability of limiting substrates or concentrations of inhibitors.

The model behaviour was generally adequate in that treatment responses in milk component production were modelled in the proper direction and fairly close in accuracy (Table 1). The exception was the response to high protein diet, in that the model predicted a greater response than was observed. The model is as yet inadequate to predict the more subtle interactions between nutrient intake, stage of lactation and milk component production. Behaviour of the model to mimic rates of body protein change was fairly robust in that body protein loss did increase in the model when low-protein rations were simulated. Sensitivity was inadequate as the model consistently predicted less body protein loss, even when the initial rate of body proteolysis was increased 75% in simulations (Table 2). However, these sensitivity analyses demonstrated that the model can respond to both dietary and 'genetic' inputs in the proper direction (behaviour). When parameter values for body proteolysis were increased, the model was able to more closely approximate observations. This result makes sense, as information available when the model parameters were originally set were from literature values of lower-producing sows, of genetic lines with less body lean, nursing pigs with smaller litters.

Behaviour of the model to mimic rates of body fat change to energy intake was in the right direction, but sensitivity was inadequate (Table 2). The model did not simulate interactions between protein and energy intake and body fat loss. Observations suggested that body fat loss increases directly with body protein loss. However the model would predict less body fat loss with increasing rates of proteolysis (Table 2).

Discussion

The modelling approach has been used with success in swine research and production (Black *et al.*, 1986; Pettigrew *et al.*, 1992). The model in this study has a research objective to demonstrate where knowledge is inadequate (Baldwin, 1995). The model nevertheless demonstrates a fair robustness at the level of intake and output. The present equations are sufficient to predict responses to dietary protein if initial parameter values are estimated precisely. The model behaviour is in the right direction for body fat and protein use; however sensitivity with present parameter values is inadequate. Increasing loss of body fat with decreasing protein intake is not stated in the present model. The present experiment suggests the following: 1) current knowledge provides an adequate basis for relating milk outputs and dietary inputs; 2) knowledge on rates of proteolysis, protein synthesis and gluconeogenesis during lactation is inadequate to predict changes in body protein and fat pools; 3) an equation relating mobilization of body protein and body fat independent of kinetic or energy balance considerations (i.e. hormonal regulation) is potentially justified.

Table 1. Litter growth and milk variables of sows fed varying protein and energy.

	LoEHiP	LoEMeP	LoELoP	HiEHiP	HiEMeP	HiELoP	SE
Day 7							
Milk Protein, %	3.7	3.1	2.9	3.0	3.6	3.0	0.09
Milk Fat, %	7.5	9.2	9.1	8.3	8.2	7.9	0.34
Day 21							
Milk Protein, %	3.2	3.3	2.9	3.3	3.1	3.1	0.09
Milk Fat, %	9.7	7.2	8.7	6.6	7.7	7.2	0.38
Days 1-20							
Litter Growth, kg	51.4	54.0	48.8	50.2	52.9	49.1	0.73
Milk Production g/d	9682	10094	9264	9452	9862	9375	115
Simulated Milk, g/d	10652	10544	9766	10500	10423	9576	
Diff[1], Sim - Obs	970	450	502	1049	561	201	
TRO[2], g/d	-412	0	-830	-642	-232	-719	
TRS[3], g/d	108	0	-778	-44	-121	-968	
Difference g/d	520	0	51	598	111	-249	

[1] Diff = Difference; [2] TRO = Treatment response observed; [3] TRS = Treatment response simulated.

Table 2. Observed and simulated changes in body composition in sows fed varying protein and energy.

	LoEHiP	LoEMeP	LoELoP	HiEHiP	HiEMeP	HiELoP	SE
Protein, kg							
Observed	-1.35	-1.29	-2.05	0.29	-1.50	-2.49	0.34
Sim, +20% PbLy	0.53	0.38	0.01	0.34	0.12	-0.48	
Sim, +30% PbLy	0.21	-0.02	-0.46	0.06	-0.28	-0.98	
Sim, +50% PbLy	-0.08	-0.40	-0.91	-0.17	-0.64	-1.44	
Sim, +75% PbLy	-0.34	-0.80	-1.38	-0.37	-0.99	-1.92	
Diff, Last Sim - Obs	1.01	0.49	0.67	-0.66	0.50	0.57	
TRO, g/d[1]	-0.07	0.00	-0.76	1.58	-0.21	-1.20	
TRS, g/d[2]	0.46	0.00	-0.58	0.43	-0.20	-1.12	
Fat, kg							
Observed	1.41	-2.76	-5.03	2.79	3.51	-2.15	1.2
Sim, +20% PbLy	-1.56	-1.16	-0.30	0.37	1.00	-0.19	
Sim, +30% PbLy	-1.52	-1.07	-0.28	0.59	1.15	-0.13	
Sim, +50% PbLy	-1.39	-0.90	-0.17	0.59	1.30	-0.02	
Sim, +75% PbLy	-1.28	-0.73	-0.01	0.67	1.46	0.15	
Diff, Last Sim - Obs	-2.69	2.03	5.05	-2.05	-2.11	2.30	
TRO, g/d	4.18	0.00	-2.27	5.55	6.27	0.61	
TRS, g/d	-0.55	0.00	0.74	1.40	2.19	0.88	

[1] TRO = Treatment response observed; [2] TRS = Treatment response simulated.

References

Baldwin R.L. (1995). Modeling ruminant digestion and metabolism. Chapman & Hall, New York.

Black J.L., Campbell R.G., Williams I.H., James K.J. & Davies G.T. (1986). Simulation of energy and amino acid utilizationation in the pig. *Res. Devel. Agric.* 3:121-145.

McNamara J.P. (1995). Role and regulation of adipose tissue metabolism during lactation. *J. Nutr Biochem.* 6:120-129.

McNamara J.P., Pettigrew J.E., Baldwin R.L., Close W.H., Walker J. & Oltjen J.W. (1991). Information needed for mathematical modelling of energy metabolism. In: Energy Metabolism in Farm Animals (C. Wenk & M. Boessinger eds) *EAAP Publ. No.* 58, pp. 468-472.

Parmley K.L.S. & McNamara J.P. (1996). Lipid metabolism in adipose tissue of pigs fed varying amounts of energy. *J. Nutr.* 126:1644-1656.

Pettigrew J.E., Gill M., France J. & Close W.H. (1992). A mathematical integration of energy and amino acid metabolism. *J. Anim. Sci.* 70:3742-3761.

Noblet J. & Etienne M. (1989). Predicting milk composition in the sow. *J. Anim.* Sci. 67:3352-3359.

MODELLING TOTAL HEAT PRODUCTION OF BROILER CHICKENS AS A BASIS FOR CLIMATE CONTROL

J.-M. Aerts[1], D. Berckmans[1,3], P. Saevels[1], J. Buyse[2] and E. Decuypere[2]

[1]Laboratory for Agricultural Buildings Research, Katholieke Universiteit Leuven, Belgium, Kardinaal Mercierlaan 92, B-3001 Heverlee; [2]Laboratory for Physiology and Immunology of Domestic Animals, Katholieke Universiteit Leuven, Belgium, Kardinaal Mercierlaan 92, B-3001 Heverlee; [3] Fund for Scientific Research - Flanders, Belgium

Summary

Sixty five experiments were carried out to measure and model the dynamic response of total heat production ($W/kg^{0.75}$) of broiler chickens to variations in temperature (°C) and light (on/off) using mathematical identification techniques. Based on measurements of temperature and light from a data period of 48 hours the total heat production of the animals was predicted. It was demonstrated that by doing so, total heat production can be predicted 30 up to 75 minutes in future. The r^2 between measured and predicted total heat production ranged from 0.58 up to 0.88. By developing such modelling techniques the following step is to develop a model based controller.
Key words: Broiler chickens, climate control, heat production, temperature, dynamic model.

Introduction

An important step in the development of more optimal climate controllers for animal production houses, would be to measure and feed back relevant responses of the animal to its varying environment and to apply modern control theory to this process.

To control processes in a more optimal way using modern control theory, two conditions must be fulfilled: 1) there has to be a feedback of the process output to control and 2) a mathematical model must be available describing the dynamic response of the process output to variations in the process inputs. The success of this approach is demonstrated in a lot of applications. In most bioprocesses however, these conditions of modern control theory are not yet fulfilled.

The objective of the research reported was to investigate if the dynamic response of total heat production (latent and sensible heat production, $W/kg^{0.75}$) of broiler chickens to variations in temperature (°C) and light (lux) can be modelled using a simple dynamic black box model. The use of inside temperature and light as control inputs is widely applicable in most of modern livestock buildings.

Materials and Methods

Experimental animals. Broiler chickens (Ross, male) of 1 to 14 days old were used during the experiments. The animals were fed *ad libitum* with a commercial meal containing 13.4 MJ/kg metabolizable energy and 220 g crude protein/kg.

Dynamic measurement of total heat production. This measurement was done by using indirect calorimetry. The installation is described more in detail by Buyse *et al.* (1997). The temperature in the respiration chamber was measured with a Pt-100 (accuracy of 0.1°C), placed at the outlet side. The light (light on) in the chamber was 2.5 lux. The measurement installation has been set up to make dynamic measurements of the total heat production (relative measuring error of 5.6%). Experiments can be carried out to measure the step response of total heat production to inside temperature with a time interval of 3 up to 7.5 minutes depending on the type of experiment.

Mathematical modelling. Since the objective of this research was to develop a model of a system for future control purposes, it was not required to develop a model based on physical or physiological insights of the system.

A simple input-output ARX (Auto Regressive with eXogenous variables) black box model was used (Ljung, 1987) with the following model structure:

$$HP(t) = a.HP(t-1) + b1.T(t-1) + b2.LI(t-1) + e(t) \quad \text{(Eq. 1)}$$

where $HP(t)$ = total heat production at time t (W/kg),
$T(t-1)$ = temperature at time t-1(°C)(time delay of 1 sample),
$Li(t-1)$ = light intensity at time t-1(lux)(time delay of 1 sample),
a, b1, b2 = model parameters.

The unknown model parameters (a, b1, b2 in Eq. 1) of the ARX model structure were estimated using a least squares approach (Ljung, 1993) based on all the data of the measurement sets. Measurement sets of 48 hours were used because, over longer periods, the response of the animals will change due to growth of the animal, possible occurrence of infections, etc. and consequently the model parameters cannot be assumed to be constant.

The resulting models were evaluated by examining the r^2 between measured and predicted total heat production.

Experiments. Sixty-five experiments, with a measuring period of 24 or 48 hours depending on the experiment, were carried out with 1 to 8 broiler chickens depending on the age and the weight of the chickens. Three different kinds of experiment were carried out (Table 1).

Table 1. Overview of the three different types of experiments.

Experiment	Number of experiments	Light		Temperature	
Type I	15	on		35°C 18°C	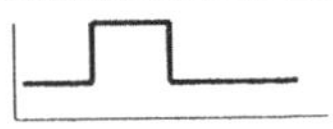
Type II	20	on off	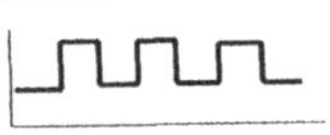	25°C	
Type III	30	on off	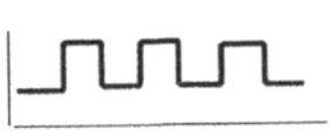	35°C 18°C	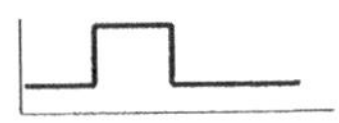

Results

For all the experiments the 10-step ahead predictions were calculated (prediction horizon of 30 up to 75 minutes depending on the sample interval).

In Fig. 1 results are shown for the response of total heat production to a combination of temperature and light (on/off) (type III experiment).

For experiments of type I r^2 values ranged from 0.58 up to 0.80, for type II experiments from 0.70 up to 0.86 and for type III experiments from 0.74 up to 0.88.

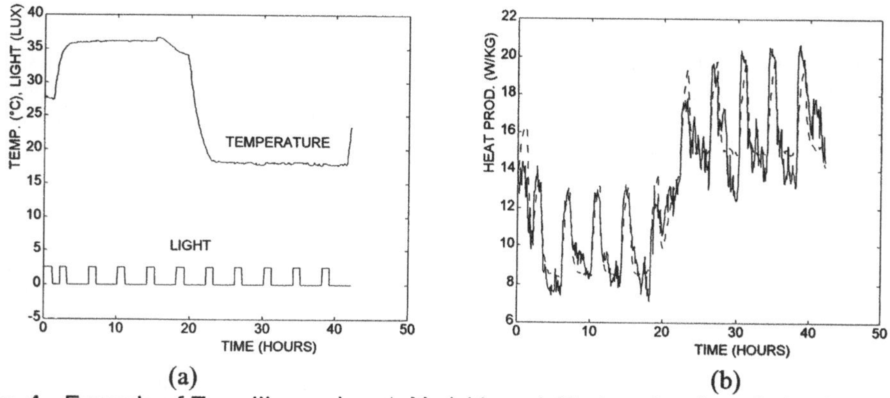

Fig. 1. Example of Type III experiment: Model-based 10-step ahead prediction (solid line: measured, dotted line: predicted) of the response of total heat production of broiler chickens (1b) to variations in the process inputs temperature and light (on/off) (1a).

Discussion

Most references in the literature describe models of static responses of heat production (Richards, 1976; Jurkschat *et al.*, 1986; Kettlewell and Moran, 1992). In this research, the total heat production can be predicted 30 up to 75 minutes in the future based on the measured value of temperature and light (on/off) one time instant ago using black box models (Aerts *et al.*, 1996).

The accuracy of the black box model still could be ameliorated by taking into account the difference in dynamics between responses to temperature and light. In the model used it is assumed that there is no difference in dynamics (only one "a" parameter in Eq. 1). Also a difference could be made between dynamics of responses to an increase or a decrease of temperature or light. Especially for light, the response of total heat production to light going on is faster than to light going off (see Fig. 1).

For longer periods (> 48 hours) the time variant character of the response of total heat production will become more important. One way to solve this problem is to estimate the model parameters in a recursive way during the production process. In this way, the model is adapted continuously to the time varying process dynamics.

Based on such compact black box models, it might become possible to control the total heat production during the production process by means of the control inputs temperature and light (on/off).

Conclusions

Sixty-five experiments were carried out to measure the dynamic step response of the total heat production of broiler chickens to variations in temperature and light (on/off). Although this is a complex biological process, the response of the total heat production of broiler chickens to variations in temperature and light can be predicted with a simple ARX model structure with constant model parameters over a 48 hour period when used in a recursive way with a prediction horizon of 10 time intervals (30 up to 75 minutes). The r^2 values for the 10-step ahead prediction of total heat production ranged from 0.58 to 0.80 for responses to temperature, from 0.70 to 0.86 for responses to light (on/off) and 0.74 to 0.88 for responses to both temperature and light (on/off).

Other experiments should be performed (smaller variations of input variables, longer duration) for a more extended evaluation of the suitability of this modelling approach for developing control algorithms to control the total heat production of broiler chickens during the production process with the control inputs air temperature and light (on/off). It might be possible - based on the combination of hardware, software, sensor and modelling technology - to feed back continuously information from the animals during the production process and use this information for dynamic modelling and further more model-based control purposes of the biological production process.

References

Aerts J.M., Berckmans D. & Schurmans B. (1996). Modelling heat production and heart rate of broiler chickens as a basis for a more optimal climate control. Paper No. 964107, 1996. ASAE Annual International Meeting, July 1996, Phoenix, Arizona.

Buyse J., Michels H., Vloeberghs J., Saevels P., Aerts J.-M., Ducro B., Berckmans D. & Decuypere E. (1997). Energy and protein metabolism between 3 and 6 weeks of age of male broiler chickens selected for growth rate or for improved food efficiency. *Br. Poult. Sci.* (in press).

Jurkschat M., Burmeister, A., Nichelmann M. & Thomas E. (1986). A mathematical equation to describe the influence of age, wind speed and ambient temperature on heat production in turkeys (Meleagris Gallopavo). *J. Therm. Biol.*, 11:105-108.

Kettlewell P.J. & Moran P. (1992). A study of heat production and heat loss in crated broiler chickens: a mathematical model for a single bird. *Br. Poult. Sci.*, 33:239-252.

Ljung L. (1987). System Identification : theory for the user. Prentice Hall, New Jersey, pp. 519.

Ljung L. (1993). System Identification Toolbox, User's Guide, The Math Works Inc., pp. 204.

Richards S.A. (1976). Evaporative water loss in domestic fowls and its partition in relation to ambient temperature. *J. agric. Sci.*, 87:527-532.

SUMMARY OF THE DISCUSSION

Discussion leader: G.C. Emmans
Scientific secretaries: H. Freetly & T. Yan

MacLeod was asked how much information is required about a genotype in order to use it in the model. He responded that the protein and fat content and the amount of feathers were the primary inputs for genotype. He was further questioned if ranging his parameter estimates by ±25% for his sensitivity analyses matched biological ranges. MacLeod responded that the typical C.V. were 10% which placed the range of ±25% within the biological range. MacLeod was asked how the model had been validated. He responded that experiments using calorimetry were used where levels of protein, energy, carbohydrate, and fat were varied. In addition, literature values were used. Asked why he had chosen to estimate maintenance as a function of body weight rather than of protein mass, he pointed out that bird weight was more easily obtained than protein mass and therefore functionally made a better input for the model. The author was complimented on choosing to test internal parameters of the model rather than final model output, thus avoiding the possibility of offsetting errors giving reasonable solutions.

In the paper of Boekholt & Schreurs it was noted that the relative pattern of protein and fat accretion with increased energy intake is similar to that seen in cattle. It was observed that this general relationship holds across species as long as the diet remains balanced. It was observed that, if fat accretion is regressed on energy intake, at zero fat accretion the birds are still gaining energy. The authors were questioned as to how cell growth could occur if there was not any fat accretion. The response was that there was no net change in fat even though birds were gaining in other tissues. The point was raised that determining energy retention as a function of fat content and protein content results in autocorrelations between the estimates. It was pointed out that determination of energy content with bomb calorimetry results in an independent measurement of energy retention. It was observed that the absolute amount of energy partitioned into protein can change but the general relationship between fat and protein is conserved with increased feed intake. Furthermore, calculating carbohydrate accretion by difference results in all measurement error being folded into this estimate.

In the general discussion the point was made that, in model development, when we select our hypothesis we must ask questions with regard to the biology and not narrow the hypothesis to interpreting the data from the last experiment conducted. Concern was expressed that, at lower levels of model aggregation, mechanistic models sometimes have elements that are incorrect in theory with respect to the biology. It was pointed out that this is not unique to mechanistic models and that the theory of maintenance can be criticized in a similar manner; however, if maintenance and k_g values are used from a single data set and not mixed across data sets they are useful tools.

In relation to the work of van Milgen *et al.* it was noted that the model inputs for the estimation of heat production were anatomical in nature rather than chemical. The author reminded the meeting that the coefficients associated with anatomical inputs of muscle, viscera, and fat were statistical fits rather than biological arguments.

Discussion of the paper of Aerts *et al.* centred on concern that the manner in which the black box model operates is not explicit to the user. It was explained that the power of the black box approach is that, if the user knows how to control the inputs then the outputs can be controlled. It was further noted that these approaches allow for dynamic models as opposed to static models.

SESSION 8

FEED EVALUATION AND DIETARY ASPECTS

ILEALLY DIGESTIBLE PEA STARCH RESULTS IN A LOWER ENERGY GAIN IN GROWING PIGS COMPARED TO ILEALLY DIGESTIBLE MAIZE STARCH

G.C.M. Bakker[1], R.A. Dekker[1], H. Everts[2], J. van der Meulen[1] and B. Smits[1]

[1]ID-DLO, Nutrition of Pigs and Poultry, P.O.Box 65, 8200 AB Lelystad, The Netherlands; [2]present address: University of Utrecht, Faculty of Veterinary Medicine, P.O. Box 80152, 3508 TD Utrecht, The Netherlands

Summary

Two different sources of starch were compared regarding the energy retained in growing pigs in relation to digestibility and portal appearance of nutrients. Maize starch and pea starch (650 g/kg diet) were selected because of their different rate of *in-vitro* degradation. Results showed that, at the terminal ileum of growing pigs, both starches were nearly completely digested (on average 98%). However, retained energy was 15% higher in pigs consuming the maize starch diet. The ratio of fat to protein retention in the energy retention, was similar for both treatments. Rate of digestion of the maize starch diet was faster than with the pea starch diet. Results from pigs with blood catheters showed that the glucose appearance in the portal vein was 32% higher in pigs receiving the maize starch diet. Additional calculations could not reveal the mechanism behind the lower energy supply by pea starch. It is concluded that the efficiency of utilization of the energy from ileally digested starch depends on the origin of the starch.
Key words: Starch, pigs, energy retention, digestibility, portal flux.

Introduction

In pigs, most sources of starch are nearly completely digested prior to the terminal ileum. Therefore, in most energy evaluation systems it is assumed that ileally digested starch is absorbed and utilized with a fixed efficiency. However, the *in-vitro* degradation rate of starch can be very different between starch sources (Cone, 1991). In addition, differences are found in the *in-vivo* rate of digestion in the small intestine (Everts *et al.*, 1996), the rate of appearance of glucose in the portal vein and the net portal-drained viscera (PDV) flux of glucose (van der Meulen *et al.*, 1996). Both the degradation rate of starch *in-vitro* and *in-vivo* showed a good relationship with the portal appearance of glucose in pigs (Smits *et al.*, 1997; van der Meulen and Smits, 1997). This information leads to the hypothesis that starch sources with a different degradation rate may result in different amounts of energy retained by pigs. The hypothesis was investigated in growing pigs receiving diets with two different starch sources by a) energy balance; b) ileal digestibility and the precaecal rate of digestion in parts of the small intestine; c) PDV flux of nutrients in the vena porta. Results from these trials were used to identify stepwise the effects of two treatments in an attempt to clarify the mechanism.

Materials and Methods

<u>Diets and feeding:</u> A basal diet was formulated to supply all the required amino acids, minerals and vitamins. The digestibility of nutrients and energy of this basal diet was determined in a preliminary trial. The two experimental diets contained 350 g/kg of this basal diet and 650 g/kg starch; either maize starch (diet M) or pea starch (diet P). Starch sources were selected on difference in rate of *in-vitro* degradation, maize starch being degraded faster. The feeding level was 2 x maintenance (870 kJ DE / W0.75) and the daily

rations were offered in two equal portions. The daily water allowance was added to the diets (4.5 l per kg diet).

Energy balance experiment: Energy, nitrogen and carbon balances were measured using indirect calorimetry in an open circuit chamber. Retained energy was calculated as the mean of the energy (ME minus heat) and the carbon/nitrogen balances. Measurements were performed at 68 kg liveweight (W) using a pair of castrates in one chamber. Experimental diets were fed from 45 kg W onwards. Per treatment five replicates were performed. Gas exchange was measured during two periods of 48 hours per balance period of ten days. One week prior to the start of the balance period the pigs were placed in metabolism crates.

Ileal digestibility experiment: Ileal digestibility of nutrients in the two diets was measured in a cross-over design with four SICV-cannulated castrates (Mroz *et al.*, 1996) averaging, 60 kg W. Cr_2O_3 was used as a marker. Pigs were fed the experimental diets at least 2 weeks prior to measurements.

Slaughter experiment: The twelve castrates in this experiment were previously used in the energy balance experiment. They received the two diets with added Cr_2O_3 as a marker for another 14 d. Pigs were killed 3.5 h after feeding. Immediately, samples were collected from the digesta in the stomach and in each third part of the small intestine. The samples were analysed for starch and its digestibility up to the different parts of the gastrointestinal tract was calculated using the marker.

Flux experiment: Portal and arterial plasma concentrations of glucose, amino acids, lactic acid and volatile fatty acids were measured during 12 h after feeding in a three period, cross-over design using four gilts. Para-amino hippuric acid was used to determine portal plasma flow. Net PDV flux was calculated from the difference between portal and arterial plasma concentrations and the corresponding portal vein plasma flow. Average W during the measurements was 42 kg.

Results

Energy balance experiment: The intended DE intake of 870 kJ per $W^{0.75}$ was achieved with both diets (Table 1). It was slightly lower with diet P due to an accumulation of small differences in energy content in the diet, feed intake and energy digestibility. Retained energy (RE) was lower for diet P (Table 1). The composition of the RE, however, was identical with both diets; the ratio of retained fat to retained protein was 2.0 (Table 1).

Table 1. Energy intake (DE, ME) and retention (RE) (all in kJ per $W^{0.75}$) in growing pigs of 68 kg W.

	Diet M	Diet P	s.e.d.[1]	F-prob[2]
DE	879	864	7.7	NS
ME	846	828	7.3	*
Retained energy (RE)	250	212	9.2	**
RE as protein (REp)	85	72	9.0	NS
RE as fat (REf)	165	140	10.3	*
REf/REp	2.0	2.0	0.97	NS

[1]standard error of difference

[2]F-probability: NS: P>0.05; *: P<0.05; **: P<0.01.

Ileal digestibility experiment: Ileal digestibility of starch was not different between diets (0.983 for diet M and 0.979 for diet P; s.e.d. 0.004). Ileal protein digestibility was significantly higher on diet M (0.777) than on diet P (0.754; s.e.d. 0.003). At the terminal ileum the concentration and pattern of volatile fatty acids (VFA) in the digesta were similar on both diets.

Slaughter experiment: The rate of digestion of maize starch was faster; starch digestibility at the beginning of the small intestine was higher for diet M than for diet P. At the end of the small intestine, however, it was similar for both diets (diet M 0.99 and diet P 0.98). Pea starch reduced ileal digestibility of protein (0.78 vs. 0.75) and some amino acids. In all three sections of the small intestine concentrations and pattern of VFA in the digesta were similar for both diets. No differences in either weight or length of the stomach and small intestine were found between treatments.

Flux experiments: Compared with diet P, net PDV flux of glucose from diet M after feeding was higher in the first 6 h, not different up to 8 h after feeding and lower thereafter. The total amount of glucose available from diet M was significantly higher than from diet P (Table 2), and accounted for 97 and 72%, respectively, of ileally digested starch. PDV flux of amino acids from diet M tended to be higher 2 to 4 h after feeding and thereafter lower than the amino acids from diet P. The total amount of amino acids available from diet M was lower than from diet P, although this difference was only significant for the sum of essential amino acids and not for the sum of non-essential amino acids (Table 2). Total portal appearance of lactic acid was not different between the diets, but portal appearance of VFA was higher with the P diet (Table 2).

Table 2. PDV plasma flux of glucose, essential and non-essential amino acids, lactic acid and volatile fatty acids in kJ/kg $W^{0.75}$ per 12 h.

	Diet M	Diet P	s.e.d.[1]	F-prob.[2]
Glucose	309	211	28.1	*
Essential amino acids	31	37	2.3	*
Non-essential amino acids	11	14	3.7	NS
Lactic acid	3	5	3.3	NS
Volatile fatty acids	10	13	1.2	*

[1]standard error of difference

[2]F-probability: NS: P>0.05; *: P<0.05.

Discussion

In all experiments described, diets were identical except for the source of starch (maize starch and pea starch, respectively). Both starches had disappeared at the terminal ileum. According to the present energy evaluation systems this would result in similar predicted energy gain. However, pigs receiving diet P retained 38 kJ/$W^{0.75}$ (15%) less energy than those receiving diet M (Table 1), of which approximately 10 kJ could be explained by a lower DE intake. The results of the slaughter experiment revealed that the pea starch was absorbed more distal in the gastrointestinal tract. As a result, the glucose peak in plasma after a meal with diet P was also later. In addition, the total amount of glucose appearing in the plasma was lower with diet P (Table 2).

Speculation about the mechanism causing the reduced glucose supply from diet P includes at least three possibilities. The first one is that a relatively large proportion of pea starch had been fermented in the stomach and/or small intestine. This was indicated by the higher PDV fluxes of VFA and lactic acid with diet P (Table 2). However, the difference seemed too low to confirm this hypothesis of large scale fermentation (Table 2). Moreover, the concentrations and pattern of VFA in the digesta of the experimental pigs did not suggest differences in fermentation between the treatments. On the other hand, utilization of VFA by the intestinal wall and microbial conversion of microbial mass more than once would underestimate the production of VFA using PDV flux (Imoto and Namioka, 1978 a, b). A second explanation might be a reduced efficiency for energy gain. However, this hypothesis is unlikely because of the similar composition of energy gain with both diets (Table 1). A third explanation might be an increased use of energy in the digestion process with diet P. This includes an increased workload of the gastrointestinal tract due to the slower rate of digestion of the P diet, and the potentially higher secretion of endogenous protein, which is supported by the reduced ileal digestion of protein. On the basis of this third hypothesis, the gastrointestinal tract would account for 20% of maintenance requirement with diet M and for 52% with diet P (Table 3). The 20% with diet M corroborates with data from Pekas (1991). The high value with diet P, however, seems unrealistic. In these calculations, however, an error may have been introduced by extrapolating the data in Table 2 to 24 h and to pigs of different sex and live weight. In conclusion, various starch sources with a similar ileal digestibility may be different in rate of digestion in the gastrointestinal tract, supply of nutrients in the portal vein and energy gain by the animal. The mechanism behind this is not clear.

Table 3. The calculated amount of energy used for maintenance, and the loss of energy in the gastrointestinal tract, in $kJ/W^{0.75}/d$.

	Diet M	Diet P
ME for maintenance[1]	421	468
Total ME (Table 1)	846	828
Energy in portal blood[2]	728	560
Energy from digestible fat	32	26
Loss in gastrointestinal tract	86	242

[1] Calculated from Table 1, with $k_p = 0.45$ and $k_f = 0.7$
[2] Calculated by extrapolating the data in Table 2 from 12 to 24 h.

References

Cone J.W. (1991). Degradation of starch in feed concentrates by enzymes, rumen fluid and rumen enzymes. *J. Sci. Food Agric.* 54: 23-34.

Everts H., Dekker R.A., Smits B & Cone J.W. (1996). Digestion of maize starch and native pea starch in the small intestine of pigs. *Proc. Nutr. Soc.* 55:59A.

Imoto S. & Namioka S. (1978a). VFA metabolism in the pig. *J. Anim. Sci.* 47: 479-487.

Imoto S. & Namioka S. (1978b). VFA production in the pig large intestine. *J. Anim. Sci.* 47: 467-478.

Meulen J. van der, Bakker J.G.M., Smits B. & De Visser H. (1996). Portal appearance of glucose and amino acids in the pig after feeding maize starch and native pea starch. *Proc. Nutr. Soc.* 55: 60A.

Meulen J. van der & Smits B. (1997). Digestion and portal appearance of glucose of different sources non-resistant starch. In: J.P. Laplace, C. Février, & A. Barbeau (Eds.) Digestive physiology in pigs. INRA, Paris, France. pp. 279-282.

Mroz Z., Bakker G.C.M., Jongbloed A.W., Dekker R.A., Jongbloed R. & van Beers A. (1996). Apparent digestibility of nutrients in diets with different energy density, as estimated by direct and marker methods in pigs with or without ileo-cecal cannulas. *J. Anim. Sci.* 74:403-412.

Pekas J.C. (1991). Specific heat output (SHO) of visceral organs/tissues: the hypothesis and first estimations. *Growth, Development and Aging*, 55: 67-80.

Smits B., Cone J.W. & Meulen J. van der (1997). Prediction of net portal glucose flux by *in-vitro* digestion or *in-vivo* disappearance from stomach and small intestine of different sources of starch. In: J.P. Laplace, C. Février, & A. Barbeau (Eds.) Digestive physiology in pigs. INRA, Paris, France. pp. 417-420.

ESTIMATION OF MAINTENANCE ENERGY REQUIREMENTS AND EFFICIENCY OF UTILIZATION OF METABOLIZABLE ENERGY FOR GROWTH IN LAMBS BY INDIRECT CALORIMETRY AND COMPARATIVE SLAUGHTER

R.W.J. Steen, L.E.R. Dawson, D.E. Kirkpatrick and S.D. Johnston

Agricultural Research Institute of Northern Ireland, Hillsborough, Co. Down BT26 6DR, UK

Summary

Two experiments have been carried out in which energy retention (ER) in lambs was estimated by indirect calorimetry and comparative slaughter, to enable a comparison to be made of the estimates of maintenance requirements (ME_m) by regression and the efficiency of utilization of metabolizable energy (ME) for growth (kg) by the two methods. In the first experiment, 120 Blackface lambs were offered grass silage (730 g DOM/kg DM) alone or with varying proportions (0.21 to 0.82 of total DMI) of cereal-based concentrates. In the second experiment 72 Texel cross lambs were offered grass or grass silage (706 and 720 g DOM/kg DM respectively) from the same sward alone or with concentrates constituting 25 or 50% of total DMI. Forty estimates of ME intake and ER were made by indirect calorimetry, each involving a group of six lambs, and ER was estimated for all lambs by comparative slaughter. For all diets and the mixed diets only respectively, the best fit relationships between ME intake and ER for the data produced by comparative slaughter (R^2 0.88 and 0.89) produced estimates of ME_m of 0.49 and 0.49 MJ/kg $W^{0.75}$ and for k_g of 0.47 (se 0.016) and 0.47 (se 0.016). For all diets and the mixed diets only respectively, the best fit relationships for data produced by indirect calorimetry (R^2 0.87 and 0.88) produced estimates of ME_m of 0.46 and 0.47 MJ/kg $W^{0.75}$ and k_g of 0.51 (se 0.032) and 0.52 (se 0.035).

Key words: Lambs, indirect calorimetry, comparative slaughter, maintenance.

Introduction

Maintenance energy requirements of ruminants can be estimated either by measuring fasting heat production or from a regression relationship between metabolizable energy (ME) intake and energy retention (ER) for a number of animals given a range of ME intakes. Energy retention has frequently been estimated by both calorimetry and the comparative slaughter technique. The results of a number of studies have suggested that grass silage-based diets have a lower efficiency of utilization of ME for growth than that generally reported for other diets (e.g. Thomas *et al.*, 1988). However, the magnitude of calculated values for the efficiency of utilization of ME depend greatly on the assumed maintenance energy requirements of the animals. The results of a series of recent studies using indirect calorimetry and involving dairy cows (Yan *et al.*, 1997) beef cattle and sheep (Steen *et al.*, 1997) have indicated that the maintenance energy requirements of cattle and sheep given grass silage-based diets are higher than those given in current feeding standards in the United Kingdom (e.g. ARC, 1980; AFRC, 1990). Two experiments have been carried out in which ER in lambs was estimated by indirect calorimetry and by comparative slaughter, to enable a comparison to be made of the estimates of maintenance requirements and efficiency of utilization of energy for growth by indirect calorimetry to be compared with those made by comparative slaughter.

Materials and Methods

In the first experiment, 120 Blackface lambs which were initially 29 kg live weight were offered:

(1) high digestibility grass silage (730 g digestible organic matter/kg dry matter) *ad libitum* as the sole feed.
(2) 200 g silage dry matter (DM) plus 200 g concentrate DM/day.
(3) 200 g silage DM plus concentrates *ad libitum*.
(4) 200 g concentrate DM plus silage *ad libitum*.

Metabolizable energy intakes were 8.8, 5.1, 17.9 and 11.9 MJ/day for diets 1 to 4 respectively. Heat production of 96 lambs, 24 per diet, were determined by indirect calorimetry during the 112 day feeding period and all 120 lambs were slaughtered at the end of the experiment plus a further 15 lambs initially to enable energy retention to be estimated by comparative slaughter as described by Johnston and Steen (1997). In the second experiment, 72 Texel x Greyface (Blackface x Border Leicester) lambs which were initially 34 kg live weight were given grass (G) or grass silage (S) made from the same sward (706 and 720 g digestible organic matter/kg DM respectively) as the sole feed or supplemented with concentrates, constituting either 25%, G25, S25 or 50%, G50, S50 of total DM intake. Heat production of all 72 lambs was determined twice by indirect calorimetry during the 120 day feeding period and the 72 lambs were slaughtered at the end of the experiment plus an additional 12 lambs initially to enable energy retention to be estimated by comparative slaughter. Metabolizable energy intakes were 10.6, 9.6, 13.3, 13.0, 13.5 and 12.6 MJ/day respectively for the animals given G, S, G25, S25, G50, S50 respectively. In both experiments, each estimate of ME intake, heat production and ER by calorimetry involved a group of six lambs, giving a total of 40 estimates over the two experiments.

Best fit relationships between ME intake and ER have been developed, the data having been adjusted for differences between experiments. Separate relationships have been developed for data produced by calorimetry and by comparative slaughter and for all diets and the mixed diets only (i.e. excluding the forage only diets).

Results

The best fit relationships between ME intake and ER for the data produced by comparative slaughter were as follows:

(1) For all diets

$$ER = 0.465\ (se\ 0.0156)\ ME\ intake - 0.230\ (se\ 0.0122) \qquad (R^2 = 0.88;\ P<0.001)$$

This relationship gives estimates of maintenance energy requirements and efficiency of utilization of ME for growth of 0.495 MJ/kg $W^{0.75}$ and 0.465 respectively.

(2) For the mixed diets only

$$ER = 0.465\ (se\ 0.0164)\ ME\ intake - 0.229\ (se\ 0.0132) \qquad (R^2 = 0.89;\ P<0.001)$$

This relationship gives estimates of maintenance requirements and efficiency of utilization of ME for growth of 0.492 MJ/kg $W^{0.75}$ and 0.465 respectively.

The best fit relationships between ME intake and ER for the data produced by indirect calorimetry were:

(3) For all diets

$$ER = 0.513 \text{ (se } 0.0319) \text{ ME intake} - 0.237 \text{ (se } 0.0247) \qquad (R^2 = 0.87; P<0.001)$$

This relationship gives estimates of maintenance requirements and efficiency of utilization of ME for growth of 0.462 MJ/kg $W^{0.75}$ and 0.513 respectively.

(4) For the mixed diets only

$$ER = 0.516 \text{ (se } 0.0346) \text{ ME intake} - 0.241 \text{ (se } 0.0274) \qquad (R^2 = 0.88; P<0.001)$$

This relationship gives estimates of maintenance requirements and efficiency of utilization of ME for growth of 0.468 MJ/kg $W^{0.75}$ and 0.516 respectively.

Discussion

In these studies estimates of maintenance energy requirements from data produced by calorimetry (i.e. 0.46 and 0.47 MJ/kg $W^{0.75}$ for all diets and the mixed diets only respectively) were close to those estimated from data produced by comparative slaughter (i.e. 0.49 MJ/kg $W^{0.75}$ for all diets and the mixed diets only). Similarly the estimates of the efficiency of utilization of ME for growth from data produced by calorimetry (i.e. 0.51 and 0.52) were close to those estimated using data produced by comparative slaughter (i.e. 0.47). Thus, these data would support the use of either indirect calorimetry or slaughter independently as methods of measuring energy retention in ruminant animals. The estimates of maintenance energy requirements in these studies, by both calorimetry and comparative slaughter are close to those reported by Steen *et al.* (1997) for lambs but are approximately 30% higher than estimates given by ARC (1980) for the animals and diets used in these studies. However, the values for efficiency of utilization of ME for growth in these studies are close to those predicted by ARC (1980) and AFRC (1990) for these diets. This would suggest that the lower performance of animals given grass silage-based diets in some earlier studies may have been due more to the animals used in these studies having a higher maintenance energy requirement than that given by current feeding standards, than to a lower efficiency of utilization of ME for growth.

The higher maintenance requirements reported in this study, and by Steen *et al.* (1997) for cattle, compared with those given by ARC (1980) may have been due to animals given forage, especially grass silage, having a higher maintenance requirement than those given other diets and secondly to animals offered food *ad libitum* having a higher maintenance requirement than those offered restricted food intakes, as estimates of maintenance requirements given by ARC (1980) are based largely on measurement of fasting metabolism following periods of restricted food intakes.

It is concluded that:

(a) indirect calorimetry and comparative slaughter produced similar estimates of maintenance energy requirements and efficiencies of utilization of ME,
(b) that estimates of maintenance requirements were about 30% higher than those used in current feeding standards and hence that estimates of efficiency of utilization of ME are higher than those calculated for silage-based diets in other recent studies.

References

Agricultural and Food Research Council (AFRC) (1990). Technical Committee on Responses to Nutrients, Report No. 5. Nutrient Requirements of Ruminant Animals : Energy. *Nutr. Abs. Revs.* Series B, 60: 729-804.

Agricultural Research Council (ARC) (1980). Nutrient Requirements of Ruminant Livestock. Commonwealth Agricultural Bureaux, Slough, UK.

Johnston S.D. & Steen R.W.J. (1997). Responses in the growth of body components of finishing lambs to additional metabolizable energy supplied from either grass silage or concentrates. (Submitted to Animal Science).

Steen R.W.J., Dawson L.E.R., Lavery N., Kirkpatrick D.E. & Johnston S.D. (1997). Estimation of maintenance energy requirements of beef cattle and lambs. Proceedings of the British Society of Animal Science, Annual Meeting, p. 118.

Thomas C., Gibbs B.G., Beever D.E. & Thurnham B.R. (1988). The effect of date of cut and barley substitution on gain and on the efficiency of utilization of grass silage by growing cattle. *Br. J. Nutr.* 60:297-306.

Yan T., Gordon F.J., Agnew R.E., Porter M.G. & Patterson D.C. (1997). The metabolizable energy requirements for maintenance and the efficiency of utilization of metabolizable energy for lactation by dairy cows offered grass silage-based diets. *Livest. Prod. Sci.* (In Press).

PREDICTING NET ENERGY OF FEEDSTUFFS FOR BEEF CATTLE

E.M. Larson and D.E. Johnson

Metabolic Laboratory, Colorado State University, Fort Collins, CO, USA

Summary

Relationships between feed net energy for gain (NEg) and diet metabolizability (ME) and chemical composition were evaluated. Literature data (512 observations) measuring net energy available for gain (NEg) were compiled (cattle and sheep, slaughter balance and respiration calorimetry). Calorimetry resulted in greater NEg values at equal ME (MJ/kg) compared to slaughter balance data. Dietary nutrient composition effects on feed NEg content were evaluated by multiple regression. The best model for calorimetry data included NDF and fat adjusters in addition to ME concentration (Neg=0.43*ME-0.04*NDF +0.14*fat+1.72). As dietary NDF increased, efficiency (kr) declined (kr=-0.004*NDF+2.80). To test the hypothesis that composition of ME affects efficiency of ME use for gain, a sheep calorimetry trial was conducted using four diets (sudex hay, alfalfa hay, sudex+40% corn, and alfalfa+40% corn). Even though ME content indicated that the sudex diets would contain more NEg than the alfalfa diets, efficiency of ME utilization was not different between the hay diets, resulting in similar NEg content (4.85 MJ/kg). Dietary NDF (51.6, 45.3, 34.1, and 31.2%; sudex, alfalfa, sudex+corn, and alfalfa+corn) was linearly related to kr (kr=-0.004*NDF+2.91). These data demonstrate the need for net energy systems to account for composition of ME when predicting efficiency of ME use for gain.
Key words: Net energy, prediction, feedstuffs, cattle.

Introduction

Current energy systems (ARC, 1980; NRC, 1996) use feedstuff digestible energy (DE), commonly measured at maintenance feed intake, to predict metabolizable energy (ME) and ME to predict net energy for maintenance (NEm) and gain (NEg). Literature data were compiled to evaluate the relationship between ME and NEg. A respiration calorimetry trial was conducted to evaluate the efficiency of ME use for gain of diets with similar ME concentrations but varying nutrient composition.

Materials and Methods

Literature data measuring NEg were compiled totalling 84 trials with 512 observations (treatment means). Experiments included two techniques and two species to determine NEg values (calorimetry: cattle, n = 28, sheep, n = 27; and slaughter balance: cattle, n = 27, sheep, n = 2. Net energy for gain was regressed against ME to evaluate and compare prediction accuracy. Alternative predictions were examined by multiple regression relating NEg to dietary nutrient composition (NDF, NSC, CP, and fat) in addition to ME.

To test the hypothesis that composition of ME affects efficiency of ME use for gain, twelve wethers (45 kg BW) were used in a balanced incomplete block design with four diets (sorghum-sudangrass hay, alfalfa hay, sorghum-sudangrass+40% corn, and alfalfa+40% corn; Table 2) offered at two levels of intake (maintenance and *ad libitum*). Energy utilization of the diets was measured via respiration calorimetry.

Results

Literature data represented intakes ranging from below to over 3X maintenance. Digestibilities of the diets ranged from 38 to 95% of GE (mean = 70 ± 8%). Metabolizability ranged from 31 to 84% of GE (mean = 58±8%) and kr ranged from 0.16 to 0.83 (mean = 0.46±0.12). Net energy for gain ranged from 1.26 to 10.13 MJ/kg (mean = 5.10±1.67).

NEg:ME When NEg (MJ/kg) was regressed against ME (MJ/kg), species (cattle vs. sheep) had minimal effect (P=0.14) on feed NEg determinations; therefore, cattle and sheep data were pooled. However, technique used to determine NEg affected (P<0.05) the regression. Calorimetry (NEg = 0.85*ME -3.77; n = 162; RSD = 0.25; R^2 = 0.61) resulted in greater NEg values at equal ME compared to slaughter balance (NEg = 0.59*ME-2.18; n = 76; RSD = 0.25; R^2 = 0.41). For low quality diets, however, NEg was approximately equal. No evidence for quadratic or cubic functions of ME were evident (P>0.10). Metabolizable energy explained only 41 and 61% of the variation in NEg determined by slaughter balance and calorimetry, respectively.

Calorimetry trials reporting dietary NDF, NSC (non-structural carbohydrates = 100 - ash - CP - NDF), fat and CP were used to evaluate the effects of nutrient composition on NEg values (Table 1). It should be noted that this evaluation was with a limited data set, mostly forages, no by-products and no slaughter balance information. Non-structural carbohydrates and CP coefficients did not significantly affect NEg (P>0.05). Equation 3 (Table 1) provided the best fit of the data. Note the negative coefficient associated with fibre (NDF) and the positive coefficient of fat. Increasing dietary NDF content reduced efficiency of ME use (kr = -0.004*NDF+2.80; n = 38; R^2 = 0.41). A similar relationship (kr= 0.004*NDF+2.91; n = 4; R^2 = 0.99) was found in the sheep calorimetry trial.

Table 1. Regression equations to predict NEg from dietary composition.[ab]

	Coefficients					
Equation no.	ME	NDF	Fat	Intercept	RSD	Adj R^2
1	0.72			-2.87	0.71	0.73
2	0.59	-0.03		0.31	0.62	0.79
3	0.43	-0.04	0.14	1.72	0.60	0.81
4		-0.08		8.67	1.02	0.44
5		-0.07	0.34	6.69	0.72	0.72

[a]NSC and CP did not enter model using stepwise selection (P>0.05).
[b]n = 38.

Dietary NDF ranged from 520 g/kg (diet DM) for sudex to 310 g/kg for alfalfa+corn (Table 2). Sudex contained 19% more NDF than alfalfa, the primary difference being hemicellulose. Cellulose content was similar between the hays but lignin was twice as high for alfalfa (75 g/kg) compared to sudex (37 g/kg). Metabolizable energy was greater (P<0.05) for diets containing sudex compared to alfalfa and greater (P<0.05) for corn+hay vs. hay-only diets (Table 3). However, kr was similar (P>0.05) between the hay-only diets and was greater (P<0.05) for the corn+hay vs. hay-only diets (Table 3). This resulted in similar NEg values for sudex and alfalfa diets (4.85 MJ/kg) and for diets containing corn+hay (6.49 MJ/kg), even though metabolizability indicated that the sudex diets would contain more NEg than the alfalfa diets.

Table 2. Diet composition and nutrient concentrations (g/kg DM).

Item	Treatments			
	Sudex	Alfalfa	Sudex + Corn	Alfalfa + Corn
		% of diet DM		
Sudex hay, ground	90	-	50	-
Alfalfa hay, ground	-	100	-	60
Corn, cracked	-	-	40	40
Protein suppl., pellet[a]	10	-	10	-
Nutrient composition[b]				
NDF	516	453	341	312
Hemicellulose	213	126	149	103
ADF	304	328	191	209
Cellulose	265	253	164	160
Lignin	39	75	28	49
Starch	126	25	365	304
Crude protein	124	166	136	140
Ether extract	16	12	24	21
Ash	70	107	48	70

[a]85% dehydrated alfalfa, 13% urea, 2% molasses; DM basis.
[b]Based on individual feed values.

Table 3. Energy utilization of the diets.

Item	Treatments				SE[a]
	Sudex	Alfalfa	Sudex + Corn	Alfalfa + Corn	
ME, MJ/kg[bcd]	10.00	9.58	11.84	11.51	0.13
kr[ce]	0.49	0.51	0.56	0.57	0.07
NEg, MJ/kg[f]	4.85	4.85	6.44	6.53	

[a]Standard error of the LS means. [b]Hay type effect, $P<0.05$. [c]Corn level effect, $P<0.05$.
[d]Analyzed with GEI (MJ/kg $BW^{0.75}$) as a continuous covariate.
[e]kr=slope of regression: MEI vs. RE. [f]$NEg=ME_{high}*kr$.

Discussion

The low coefficient of determination for the relationship between NEg and ME indicates considerable error or that other important factors are involved in predicting heat increment of gain. Variation associated with different experiments such as varying fat vs. lean proportions in growth could not be separated and probably contributes to the low coefficient of determination in both techniques. Perhaps the most important factor contributing to the weak relationship between ME and NEg, however, is that the quantity of ME may not be as important as the composition of the ME. In other words, whether ME is derived from cellulose vs. some other energy yielding nutrient may be more important than the ME content per kg of feed. Considerable data have shown kr is not directly related to ME concentration of the diet. At similar ME concentrations, research shows efficiency of ME utilization for gain is less for forages compared to concentrates (Wainman *et al.*, 1970; Beever *et al.*, 1988; Reynolds *et al.*, 1991), less for mature compared to immature forages (MacRae *et al.*, 1985), and less for grasses compared to legumes (Varga *et al.*, 1990; Waldo *et al.*, 1990; Tyrrell *et al.*, 1992). A common link among the above comparisons is the inverse relationship between cell wall content and kr. Several reasons for the decline in efficiency of ME utilization for gain with an increasing proportion of dietary cell walls have been proposed and include related factors such as: 1) shift in site of digestion towards the rumen, 2) increased gut contents, 3) increased rumen degradable energy/N ratio, 4) change in profile of nutrients absorbed (increased VFA, especially acetate and decreased glucose), and 5) increased portal-drained viscera oxygen consumption.

Predicting NEg of feeds could be improved to reflect our knowledge about how urinary, methane, and heat energy varies according to dietary nutrient composition. Theoretically, the NEg of a feedstuff should be more accurately predicted from the proportions of nutrients composing ME, rather than merely the concentration of ME. To fully determine composition of ME, we must be able to predict the losses of each nutrient via the faeces, urine, and methane; as well as the site of digestion (fermentation vs. enzymatic digestion) which determines the types of digestion end-products (i.e. VFA vs. glucose). Additionally, heat loss associated with rumen events must also be taken into account.

References

ARC. (1980). The Nutritive Requirements of Ruminant Livestock. Agricultural Research Council, Commonwealth Agric. Bureaux, Slough, U.K.

Beever D.E., Cammell S.B., Thomas C., Spooner M.C., Haines M.J. & Gale D.L. (1988). The effect of date of cut and barley substitution on gain and on the efficiency of utilization of grass silage by growing cattle. 2. Nutrient supply and energy partition. *Br. J. Nutr.* 60: 297.

MacRae J.C., Smith J.S., Dewey P.J.S., Brewer A.C., Brown D.S. & Walker A. (1985). The efficiency of utilization of metabolizable energy and apparent absorption of amino acids in sheep given spring-and autumn- harvested dried grass. *Br. J. Nutr.* 54: 197.

NRC. (1996). Nutrient Requirements of Beef Cattle. National Academy Press, Washington DC.

Reynolds C.K., Tyrrell H.F. & Reynolds P.J. (1991). Effects of diet forage-to-concentrate ratio and intake on energy metabolism in growing beef heifers: whole body energy and nitrogen balance and visceral heat production. *J. Nutr.* 121: 994.

Tyrrell H.F., Thomson D.J., Waldo D.R., Gowering H.K. & Haaland G.L. (1992). Utilization of energy and nitrogen by yearling Holstein cattle fed direct-cut alfalfa or orchardgrass ensiled with formic acid plus formaldehyde. *J. Anim. Sci.* 70: 3163.

Varga G.A., Tyrrell H.F., Huntington G.B., Waldo D.R. & Glenn B.P. (1990). Utilization of nitrogen and energy by Holstein steers fed formaldehyde- and formic acid-treated alfalfa or orchardgrass silage at two intakes. *J. Anim. Sci.* 68: 3780.

Wainman F.W., Blaxter K.L., Smith J.S. & Dewey P.J.S. (1970). Calorimetric study of the nutritive value of dried grass. In : A. Schurch and C. Wenk (Ed.), Energy Metabolism of Farm Animals. *EAAP Publ No 13*: pp. 17-20.

Waldo D.R., Varga G.A., Huntington G.B., Glenn B.P. & Tyrrell H.F. (1990). Energy components of growth in Holstein steers fed formaldehyde- and formic acid-treated alfalfa or orchardgrass silages at equalized intakes of dry matter. *J. Anim. Sci.* 68: 3792.

THE EFFECT OF STAGE OF HARVEST OF MAIZE SILAGE ON THE CONCENTRATION AND EFFICIENCY OF UTILIZATION OF METABOLIZABLE ENERGY BY LACTATING DAIRY COWS

D.E Beever, S.B. Cammell, J.D. Sutton and D.J. Humphries

Centre for Dairy Research (CEDAR), Department of Agriculture, The University of Reading, Earley Gate, Reading RG6 6AT, UK

Summary

Maize was harvested at two stages of maturity (low and high dry matter (DM)) in consecutive years to provide four silages and fed to cows in total mixed rations (TMR) comprising maize silage and concentrate blend (60:40 DM basis) with a minimal amount of chopped barley straw. Using open circuit calorimetry, full energy balances were obtained. In year 1 the silages had ME contents of 10.9 and 10.1 MJ/kg DM respectively, the higher value being related to the higher gross energy content of the less mature crop. However, both values were generally lower than those obtained when the same silages were analysed by six commercial laboratories. In year 2 both silages had ME contents of 11.2 MJ/kg DM, whilst in both studies, with average milk yields of between 27 and 34 kg/d, the efficiency of ME utilization (k_l) ranged between 536 and 552 kJ/MJ, lower than predicted by AFRC (1993). This suggests that current estimates of k_l are too high or ME maintenance requirements too low. Both this and discrepancies in prediction of ME content require urgent reconsideration.

Key words: Milk production, energy efficiency, maize silage.

Introduction

Maize silage is an important forage for dairy cows due to its capacity to improve DM intake and milk yield and increase milk protein/fat ratio compared with grass silage (Phipps *et al.*, 1995). The area of maize in the UK has increased five-fold in the last decade and is now grown in more diverse environments. In favourable areas it is relatively easy to harvest maize at DM contents of >300 g/kg fresh weight and starch contents of >250 g/kg DM, but in marginal areas much lower values of 240 g DM/kg fresh weight and 120 g starch/kg DM or less are quite common. Such changes are likely to affect both intake and nutritional characteristics of maize silage for milk production, but there has been no systematic evaluation of maize silage in dairy cow rations. Two studies on the evaluation of maize silages are reported in the present paper.

Materials and Methods

Maize (var. Energy) was harvested in early October 1995, using a precision chop forage harvester with corn crackers, and ensiled in concrete walled silos (low DM, year 1, (LDM1)). A further harvest was made 7 d later, after two night frosts had occurred, and ensiled in a similar clamp (HDM1). In 1996 maize (var. LG2080) was cut in late August (LDM2), with a further harvest in mid September (HDM2), using the same procedures, the earlier dates being due to the more advanced growing season in 1996. Thereafter all silages were left for at least 6 weeks before being opened for feeding.

Using the two silages from each year, total mixed rations based on a 60:40 DM ratio of maize silage to concentrate were formulated and offered *ad libitum* with 0.7 kg DM/d chopped barley straw. Concentrate composition (kg/t) was similar for both years (HiPro rape seed meal 373; '48%' soya 248; rolled barley 124; molassed sugar beet 100; white fishmeal 81; beet molasses, 62; minerals/vitamins 12). It was provided as a meal and contained 375 g crude protein/kg DM.

Within each year, a duplicated 2x2 Latin square design was adopted to examine the effect of maize maturity on energy utilization. Each experimental period lasted 5 weeks with weeks 1-4 for diet adaptation, followed in week 5 by full energy balances using open circuit indirect respiration calorimetry.

In both studies, 4 multiparous Holstein/Friesian cows were introduced to the experimental rations in lactation week 6. All cows were normally held in individual standings and bedded on shavings but were transferred to the calorimeters for estimation of feed intake and milk, faeces and urine output (6 d) and oxygen consumption and carbon dioxide and methane production (3 d) (Beever *et al.*, 1991).

Results and Discussion

Both harvest date and season had appreciable effects on maize silage composition (Table 1). In year 1 delayed harvest resulted in only small increases in starch and neutral detergent fibre (NDF) contents despite a large increase in DM content, whilst both silages had similar contents of fermentation products but quite different gross energy (GE) contents. In contrast, LDM2 had lower starch and higher NDF contents than HDM2 but similar GE content. Lower contents of lactate+VFA in the later crop indicated a reduced extent of fermentation.

Table 1. Chemical composition of the four maize silages (g/kg volatile corrected DM (VCDM) unless stated).

	LDM1	HDM1	LDM2	HDM2
VCDM (g/kg fresh wt)	270	329	329	388
NDF	439	478	487	421
ADF	226	226	259	208
Starch	241	262	161	291
Total N	15.6	15.4	11.9	13.7
Gross energy (MJ/kg VCDM)	18.9	17.6	18.0	17.8
Lactate + VFA	90.1	94.2	88.1	65.9

No significant differences in intake or milk production were established between year 1 silages (Table 2). Intake on HDM2 was similar to year 1 silages, with slightly lower milk and milk fat yields, whilst protein yield was maintained. Compared with HDM2, intake of LDM2 was lower ($P<0.05$) as were yields of milk, milk fat and milk protein ($P<0.10$). Some of this effect may have been due to the lower starch content of LDM2 maize silage.

GE intakes were lower in year 2 than year 1 (Table 3), but apparent digestibility of dietary energy was unaffected by stage or year of harvest. Urine and methane energy accounted for between 120 and 150 kJ/MJ DE, with a mean ME/DE ratio for all diets of 863 kJ/MJ. ME intakes were about 245 MJ/d for all diets except LDM2 where intake was only 218 MJ/d. Diets based on LDM1 and 2 and HDM2 had similar ME contents (about 11 MJ/kg DM), whilst HDM1 had a value of 10.4 MJ/kg DM. However ME/GE values were unaffected by crop maturity results for year 1 (579 and 576 kJ/MJ respectively) being marginally lower than those for year 2 (608 and 607 kJ/MJ respectively), suggesting that the lower ME/DM value for HDM1 was largely due to its lower GE content though this did not appear to be related to the respective contents of fermentation products.

Table 2. DM intake and lactational performance (kg/d).

	LDM1	HDM1	se	LDM2	HDM2	se
Intake:						
Maize silage	13.4	14.2		11.7	13.1	0.37
Total	22.5	23.5	0.32	19.6	22.5	0.56
Milk yield:						
Total	33.7	34.1	0.51	27.4	31.1	0.73
Fat	1.41	1.42	0.006	1.25	1.32	0.015
Protein	1.08	1.11	0.010	0.96	1.12	0.033

After allowance for calculated ME content of the concentrates and straw plus level of feeding corrections (AFRC, 1993), ME contents of 10.9, 10.1, 11.2 and 11.2 MJ/kg VCDM were derived for LDM1, HDM1, LDM2 and HDM2 silages respectively. Based on analyses by six independent laboratories, mean ME values of 11.2 and 11.5 MJ/kg DM were obtained for LDM1 and HDM1 silages respectively. Thus whilst the value for LDM1 agreed closely with the *in vivo* derived value, the suggestion that HDM1 had a higher ME value was contrary to the data derived using calorimetric procedures, possibly indicating that a major failure of the laboratory procedures was to take account of the lower GE content of the later cut maize. Furthermore, the between-laboratory variation for LDM1 and HDM1 respectively ranged between 10.5 and 12.1 MJ/kg DM and 10.7 and 12.3 MJ/kg DM.

Table 3. Partition of gross energy.

	LDM1	HDM1	se	LDM2	HDM2	se
Intake	424	425	8.7	359	409	13.5
Faeces	136	138	1.6	109	126	3.5
Urine	16	16	0.2	9	9	0.1
Methane	27	27	0.2	24	25	1.1
ME/DE	850	851	4.9	872	879	2.7
ME(MJ/kg DM)	10.9	10.4	0.30	11.1	11.1	0.33

For LDM1 and HDM1, daily ME intake averaged 1893 $kJ/kg^{0.75}/d$, with a total energy availability of 1985 $kJ/kg^{0.75}/d$ as all cows were in slight negative energy balance (mean 12 MJ/d) (Table 4). From AFRC (1993) a mean maintenance ME requirement of 484 $kJ/kg^{0.75}/d$ was estimated, leaving 1500 kJ $ME/kg^{0.75}/d$ on both diets for the synthesis of milk constituents (MEp).

Table 4. Efficiency of ME utilization ($kJ/kg^{0.75}/d$ unless stated).

	LDM1	HDM1	se	LDM2	HDM2	se
Energy availability:						
ME intake	1897	1889	81.0	1713	1955	70.1
Tissue mobilization	81	102	39.9	97	14	103.5
Total	1978	1991	42.5	1810	1969	62.6
Energy expenditure:						
Maintenance	483	484	-	477	477	2.7
Production (MEp)	1495	1507	42.5	1334	1492	60.5
Milk energy	819	832	9.9	719	803	24.2
k_l (kJ/MJ MEp)	548	552	10.5	539	536	13.1

Furthermore both diets had similar milk energy outputs (mean 107 MJ/d, 826 $kJ/kg^{0.75}/d$), with similar values for k_l. In year 2, ME intake and total ME available were lower ($P<0.10$) on LDM2 than HDM2. Consequently MEp was lower on LDM2 resulting in a significant depression in milk energy output (92 vs. 102 MJ/d; 719 vs. 803 kJ/kg$^{0.75}$/d, $P<0.10$). However k_l was similar on both diets and marginally lower than the values obtained the previous year.

Conclusion

ME content of the four contrasting silages varied between 10.1 and 11.2 MJ/kg VCDM, indicating that maize silage is unlikely to have ME contents greater than those for grass silage. This suggests that it is the intake characteristics of maize silage rather than its energy density that influences observed responses in milk yield and composition. Furthermore, the data indicate that current laboratory procedures for estimating the ME content of maize silages are not adequate, implying a need for a more extensive database of *in vivo* estimates before reliable predictive procedures can be developed. The consistency in the estimates of k_l suggests that energy availability rather than energy type was more important in determining the yield of milk energy. If the low estimates for k_l are confirmed, there will be a need to revise current energy recommendations.

Acknowledgements

Funded by the Milk Development Council, Maize Growers Association and Force LimaGrain.

References

Agricultural and Food Research Council (1993). Energy and Protein Requirements of Ruminants. CABI, Wallingford.

Beever D.E., Cammell S.B., Spooner M.C., Humphries D.J. & Haines M.J. (1991). Effect of defined manipulations of nutrient supply on energy and protein utilization by dairy cows. In: C.Wenk & M Boessinger (Ed.) Energy Metabolism of Farm Animals. *EAAP Publ. No* 58:325-328.

Phipps R.H., Sutton J.D. & Jones A.K. (1995). Forage mixtures for dairy cows: the effect on dry matter intake and milk production of incorporating either fermented or urea treated whole crop wheat, brewers grains, fodder beet or maize silage into diets based on grass silage. *Anim. Sci.* 61: 491-496.

PARTITION OF ENERGY DIGESTION IN PIGS FED BARLEY-BASED DIETS WITH FORAGE MEAL INCLUSION

J.E. Lindberg and C. Andersson

Swedish University of Agricultural Sciences, Department of Animal Nutrition and Mangement, Box 7024, S-750 07 Uppsala, Sweden

Summary

PVTC-cannulated growing pigs were used to determine the ileal, hindgut and total tract digestibility of dietary components and energy in barley-based diets with forage meals (lucerne, white clover, red clover and perennial ryegrass) at three levels of inclusion (0, 10 and 20%). The ileal and total tract digestibilities of organic matter, crude protein and energy were significantly ($P<0.05$) reduced with increasing inclusion of forage meals in the diet. The ileal digestibility of energy decreased by approximately 1.4, 2.2 and 2.9% units per percentage increase of dietary TF, NDF and CF in DM, respectively. This was considerably faster (+41 to +47%) than the decrease in dE measured over the total tract. Significant differences were found between treatments in the quantity of dietary DE utilized in the hindgut. The present data suggest that the effect of dietary fibre on dE is most pronounced at the ileal level and that the decrease in dE with forage fibre inclusion is more limited than with cereal fibre inclusion.
Key words: Fibre, energy digestibility, hindgut digestion, digestible energy.

Introduction

Forages represent significant potential as protein and energy sources for domestic animals and there has, in recent years, been a renewed and increasing interest for using forages in modern pig production (Wiesemüller & Poppe, 1990). However, the high fibre content of forages is of major concern since plant fibre has been shown to reduce nutrient and energy digestibility in pigs (Just, 1982a, b). The total tract digestibility of dietary components and energy in lucerne has been studied extensively in the pig while more limited information exists on other forages (Andersson, 1997). However, only a few papers have been published on the ileal digestibility of dietary components in lucerne and there was virtually no information available on the ileal digestibility of other forages until recently (Andersson & Lindberg, 1997 a, b). Also recently, the digestion and nutritive value of lucerne leaf meal has been studied in pigs (Lindberg & Cortova, 1995).

The present experiments were conducted to determine the ileal and total tract apparent digestibility, and hindgut disappearance, of nutrients and energy in a barley-based diet with increasing inclusion of either lucerne, white clover, red clover or perennial ryegrass meal when fed to growing pigs.

Materials and Methods

Castrated male pigs (LxY) were surgically fitted with post valvular t-cannulas (PVTC; Leeuwen *et al.*, 1991) and were housed individually in 3.5 m^2 pens with rubber mats. The average live weights (LW) of the pigs at the start of the experiment were 41 (SD 4.9) kg in Exp 1 and 43 (SD 1.5) kg in Exp 2, respectively.

The treatments were a barley-based diet (C) and forage meal diets which were composed of two levels (10% and 20%) of lucerne (L10 and L20), white clover (W10 and W20), red clover (R10 and R20) or perennial ryegrass (PR10 and PR20), respectively (Table 1). The diets contained 0.3% chromium oxide (Cr_2O_3) as an indigestible marker.

The pigs were fed daily in three equal meals, at 8:00, 16:00 and 24:00 h. The feed allowances were restricted to 4% of the mean LW up to 75 kg and thereafter 3 kg per pig per day. Water was provided *ad libitum*.

The experiments were conducted as 5x5 change-over designs comprising 9 d of adaptation, 2 d of ileal digesta and 5 d of faeces collection. Ileal digesta flow and faecal excretion were estimated assuming a complete recovery of chromium oxide at the two sites of sampling. Samples were analysed with conventional methods (see Andersson & Lindberg, 1997a) and the content of total fibre (TF) was estimated as a residual fraction by subtraction of the analysed content of sugars, starch, crude protein (CP), crude fat (EE) and ash from the DM content (Åman & Hesselman, 1984). Gross energy was determined using an iso-thermal calorimeter. All results are presented on a DM basis.

Analysis of variance was performed using the General Linear Models procedure and linear regression analysis using the Regression procedure in SAS (SAS, 1988). The results are presented as least square means with their standard error (SEM).

Results

The intake of starch decreased significantly (P<0.05) with increasing inclusion of the forage meal, whereas the intake of CP and fibre (TF, NDF, ADF and crude fibre) increased significantly (P<0.05). The ileal and total tract flow of OM, TF, NDF, ADF, crude fibre (CF) and energy increased significantly (P<0.05) with inclusion of forage meal in the diet (data not shown). No significant effect on the ileal flow of starch and sugars was observed. The ileal digestibility of OM, CP, EE and energy were significantly (P<0.05) reduced with increased inclusion of forage. In contrast, the ileal digestibility of starch and fibre (TF, NDF, ADF and CF) were not significantly affected by forage meal inclusion.

Table 1. Hindgut disappearance (%) of organic matter (hdOM) and energy (hdE), and dietary content (MJ/kg DM) of digestible energy (DE) at the ileum, in the hindgut and in the total tract in growing pigs fed barley-based diets with forage meal inclusion.

	Diets in Exp 1					
	Control	L10	L20	W10	W20	SEM
hdOM	47[ab]	43[c]	43[bc]	48[a]	47[a]	1.1
hdE	43	40	40	43	42	1.2
DE ileum	12.2[a]	11.3[b]	9.8[c]	11.4[b]	10.1[c]	0.17
DE hindgut	2.3[c]	2.5[c]	3.1[ab]	2.8[bc]	3.2[a]	0.14
DE total tract	14.5[a]	13.8[c]	12.9[e]	14.2[b]	13.3[d]	0.07
	Diets in Exp 2					
	Control	R10	R20	PR10	PR20	SEM
hdOM	51[a]	48[ab]	49[ab]	46[bc]	43[c]	1.3
hdE	47[a]	43[ab]	45[ab]	41[bc]	39[c]	1.4
DE ileum	12.1[a]	10.4[b]	9.0[c]	9.8[b]	8.8[c]	0.23
DE hindgut	3.0	3.2	3.9	3.2	3.4	0.20
DE total tract	15.1[a]	13.6[b]	13.0[c]	13.1[c]	12.2[d]	0.10

[a, b, c, d] Means with different superscripts within rows are significantly different at P<0.05

The hindgut disappearance of OM (hdOM) was significantly (P<0.05) higher in the white clover meal diets than in the lucerne meal diets (+ 4.2%-units), while the hindgut disappearance of fibre (TF, NDF, ADF and CF; data not shown) and energy (hdE) was not significantly different between diets (Table 1). The hdOM and hdE was reduced (P<0.05) in the perennial ryegrass meal diets compared with both the control diet and the red clover meal diets (Table 1), which were similar. The hdOM was significantly (P<0.05) higher in the red clover meal diets than in the perennial ryegrass meal diets (+4.5%-units). The disappearance of fibre (NDF, ADF and CF; data not shown) tended to increase, although not significantly, with inclusion of red clover and perennial ryegrass meal.

There was a significant (P<0.05) reduction in total tract digestibility of energy (dE) with increasing inclusion of forage meal (Table 2). In Exp 1 mean dE was significantly (P<0.05) higher for the white clover meal diets as compared with the lucerne meal diets (+ 1.4%-units). In Exp 2 the reduction was greater for the perennial ryegrass meal diets than for the red clover meal diets (P<0.05). Average dE in the red clover meal diets was significantly (P<0.05) higher as compared with the perennial ryegrass meal diets (+3.3%-units). The dietary content of digestible energy (DE) was significantly (P<0.05) reduced at both the ileum and the total tract with inclusion of forage meal (Table 1). However, the proportion of DE digested in the hindgut increased significantly in Exp 1 (P<0.05).

Table 2. Relationship between the ileal and total tract digestibility of energy (dE, %), and the dietary content (% in DM) of total fibre (TF), neutral detergent fibre (NDF) and crude fibre (CF) in growing pigs fed barley-based diets with forage meal inclusion.

	dE (%), ileum			dE (%), total tract		
	TF	NDF	CF	TF	NDF	CF
Intercept	99.4	98.1	78.1	103.9	104.0	89.9
SEM	11.8	8.6	7.9	7.5	4.6	5.0
Slope	-1.38	-2.23	-2.93	-0.94	-1.57	-2.08
SEM	0.40	0.48	1.20	0.26	0.26	0.75
R^2	0.60	0.72	0.43	0.63	0.82	0.49
CV	7.3	6.0	8.7	3.6	2.5	4.2
Sign. level	0.01	0.01	0.05	0.01	0.001	0.05

The content of total tract DE was significantly (P<0.05) higher for white clover meal diets as compared with lucerne meal diets (+ 0.4 MJ/kg DM). In contrast, there was no significant difference in the dietary content of ileal and hindgut DE between lucerne and white clover meal diets. There was also a significant (P<0.05) reduction in dietary content of DE in the total tract for perennial ryegrass meal diets compared with the red clover meal diets (-0.7 MJ/kg DM). In contrast, there was no significant difference between the experimental diets at the ileum. The amount of DE disappearing in the hindgut tended (P>0.05) to increase with inclusion of red clover and perennial ryegrass meal as compared with the control diet.

Discussion

The present study shows that increasing inclusion of forage meals in a barley-based diet for growing pigs will result in a decrease in the digestibility of OM and energy. This can be attributed to the increase in fibre intake and was in agreement with others (Just, 1982a, b; Lindberg & Cortova, 1995). The decrease in dE was more pronounced at the ileal level than in the total tract, as also shown by Lindberg & Cortova (1995). This effect was consistent irrespective of fibre measure (TF, NDF, CF) and amounted to (ileal vs. total tract) +47% for TF, +42% for NDF and +41% for CF. Also with increasing forage meal intake the dietary DE content will decrease significantly both at the ileal and total tract level. The decrease in DE at the total tract level will, however, be less pronounced as compared with the ileal level due to a compensation by hindgut digestion.

The data on OM and energy disappearance in the hindgut suggests that the inclusion of forage fibre in the diet of growing pigs has only marginal effects on their utilization at this site of the GI-tract. The maximum increase in daily fibre intake was +34% for TF, +44% for NDF and +99% for CF, respectively, in the lucerne meal diets. Correspondingly fibre intake increased by +26% for TF, +21% for NDF and +41% for CF, respectively, in the white clover meal diets, +37% for TF, +27% for NDF and +41% for CF, respectively, in the red clover meal diets and +41% for TF, +49% for NDF and +60% for CF, respectively, in the perennial ryegrass meal diets. This shows that, despite a very large increase in fibre intake, the disappearance of OM and energy in the hindgut was only affected to a limited extent, which was in agreement with Lindberg & Cortova (1995). The observed differences in hindgut disappearance appeared to be related to differences in chemical composition between forage meals.

The decrease in dE in the total tract as a result of increasing fibre content in the diet was lower in the present study (-2.1%-units per % CF) than that reported by Just (1982a) for cereal fibre (-2.9%-units per % CF) and of similar magnitude to that reported by Just (1982b) for barley straw (-2.1%-units per % CF). In earlier studies on lucerne leaf meal inclusion in barley-based diets for growing pigs (Lindberg & Cortova, 1995) a decrease in dE considerably lower than 2%-units per unit increase in CF could be calculated. This suggests that differences in fibre quality in the diet can have a considerable influence on predictions of dE. It was also apparent from the present study that the precision in predicting dE could be markedly improved by using NDF as a measure of dietary fibre content. This is in agreement with Noblet & Perez (1993).

The present data suggest that the effect of dietary fibre on dE is most pronounced at the ileal level and that the decrease in dE with forage fibre inclusion is more limited than with cereal fibre inclusion. It is further suggested that the hindgut in the growing pig has a substantial, but poorly utilized capacity to digest fibre and other dietary components.

References

Åman P. and Hesselman K. (1984). Analysis of starch and other main constituents of cereal grains. *Swedish J. agric. Res.* 14:135-139.

Andersson C. (1997). Forages for growing pigs-partition of digestion and nutritive value. Swedish University of Agricultural Sciences, Acta Universitatis Agriculturae Sueciae, Agraria 31.

Andersson C. & Lindberg J.E. (1997). Forages in diets for growing pigs. 1. Digestibilities and partition of digestion in barley-based diets with inclusion of lucerne and white clover meal. *Anim. Sci.* (In press).

Andersson C. & Lindberg J.E. (1997). Forages in diets for growing pigs. 2. Digestibilities and partition of digestion in barley-based diets with inclusion of red clover and perennial ryegrass meal. *Anim. Sci.* (In press).

Just A. (1982a). The influence of crude fibre from cereals on the net energy value of diets for growth in pigs. *Livest. Prod. Sci.* 9: 569-580.

Just A. (1982b). The influence of ground barley straw on the net energy value of diets for growth in pigs. *Livest. Prod. Sci.* 9: 717-729.

Leeuwen van P., van Kleef D.J., van Kempen G.J.M., Huisman J. & Verstegen M.W.A. (1991). The post valve T-caecum cannulation technique in pigs applicated to determine the digestibility of amino acid in maize, groundnut and sunflower meal. *J. Anim. Physiol. Anim. Nutr.* 65: 183-193.

Lindberg J.E. & Cortova Z. (1995). The effect of increasing inclusion of lucerne leaf meal in a barley-based diet on the partition of digestion and on nutrient utilization in pigs. *Anim. Feed Sci. Technol.* 56: 11-20.

Noblet J. & Perez J.M. (1993). Prediction of digestibility of nutrients and energy value of pig diets from chemical composition. *J. Anim. Sci.* 71: 3389-3398.

SAS (1988). SAS Institute Inc. SAS/STAT Users Guide, Release 6.03 Edition. Cary, NC:SAS Institute Inc.

Wiesemüller W. & Poppe S. (1990). Lucerne silage as a feedstuff for sows. *World Rev. Anim. Prod.* XXV:47-53.

ENERGY UTILIZATION OF DIETS WITH DIFFERENT SOURCES OF DIETARY FIBRE IN GROWING PIGS

H. Jørgensen

Danish Institute of Agricultural Sciences, Research Centre Foulum,
P.O. Box 50, DK 8830 Tjele, Denmark

Summary

The influence of dietary fibre on digestibility, microbial fermentation, energy disappearing in the hind-gut and utilization of ME in growing pigs was investigated. Dietary fibre (0, 30, 60, 90 and 120 g/kg) from either soya-bean or potato was used as the sole fibre source and added to wheat starch and fish meal based diets. Ten castrated male pigs, fitted at 35 kg with a simple T-cannula at the terminal ileum, were used. The experiment was carried out in the weight range 50-70 kg. Three balance experiments were performed with each pig in a randomized design yielding 6 replicates of the control diet and 3 replicates of each of the 8 fibre diets. Each balance experiment consisted of an 8-day initial period followed by a 4-day collection of faeces and urine and a 2-day collection of ileal digesta. A 24-h respiration experiment was performed using indirect calorimetry in each collection period.

With increasing dietary fibre level from both fibre sources, CH_4 production increased linearly. H_2 production was constant and only contributed less than 0.5 l/day. The added fibre sources had no influence on the faecal energy digestibility; however the ileal energy digestibility decreased. This shows that the fibre sources were not digested/fermented in the small intestine but to a high degree fermented in the hind-gut. The amount of RE as a portion of ME decreased in relation to the amount of energy fermented in the hind-gut. A regression analysis indicated that 4-5% of GE in the fibre sources is lost as CH_4 due to the fermentation in the hind-gut. Potato fibre, which is more water-soluble than soya fibre, was also more fermentable (86% compared to 71% for soya fibre).

Key words: Pigs, cannulated, ileal digestibility, fibre, energy retention, CH_4.

Introduction

There is increasing interest in elucidating the effects of dietary fibres on the digestion, absorption and utilization of nutrients in humans and growing pigs. These effects include increased volatile fatty acid production, alteration of bacterial populations in the small and large intestine, prevention of the proliferation of pathogenic bacteria, increased viscosity, increased water holding capacity, increased rate of passage, modification of digestive enzyme activities and effects on intestinal enterocytes. Several studies have demonstrated that the net energy value of different fibre sources depends, to a large extent, on the composition of the dietary fibre i.e. soluble or insoluble non-starch polysaccharides (NSP) (Livesey *et al.*, 1995; Just *et al.*, 1983; Jørgensen *et al.*, 1996a, b).

The objective of this experiment was to study the effect of two sources of dietary fibre (from potato and soya bean) on digestibility, microbial fermentation, energy disappearance and utilization of metabolizable energy (ME) in growing pigs.

Materials and Methods

Dietary fibre (0, 30, 60, 90 and 120 g/kg) from either soybean or potato was used as the sole fibre source and added at the expense of wheat starch. Fish meal (223 g/kg) was used as protein source. Sugar (50 g/kg) and soya oil (30 g/kg) were added to improve palatability, in addition to the necessary minerals and vitamins and, for calculating digestibility, chromic oxide (2 g/kg). Ten castrated male pigs, fitted at 35 kg with a simple T-cannula at the terminal ileum, were used. The experiment was carried out in the weight range 50-70 kg. Three balance experiments were performed with each pig in a randomized design yielding 6 replicates of the control diet and 3 replicates of each of the 8 fibre diets. Each balance experiment consisted of an 8-day initial period followed by a 4-day collection of faeces and urine and a 2-day collection of ileal digesta. During the collection and balance periods the pigs were kept in stainless steel metabolic cages. Heat production was measured in each collection period for 24 h using an open circuit respiration unit (Jørgensen *et al.*, 1996a).

Results and Discussion

The two fibre sources potato or soya were both rich in dietary fibre, 615 and 726 g/kg DM, respectively (Table 1). However, the chemical composition of the fibre fraction differed between the sources. Potato fibre contained nearly as much water-soluble-NSP (46%) as insoluble-NSP whereas in soya fibre only 27% of the NSP fraction was soluble. Fibre from potato had a stronger influence on the ileal energy digestibility than fibre derived from soya. The ileal energy digestibility decreased 1.7 and 1.3 percentage units per percentage dietary fibre from potato and soya fibre, respectively. At the faecal level, fibre from soya exerted a stronger influence on energy digestibility than fibre from potato (0.27 and 0.32) the difference being small although significant ($p=0.004$).

Table 1. Composition (g/kg DM) of the basal diet and the fibre sources.

	Basal diet	Potato fibre	Soya fibre
Crude protein	178	76	116
Total-NSP	3	593	711
Soluble-NSP	1	275	190
Insoluble-NSP	2	318	520
Dietary Fibre	12	615	726

The methane (CH_4) production increased with increasing dietary fibre in the experimental diets (Figure 1). Hydrogen (H_2) production was constant and only contributed less than 0.5 l/day. In experiments with chickens H_2 production increased with increased fibre level in the diet (Jørgensen *et al.*, 1996b). As H_2 mainly is present in the small intestine (Jensen & Jørgensen, 1994) the present results suggest that fermentation in the small intestine was of minor importance. The energy from CH_4 production did not exceed 0.6% of DE in the experimental diets (Table 2). However, a regression analysis showed that 3.6 and 4.6% of GE from potato and soya fibre was lost as CH_4, respectively (Table 3).

The low or negative ileal digestibility values found by regression analysis for the fibre sources (Table 3) are presumably due to increased endogenous matter. The energy in potato fibre was more fermentable compared to energy in soya fibre. This is in agreement with potato fibre having a higher proportion of soluble-NSP in the fibre fraction. However, both fibre sources were to a large extent fermented in the hind-gut (86 and 71% of GE in potato and soya fibre, respectively). The fermentation of soya fibre was lower than that found by Livesey *et al.* (1995) in a study using laboratory rats. The end-products of the microbial degradation of dietary fibre are various gases (H_2, CO_2, CH_4), lactic acid and short-chain fatty acids (SCFA) of which acetate, propionate and butyrate are dominant. The SCFA produced are rapidly absorbed from the gut lumen. SCFA in the circulatory system have been ascribed various effects. The water-soluble fibre can lower serum lipids. The energy efficiency of short chain fatty acids (SCFA) derived from the microbial fermentation is lower than from energy absorbed as glucose in the small intestine (Jørgensen *et al.*, 1997). Whereas acetate can be used for biosynthesis of higher lipids, propionate for gluconeogenesis, and butyrate for the formation of ketone bodies, the eventual fate of these SCFA is oxidation for energy.

Table 2. Energy metabolism of pigs fed diets with either potato or soya-bean fibre.

Fibre source		Potato				Soya			
Fibre inclusion, g/kg	0	30	60	90	120	30	60	90	120
ME intake, MJ/d	27.5	27.9	27.5	27.5	27.4	27.8	27.6	27.4	27.1
DE ileal, % of GE	93	92	83	85	81	91	86	85	82***
DE faecal, % of GE	96	96	96	95	94	96	95	94	93
ME, % of DE	96.9	97.4	96.9	96.9	97.2	97.2	96.8	96.8	96.7
CH_4-energy, % of DE	0.1	0.2	0.2	0.4	0.4	0.2	0.4	0.5	0.6***
HP, % of ME	51.9	49.9	53.4	56.7	52.6	51.4	52.0	54.2	54.9
RE, % of ME	48.1	50.1	46.5	43.3	47.4	48.6	48.0	45.8	45.1
RE-protein, % of RE	23.2	28.4	24.6	30.5	27.5	26.8	26.3	29.0	30.3
RE-fat, % of RE	76.8	71.6	75.4	69.5	72.5	73.2	73.7	71.0	69.7

*** significant at $p < 0.001$.

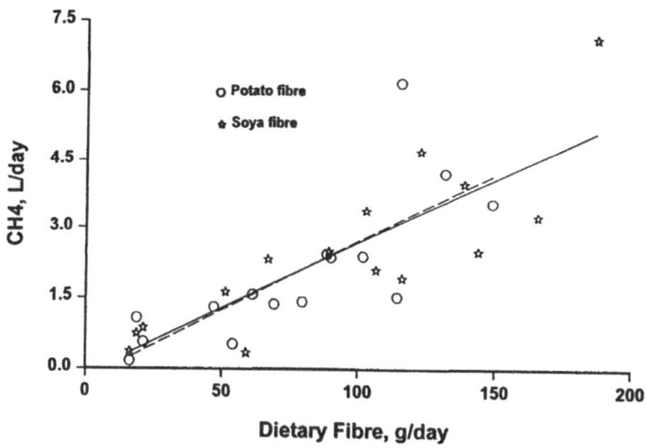

Fig. 1. Methane production depending on intake of dietary fibre in growing pigs.

Table 3. Energy efficiency of the basal diet and the supplemented fibre sources estimated as the regression coefficient from the equation $Y = b_1{*}GE_{basal} + b_i{*}GE_{fibresource}$.

	Basal diet	Potato fibre	Soya fibre	Fibre source
DE ileal / GE	0.93	-0.17	-0.13	ns
DE faecal / GE	0.96	0.86	0.71	***
CH_4-energy / GE	0.001	0.036	0.046	***
Urine-energy / GE	0.028	-0.012	0.003	ns
ME / GE	0.93	0.84	0.66	***
HP / GE	0.48	0.71	0.62	**
RE / GE	0.46	0.13	0.04	ns

ns = non-significant; **, *** significant at $p<0.01$ and $p<0.001$, respectively.

When diets high in fibre are given, there is hypertrophy and enlargement of the digestive tract, especially the hind-gut (Jørgensen *et al.*, 1996a, b), which might increase energy expenditure. A direct physical effect of fibres cannot be ruled out, but SCFA themselves could affect intestinal function. Butyrate is considered to be the preferred respiratory fuel for colonocytes and the primary trophic factor in the colon. In the present study only 15 and 6% of ME contributed to energy retention from potato and soya fibre, respectively, whereas the remainder was metabolized and lost as heat.

References

Livesey G., Smith T., Eggum B.O., Tetens I.H., Nyman M., Roberfroid M., Delzenne N., Schweizer T.F. & Decombaz J. (1995). Determination of digestible energy values and fermentabilities of dietary fibre supplements: a European interlaboratory study *in vivo*. *Br. J. Nutr.* 74:289-302.

Jensen B.B. & Jørgensen H. (1994). Effect of dietary fiber on microbial activity and microbial gas production in various regions of the gastrointestinal tract of pigs. *Appl. Environ. Microbiol.* 60:1897-1904.

Just A., Fernández J.A. & Jørgensen H. (1983). The net energy value of diets for growth in pigs in relation to the fermentative processes in the digestive tract and the site of absorption of the nutrients. *Livest. Prod. Sci.* 10:171-186.

Jørgensen H., Zhao X.Q. &. Eggum B.O. (1996a). The influence of dietary fibre and environmental temperature on the development of the gastrointestinal tract, digestibility, degree of fermentation in the hind-gut and energy metabolism in pigs. *Br. J. Nutr.* 75:365-378.

Jørgensen H., Zhao X.Q., Bach Knudsen K.E. & Eggum B.O. (1996a). The influence of dietary fibre source and level on the development of the gastrointestinal tract, digestibility and energy metabolism in broiler chickens. *Br. J. Nutr.* 75:379-395.

Jørgensen H., Larsen T., Zhao X.Q. & Eggum B.O. (1997). The energy value of short-chain fatty acids infused into the caecum of pigs. *Br. J. Nutr.* 77:745-756.

THE EFFECTIVE ENERGY OF SIX POULTRY DIETS MEASURED IN YOUNG AND ADULT BIRDS

D.J. Farrell[1,2] A. Smulders[1], P.F. Mannion[1], M. Smith[2] and J. Priest[2]

[1]The Queensland Poultry Research and Development Centre, Alexandra Hills Qld 4163, Australia; [2]The Department of Agriculture, The University of Queensland, St Lucia, Qld 4072, Australia

Summary

An experiment was undertaken to determine the effective energy content of a basal diet and of five ingredients when included individually at 0.33 in the basal diet using growing broilers and adult hens. There were large differences in the apparent metabolizable energy of hens and broilers resulting in substantially higher effective energy values for hens. When compared with values used to construct diets in practice and based on the effective energy system, these values were in better agreement with those determined here with adult birds. It was concluded that it is extremely time consuming to determine effective energy values for feed ingredients especially as they differ between young and old birds.
Key words: Poultry, broiler, layer, effective energy, digestibility.

Introduction

The concept of effective energy of domestic livestocks has appeared in various publications and recently in more detail and in its final form by Emmans (1994). This and other energy systems were reviewed recently by Farrell (1996).

The effective energy system and its components can be applied across domestic species and the same coefficients used to predict their effective energy requirements as well as the effective energy content of feedingstuffs. However much of the data on requirements are derived from selected, published and unpublished experiments that meet certain criteria identified by Emmans (1994). The measurement of the effective energy content of a feedingstuff or diet requires a number of chemical analyses and the need to use biological data. Effective energy values are normally derived from published tables which may or may not be appropriate and, to our knowledge, no one has actually derived data for ingredients or diets using livestock in order to compute specifically the effective energy of a feedstuff or diet.

The purpose of this paper was to determine with broiler chickens and adult hens the effective energy content of six diets. It is known that young and old birds differ in their ability to metabolize the same diet (Johnson, 1987).

Materials and Methods

The effective energy content of a feedstuff (kJ/g) =
$AME_n - W_d * FOM - 0.16\ W_u * DCP + 12\ Z * DCL$

AME_n = apparent metabolizable energy of the feedstuff corrected to zero nitrogen (n) balance.
FOM = faecal organic matter (g) and $W_d = 3.8$ kJ/g.
DCP = digestible crude protein (g) and $W_u = 29.2$ kJ/gN.
DCL = digestible crude lipid (g).
Z = proportion of dietary fat that is deposited directly into tissue lipid and is a positive characteristic of the feed.

For broiler chickens, Z is normally 0.33 but may vary, particularly in laying hens when incorporated into egg lipid. The coefficients are 4.4 kJ/g and 16.6 kJ/g for direct incorporation and for synthesis of lipid and from non-dietary fat substrate respectively.

The basal diet (g/kg) consisted of maize 863, fish meal 68, soybean meal 24, minerals and vitamins 10 and bentonite 10. The six diets used were (i) the basal diet; 0.33 of this was replaced with either (ii) millrun, (iii) meat and bone meal, (iv) grain sorghum, (v) sunflower meal or (vi) field peas. Each diet was fed to five individually caged broiler chickens (14-21 d of age) for a period of six days; this was followed by a similar period of feeding to five individually caged laying hens. Excreta samples were collected during the last two days, pooled, dried, milled and subsampled for analyses.

At the end of the experiment birds were killed by cervical dislocation; the digestive tract removed and the contents of the last 15 cm of the ileum squeezed into containers. These were cooled immediately in ice, then frozen and later dried for subsequent analyses.

Acid insoluble ash (AIA) was used to determine dry matter digestibility of the diets at the terminal ileum and at the cloaca following the method of Van Keulen & Young (1977). Analyses for dry matter, ash and lipid in diets and excreta were undertaken according to the AOAC (1994). Nitrogen (N) was measured in subsamples of feed, ileal digesta and excreta using the Leco Nitrogen Analyser (Leco Corp., Michigan). AME values were adjusted to zero N retention using the factor of 34 kJ/gN. The data for each parameter were analysed using an analysis of variance on the raw values. In some cases data were missing, mainly because of insufficient material for analysis. In other cases digesta were combined to provide sufficient quantity for analysis.

Results

The results of the measurements made are given in Table 1. These are for the diets and not the individual ingredients. It can be seen that AME_n of each diet was consistently lower for broilers than for laying hens by on average 1.9 MJ/kg. Only the basal diet and the diet with field peas gave similar ($P>0.05$) values for AME_n for young and old birds. Overall effective energy was also reduced in broilers by 2.2 MJ/kg on a dry matter basis. This difference was due to the consistently lower dry matter digestibility in broilers compared to layers but the digestibility of crude protein was very similar. Differences were larger for byproducts, particularly sunflower seed meal, which contains high concentrations of dietary fibre.

Discussion

There was considerable variability in dry matter digestibility between birds; this seemed to be higher among individual broilers than hens (see Table 1). Small sample size and some missing values due to insufficient sample size added to the variation.

Since the starting point of these calculations is AME_n, differences between broilers and layers for this parameter must necessarily carry over to effective energy. The AME_n value for the basal diet is close to that expected, given its ingredient composition.

In many laboratories adult birds are used to measure AME_n and other characteristics of a feed ingredient or diet, e.g.. The European Table of Energy Values for Poultry Feedstuffs (1986) is based on data mainly from adult cockerels. It is clear that such AME_n values do not always have application to broilers and the magnitude of the differences is not constant (Table 1). Effective energy values for young and old birds may be further affected because faecal organic matter was generally greater for young than older birds although DCP was similar; DCL makes only a small contribution to effective energy calculations but was lower for young than adult birds.

Table 1. The apparent metabolizable energy adjusted to nitrogen (n) balance (AME_n), faecal organic matter (FOM), digestible crude fat (DCL), digestible crude protein (DCP) and the effective energy (EE) of diets containing 0.66 basal and 0.33 single ingredient when fed to broilers (B) and adult hens (H). The total number of observations (n) and the overall mean ± standard deviation (SD) are also given†.

Diet		AME_n (MJ/kg DM)		FOM (g/100 g)		DCL (g/100 g)		DCP (g/100 g)		EE MJ/kg DM)	
		B	H	B	H	B	H	B	H	B	H
1.	Basal	12.71^{a}	13.61^{ab}	18.6^{a}	13.4^{a}	4.3^{a}	5.5^{a}	10.8^{a}	11.1^{a}	11.74^{a}	13.05^{a}
2.	Basal + Millrun	10.44^{b}	12.81^{bc}	30.8^{bd}	18.7^{b}	1.5^{bc}	3.2^{b}	12.9^{b}	12.4^{ab}	8.53^{b}	11.63^{abc}
3.	Basal + Meat and bone meal	10.92^{ab}	12.95^{b}	27.7^{bd}	21.1^{bc}	4.1^{ad}	6.5^{c}	19.4^{d}	19.0^{d}	9.12^{b}	11.98^{abc}
4.	Basal + Sorghum	12.04^{a}	14.19^{a}	23.9^{ad}	18.9^{b}	1.0^{b}	2.4^{d}	9.9^{a}	10.0^{a}	10.71^{a}	12.81^{a}
5.	Basal + Sunflower meal	8.77^{c}	12.09^{c}	44.1^{c}	25.5^{c}	2.1^{cd}	2.5^{d}	17.7^{c}	16.9^{d}	6.35^{c}	10.21^{bc}
6.	Basal + Field peas	12.91^{a}	13.19^{b}	24.0^{ad}	20.6^{b}	1.4^{bc}	2.3^{d}	14.0^{b}	14.2^{b}	11.39^{a}	12.03^{a}
Mean		11.27^{A}	13.17^{B}	28.0^{A}	19.7^{B}	2.4^{A}	3.5^{B}	14.1^{A}	13.8^{A}	9.64^{A}	11.90^{B}
n		28	26	28	26	29	25	30	19	27	15
SD		1.03	0.57	5.1	2.6	0.58	0.16	0.57	0.78	1.2	0.73

† Values within a column (a - d) or pair of means (A-B) without the same superscript are significantly different ($P<0.05$).

A comparison is made in Table 2 between AME_n and effective energy values measured here for the six diets with those used in diet formulation. The latter data are normally used in diet formulation of broiler chickens when using the effective energy system but are in better agreement with those determined here with adult birds for both AME_n and effective energy.

Table 2. A comparison of the apparent metabolizable energy (AME_n MJ/kg DM) and effective energy (EE MJ/kg DM). Values determined with broilers and layers compared with table values used in formulation of diets in practice for broilers by R. Gous, University of Natal (1997).

	Broilers		Hens		Tables	
	AME_n	EE	AME_n	EE	AME_n	EE
Basal (B)	12.71	11.74	13.61	13.05	13.49	12.73
B + millrun	10.44	8.53	12.81	11.63	11.90	10.81
B + meat & bone	10.92	9.12	12.95	11.98	12.29	10.94
B + sorghum	12.04	10.71	14.19	12.81	13.39	12.67
B + sunflower meal	8.77	6.35	12.09	10.21	11.57	10.12
B + field peas	12.91	11.39	13.19	12.03	12.56	11.53

The effective energy system is unquestionably more refined than the AME system and is likely to predict bird performance more accurately. This advantage is offset to some extent by the laborious and time-consuming measurements needed to determine effective energy of diets or feedstuffs particularly if different sets of values are needed for growing chickens and adult birds.

References

AOAC (1994). Association of Official Analytical Chemists. Official Methods of Analysis 16th edn. Washington, DC, Association of Official Analytical Chemists.

Emmans, G.C. (1994). Effective energy: a concept of energy utilization across species. *Br. J. Nutr.* 171:801-821.

Farrell D.J. (1996). Proceedings XX World's Poultry Congress Vol 11. World's Poultry Science Association, New Delhi, pp. 225-235.

Johnson R.J. (1987). Metabolizable energy, recent research in poultry. In: D.J. Farrell (Ed) Recent Advances in Animal Nutrition in Australia 1987. University of New England, Armidale, NSW. pp. 228-243.

Van Keulen J. & Young B.A. (1977). The evaluation of acid-insoluble ash as a natural marker in ruminant digestibility studies. *Canad. J. Anim. Sci.* 44:282-287.

THE EFFECT OF GRASS SILAGE:CONCENTRATE RATIO IN THE DIET OF BEEF CATTLE AND LAMBS ON THE EFFICIENCY OF UTILIZATION OF METABOLIZABLE ENERGY FOR GROWTH

R.W.J. Steen, L.E.R. Dawson, N. Lavery, D.E. Kirkpatrick and S.D. Johnston

Agricultural Research Institute of Northern Ireland, Hillsborough, Co. Down BT26 6DR, UK

Summary

A series of nine experiments was carried out at this Institute involving diets containing a wide range of grass silage:concentrate ratios (zero to 86% concentrates). Data from these studies have been analysed to examine the effect of silage:concentrate ratio in the diet on the efficiency of utilization of metabolizable energy (ME) for growth. The studies involved both beef cattle (four studies) and lambs (five studies) and energy retention was estimated by indirect calorimetry (six studies) or by comparative slaughter (three studies). Estimates of the efficiency of utilization of ME for silage were 8% lower than those for concentrates, indicating that the net energy content of the high digestibility silages used in these studies as a proportion of their ME content was 8% lower than that for a cereal-based concentrate.

Key words: Beef cattle, grass silage, concentrates, efficiency of utilization of metabolizable energy.

Introduction

The results of a number of metabolism and comparative slaughter experiments (e.g. Thomas *et al*, 1988) have suggested that the metabolizable energy (ME) in silage-based diets is utilized with lower efficiency for growth than that in other diets. However, the magnitude of calculated values for the efficiency of utilization of ME may depend greatly on factors such as the assumed maintenance energy requirements of the animals and the digestibility or ME concentration of the forages involved. For experiments carried out in the United Kingdom over the past 15 years, efficiencies of utilization of ME for growth have generally been calculated using assumed maintenance energy requirements given by the Agricultural Research Council (1980). However, the results of recent metabolism studies involving dairy cows (Yan *et al*, 1997) beef cattle and sheep (Steen *et al*, 1997) have indicated that the maintenance energy requirements of cattle and sheep given grass silage-based diets are higher than those given by ARC (1980) and AFRC (1990). Nine experiments have been carried out at this Institute involving diets containing a wide range of grass silage:concentrate ratios. Data from these studies have been analysed to examine the effect of silage:concentrate ratio in the diet on the efficiency of utilization of ME for growth, using maintenance energy requirements for the animals which have been estimated from the results of the same experiments. These were 0.61 and 0.46 MJ/kg $W^{0.75}$ for cattle and sheep respectively (Steen *et al*, 1997).

Materials and Methods

High digestibility grass silages containing 716 to 742 (mean 726) g digestible organic matter/kg dry matter, were used in all nine experiments. All silages were well preserved with ammonia N concentrations of 52 to 97 (mean 77) g/kg total nitrogen, and dry matter and crude protein concentrations of 151 to 229 (mean 199) g/kg and 127 to 172 (mean 147) g/kg DM respectively. The concentrates used in experiments 1 to 7 and experiment 9 were composed of rolled or ground barley, soyabean meal and minerals and vitamins while that used in experiment 8 was also based on barley and soyabean meal but also contained 10% maize meal and 20% citrus pulp. Four of the experiments involved finishing beef cattle between 405 to 645 kg live weight and five involved lambs between 29 to 58 kg live weight. Metabolizable energy intakes were measured by indirect calorimetry and energy retention by indirect calorimetry or by comparative slaughter. Further details of the individual experiments are given in Table 1. Relationships have been developed between efficiency of utilization of ME for growth and proportion of concentrates in the diet, the following variables having been included in the analyses:

1. differences between experiments only.
2. differences between experiments and the lipid:protein ratio in the gain.
3. differences between experiments and the energy concentration in the diets (ME/GE).
4. differences between experiments, the energy concentration in the diets and lipid:protein ratio in the gain.
5. the energy concentration in the diet and lipid:protein ratio in the gain.

Table 1. Description of experiments.

Expt no.	Species of animal	Method of estimating energy retention	Proportions of concentrates in diet (%)	DOM concentrations in silage (g/kg DM)
1	Cattle	Calorimetry	0, 25, 43, 68	725
2	Cattle	Calorimetry	0, 52	742
3	Sheep	Slaughter	0, 25, 50	720
4	Sheep	Calorimetry	0, 25, 50	720
5	Sheep	Calorimetry	34, 35, 70, 72	716
6	Sheep	Slaughter	0, 21, 82	730
7	Sheep	Calorimetry	0, 21, 86	730
8	Cattle	Calorimetry	21, 78, 78	732
9	Cattle	Slaughter	0, 35, 36, 61	721

For studies in which energy retention was estimated by calorimetry the composition of the gains in corresponding animal production studies involving the same or similar diets were used.

Results

The best fit relationship between efficiency of utilization of ME for growth and the proportion of concentrates in the diet when only differences between experiments were removed (Equation 1) was:

$$y = 0.519 \text{ (se } 0.0415) + 0.00039 \text{ (se } 0.000411) x \qquad (R^2{=}0.05; \text{NS})$$

where y = efficiency of utilization of ME and x = % concentrates in diet.

Including either the energy concentration in the diet or the composition of the gain or both in the relationship between the efficiency of utilization of energy and the proportion of concentrates in the diet, in addition to differences between experiments, did not significantly improve the relationship. However, when differences between experiments were excluded from the analysis, the effects of proportion of forage in the diet, energy concentration in the diet and the composition of the gain on the efficiency of utilization of energy were all significant ($P<0.05$; $P<0.01$ and $P<0.001$ respectively) when all three variables were included in a single relationship. This relationship (Equation 2) was:

$$y = -2.557 \text{ (se 0.864)} - 0.00346 \text{ (se 0.00135)}\ x_1 + 4.36 \text{ (se 1.33)}\ x_2 + 0.109 \text{ (se 0.0268)}\ x_3 \quad (R^2 = 0.50;\ P<0.001).$$

where y = efficiency of utilization of ME, x_1 = % concentrates in diet, x_2 = ME/GE ratio in diet and x_3 = lipid:protein ratio in the gain.

Discussion

On the basis of equation 1, the proportion of concentrates in the diet did not significantly affect the efficiency of utilization of ME. Nevertheless using equation 1 to calculate efficiencies gives values of 0.519 and 0.558 for silage only and concentrate only diets respectively. This would suggest that the efficiency of utilization of ME from silage was 7% lower than that from concentrates. To enable a similar estimate to be made using equation 2, the relationship between the proportion of concentrates and ME/GE ratio in the diets has been developed. Estimated ME/GE values for all silage and all concentrate diets from this relationship are 0.64 and 0.73 respectively, which with a value of 2.7 for the lipid:protein ratio in the gain (i.e. the mean value for all the data) gives estimated efficiencies of utilization of ME of 0.528 and 0.574 for silage only and concentrate only diets respectively using equation 2. The value for silage is 8% lower than that for concentrates, indicating that the net energy content of the high digestibility grass silages used in these studies as a proportion of their ME concentration was 8% lower than that of cereal-based concentrates.

This value is close to those reported by Johnston and Steen (1997) who found that the efficiency of utilization of ME for growth was 9% lower for high digestibility silage than for concentrates in each of two experiments. The higher estimate of the efficiency for silage compared to concentrates in the present studies compared to the low efficiencies reported for silage in earlier studies may be partly due to the lower maintenance energy requirements which were assumed for the animals in the earlier studies, and the fact that ME intake was higher for the high-concentrate diets than for the high-silage diets in some studies. In the present studies ME intakes were equalized across most of the treatments in seven of the nine studies. The high digestibility of the silages used in the present studies is also likely to have contributed to the high efficiencies of utilization of ME from them.

References

Agricultural and Food Research Council (AFRC) (1990). Technical Committee on Responses to Nutrients, Report No. 5. Nutrient Requirements of Ruminant Animals : Energy. *Nutr. Absts. and Revs.* Series B 60: 729-804.

Agricultural Research Council (ARC) (1980). Nutrient Requirements of Ruminant Livestock. Commonwealth Agricultural Bureaux, Slough, UK.

Johnston S.D. & Steen R.W.J. (1997). Responses in the growth of body components of finishing lambs to additional metabolizable energy supplied from either grass silage or concentrates. Submitted to Animal Science.

Steen R.W.J., Dawson L.E.R., Lavery N., Kirkpatrick D.E & Johnston S.D. (1997). Estimation of maintenance energy requirements of beef cattle and lambs. Proceedings of the British Society of Animal Science, Annual Meeting p. 118.

Thomas C., Gibbs B.G., Beever D.E. & Thurnham B.R. (1988). The effect of date of cut and barley substitution on gain and on the efficiency of utilization of grass silage by growing cattle. *Br. J. Nutr.* 60:297-306.

Yan T., Gordon F.J., Agnew R.E., Porter M.G. & Patterson D.C. (1997). The metabolizable energy requirements for maintenance and the efficiency of utilization of metabolizable energy for lactation by dairy cows offered grass silage-based diets. *Livest. Prod. Sci.* (In Press).

METHANE SUPPRESSION BY CALCIUM SOAPS OF STEARIC, OLEIC AND LINOLEIC ACID MIXTURES IN CATTLE

T. Nishida[1], M. Kurihara[1], A. Purnomoadi[2], F. Terada[1] and M. Shibata[3]

[1]National Institute of Animal Industry, Tsukuba Norindanchi P.O. Box 5, Ibaraki, 305 Japan; [2]Diponegoro University, Semarang, 50241 Indonesia; [3]Agriculture, Forestry and Fisheries Research Council Secretariat, Chiyoda-ku Kasumigaseki 1-2-1, Tokyo, 100 Japan

Summary

Two types of calcium soaps of fatty acids (Ca-FA) were used; Ca-FA-1 (C18:0 17.0%, C18:1 65.0%) and Ca-FA-2 (C18:0 20.0%, C18:1 40.0%, C18:2 35.0%). In Expt 1 four dry cows were fed Italian ryegrass hay and concentrate as basal diet (BD). Treatments were BD, BD + Ca-FA-1 (300g), BD + Ca-FA-1 (600g) or BD + Ca-FA-2 (600g). In Expt 2 four lactating cows were fed bahiagrass hay and concentrates with or without the Ca-FA-2 supplements (500g). In Expt 1 there were no significant differences in NDF digestibility by feeding Ca-FA. Methane production per dry matter intake (DMI) in dry cows tended to decrease about 7-9% (17-23 litres/day) by feeding Ca-FA but no effects of degree of saturation and amount of Ca-FA fed were observed. In Expt 2 DMI and digestibilities of dry matter (DM) and crude protein (CP) were not affected by feeding Ca-FA. Methane suppression in lactating cows was about 4% (22 litres/day/kg DMI).

From the above results, it is concluded that feeding 300-600 g/day of Ca-FA is effective in suppressing methane production without reduction of productivity.

Key words: Methane production, calcium soaps, cows, digestibility.

Introduction

It is important now to reduce methane production from ruminants because it is recognized to be one of the major greenhouse gases. Methane excretion by ruminant animals is about 8% of total energy intake of the animal (Blaxter, 1967). The methanogenic bacteria produce methane in the rumen mainly from hydrogen and carbon dioxide. Hydrogenation of the double bonds of unsaturated fatty acids might share hydrogen with the methanogenesis process in the rumen (Blaxter & Czerkawski, 1966). Therefore, the addition of unsaturated fatty acids to the diet may result in an inhibition of methane production.

Increasing energy density by feeding supplemental Ca-FA can increase production, especially in high producing lactating cows (Schneider *et al.*, 1988, Sklan *et al.*, 1989, Andrew *et al.*, 1991). Preformed calcium soaps are easily mixed with rations and do not affect normal rumen fermentation (Jenkins & Palmquist, 1984) and digestibilities of most nutrients (Schauff & Clark, 1992). For that reason it is practical for use by farmers to suppress methane production from ruminants without reducing production performance.

The objectives of the experiments were to evaluate the effect of unsaturated Ca-FA supplementation on the methane production and digestion of nutrients in cattle.

Materials and Methods

Two experiments were conducted using 4 dry cows (Expt 1, average live weight 643kg) in one way classification and 8 lactating cows (Expt 2, average live weight 626kg) in one way classification with blocks (protein source). Two types of Ca-FA were used; Ca-FA-1 (C18:0 17.0%, C18:1 65.0%; Fine feed I, Miyoshi Oils and Fat Co., Ltd.) and Ca-FA-2 (C18:0 20.0%, C18:1 40.0%, C18:2 35.0%; Fine feed B). The chemical composition of the hay, concentrate and Ca-FA used in the trial are shown in Table 1. In Expt 1 dry cows were fed a diet of 60% Italian ryegrass hay and 40% concentrate on DM basis to meet metabolizable energy (ME) requirement (AFFRC, 1994) as basal diet (BD). Treatments were BD (control), BD + Ca-FA-1 (300g, C18:1L), BD + Ca-FA-1 (600g, C18:1H) or BD + Ca-FA-2 (600g, C18:2H) with 14 days feeding period. Faeces and urine were collected for the last 4 days, and each cow spent 2 days of each collection period in open-circuit respiration chambers for the measurement of respiratory exchange (Iwasaki *et al.*, 1982). In Expt. 2 lactating cows were fed bahiagrass hay *ad libitum* and concentrates with or without the Ca-FA-2 supplement (500g, C18:2) to meet 80% of ME requirement (AFFRC, 1994) for 14 days. Faeces and urine collections were done for the last 7 days, and studies of energy balance were conducted during the last 5 days of the collection period using open circuit respiration chambers (Mukai *et al.*, 1989).

All data were analyzed by the GLM procedure (SAS, 1988) and treatment means were tested with Tukey's multiple range test if F tests were significant ($P<0.05$).

Table 1. Chemical composition of hay, concentrate and calcium soaps of fatty acids (Ca-FA) (g/kg or MJ/kg DM).

	DM	CP	EE	NDF	GE
Expt 1 (dry cows)					
Italian ryegrass hay	884	99	29	620	17.8
Concentrate[1)]	902	173	40	188	18.3
Ca-FA-1	971		870		33.3
Ca-FA-2	970		880		34.0
Expt 2 (lactating cows)					
Bahiagrass hay	888	73	15	709	17.9
Concentrate (CaFA2+)[2)]	888	148	63	171	16.9
Concentrate (CaFA2-)[3)]	886	178	29	193	16.4

DM: Dry matter, CP: Crude protein, EE: Ether extract, NDF: Neutral detergent fibre, GE: Gross energy.

1) Ingredients (ADM%): corn 30%, barley 25%, wheat bran 9%, defatted rice bran 7%, soybean meal 12%, molasses 3.7%, alfalfa meal 7%, beat pulp 5%, $CaCO_3$ 1.3%.

2) Ingredients (ADM%): corn 41.9%, barley bran 2.2%, soybean meal 18.2-6.8%, fish meal 0-6.8%, alfalfa meal 10.6%, soybean hull 7.1%, beet pulp 7.3-13.7%, molasses 7.1%, calcium soaps of fatty acids 3.5%, $Ca_3(PO_4)_2$ 0.9%, mineral suppl. 0.7%.

3) Ingredients (ADM%): corn 50.3%, barley bran 2.7%, soybean meal 17.5-7.1%, fish meal 0-7.1%, alfalfa meal 7.1%, soybean hull 7.1%, beet pulp 14.0-17.7%, $Ca_3(PO_4)_2$ 0.5%, $CaCO_3$ 0.5%, mineral suppl. 0.7%.

Results and Discussion

The least square means of BW, milk yield, milk fat, digestibility, methane production, nitrogen balance and energy balance with dry and lactating cows are shown in Table 2. Milk fat concentration was lower in the Ca-FA period, but milk fat yield was more (1.01 kg/day) than the control (0.99kg/day) because of increased milk production. There were no significant differences in DM, CP and NDF digestibilities by feeding Ca-FA with dry cows or in DM and CP digestibility with lactating cows. These results were similar to previous experiments (Klusmeyer *et al.*, 1991; Schauff & Clark, 1992). DMI was not affected by feeding Ca-FA. Methane production tended to decrease when Ca-FA was fed to dry cows, but was increased by feeding Ca-FA (552.9 litres/day) compared to the control diet (540.4 litres/day) in lactating cows. Methane production per DMI in dry cows was tended to decrease about 6.7-8.7% (17-23 litres/day) by feeding Ca-FA but no effect of the degree of saturation nor amount of Ca-FA was observed. Methane production per kg DMI in lactating cows was reduced by feeding Ca-FA-2 (30.0 litres/kg) compared to the control (31.2 litres/kg), and methane suppression per DMI was about 3.8% (22.4 litres/day).

Table 2. Least square means of live weight (LW), milk yield, milk fat, digestibility, methane production, nitrogen balance and energy balance in dry and lactating cows.

	Expt 1 (dry cows)					Expt 2 (lactating cows)		
Treatment	Control	C18:1L	C18:1H	C18:2H	SE	Control	C18:2	SE
Number of cows	4	4	4	4		4	4	
LW (kg)	677	650	651	668	18	597	653	19
DMI (kg/day)	6.5	6.5^{ab}	6.5	6.5	0.1	17.4	18.4	0.9
Ca-FA intake(kg/day)		0.3	0.6	0.6			0.5	
Milk yield (kg/day)						25.6	27.3	1.2
Milk fat (%)						3.9	3.7	0.1
Digestibility (%)								
Dry matter	71.2	71.6	69.7	70.5	0.8	68.1	68.0	0.7
Crude protein	62.4	62.8	62.3	61.5	1.7	57.7	59.5	0.6
NDF[1]	66.9	66.9	66.4	67.0	1.1			
CH_4 (l/day)	263	240	245	243	5.1	540	553	22
CH_4/DMI (l/kg)	40.6	37.1	37.9	37.5	0.9	31.2	30.0	0.7
% of control	$(100)^{a}$	$(91.3)^{b}$	(93.4)	(92.4)		(100)	(96.2)	
CH_4/DDMI (l/kg)	57.0^{a}	51.8^{b}	54.3	53.2	1.0	46.0^{a}	43.1^{b}	0.7
N balance (g/day)								
Intake	134	134	134	134	2.6	402	436	21
Faeces	50.4	49.8	50.6	51.6	3.1	168	176	9.2
Urine	72.1	68.5	60.2	66.8	4.3	54.4	82.1	8.9
Milk						129	132	4.0
Retention	11.3	15.4	22.9	15.3	3.7	50.3	45.4	10
Energy balance ($kJ/kg^{0.75}$)								
ME intake	538^{A}	585^{A}	667^{B}	655^{B}	4.9	1580	1654	57
Milk						650	615	16
Heat production	488^{A}	512^{A}	546^{B}	529^{B}	11	862	913	21
Retention	50	73	121	127	23	69	126	30

a, b: Means in the same row within dry cows or lactating cows with different superscripts differ ($P<0.05$) by Tukey's multiple range test.

A, B: Means in the same row within dry cows or lactating cows with different superscripts differ ($P<0.01$) by Tukey's multiple range test.

1) NDF: Neutral detergent fibre.

Nitrogen balance was not affected by Ca-FA feeding in Expt 1 but urinary nitrogen excretion tended to increase with addition of Ca-FA in Expt 2. Metabolizable energy intake and heat production were significantly increased with dry cows and retention energy tended to increase when 500-600 g of Ca-FA were fed with dry and lactating cows.

From the results of these experiments we calculated the decrease of methane production C18:1-L; 1.36, C18:1-H; 0.80, C18:2-H; 0.86 and C18:2-H (lactating cows) 0.89 moles/mole Ca-FA given. Methane production per mole of Ca-FA was not affected by unsaturation of the acid 500-600 g of Ca-FA feeding periods. Blaxter & Czerkawski (1966) observed the effect of stearic, oleic and linolenic acid for methane depression were 3.1, 1.7 and 1.8 moles/mole respectively with sheep, but they reported loss of appetite and decrease of DMI and digestibility due to the effects on rumen microorganisms. In our study smaller effects on methane suppression compared with their experiments were confirmed by feeding Ca-FA with cows but no negative effects of DMI, digestibility, milk production, nitrogen and energy balance were observed. This is probably because Ca soaps protect ruminal microbes from the adverse effects of fat (Chalupa *et al.*, 1986). However, they observed a reduction in methane production per mole of unsaturated fatty acid increased with increasing unsaturation of the acid at a rate of 0.24 ± 0.09 moles/mole double bond. In our experiments Ca-FA-2 contained only 35.0% unsaturated fatty acid (C18:2) so the effects of degree of unsaturation might not be observed clearly.

From the above results, it is concluded that feeding 300-600 g/day of Ca-FA is effective in suppressing methane production without reduction of productivity.

References

Agriculture, Forestry and Fisheries Research Council Secretariat (AFFRC) (1994). Japanese Feeding Standards for Dairy Cattle. AFFRC. Tokyo.

Andrew S.M., Tyrrell H.F., Reynolds C.K. & Erdman R.A. (1991). Net energy for lactation of calcium salts of long-chain fatty acids for cows fed silage-based diets. *J. Dairy Sci.* 74:2588-2600.

Blaxter K.L. (1967). Energy metabolism of ruminant. Hutchinson Scientific and Technical. London.

Blaxter K.L. & Czerkawski J. (1966). Modifications of the methane production of the sheep by supplementation of its diet. *J. Sci. Food Agric.* 17:417-421.

Chalupa W., Vecchiarelli B., Elser A.E., Kronfeld D.S., Sklan D. & Palmquist D.L. (1986). Ruminal fermentation in vivo as influenced by long chain fatty acids. *J. Dairy Sci.* 69:1293-1301.

Iwasaki K., Haryu T., Tano R., Terada F., Itoh M. & Kameoka K. (1982). New animal metabolism facility especially the description of respirational apparatus. *Bull. Nat. Inst. Anim. Indust.* 39:41-78.

Jenkins T.C. & Palmquist D.L. 1984. Effect of fatty acids or calcium soap on rumen and total nutrient digestibility of dairy rations. *J. Dairy Sci.* 67:978-986.

Klusmeyer T.H., Lynch G.L., Clark J.H. & Nelson D.R. (1991). Effects of calcium salts of fatty acids and protein source on ruminal fermentation and nutrient flow to duodenum of cows. *J. Dairy Sci.* 74:2206-2219.

Mukai A., Shibata M. & Kurihara M. (1989). Energy metabolism of dairy cattle under high environmental temperature 1. Description of the energy metabolism laboratory at the Kyushu National Agricultural Experimental Station. *Bull. Kyushu Natl. Agric. Exp. Stn.* 26:27-69.

SAS (1988). SAS/STAT™ user's guide, release 6.03 ed. SAS Inst. Inc. Cary, NC, USA.

Schauff D.J. & Clark J.H. (1992). Effects of feeding diets containing calcium salts of long-chain fatty acids to lactating dairy cows. *J. Dairy Sci.* 75:2990-3002.

Schneider P., Sklan D., Chalupa W. & Kronfeld D.S. (1988). Feeding calcium salts of fatty acids to lactating cows. *J. Dairy Sci.* 71:2143-2150.

Sklan D., Bogin E., Avidar Y. & Gur-arie S. (1989). Feeding calcium soaps of fatty acids to lactating cows: effect on production, body condition and blood lipids. *J. Dairy Res.* 56:675-681.

EXTRAPOLATING FROM ME TO NE: UNINTENDED CONSEQUENCES

D.E. Johnson. E.M. Larson and M.J. Jarosz

Metabolic Laboratory, Colorado State University, Ft. Collins, CO, USA

Summary

The assumption of NRC (1996) that NE increases at a decreasing rate as ME of the feedstuff increases may be under-valuing feedstuffs at high and low ME concentrations. Errors of measurement, particularly of the independent term of predictive relationships can result in biased low coefficients. Energy retention efficiencies determined by respiration calorimetry are biased high and those arrived at by slaughter balance technique biased appositely. A reduced major axis regression should be explored to derive less biased constants.

Key words: Cattle, net energy, energy systems, regression.

Introduction

The necessarily empirical prediction of the NE value of diets is centrally important to energy requirement standards for animals. All systems (UK, US, Dutch, German, Swiss, French, Australian) start with a similar base of information on the feeds in their country, the digestible energy value of the feed. All but the US beef and sheep systems progress similarly to NE from there. Why the US system is different and its potential consequences is the subject of this paper.

Discussion

Nearly all world net (NE) systems base the conversion of metabolizable energy (ME) to net energy on a simple linear equation relating the efficiency of conversion (k) to the concentration of ME in the diet expressed as a ratio of joules (ME/GE) or ME/kg DM. The ARC (1980) for example predicts the efficiency of ME use for NE as growth (NEg) to vary as: $k_f = 0.78q + 0.006$ (see also Figure 1), where q = the ratio of ME to GE of the diet measured at maintenance and k_f = the ratio of NEg/ME. This procedure predicts a linearly increasing efficiency of ME conversion to NE as the energy availability in the diet increases. It follows that the NEg value of the diet per unit of DM increases in an ever increasing curvilinear manner as the ME per unit of DM increases (Figure 2).

The US prediction system for growing ruminants (NRC, 1996) however, results in a strikingly different pattern of increasing NE as diet digestibility increases. Both use feedstuff energy digestibility (DE) as the base and a similar conversion (i.e. x 0.82) to convert DE to ME and both result in a curvilinearly increasing NE with increasing ME. The foci of the curves are opposite, however, with the NE values in the US system increasing at an ever decreasing rate. As a consequence of this approach, the NEg values become negative at DE values of 0.40 or lower and approach a plateau when DE values increase to high levels with the US system, very unlike the other world systems at both high and low DE extremes.

Given this disparate picture of NE vs. ME of both high and low quality diets or feedstuffs, a critical examination of the relationship is needed. The data set used to develop the quadratic relationship between NE and ME for the US system has been described

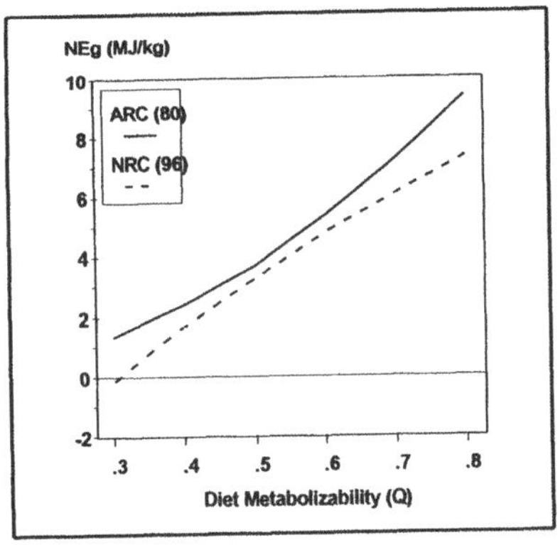

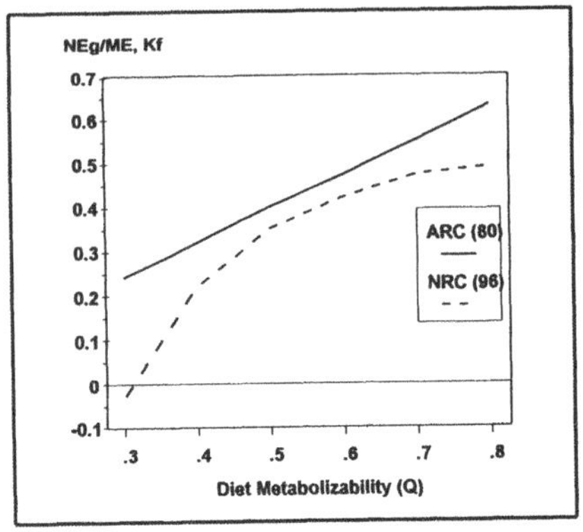

Fig. 1. ME use efficiency vs. ME/GE.

Fig. 2. Neg vs. Diet ME (ARC vs. NRC).

(Garrett, 1980). The observations were made from slaughter balance trials with the vast majority of diets of medium quality (i.e. 65% between 72 and 79% DE and 22% between 51 and 72% DE).

Less than 1% of the observations were with diets <51% DE and only 12% were with diets >79% DE. Using this relationship to predict the value of either straws on the low end or fats on the upper end will predict feed values beyond the range of the equations solution set. The energy values of diets at these extremes are where deviations from other world systems are greatest. Medium quality diet NE values are predicted to be similar by both US and the other world schemes.

As noted in a companion paper (Larson and Johnson, 1998), the body of evidence in the world literature relating NE to ME, shows no support for a curvilinear relationship. This holds for the data derived from comparative slaughter as well as respiration calorimetry methodology. Both data sets show a large degree of scatter, however, and it is not possible either to conclude that there is no curvilinearity. The lack of precision of these regression relationships causes further concern. Regression coefficients are biased low if the independent variable is measured with error. Other things being equal, low quality diets are expected to be biased high and conversely for high quality diets.

The use of regression techniques to establish biological relationships introduces a bias in proportion to the error of measurement of the independent variable. The change in Y in response to a change in independent variable X will be underestimated. Absolute values of Y at mean values of X will be unaffected, values of Y at upper ranges of X will be underestimated and values of Y at lower ranges of measured X's will be overestimated. These errors can be minimised by high accuracy of measurement of X and by choosing the variable measured with the most accuracy as the X variable. However, these conditions are not always available in biological circumstances.

Extrapolation or determination problems occur for many net energy system components. For example, the relationship of retained energy to ME intake can show different results depending on the mathematical technique. A problem can be encountered when the partial efficiency of ME use for a given diet is determined with varying error across levels of intake. In two example experiments, heat production was measured by indirect respiration calorimetry while feeding steers at near *ad libitum* and at approximately 60% of this intake (Wedegaertner, 1980; Benz, 1982). The regression of individual animal HP on ME as the independent found high r^2 relationships (0.99) in the first experiment but lower r^2 in the latter case (0.66 and 0.86). The regression of individual animal data indicated that 32 to 38% of ME intake above maintenance was given off as heat by steers in the first experiment and 43 to 52% in the second (Table 1). If, however, the high and low intake group means are used to calculate this heat vs. ME response, a nearly identical result is obtained for the first but not the second experiment.

Table 1. Apparent ME use efficiency vs. calculation method.

Expt	Trt	r^2	HP vs. ME by individual regression slope (inefficiency)	HP high-low group/ ME high-low group slope (inefficiency)
I	1	0.99	0.38	0.38
	2	0.99	0.32	0.33
II	1	0.66	0.52	0.69
	2	0.86	0.43	0.57

The slopes agree for HP response to ME for the first experiment, where the statistical relationship is strong, but the two methods of calculation do not agree for the second experiment when r^2 is lower. In this case, the regression across the entire individual animal data results in slopes that are biased low. Heat increments are thus biased low (and ME use efficiencies biased high) in relative proportion to independent variable measurement error for respiration calorimetry experiments.

The opposite occurs for slaughter balance experiments, where retained energy is the primary dependent and biased low and thus ME use efficiencies are biased low. Is it possible that the widely observed (Larson and Johnson, 1997; Geay, 1984; NRC, 1996; ARC, 1980; etc.) lower slaughter balance determined efficiencies as compared to those determined by respiration calorimetry result partially, at least, from a differential regression bias?

An analogous problem can occur when relating NE to ME concentration, J/kg DM in diet where results vary with error of measurement and/or choice of variable as the dependent. For example, the analysis of 73 slaughter balance experiments reported in the literature (Larson *et al.*, 1997) found; NEg = 0.61ME - 2.34, r^2=0.41, when choosing NE as dependent. The solution with ME as dependent was; ME = 0.667NEg + 8.08, which solves to NEg = 1.50ME -12.1. The latter equation is clearly untenable, biologically and leaves both equations in question. In this example using slaughter balance data, it can be argued that NE is determined with least error and thus should be the independent variable.

Given these difficulties, another form of regression analysis, i.e. reduced major axis analysis, should be explored when assessing central tendency of biological relationships as opposed to statistical variance components. Such a technique has been proposed for estimating the appropriate exponent of body weight to relate to metabolic rate (Calder, 1987). If the regression coefficients for the world NEg vs. ME data, noted above, are adjusted to a reduced major axis (RMA) basis by dividing by the correlation coefficient, the dependent vs. independent anomalies disappear and the respiration calorimetry vs. slaughter balance differences are greatly reduced:

NEg = 1.08ME - 6.32 (RMA for respiration calorimetry)

NEg = 0.93ME - 5.15 (RMA for slaughter balance data).

The least squares regression coefficient by slaughter balance is 31% lower than by respiration calorimetry. The RMA analysis reduces this to a 15% lower value for slaughter balance. Analysis of the same data set for k_g predicted from measures of ME respond to the RMA differently since correlations are lower and not always higher with respiration calorimetry (Table 2).

Table 2. Comparison of least squares to reduced major axis (RMA) regression.

Technique	Y	B*X	I0	r^2
Least squares regression:				
RC[a]	NEg =	0.85ME	-3.77	0.61
SB[b]	NEg =	0.59ME	-2.18	0.41
RC	Kr =	0.63Q[c]	+0.12	0.21
SB	Kr =	0.57Q	+0.07	0.23
RMA regress:				
RC	NEg =	1.08ME	-6.32	0.61
SB	NEg =	0.93ME	-5.15	0.41
RC	Kr =	1.36Q	-0.31	0.21
SB	Kr =	1.21Q	-0.32	0.23

[a]Respiration calorimetry; [b]slaughter balance; [c]Q=ME/GE.

Conclusions

The NE and economic value of feedstuffs in the upper range of ME concentrations in use by the ruminant livestock industry are probably being underestimated as compared to average feedstuffs when using NRC (1996) beef and NRC (1989) growing dairy nutrient requirement schemes. Low quality feedstuffs are probably being undervalued. The use of nutrient content, digestion kinetics, end product absorption and alternative regression expressions of variations should be examined as potential methods to determine the productive value of feedstuffs/nutrients.

References

ARC (1980). Nutrient requirements of ruminant livestock. Commonwealth Agric. Bureau, London.

Benz D.A. (1982). Energy partitioning of steers fed bromgrass hay with monensin. M.S. Thesis, Colorado State University, Fort Collins.

Calder W.A. (1987). Scaling energetics of homeothermic vertebrates: An operational allometry. *Ann. Rev. Physiol.* 49:107-20.

Garrett W.N. (1980). Energy utilization of growing cattle as determined in 72 comparative slaughter experiments. In: L.E. Mount (Ed) Energy Metabolism, *EAAP Publ. No.* 26: 3-7.

Geay Y. (1984). Energy and protein utilization in growing cattle. *J. Anim. Sci.* 58:766-773.

Larson E.M. & Johnson D.E. (1998). Predicting net energy of feedstuffs for beef cattle. In: K.J. McCracken, E.F. Unsworth and A.R.G. Wylie (Eds) Proceedings 14th Symposium on Energy Metabolism, Newcastle. CAB International, Wallingford UK, pp. 355-358.

NRC (1989). Nutrient requirements of dairy cattle, 6th edition, NRC, National Academy Press.

NRC (1996). Nutrient requirements of Beef Cattle, NRC, 7th edition, National Academy Press.

Wedegaertner T.C. (1980). Monensin and level of intake effects on energy partitioning by steers. M. S. Thesis, Colorado State University, Fort Collins.

TREATMENT OF UREA-TREATED WHOLE CROP WHEAT TO IMPROVE ITS ENERGY VALUE FOR LACTATING DAIRY COWS

J.D. Sutton, S.B. Cammell, D.E. Beever, D.J. Humphries and R.H. Phipps

Centre for Dairy Research, Department of Agriculture, The University of Reading, Earley Gate, Reading RG6 6AT, UK

Summary

In an attempt to improve the energy value of urea-treated whole crop wheat (WCW) for milk production, energy and nitrogen balance was measured in four lactating cows given 1:2 grass silage:WCW (DM basis) *ad libitum* with concentrates. The WCW was fed untreated or treated with caustic soda to improve digestibility or supplemented with molasses on the forage or concentrates to reduce rumen ammonia concentrations. None of the treatments increased gross energy intake significantly but all increased milk energy and fat output. Caustic treatment increased fibre digestibility but reduced starch digestibility while molasses reduced fibre digestibility. There were no significant effects on energy digestibility or the partition of digestible or metabolizablesable energy. It is concluded that the benefits of the treatments are unlikely to be of practical importance to the farmer.

Key words: milk production, energy efficiency, whole crop wheat.

Introduction

In the United Kingdom whole crop wheat (WCW) is frequently harvested at a mature stage (about 60% DM) and preserved by the addition of urea. In feeding trials, milk production responses to the partial replacement of grass silage by WCW were found to be less than expected according to the conventional evaluation of the crop at about 10 MJ metabolizable energy (ME)/kg corrected DM (CDM). Leaver and Hill (1992) suggested, on the basis of indirect evidence, that this reflected inefficient utilizationation of ME possibly caused by the high intakes of ammonia in the WCW. However energy balance studies with lactating dairy cows (Cammell *et al.*, 1996) showed that a major problem with WCW was poor digestibility of both the starch and fibre fractions of the crop giving rise to low energy digestibility and an ME value of only 8 MJ/kg CDM. There was no evidence of unusually low efficiency of utilization of the ME.

The present experiment was an attempt to improve energy utilization from WCW for milk production either by improving digestibility by pre-treatment with caustic soda or by reducing rumen ammonia concentrations and increasing microbial protein synthesis by the addition of molasses.

Materials and Methods

Four multiparous Holstein-Friesian cows (mean live weight during the experiment 621 kg), at an average of week 11 of lactation, were allocated to a 4x4 Latin square experiment. Each period consisted of 5 weeks with measurements of feed intake and milk, faeces and urine output being made in respiration chambers for 6 days and gas exchange for 4 days during the last week. The Control diet consisted of a 1:2 grass silage:WCW mix (DM basis) fed *ad libitum* with 9 kg fresh weight of concentrates (234 g crude protein/kg DM) daily. WCW treatments were: Caustic - pre-treatment with 50 kg caustic soda/t DM followed by 60 l water; Mol/For - addition of 2 kg/day of molasses to the forage mix at feeding; Mol/Conc - addition of 2 kg/day of molasses to the concentrates at feeding. The concentrate ration was reduced and re-formulated as necessary to maintain a constant energy and nitrogen intake from concentrates and molasses combined. The WCW (cv Hussar) was a winter variety and was harvested at 550-590 g DM/kg on 26 July. It was treated with 40 kg urea/t DM before being stored in clamps under plastic sheeting. Grass silage was a first-cut perennial ryegrass treated with AddSafe (ammonium tetraformate). Daily rations were offered in two equal meals at 8.30 h and 16.30 h. Cows were milked twice daily.

Results

The WCW had a high starch content (Table 1). Caustic treatment increased the ash content and pH and reduced the energy and ammonia contents.

Table 1. Composition of the forages (g/kg corrected DM unless stated).

	Grass silage	WCW	Caustic WCW
Corrected DM (g/kg fresh)	289	587	622
Ash	90	43	93
NDF	537	467	411
Starch	-	311	250
Energy (MJ/kg CDM)	18.8	18.4	17.6
Crude protein	101	153	103
Ammonia N (g/kg total N)	106	282	49
pH	3.9	8.0	9.4
Lactic acid + VFA	89	13	13

Caustic increased fat yield ($P<0.05$) and tended to increase milk and protein yields ($P<0.10$) (Table 2). Both molasses treatments increased fat yield ($P<0.01$). Mol/For increased milk fat content ($P<0.05$) but tended to reduce protein content ($P<0.10$) while Mol/Conc tended to increase milk yield ($P<0.10$) but reduced protein content ($P<0.05$).

Table 2. Milk yield and composition.

	Control	Caustic	Mol/For	Mol/Conc	SE
Milk yield (kg/day)	26.3	27.6	26.5	27.6	0.39
Fat (g/kg)	41.6	42.4	43.7	41.9	0.60
Protein (g/kg)	31.6	31.7	30.8	30.2	0.38
Fat (g/day)	1093	1164	1161	1155	13.3
Protein (g/day)	828	867	815	828	12.3

Table 3. Intake and digestibility.

	Control	Caustic	Mol/For	Mol/Conc	SE
DM intake	19.4	20.3	20.0	19.6	0.33
OM intake	18.0	18.4	18.5	18.1	0.32
Digestibility					
OM	0.701	0.701	0.687	0.690	0.0058
Energy	0.674	0.676	0.659	0.664	0.0056
NDF	0.593	0.628	0.564	0.557	0.0085
Starch	0.909	0.862	0.880	0.899	0.0039

Treatments had no significant effects on intake or digestibility of DM or OM or on energy digestibility (Table 3). Caustic increased NDF digestibility ($P<0.05$) but reduced starch digestibility ($P<0.001$). Mol/For reduced starch ($P<0.01$) and tended to reduce NDF ($P>0.01$) digestibility while Mol/Conc only reduced the digestibility of NDF ($P<0.05$).

None of the treatments increased intakes of gross energy (GE), DE or ME significantly but all increased milk energy output (Caustic P<0.01; Molasses P<0.05) (Table 4). Molasses tended to increase heat losses (P<0.10) but the increase in energy mobilizationation was not significant (P>0.10).

Table 4. Energy balance (MJ/day) and metabolizable energy partition.

	Control	Caustic	Mol/For	Mol/Conc	SE
Intake	360	370	370	362	6.3
Faeces	117	120	126	122	1.4
Methane	26	26	27	26	0.4
Urine	11	10	11	12	0.2
Milk	82	87	85	86	0.9
Heat	123	127	135	129	3.1
Retained	0	-1	-15	-12	5.9
DE	243	250	244	241	6.0
ME	205	213	206	202	6.4
Partition of ME					
Heat	0.599	0.596	0.657	0.637	0.0202
Milk	0.402	0.409	0.415	0.425	0.0126
Retained	-0.001	-0.005	-0.073	-0.062	0.0304

There were no effects of treatments on partition of DE or ME. Of the DE, a mean of 0.046 (SE 0.0009) was partitioned to urine, 0.108 (SE 0.0021) to methane, and 0.846 (SE 0.0039) to ME. The partition of ME on Caustic was almost identical to that on Control (Table 4). The tendencies for higher heat losses and energy mobilization and a greater partition of ME to milk with the Molasses treatments were not significant.

The M/D values of Control, Caustic, Mols/For and Mols/Conc diets respectively were 10.6, 10.5, 10.3 and 10.3 (SE 0.17) MJ ME/kg CDM and were not significantly different.

Table 5. Nitrogen balance (g/day).

	Control	Caustic	Mol/For	Mol/Conc	SE
Intake	537	505	551	542	5.5
Faeces	203	212	217	204	5.4
Urine	211	153	204	214	4.9
Milk	130	136	128	130	1.9
Digested	335	292	334	338	9.3
Allantoin (mmol/day)	395	388	428	461	19.7

Caustic reduced N intake (P<0.01) by reducing the ammonia content of the WCW (Table 5). This was associated with a large reduction in N digested (P<0.05) and in urine N output (P<0.001). Molasses had no significant effect on N balance but tended to increase allantoin excretion (P<0.10). None of the treatments affected N retention (mean -1 g/day SE 4.3).

Discussion

The decision to treat WCW with caustic soda was based on observations in earlier studies that a major reason for the low energy value of urea-treated WCW was the poor digestibility of the starch in the wheat grains (Sutton *et al.*, 1997). Caustic soda is well-known as being effective in improving the digestibility of fibre and it was hoped that it might attack the husk of the grains so making the starch more digestible. However, while the caustic treatment did improve fibre digestibility, it actually reduced starch digestibility with no net effect on energy digestibility or partition of DE or ME. Milk energy output did increase by 5 MJ/day (6.1%) but this mainly reflected a small, non-significant increase of 10 MJ/day in GE intake rather than any improvement in efficiency of energy utilization. The small benefits in milk production must be balanced against the problems of using caustic soda on the farm.

The choice of molasses as a supplement was to attempt to utilize the high concentrations of ammonia found in the rumen when urea-treated WCW is fed to cows (Abdalla *et al.*, 1996). Further, it was argued that greater benefits might result from inclusion of the molasses in the forage, which was the source of the extra ammonia, than with the concentrates. Evidence regarding the effect of the molasses on microbial protein synthesis was conflicting. Allantoin excretion in urine tended to increase, suggesting an increase in microbial synthesis, but this conflicted with other indirect evidence based on the reduction in milk protein concentration and, in a parallel experiment with fistulated cows (unpublished), the absence of any effect on rumen ammonia concentrations. Molasses did increase milk energy output slightly though significantly but this reflected the effects of a variety of small non-significant positive and negative changes in energy intake and utilization. The higher milk energy output appeared to reflect primarily a tendency for increased energy mobilization with molasses. There was a tendency for dietary M/D values to be reduced by molasses supplements though this did not reach significance.

The present measurements of the energy utilization of diets containing high proportions of urea-treated WCW are very similar to those of the previous experiment (Cammell *et al.*, 1996). Neither experiment provides any evidence that ME utilization is unusually low with such diets.

Acknowledgements

This experiment constituted part of a project funded within the LINK programme "Technologies for Sustainable Farming Systems" by Agricultural Genetics Co, BOCM Pauls, Dalgety Agriculture, Hi Spec Engineering, ICI Nutrition, J Bibby Agriculture, Maize Growers Association, Milk Marketing Board of England and Wales, Ministry of Agriculture, Fisheries and Food, Rumenco, Trouw Nutrition UK and Zeneca Seeds UK.

References

Abdalla A.L., Sutton J.D., Humphries D.J. & Phipps R.H. (1996). Digestibility of diets of grass silage and whole-crop wheat in the rumen of lactating cows. *Anim. Sci.* 62:631.

Cammell S.B., Sutton J.D., Humphries D.J., Phipps R.H. & Beever D.E. (1996). Energy balance of lactating cows given diets of grass silage and whole-crop wheat. *Anim. Sci.* 62:632.

Leaver J.D. & Hill J. (1992). Feeding cattle on whole-crop cereals. In: B.A. Stark & J.M.Wilkinson (Eds)Whole-crop cereals.Chalcombe Publications, Canterbury pp. 59-69.

Sutton J.D., Abdalla A.L., Phipps R.H., Cammell S.B. & Humphries D.J. (1997). The effect of the replacement of grass silage by increasing proportions of urea-treated whole crop wheat on feed intake and digestibility and milk production by dairy cows. *Anim. Sci.* (in press).

THE INFLUENCE OF DIETARY CARBOHYDRATE AND PROTEIN LEVELS ON ENERGY AND NITROGEN UTILIZATION OF ATLANTIC SALMON IN SEAWATER

S.J. Helland and B. Grisdale-Helland

AkvaForsk, Institute of Aquaculture Research AS, N-6600 Sunndalsøra, Norway

Summary

A 10-week comparative slaughter and metabolism trial was conducted with Atlantic salmon. Four experimental diets were used in which the protein and carbohydrate contents were altered by substituting fish meal with wheat starch. Feed intake was similar in all groups, but both specific growth rate and feed efficiency ratio improved as the dietary protein content increased. The nitrogen budget showed that, per kg gain, nitrogen intake was positively related to the dietary protein content. This was again reflected in the content in the whole body, resulting in no change in percentage nitrogen retention. Heat increment increased linearly with increasing carbohydrate content in the diet. There was no effect of treatment on the level of non-faecal nitrogenous losses and, thus, the results indicate that the increase in heat increment was caused by the increase in dietary carbohydrate level.
Key words: Atlantic salmon, protein, carbohydrate, net energy, heat increment.

Introduction

The availability of dietary nutrients for essential metabolic processes and growth is a key factor in the economy of aquaculture production. Information about how dietary manipulations influence the net utilization of the macronutrients and energy for Atlantic salmon is limited. The objective of this study was to investigate the effect of substituting dietary protein for carbohydrate on net utilization of the diet.

Materials and Methods

A 10-week comparative slaughter and metabolism trial was conducted with Atlantic salmon. The fish were transferred to saltwater (9.4±0.3°C) 8 weeks prior to the experiment, and were less than a year of age at the start (0-year smolt). Groups of salmon (weight, 100±1.6 g; length, 20.4±0.7 cm) were randomly allotted to each of 12 combined growth and metabolism tanks (Grisdale-Helland & Helland, 1995). The water flow rate was 3.5 l/min and the water speed, measured 5 cm from the edge, was 12±0.2 cm/sec.

Four experimental diets were used in which the protein and carbohydrate contents were altered by replacing fish meal with wheat starch (Table 1). Ytterbium oxide was included in the diets (0.01%) as a digestibility marker. The diets (4 mm pellets) were produced by Dansk Ørredfoder, Brande, Denmark, using an experimental extruder.

The fish were fed in excess, six times per hour during the 16 hour light period. Feed waste was collected daily and feed intake determined (Helland *et al.*, 1996). An initial slaughter group of 15 fish was sampled after a two day fast. The weights of the whole fish, liver and the remaining viscera were recorded, as well as fish length. Fifteen fish from each tank were slaughtered in a similar manner at the termination of the experiment. During the ninth week of the experiment, faeces were collected from the effluent water using an automatic straining system (Choubert *et al.*, 1982).

Ytterbium oxide in diets and faeces was measured as described by Refstie *et al.* (1997).

The analysis of the carbohydrate content of the diets and faeces, together with the analyses of dry matter, nitrogen, ash, fat and energy (GE) in the diets, fish and faeces were measured as described by Grisdale-Helland & Helland (1997).

Table 1. Chemical analysis of diets (g or MJ/kg diet).

Diet	1	2	3	4
Dry matter	973	966	976	980
Dry matter basis:				
Protein	422	455	490	520
Fat	262	273	267	271
Carbohydrate	242	202	161	126
Ash	74	80	87	91
Gross energy	24.0	24.3	24.3	24.5

The branchial and urinary nitrogenous excretions were calculated by difference from the nitrogen balance. The energy content of these excretions was estimated by employing the ammonia to urea ratio obtained during a 4 day measurement in week 4 and the caloric constants for these metabolites reported by Elliot & Davison (1975). Continuous measurements of oxygen consumption of each group of fish were carried out over 14 days (weeks 3 and 4), and extrapolated for the entire experimental period. Total heat loss (energy for maintenance, activity and heat increment (HI)) was calculated based on oxygen consumption (13.6 kJ/g O_2; Elliot & Davison, 1975). The energy costs of maintenance (fasted fish swimming at zero speed) and activity (extra energy used for swimming against the actual water current) were estimated using data from an experiment with 50 g sockeye salmon at 10°C (Brett, 1964). HI was calculated as the differance between the total heat production and the energy used for maintenance and activity.

The data were analysed by linear regression using the dietary levels of protein or carbohydrate as the independent variable.

Results and Discussion

Both specific growth rate (SGR) and the feed efficiency ratio (FER) (Table 2) improved significantly as the dietary protein content increased and carbohydrate content decreased. The growth rate for the fish fed Diet 4 was similar to that expected for salmon of this size in seawater (Austreng *et al.*, 1987). The difference in growth rate was not a reflection of increased appetite, since feed intake, measured as a percentage of body weight, was not influenced by treatment. The hepatosomatic index of the fish increased with increasing dietary carbohydrate content, probably reflecting increased deposition of glycogen in the liver.

As expected, carbohydrate digestibility was negatively related to the carbohydrate inclusion level and decreased linearly from 0.79 to 0.73. Together with a 1% drop in fat digestibility, these changes resulted in a decrease in energy digestibility from 0.92 to 0.89 ($P<0.05$) when the dietary carbohydrate content was increased.

The increase in the protein content of the diets had only small effects on body composition, and only the change in protein content, from 17 to 18%, was significant. Nitrogen intake per kg gain was positively related ($P<0.05$) to the protein content in the diets (Table 3) and subsequently, to the content in the whole body. As a result, there was no change in percentage nitrogen retention (56±1%) and, in contrast to the results of Grisdale-Helland & Helland (1997) and Hemre *et al.* (1995) with Atlantic salmon, no protein sparing effect of carbohydrate was found. There were no treatment effects on the levels of nitrogenous losses in faeces or metabolic excretions.

Table 2. Performance of Atlantic salmon fed the experimental diets (mean±SE).

Diet	1	2	3	4	Sig.
SGR[1]	1.02±0.01	0.99±0.03	1.12±0.03	1.20±0.03	*
Intake[2]	0.69±0.01	0.66±0.02	0.68±0.02	0.69±0.01	NS
FER[3]	1.49±0.01	1.52±0.02	1.65±0.01	1.73±0.02	*
Hepatosomatic index	1.55±0.06	1.34±0.01	1.21±0.01	1.09±0.04	*

[1]Specific growth rate (% body weight/d) [100*(ln(end weight) - ln(start weight))/days fed].
[2]Daily feed intake as a percentage of body weight, dry matter basis.
[3]Feed efficiency ratio (g wet gain/g dry feed intake).
*Linear effect of dietary protein content (P≤0.05); NS, non-significant

The GE content of the diets was similar and thus, energy intake per kg gain (Table 3) reflected the differences in SGR and FER. Both in terms of per kg gain and as a percentage of GE intake, HI increased linearly with increasing carbohydrate content in the diet (P<0.05). The percentage of GE lost as HI increased from 10 to 12% with the increase in dietary carbohydrate content. In a trial in which rainbow trout were fed isoenergetic (digestible energy (DE) basis) and isonitrogenous diets, it was found that HI as a percentage of calculated levels of GE, DE and metabolizable energy (ME) was higher for the diet containing a high level of glucose in comparison with the diet containing a high level of lipid (Beamish *et al.*, 1986). It was noted by these authors that the higher HI may be a reflection of transforming or excreting excess glucose, or the result of a possible increase in metabolism of amino acids due to poor utilization of glucose for energy. There was no effect of treatment on the level of non-faecal nitrogenous losses in the present study, indicating that the change in HI was caused by the dietary carbohydrate level.

The results of Brett (1964) for the energy cost of swimming against a current for sockeye salmon, which were used in the calculation of maintenance and activity, are close to our unpublished results from a trial done with Atlantic salmon at 9.4°C. Both of these variables are directly related to the differences in growth rate (fish size).

Table 3. Partitioning of nitrogen and energy consumed by the fish, g or MJ/kg gain (mean±SE).

Diet	1	2	3	4	Sig.
Nitrogen intake	45.4±0.3	47.8±0.6	47.4±0.3	48.2±0.5	*
Faeces	3.6±0.1	4.0±0.1	3.7±0.2	4.0±0.1	NS
Metabolic[1]	16.3±0.8	17.9±0.6	16.8±0.1	16.5±0.4	NS
Deposited	25.5±0.7	25.9±1.0	26.9±0.3	27.7±0.6	*
Energy intake	16.2±0.1	16.0±0.2	14.7±0.1	14.2±0.1	*
Faeces	1.7±0.04	1.5±0.1	1.2±0.02	1.1±0.04	*
Metabolic[2]	0.4±0.02	0.4±0.01	0.4±0.003	0.4±0.01	NS
Heat	4.8±0.2	4.7±0.2	4.0±0.02	3.8±0.1	*
HI[3]	2.0±0.2	1.8±0.1	1.5±0.04	1.4±0.1	*
Activity	0.7±0.02	0.7±0.02	0.6±0.01	0.6±0.03	*
Maintenance	2.1±0.02	2.2±0.1	1.9±0.1	1.8±0.04	*
Deposited	8.3±0.3	8.8±0.2	8.5±0.2	8.0±0.2	NS
Difference	1.0±0.2	0.5±0.2	0.5±0.2	0.8±0.2	NS

[1]Calculated by difference.
[2]Calculated from metabolic nitrogenous losses in the nitrogen budget and the fraction between the measured excretion levels of ammonia and urea.
[3]Heat increment and feeding activity.
*Linear effect of dietary protein/carbohydrate content (P≤0.05); NS, non-significant.

The amount of energy deposited in the fish per kg gain (Table 3) was not linearly related to dietary protein content. As a percentage of GE intake though, energy deposition (NEp) increased linearly ($P<0.05$) with increasing dietary protein. This also tended to be the case for NEp as a percentage of DE and ME ($P<0.1$). The proportion of GE in the diets which the fish used for total net energy increased from 0.69 to 0.73 ($P<0.05$) when the protein level was increased.

The effects of increasing dietary carbohydrate level included increases in faecal losses, hepatosomatic index and HI. Furthermore, the amount of protein available for growth and the net energy level of the diet were both reduced by this change in diet composition. The parallel changes in protein and net energy contents in the diets preclude a determination of whether one, or both, of these factors was limiting for growth.

References

Austreng E., Storebakken T. & Åsgård T. (1987). Growth rate estimates for cultured Atlantic salmon and rainbow trout. *Aquaculture* 60:157-160.

Beamish F.W.H., Hilton J.W., Niimi E. & Slinger S.J. (1986). Dietary carbohydrate and growth, body composition and heat increment in rainbow trout (*Salmo gairdneri*). *Fish Physiol. Biochem.* 1:85-91.

Brett J.R. (1964). The respiratory metabolism and swimming performance of young sockeye salmon. *J. Fish. Res. Bd. Canada* 21:1183-1226.

Choubert G., Noüe De la J. & Luquet P. (1982). Digestibility in fish: improved device for the automatic collection of faeces. *Aquaculture* 29:185-189.

Elliott J.M. & Davison W. (1975). Energy equivalents of oxygen consumption in animal energetics. *Oecologia* (Berl.) 19:195-201.

Grisdale-Helland B. & Helland S.J. (1995). Methods to assess optimal nutrient composition of diets for Atlantic salmon. *J. Appl. Ichthyol.* 11:359-362.

Grisdale-Helland B. & Helland S.J. (1997). Replacement of protein by fat and carbohydrate in diets for Atlantic salmon (*Salmo salar*) at the end of the freshwater stage. *Aquaculture* (In Press).

Helland S.J., Grisdale-Helland B., & Nerland S. (1996). A simple method for the measurement of daily feed intake of groups of fish in tanks. *Aquaculture* 139:157-163.

Hemre G.-I., Sandnes K., Lie Ø., Torrissen O. & Waagbø R. (1995). Carbohydrate nutrition in Atlantic salmon, *Salmo salar* L.: growth and feed utilization. *Aquacult. Res.* 26:149-154.

Refstie S., Helland S.J. & Storebakken T. (1997). Adaptation to soybean meal in diets for rainbow trout, *Oncorhynchus mykiss*. *Aquaculture* (in press).

SUMMARY OF THE DISCUSSION

Discussion leader J.D. Sutton
Compiled by J.W. Schrama and J.P. Hutcheson

The discussion was divided into five sections reflecting the diverse contents of the session.

Maintenance

The opening discussion centred on the recent estimates on ME_m from ARINI, Hillsborough presented at the current meeting and also at the 1997 Winter meeting of the British Society of Animal Science. In all of these papers estimates of ME_m for ruminants, offered diets based primarily on grass silage, were markedly higher than standard ARC (1980) values. Values were 1.28 times those in ARC (1980) for lambs, 1.38 times for beef cattle and 1.40 times for lactating dairy cows. In the discussion the importance of distinguishing between practical advisory models and more basic scientific models was stressed. Calculations of ME_m and k values must be within a compatible data set, not from different sources. The Hillsborough results, which involve little extrapolation, indicate that the ARC (1980) factors do not stand up in practice. The factors established by the Beltsville team had been based on the best data available at the time, but were not expected to be universal. They had calculated that 80% of the error in the system lay in estimates of the metabolizability values for the feeds. For scientific purposes and progress, mechanistic models were more appropriate. These must be based on sound chemical parameters including reliable values for feed composition. The importance of basing such models on animal results rather than pure stoichiometry alone was stressed.

Efficiency of gain

Four papers had dealt with aspects of k_g. Discussion focused mainly on curvilinearity of the RE/ME relationship and was notable for considerable differences of opinion. Principal factors identified as contributing to curvilinearity were level of intake, type of diet (e.g. q value) and nature of the gain (i.e. protein/fat ratio). Once again the dangers of mixing data types (e.g. animal species) and extrapolation of data were stressed. In relation to broilers, there was some debate as to whether the relationship was actually curvilinear and as to the degree of constancy of the fat:protein ratio in gain at high levels of intake.

Effective energy

One paper had discussed the practical aspects of the feed evaluation for poultry under the effective energy system recently proposed by Emmans (1994). This system is based on a new approach to predicting heat increment. A central factor is determination of digestible protein which is difficult and time-consuming in poultry. It was suggested that the new system was an advance on the AME system towards an NE system for poultry but there appear to be no data available yet to test the validity of this opinion.

Fish nutrition

The report on the nutrition of Atlantic salmon showed their response to very high levels of fish protein (up to 520 g crude protein/kg diet). A long-term aim of the programme was to attempt to reduce reliance on high levels of fish meal in fish diets.

Pig nutrition

Two papers from Scandinavia had studied the value of fibre in pig diets. The question of definition and measurement of "fibre" was raised. For practical purposes NDF was considered to be more useful than CF. The division into insoluble and soluble non-starch polysaccharides (NSP) in the Danish study was primarily for scientific rather than practical purposes, the soluble NSP being more fermentable. However the problems arising from the increasingly complex methods of fibre determination were stressed. It was claimed that pigs have a very considerable capacity to digest fibre which should be exploited in appropriate situations. In a comparison of pea and maize starch for pigs, the authors reported that they could find no evidence of hind-gut fermentation of pea starch. However this conclusion was challenged: the low flux rate of VFA in the portal might simply reflect extensive utilization by the gut wall and the low faecal N digestibility could reflect microbial protein synthesis in the caecum. The question was left unresolved.

Reference

Emmans G.C. (1994). Effective energy: a concept of energy utilization applied across species. *Br. J. Nutr.* 71:801-821.

LIST OF PARTICIPANTS

Dr R. Agnew
Agricultural Research Institute
Large Park
Hillsborough
BT26 6DP
NORTHERN IRELAND

Dr J.F. Aguilera
Estación Experimental del Zaidín (CSIC)
Departamento de Nutrición Animal
Profesor Albareda, 1,
18008
Granada
SPAIN

Dr Y. Aharoni
Newe Ya-ar Research Centre
PO 90000
Haifa 31900
ISRAEL

Dr S. Andrew
Department of Animal Science
3636 Horsebarn Road Extension
U Box 40, Storrs
Connecticut
06269-04040
USA

Dr A. Araba
Dept. of Animal Sciences
Institut Agromique et Vétérinaire
Hassan II
PO Box 6202
Rabit-Instituts 10100
MOROCCO

Dr D.A. Astuti
Laboratory of Physiological Chemistry
Department of Physiology
Pharmacology
Faculty of Veterinary Medicine
Bogor Agricultural University
Jalan Taman Kencana 1 No. 3
Bogor 16151
INDONESIA

Dr G.C.M. Bakker
Research Institute for Livestock Feeding and Nutrition
Runderweg 2
Lelystad PO Box 160
8200 AD Lelystad
THE NETHERLANDS

Dr R.L.Baldwin
USDA/ARS
Room 211, Bldg 163
BARC-EAST
Beltsville
MD 20706
USA

Mr J. Birnie
Agricultural Research Institute
Large Park
Hillsborough
BT26 6DP
NORTHERN IRELAND

Dr F. Bocquier
Laboratoire Sous-Nutrition de Ruminants
INRA-Theix
St Genes Champanelle
FRANCE

Dr H.A. Boekholt
Agricultural University Wageningen
Department of Animal Physiology
Haarweg 10
6709 PJ Wageningen
THE NETHERLANDS

Miss H. Bray
Institute of Animal Science, Physiology
Edelhertweg 15,
Postbus 65
8200 AB Lelystad
THE NETHERLANDS

Dr A. Brosh
Agricultural Research Organization
Ministry of Agriculture
Institute of Animal Science
Newe Yaar Research Center
PO Haife 31999
ISRAEL

Dr D.P. Bureau
Dept. of Human Biology and Nutrional Science
University of Guelph
Guelph
Ontario NIG 2WI
CANADA

Dr F.M. Byers
Texas A & M University
Department of Animal Science
College Station
Texas 77843
USA

Dr A. Carson	Agricultural Research Institute Large Park Hillsborough BT26 6DP NORTHERN IRELAND
Dr G.E. Carstens	Texas A & M University Department of Animal Science College Station Texas 77843 USA
Dr Y. Chilliard	Institut National de la Recherche Agronomique Centre de Clermont-Ferrand/Theix 63125 Saint-Gènes Champanelle FRANCE
Dr A. Chudy	Oskar-Kellner-Institut für Tierernährung Justus von Liebigs-Weg 2 18059 Rostock GERMANY
Dr A. Chwalibog	Royal Veterinary and Agricultural University Dept. of Animal Science & Animal Health Bülowsvej 13 DK-1870 Frederiksberg DENMARK
Dr C.N. Coon	Center of Excellence for Poultry Science Room 0-211, Poultry Science Center University of Arkansas Fayetteville AR 72701 USA
Dr B.A. Crooker	University of Minnesota Department of Animal Science St-Paul Minnesota 55108 USA
Dr G.M. Crovetto	Istituto di Zootecnia Generale Facolta di Agraria Universita degli Studi di Milano Vía Celoria 2 20133 Milano ITALY

Dr K. Cummings	Church & Dwight Company 469 N Harrison Street Princeton New Jersey 08543-5297 USA
Dr M. Decoux	Ralston Purina 1 Plau Ch de Gaulle BP 301 78054 St Quentin en Yuelines FRANCE
Mr B. de Pauw	University of Leuven Laboratory for Agricultural Research Kardinaal Mercierlaan, 92 3001 Leuven-Heverlee BELGIUM
Dr M. Derno	Oskar-Kellner-Institut für Tierernährung Justus von Liebigs-Weg 2 18059 Rostock GERMANY
Dr G. Emmans	SAC Genetics/Behavioural Science Dept Bush Estate Penicuik Midlothian EH26 0QE UK
Dr H. Fandrejewski	Polish Academy of Sciences The Kielanowski Institute of Animal Physiology and Nutrition 05-110 Jablonna near Warsaw POLAND
Dr D.J. Farrell	University of Queensland Department of Agriculture Brisbane Qld 4072 AUSTRALIA
Dr J.A. Fernández	The National Institute of Animal Science Department of Animal Physiology and Biochemistry Foulum PO Box 39 8830 Tjele DENMARK

Dr C.P. Ferris	Agricultural Research Institute Large Park Hillsborough BT26 6DP NORTHERN IRELAND
Dr H. Freetly	US Department of Agriculture, ARS Roman L. Hruska US Meat Animal Research Center Clay Center PO Box 166 Nebraska 68933 USA
Prof F.J. Gordon	Agricultural Research Institute Large Park Hillsborough Co Down NORTHERN IRELAND
Dr B. Grisdale-Helland	Institut for Akvakulturforskning Akvaforsk PO Box 203 6600 Sunndalsöra NORWAY
Dr A. Guerouali	Laboratory of Animal Physiology Institut Agronomique et Vétérinaire Hassan II PO Box 6202 Rabat MOROCCO
Dr A.B. Hall	Dept. of Animal Production University of Queensland Gatton College 4345 Queensland AUSTRALIA
Dr M.J.W. Heetkamp	Department of Animal Husbandry Agricultural University Wageningen PO Box 338 6700 AH Wageningen THE NETHERLANDS
Dr S.J. Helland	Institut fur Akvakulturforskning Akvaforsk PO Box 203 6600 Sunndalsöra NORWAY

Dr W. Henry
Nutec Ltd
Greenhills Centre
Greenhills Road
Tallaght
Dublin 24
REPUBLIC OF IRELAND

Dr P. Herpin
INRA, Centre de Recherches de Rennes
St-Gilles, Station de Recherches
Porcines
35590 L'Hermitage
FRANCE

Dr F. Hocquette
INRA Centre de Clermont-Ferrand/Theix
UR Metabolismes Energètique et Lipidique
63122 Saint-Gènes Champanelle
FRANCE

Dr J. Hutcheson
Pfizer Animal Health
6908 Columbia Lane
Armarillo
Texas 79109
USA

Dr Kirsten Jakobsen
National Institute of Animal Science
Department of Animal Physiology and Biochemistry
Foulum
PO Box 39
8830 Tjele
DENMARK

Dr G. Janssens
Agricultural Research Centre
Ghent-Rijksstation voor Kleinveeteelt
Burgemeesters van Gansberghelaan, 92
89820 Merelbeke
BELGIUM

Dr W. Jentsch
Oskar-Kellner-Institut für Tierernährung
Justus von Liebigs-Weg 2
18059 Rostock
GERMANY

Dr D.E. Johnson
Colorado State University
209 Animal Science Bldg
Fort Collins
Colorado 80523
USA

Dr K.A. Johnson

Washington State University
Department of Animal Sciences
231 Clark Hall
Pullman
Washington, DC 99164-6320
USA

Dr H. Jorgensen

The National Institute of Animal Science
Department of Animal Physiology & Biochemistry
Foulum
PO Box 39 8830 Tjele
DENMARK

Dr P. Junghans

Forschungssinstitut fur die Biologie
FB Ernahrungsphysiologie
"Oskar Kelyser"
Justus-von-Liebig-Weg 2
Rostock-18059
GERMANY

Dr V. Khummirdpetch

Human Biology, Natural Sciences
Animal Science Building
University of Guelph
Guelph
Ontario N1G 2W1
CANADA

Mr E.J. Kim

IGER
Plas Gogerddan
Aberystwyth
Dyfed
SY23 3EB
UK

Dr M. Klein

Oskar-Kellner-Institut für Tierernährung
Justus von Liebigs-Weg 2
Rostock 18059
GERMANY

Dr M. Kreuzer

Institut für Nutzierwissenschaften
Gruppe Ernährung, ETH-Zentrum
CH-8092 Zürich
SWITZERLAND

Dr M. Lachica

Estación Experimental del Zaidín (CSIC)
Departamento de Nutrición Animal
Profesor Albareda, 1
18008 Granada
SPAIN

Dr E.M. Larson | Larson Nutrition Services
400 E. Park Avenue
Fowler
Colorado 81039
USA

Miss F. Lewis | Agricultural and Environmental Science
The Queen's University of Belfast
Newforge Lane
Belfast BT9 5PX
NORTHERN IRELAND

Dr J.B. Liang | Livestock Research Division
Malaysian, Agricultural Research & Development Institute
PO Box 12301
GPO 50744
Kuala Lumpur
MALAYSIA

Dr J.E. Lindberg | Swedish University of Agric. Sciences
Dept. of Animal Nutrition & Management
Kungsängens Försöksgard
S-775 90 Uppsala
SWEDEN

Dr A. Machmuller | Institut für Nutzierwissenschaften
Gruppe Ernährung, ETH-Zentrum
CH-8092 Zürich
SWITZERLAND

Dr M.G. MacLeod | Agricultural and Food Research Council
Roslin Institute (Edinburgh)
Roslin
Midlothian
EH25 9PS
UK

Dr K.J. McCracken | Agricultural and Environmental Science
Department of Agriculture
Newforge Lane
Belfast
BT9 5PX
NORTHERN IRELAND

Mr P. McGrattan | Agricultural and Environmental Science
The Queen's University of Belfast
Newforge Lane
Belfast BT9 5PX
NORTHERN IRELAND

Dr K. McLeod	USDA/ARS Room 211, Bldg 163 BARC-EAST Beltsville MD 20706 USA
Dr J.P. McNamara	Washington State University Department of Animal Science 231 Clark Hall Pullman Washington, DC 99163 USA
Dr J.G. Mahardika	Department of Animal Nutrition Udayana University Jalan PB Sudiman Denpasar Bali INDONESIA
Dr J. Michal	Washington State University Department of Animal Sciences 231 Clark Hall Pullman Washington, DC 99164-6320 USA
Dr A. Mitchell	USDA Agricultural Research Service Blgd 200, Rm 205 BARC-East Beltsville MD 20705 USA
Dr A. Moloney	Grange Research Centre TEAGASC Dunsany Meath REPUBLIC OF IRELAND
Dr A. Moss	ADAS Drayton Stratford upon Avon UK
Dr H.L. Müller	Technische Universität München Institut für Ernährungsphysiologie 85354 Fresing-Weihenstephan GERMANY

Dr T. Nishida
National Institute of Animal Industry
Department of Animal Nutrition
Tsukuba Norindanchi, PO Box 5
Ibaraki 305
JAPAN

Dr J.W. Oltjen
University of California
Dept of Animal Science
Davis
California 95616-8521
USA

Dr I. Ortigues
Laboratoire Croissance et Metabolisme
INRA - Theix
St-Genes Champanelles
LA FRANCE

Dr D.L. Palmquist
The Ohio State University
Dept of Dairy Science
1680 Madison Avenue
Wooster
Ohio 44691
USA

Dr D.S. Parker
University of Newcastle-Upon-Tyne
The Dept of Biological & Nutritional Sciences
Newcastle Upon Tyne NE1 7RU
UK

Mr V. Pirgozliev
Harper Adams College
Edgmond
Newport
Shropshire TF10 8NB
UK

Mr M. Porter
The Agriculture Research Institute
Large Park
Hillsborough
BT26 6DP
NORTHERN IRELAND

Dr A. Purnomoadi
Laboratory of Energy Metabolism
National Institute of Animal Industry
Tsukuba Norindanchi, PO Box 5
Ibaraki 305
JAPAN

Dr J.J. Ramsey
Wisconsin Regional Primate Center
University of Wisconsin
Madison
Wisconsin 53715-1299
USA

Dr L. Rapetti
Instituto di Zootecnia Generale
Facolta di Agraria
Universita degli Studi di Milano
Vía Celoria 2
20133 Milano
ITALY

Dr P.J. Reeds
Children's Nutrition Research Center
Department of Pediatrics
1100 Bates Street
Houston, Texas 77030
USA

Dr C.K. Reynolds
Centre for Dairy Research
University of Reading
Early Gate
Reading RG6 6AT
UK

Miss K. Robertshaw
BOCM Pauls Ltd
47 Key Street
Ipswich
IP4 1BX
UK

Prof D. Romsos
Dept of Food Science and Nutrition
Michigan State University
106 GM Trout Building
East Lansing
MI 48824-1224
USA

Dr P. Rose
Harper Adams College
Edgmond
Newport
Shropshire TF10 8NB
UK

Dr R. Rosebrough
USDA
Non-Ruminant Animal Nutrition
LPSI
Bldg 200, Room 201 BARC-East
Beltsville MD 20705
USA

Dr W.V. Rumpler
US Dept of Agriculture, ARS
Beltsville Human Nutrition Center
Energy and Protein Nutrition Lab
B-308 BARC-East
Maryland 20205
USA

Dr P. Saevels
Laboratory for Agricultural Building Research
Katholic University, Leuven
Kard Mercierlaan 92
B 3007 Heverlee
BELGIUM

Ms E. Saggau
Forschungssinstitut fur die Biologie
FB Ernahrungsphysiologie
"Oskar Kellner"
Justus-von-Liebig-Weg 2
Rostock-18059
GERMANY

Dr D. Sastradipradja
Bogor Agricultural University
Department of Physiology
Pharmacology
Jalan Taman Kencana 1 No. 3
Bogor 16151
INDONESIA

Dr J. Schrama
Agricultural University Wageningen
Department of Animal Husbandry
PO Box 338
6700 AH Wageningen
THE NETHERLANDS

Dr N.D. Scollan
AFRC Institute of Grassland & Environmental Research
Plas Gogerddan, Aberystwyth
Dyfed
SY23 3EB
UK

Dr Gregorz Skiba
Institute of Animal Physiology & Nutrition
05-100 Jablonna near Warsaw
POLAND

Mr J. Sloss
Moy Park Ltd
39 Seagoe Industrial Estate
Craigavon
NORTHERN IRELAND

Dr R.W.J. Steen
Agricultural Research Institute
Large Park
Hillsborough
BT26 6DP
NORTHERN IRELAND

Prof A. Susenbeth
Institute of Animal Nutrition
University of Kiel
24098 Kiel
GERMANY

Dr J.D. Sutton
Centre for Dairy Research
The University of Reading
Early Gate
Reading RG6 6AT
UK

Dr J. Takahashi
Obihiro University of Agriculture and Veterinary Medicine
Laboratory of Animal Nutrition
Obihiro 080
JAPAN

Dr A.H. Tauson
Royal Veterinary & Agricultural University
Department of Animal Nutrition
Bülowsvej 13
1870 Frederiksberg C
DENMARK

Dr G. Thorbek
Hegnskrogen 12
2800, Lyngby
DENMARK

Dr H.F. Tyrrell
US Department of Agriculture
Comparative State Research Service
AG Box 2220
SW Washington, DC 20250-2220
USA

Dr E.F. Unsworth
Agricultural and Environmental Science
Department of Agriculture
Newforge Lane
Belfast
BT9 5PX
NORTHERN IRELAND

Dr H. van den Brand
Dept. of Animal Husbandry
Wageningen Agricultural University
Wageningen 6700 AH
THE NETHERLANDS

Dr Y. van der Honing
Research Institute for Livestock
Feeding & Nutrition
Runderweg 2
Lelystad PO Box 65
8200 AD Lelystad
THE NETHERLANDS

Dr J. van Milgen
INRA-Station de Recherches Porcines
Domaine de la Prise
35596 Saint-Gilles
FRANCE

Dr M. Vermorel
Institut National de la Recherche Agronomique
Centre de Clermont-Ferrand/Theix
63122 Saint-Gènes Champanelle
FRANCE

Miss W. Wakeman
BOCM Pauls Ltd
47 Key Street
Ipswich
IP4 1BX
UK

Dr W. Weber
Dept. of Animal Science
130 Haecker Hall
University of Minnesota
St Paul
MN 55108
USA

Prof C. Wenk
Institut für Nutzierwissenschaften
Gruppe Ernährung, ETH-Zentrum
CH-8092 Zürich
SWITZERLAND

Dr A.R.G. Wylie
Agricultural and Environmental Science
Department of Agriculture
Newforge Lane
Belfast
BT9 5PX
NORTHERN IRELAND

Dr T. Yan
Agricultural Research Institute
Large Park
Hillsborough
BT26 6DP
NORTHERN IRELAND

INDEX OF LABORATORIES STUDYING
ENERGY METABOLISM IN FARM ANIMALS

(SEPTEMBER 1997)

Compiled by **Y. van der Honing**
ID - DLO,
PO Box 65
8200 AB Lelystad
The Netherlands

Tel: +31.320.237314
Fax: +31.320.237320
E-mail: Y.vanderHoning@id.dlo.nl

AUSTRALIA

1. **Research Institute, Department of Agriculture, R.M.B. 8010, Kyabran, Victoria 8620**
   Phone Number :
   Investigators : D. DELLOW
   Main research topics :
   Techniques : CO_2 entry rate,
   Species : Sheep, cattle

2. **University of Melbourne, School of Agriculture and Forestry, Parkville 3052, Victoria**
   Phone Number : 61.3.344.50.07 Fax number : 61.3.344.55.70
   Investigators : A.R. EGAN
   Main research topics :
   Techniques : Slaughter
   Species : Sheep, pigs, beef cattle

3. **La Trobe University, School of Agriculture, Bundoora 3083, Victoria**
   Phone Number :
   Investigators : P. CRANWELL
   Main research topics :
   Techniques : Isotope dilution, Slaughter
   Species : Pigs, sheep, cattle

4. **University of New England, Department of Biochemistry, Microbiology & Nutrition, Armidale, N.S.W., 2351.**
   Phone Number : 067.73.22.70 Fax number : 067.72.82.35
   Investigators : P.W. WARWICK, G.D.P. JONES, M. CHOCT
   Main research topics : Genotype/environment interactions in poultry
   Techniques : Closed circuit respiration chambers ; Ventilation hood
   Species : Poultry, humans

5. **University of Sydney, Department of Animal Science, Dairy Research Unit, PMB 3, CAMDEN NSW 2570**
   Phone Number : + 61.46.552.300 Fax number : 61.46.552.374
   Investigators : J.GOODEN, P.C. WYNN
   Main research topics : Influence of high temperatures on energy expenditure in growing pigs and lactating sows ; Energy expenditure in exercising sheep ; Oxygen consumption of the mammary gland in lactating dairy cows
   Techniques : Blood flow using transit-time ultrasound; Blood oxygen saturation using fibre optic oximetry
   Species : Pigs, sheep and dairy cattle

6. **CSIRO, Locked Bag 1, Deliviry Centre, BLACKTOWN, N.S.W. 2148.**
   Phone Number : 02.840.27.00 Fax number : 02.840.29.40
   E. Mail : D. MARGAN@prospect.anprod.CSIRO.av
   Investigators : D.E. MARGAN
   Main research topics : Modelling energy + protein metabolism in beef cattle; Feed evaluation in sheep
   Techniques : Closed-circuit respiration chambers currently not in use. Facility no longer being supported
   Species : Sheep

7. **Tropical Cattle Research Centre, CSIRO, Rockhampton, Queensland 4700.**
Phone Number :
Investigators : J.E. FRISCH, G.D. INDER
Main research topics : Tropical animal production
Techniques : Open circuit respiration chambers
Species : Cattle, mice, rats

8. **University of Queensland, Department of Animal Production, Gatton College, UQG, LAWES Qld 4343**
Phone Number : 0754.601.251 Fax number : 0754.601.444
E. Mail : bay@warigal.uqg.uq.oz.au
Investigators : B.A. YOUNG, Anita B. HALL
Main research topics : Environmental influence on farm animals; Modifying rumen metabolism
Techniques : Open-circuit respiratory ; Heart rate telemetry for field ; Animal mounted field equipment, Dairy cow environment heat load body temperature measurements
Species : Cattle, sheep, pigs, poultry

9. **University of Queensland, Department of Agriculture, Brisbane Qld 4072**
Phone Number : 07.365.11.77 Fax number : 07.366.09.73/07.824.43.16
Investigators : D.J. FARRELL, R.A.E. PYM
Main research topics : Environmental effects on poultry
Techniques : Closed circuit calorimeters ; Slaughter method
Species : Poultry

BELGIUM

1. **University of Ghent, Laboratory of Animal Nutrition, Heidestraat 19, 9220 MERELBEKE**
Phone Number : +32 9 252 19 41 Fax number : +32 9 252 53 04
E. Mail: Roland.DeWilde@rug.ac.be
Investigators : R. DE WILDE, G. JANSSENS, M. SEYNAEVE, M. HESTA, H. DE RIJCKE
Main research topics : Carnitine metabolism in racing pigeons; Phosphorus digestibility in pigs, ostriches and dogs in relation to phytate; Carbohydrases in the feeds of dogs and cats; Inhibition of legume lectins by glycoproteins in animal byproducts
Techniques : HPLC, colorimetry, electrophoresis and western blotting, ELISA, slaughter method
Species : Pigs, cats, dogs, pigeons (ostrich experiments on commercial farms)

2. **Katholic University of Leuven, Laboratory for Physiology and Immunology of Domestic Animals, Kardinaal Mercierlaan, 92, 3001 HEVERLEE**
Phone Number : 32.16.321436 Fax number : 32.16.321994 E. Mail : eddy.decuypere@agr.kuleuven.ac.be
Investigators : E. DECUYPERE
Main research topics : Endocrinology of growth in chickens; Energy metabolism in growing chicks in relation to endocrine parameters and as a function of genotype, treatment and feeding regimes; Inhibin in the chicken embryo and laying hen; Reproduction in broiler breeders; Ascites in broilers ; Fecundity gene in sheep

Techniques : Open circuit respiration chambers; radioimmunoassay (α- and β-counter); Enzyme-linked immunosorbent assay; Polymerase chain reaction techniques; Cell culture and organ or tissue perfusion techniques; Flow cytometry
Species : Chicken and sheep

3. **Agricultural Research Centre, Ghent - Rijksstation voor Kleinveeteelt, Burgemeesters van Gansberghelaan 92, 9220 MERELBEKE**
Phone Number : 32.92.52.19.71 Fax number : 32.92.52.53.98
Investigators : G. DE GROOTE, G. HUYGHEBAERT, G. DE MUNTER
Main research topics : Nutrition, reproduction poultry, rabis - Energy evaluation feedstuffs ; Energy metabolism studies
Techniques : Open-circuit respiration, Slaughter method
Species : Poultry, rabbits

4. **Katholic University of Leuven, Laboratory for Agricultural Buildings Research, Kardinaal Mercierlaan, 92, 3007 LEUVEN-HEVERLEE**
Phone Number : 32.16.321434 Fax number : 32.16.321977
E. Mail : Rony Geers@AGR.KULEUVEN.AC.BE
Investigators : R. GEERS, S. PERREMANS, B. DE PAUW
Main research topics : Stress physiology during housing and transport of pigs
Techniques : Climatic rooms, I.R. thermography, telemetry, echography, DEXA (Dual Energy X-ray Absorptiometry), animal behaviour by image analysis
Species : Pigs

5. **Katholic University of Leuven, Laboratory for Agricultural Buildings Resarch, Kardinaal Mercierlaan 92, B-hoon, LEUVEN-HEVERLEE**
Phone Number : ++32.16.32.14.36 Fax : ++32.16.32.14.80
E. Mail : daniel.berchmans@agr.pheleuven.ac.be
Investigators : D. BERCHMANS, P. SAEVELS, J-M. AERTS.
Main research topics : heat production in relation to climate, modelling, controlling techniques: open circuit respiration chambers, telemetry of heart rate, body temperature and activity, soxlet, kjeldahl.
Species : Broiler chickens.

CANADA

1. **University of Alberta, Department of Animal Science, Edmonton, Alberta, T6G 2P5.**
Phone Number :
Investigators : R.J. CHRISTOPHERSON, G.W. MATHISON, R.J. HUDSON
Main research topics :
Techniques : Open-circuit respiration ; Modelling cell and tissue respiration ; Isotopic tracers; Unrestrained animal field techniques
Species : Cattle, sheep, pigs

2. **University of Guelph, Department of Animal and Poultry Science, Guelph, Ontario, N1G 2W1**
Phone Number : 1.519.824.41.20 Fax number : 519.836.98.73
Investigators : L.P. MILLIGAN, B.W. McBRIDE
Main research topics :

Techniques : Cell and tissue respiration, Na^+/K^+ transport, Nuclear magnetic spectro. ^{31}P, Isotopic tracers
Species : Pigs, sheep, cattle, chickens

3. **Department of Human Biology and Nutritional Sciences, University of Guelph, Ontario, N1G 2W1**
Phone Number : 1.519.824.41.20 Fax number : 1.519.763.59.02
E. Mail : HBayley.ns@aps.voguelph.ca
Investigators : H.S. BAYLEY
Main research topics : Energy metabolism in swine
Techniques : Open circuit indirect calorimetry using ^3H and ^{14}C labelled compounds
Species : Pigs, poultry, (associated studies with human infants)

4. **Fish Nutrition Research Laboratory, University of Guelph, and Ontario Ministry of Natural Resources (OMNR), GUELPH, Ontario, N1G 2W1**
Phone Number : 1-519-836-1338 Fax number : 1-519-763-5902
E. Mail : ycho@uoguelph.ca
dbureau.ns@aps.uoguelph.ca
Investigators : C.Y. CHO, D.P. BUREAU
Main research topics : Energy metabolism in fish; Nutrient requirements and metabolism of fish, Feed ingredients evaluation. Feeding strategies. Modelling of energy requirement of poikilothermic animals
Techniques : Open circuit indirect calorimetry; Comparative carcass analysis, apparent digestibility.
Species : Salmonid fish

5. **University of Guelph, Magnetic Resonance Imaging Centre, Guelph, Ontario, NG 2W1**
Phone Number :
Investigators : E. JANZEN
Techniques : Magnetic Resonance Imaging, ^{31}P, H, ^{13}C spectroscopy
Species : Poultry

6. **University of Manitoba, Department of Animal Science, Winnipeg, Manitoba, R3T2N2**
Phone Number :
Investigators : L.D. CAMPBELL, A.D. KENNEDY
Main research topics :
Techniques : Comparative slaughter, Temperature telemetry, Hormone binding techniques
Species : Chickens, sheep, pigs, cattle

CHINA

1. **Ji Lin Academy of Agricultural Science, Institute of Animal Nutrition, Energy Metabolism Research Section, 186 East Xinghua Street, GONGZHULING CITY, Jilin Province 136100**
   Phone Number : +86.0434 6229167 Fax number : +86.0434 6214884
   E. Mail: ghtc@public.sta.net.cn
   Investigators : MA XIYI, JIN SHIHOU, YU XIUFANG, YOUNG LIBIN, QI HONGWEI
   Main research topics : Studies on efficieny of energy utilization of farm and economic animals ; Metabolic mechanism of heat stress
   Techniques : Automatic open circuit respiration chambers; open and shut respiration chambers; closed-circuit respiration chambers; molecular study
   Species : Pigs, broilers, layer, sheep/goat, economic animals

2. **Department of Animal Science, Laboratory of Ruminant Nutrition and Energy Metabolism, Beijing Agricultural University, Yuanmingyuan xi road, Haidian, BEIJING 100094**
   Phone Number : 2582244-406 Fax number : 86-1-2582332
   Investigators : FENG YANGLIAN, ZHANG XIAOMING, MO FANG, ZHAO GUANGYONG
   Main research topics : Small intestine energy nutrition of ruminants; Manipulation of rumen fermentation; VFA production and its manipulation; Lipid metabolism and utilization ; Effect of undegradable nutrients on energy conversion; Evaluation of feeds nutritive value for ruminants; Technology of coated feeds
   Techniques : Automated large and small respiration chambers; Cannulae technique; Isotope technique; Infusion technique; Rumen simulation technique
   Species : Cattle, sheep, deer

3. **Feed Research Centre, Academy of the Ministry of Commerce, Beijing 100037**
   Phone Number :
   Investigators : LI AIKE
   Main research topics :
   Techniques : Comparative open-shut and closed circuit respiration chambers
   Species : Swine, poultry, beef cattle

CZECH REPUBLIC

1. **Veterinary Research Institute, Hudcova 70, BRNO 21**
   Phone Number :
   Investigators : J. PRINC
   Main research topics :
   Techniques : Open-circuit respiration chambers
   Species :

2. **Veterinary Faculty, Department of Physiology, Palackeho, 1 621200 BRNO**
   Phone Number :
   Investigators : A. HOLUB
   Main research topics :
   Techniques :
   Species :

DENMARK

1. **Danish Institute of Animal Science, Department of Nutrition, Research Centre Foulum, PO Box 50, 8830 TJELE**
   Phone Number : +45.89.99.19.00 Fax number : +45.89.99.1166
   E. Mail : Kirsten.Jakobsen@sh.dk
   Henry.Jorgensen@sh.dk
   JoseA.Fernandez@sh.dk
   Investigators : H. JØRGENSEN, J.A. FERNANDEZ, K. JAKOBSEN
   Main research topics : Nutritional physiology; GIT metabolism; thermal environment techniques : indirect calorimetry; open circuit respiration chambers; tissue respiration, stable and radio isotopes,
   Species : Calves, sheep, pigs, rabbits, poultry, rats

2. **Royal Veterinary and Agricultural University, Department of Animal Science and Animal Health, Bülowsvej 13, 1870 Frederiksberg C**
   Phone Number : 45 35283039 Fax number : 45 35283020
   E. Mail : ac@kvl.dk
   aht@kvl.dk
   Investigators : A. CHWALIBOG, A.H. TAUSON
   Main research topics : Substrate oxidation; Simulation of energy metabolism; Nutrient utilization
   Techniques : Indirect calorimetry, mathematical modelling, telemetry, stable isotopes
   Species : Cattle, pigs, poultry, mink, foxes, dogs

3. **Royal Veterinary and Agricultural University, Research Department of Human Nutrition, Rolighedsvej 25, 1958 Frederiksberg C**
   Phone Number : 4535284070 Fax number : 4531351738
   Investigators : A. ASTRUP, S. TOUBRO, G. THORBEK
   Main research topics : Human nutrition
   Techniques : Indirect calorimetry; Open-circuit respiration chambers
   Species : Humans

FRANCE

1. **INRA, Laboratoire Croissance et Métabolismes des Herbivores, Centre de Clermont-Ferrand/Theix, 63122 Saint-Genès Champanelle**
   Phone Number : 0473.62.40.00 Fax number : 0473.62.46.39
   E. Mail : Michel.Vermorel@clermont.inra.fr
   Hocquet@clermont.inra.fr
   Isabelle.Ortigues@clermont.inra.fr
   Investigators : M. VERMOREL, B. AUROUSSEAU, J.F. HOCQUETTE, I. ORTIGUES, J. VERNET
   Main research topics : Energy metabolism of growing ruminants; Organ and tissue metabolism; Nutritional and hormonal regulation of key-enzymes of energy metabolism
   Techniques : Open-circuit respiration chambers; Multicatheterization; A/V differences; Net nutrient flux ; Cellular and molecular biology
   Species : Sheep, cattle

2. **Centre de Recherches en Nutrition Humaine, Laboratoire de Nutrition Humaine, BP 321, 58, Rue Montalembert, 63000 Clermont-Ferrand**
   Phone Number : 73.60.82.50 Fax number : 73.60.82.55
   Investigators : M. VERMOREL, P. RITZ, B. MORIO, C. MONTAURIER
   Main research topics : Energy metabolism in elderly people
   Techniques : Open-circuit respiration chambers; Doubly labelled water
   Species : Humans

3. **INRA, Centre de Recherches de Tours, Station de Recherches Avicoles, 37380 NOUZILLY**
   Phone Number : (33) 47.42.79.25 Fax number : (33) 47.42.77.78
   Investigators :
   Main research topics : Energy expenditure: components and regulation particularly diet-induced thermonegenesis; Heat-induced changes in energy metabolism
   Techniques : Indirect calorimetry; Respiratory chambers; Telemetry
   Species : Poultry

4. **INRA, Station de Recherches Porcines, 35590 ST-GILLES**
   Phone Number : 0299.28.50.87/0299.28.50.49 Fax number : 0299.28.50.80
   E. Mail : Jean.Noblet@rennes.inra.fr
   Patrick.Herpin@St-gilles.rennes.inra.fr
   Jacob.Van.Milgen@St-gilles.rennes.inra.fr
   Marie.Damon@St-gilles.rennes.inra.fr
   Jean.LeDevodoch@St-gilles.rennes.inra.fr
   Investigators : P. HERPIN, J. NOBLET, J. VAN MILGEN, J. LE DIVIDICH, MARIE DAMON
   Main research topics : Thermoregulation, energy metabolism and nutrition of the neonatal pig; Thermogenic effect and digestive and metabolic utilization of colostrum; Energy requirements of growing pigs and sows; Components of heat production; Net energy value of feeds
   Techniques : Indirect calorimetry; Comparative slaughter technique; lectromyography; Tissue and mitochondrial respiration; Thyroid hormone metabolism (receptors,...), modelling, molecular biology
   Species : Newborn pigs, growing pigs, sows

5. **INRA, Unité mixte INRA-IFREMER de Nutrition des Poissons, Station d'Hydrobiologie, St Pée-sur-Nivelle, B.P. 3, 64310 Ascain**
   Phone Number : 59.51.59.51 Fax number :59.54.51.52
   E. Mail : medale@st-pee.inra.fr
   Investigators : F. MEDALE, G. CORRAZE
   Main research topics : Energy expenditure, substrates oxidation and lipogenesis in relation to dietary supply and environmental factors.
   Techniques : Slaughter technique, Indirect calorimetry, Enzymes activity
   Species : Fish (Freshwater and seawater)

6. **INRA, Laboratoire de Recerches sur la Sous-Nutrition des Ruminants, Centre de Clermont-Ferrand - Theix, 63122 St Genés Champanelle, France**
   Phone Number : (33) 04 73 62 14 Fax number : (33) 04 73 62 45 19
   E. Mail : chilliar@clermont.inra.fr
   Investigators : Y. CHILLIARD, M. DOREAU, F. BOCQUIER, A. FERLAY, P. NOZIERE, Y. FAULCONNIER and M. BONNET

Main research topics : Energy metabolism of underfed ruminants, Lactation, Reproduction, Photoperiod, Adipose tissue metabolism (lipogenesis and lipolysis), Endocrinology, Digestion and Absorption
Techniques : *in vivo* dilution technique, adipose tissue cellularity, blood metabolites and hormones, adipose tissue microdialysis, multicatheterization and A/V differences, lipogenic enzyme activities, cellular and molecular biology, milk fatty acids
Species : Cattle, sheep and goats

GERMANY

1. **Institut für Tierernährung, Bundesforschungsanstalt für Landwirtschaft, Bundesallee 50, 38116 BRAUNSCHWEIG**
Phone Number : 0531.59.64.33 Fax number : 0531.59.63.76
E. Mail : gaedekenàte.fal.de
Investigators : D. GÄDEKEN,
Main research topics :
Techniques : Open-circuit respiration chambers; Comparative slaughter technique
Species : Cattle, pig, sheep

2. **University of Hohenheim, Institute of Animal Nutrition, 450, Emil-Wolff-Str. 10, D-70599 STUTTGART**
Phone Number : 0711.459.2420 Fax number : 0711.459.2421
Investigators : W. DROCHNER, H. STEINGASS
Main research topics : Partition and utilization of energy in animals; Growth, lactation, physical work
Techniques : Open-circuit respiration chambers; Mask technique
Species : Cattle, pigs, sheep

3. **Technische Universität München, Institut für Ernährungsphysiologie, 85350 FREISING-WEIHENSTEPHAN**
Phone Number : 08161/713400 Fax number : 08161/713999
Investigators : M. KIRCHGESSNER, H.L. MULLER, F. SCHWARZ
Main research topics : Dietary-induced thermogenesis; Hintgut fermentation; Reproduction and lactation; Energy requirements of carp
Techniques : Open-circuit respiration; Slaughter technique
Species : Cattle, pig, fish

4. **Universität München, Institut für Physiologie, Physiologische Chemie und Tierrernährung, Veterinärstrasse 13, 80539 MÜNCHEN**
Phone Number : 089.21.80.34.48
Investigators : H. PETRY
Main research topics :
Techniques : Open-circuit respiration chambers
Species : Rats, marmosets, chinchillas

5. **Research Institute for the Biology of Farm Animals, Dummerstorf-Rostock, Department of Nutritional Physiology « Oskar Kellner », Justus-von-Liebig-Weg 2, D-18059 ROSTOCK**
Phone Number : 0381/4020200 Fax number : 0381/4020299
E. Mail : fbn@fbn.uni.rostock.de

Investigators : M. BEYER, A. CHUDY, M. DERNO, P. JUNGHANS, M. KLEIN, H. SCHOLZE
Main research topics : Energy and protein metabolism of farm and laboratory animals during growth, pregnancy and lactation in dependence on feeding and environmental temperature including endocrine and immunological parameters
Techniques : Open-circuit indirect calorimetry; Climate respiration chambers; Locomotor activity; Isotope techniques; Slaughter technique
Species : Cattle, pig, sheep, poultry, rodents

6. **University of Kiel; Institute of Animal Nutrition, Physiology and Metabolism; D-24098 KIEL**
Phone Number: +49 431 880 2013 Fax number: +49 431 880 1528
E Mail: susenbeth@aninut.uni-kiel.de
Investigators : A. SUSENBETH
Main research topic : Partition and utilization of energy in animals; growth; physical work of draught animals; chewing activity; body composition; feed evaluation
Techniques : Indirect calorimetry, D_2O-dilution technique
Species : Cattle, pigs

ICELAND

1. **Rannsoknastofnun Iandbunadarins, Agricultural Research Institute, Keldnaholt, IS-112-Reykajvik**
Phone Number :
Investigators : O. GUDMUNDSSON, T. LARUSSON, B. L. OLAFSSON
Main research topics :
Techniques : Slaughter method
Species : Sheep, cattle, fur animals and fish

INDIA

1. **National Dairy Research Institute, Karnal, 132001 Haryana**
Phone Number :
Investigators : R.C. UPADHYAY
Main research topics :
Techniques : Indirect calorimetry, tissue, organ and cell metabolism
Species : Cattle, buffalo, goat

2. **Indian Veterinary Research, Institute Izatnagar, Bareilley (U.P.) 243122**
Phone Number :
Investigators : :D.C. SHUKLA
Main research topics :
Techniques : Indirect calorimetry, tissue, organ and cell metabolism
Species : cattle, buffalo , goat, sheep.

3. **Indian Veterinary Research, Animal Nutrition Division, Institute Izatnagar, Bareilley (U.P.) 243122**
Phone Number : Fax number : E. Mail:
Investigators : M.Y. KHAN, Jai KISHAN, MURARI LAL
Main research topics :

Techniques : Open circuit respiration chambers
Species : Cattle, buffalo, goat and sheep.

4. **Central Institute for Research on Goat, P.O. Makhdoom via Farah, Mathura (U.P.)**
Phone Number :
Investigators : KHUB SINGH
Main research topics :
Techniques : Indirect calorimetry, tissue, organ and cell metabolism
Species : Cattle, goat

INDONESIA

1. **Center for Metabolic Research, Institut Pertanian Bogor, (Bogor Agricultural University), Jalan Taman Kencana 1 no. 3, BOGOR 16151**
Phone Number : 62 251-328487 Fax number : 62 251-323161
E. Mail : d@indo.net.id
Investigators : D. SASTRADIPRADJA, B. KIRANADI, D.A. ASTUTI, D.R. EKASTUTI, T. SUTARDI
Main research topics : Metabolism of indigenous animals; dietary factors, endogenous and exogenous manipulation of animal metabolism
Techniques : Respirocalorimetry, heart rate telemetry, carbon dioxide entry rate technique, nutrient kinetics using radiotracers, organ metabolism using Fick principle, *in vivo* body composition
Species : Goats, sheep, swamp buffaloes, cattle, poultry, aquatic animals, silk worm (*Bombyx mori*)

2. **Research Program in Animal Metabolism, Faculty of Animal Husbandry, Udayana University, Jalan P.B. Sudirman, DENPASAR - BALI**
Phone Number : 62 361 222096
Investigators : I.G. MAHARDIKA, I.K. SAKA, I.K. SUMADI, I.A. SUKARINI, I.G.A. PUTRA, S. PUTRA, I.W. KASA, I.K. LANA, D. SASTRADIPRADJA (consultant)
Main research topics : Metabolism of indigenous animals; dietary and exogenous factors affecting metabolism
Techniques : *In vivo* body composition based on body densitometry and water space, heart rate telemetry, organ metabolism using Fick principle.
Species : Bali cattle (*Bos banteng*), swamp buffaloes, swine

ISRAEL

1. **Hebrew University, Dep. Zoology, Jerusalem**
Phone Number :
Investigators : A. BORUT, Y. MERDER
Main research topics :
Techniques : Open circuit respiration chambers, Temperature and metabolic rooms
Species : Birds, rodents

2. **Hebrew University, School of Medicine, Department of Physiology, Jerusalem**
Phone Number :
Investigators : M. HOROWITZ, S. SAMUELOFF
Main research topics :
Techniques : Open circuit respiration chambers, Climatic rooms
Species : Rodents

3. **Hebrew University, Dept. Animal Science of Faculty of Agriculture, Rehovot 76100.**
Phone Number : 972.84.81.307 Fax number : 972.894.65.763
Investigators : A. ARIELI, A. BERMAN,
Main research topics :
Techniques : Open circuit respiration chambers, Climatic rooms
Species : Small ruminants, poultry

4. **Agricultural Research Organization, Beit Dagan**
Phone Number :
Investigators : S. HURVITZ, I. PLAVNIK, I. BARTOV
Main research topics :
Techniques : Climatic chambers
Species : Birds

5. **Beer Sheva University, Department of Biology, Beer Sheva**
Phone Number :
Investigators : A. DEGEN, B. PINSHOW, A. ZARIZKI
Main research topics :
Techniques : Open circuit respiration chambers
Species : Birds, small ruminants

6. **Migal, Animal Science Department, Kiriat-Shmona**
Phone Number :
Investigators : G. DGANI, I. PIATOVSKI
Main research topics :
Techniques : Open circuit respiration chambers
Species : Poultry, fishes

7. **Tel-Aviv University, Zoology Department, Tel-Aviv**
Phone Number :
Investigators : A. AR, I. CHOSHNIAK, Y. LOYA, A. SHKOLNIK, R. DMIEL
Main research topics :
Techniques : Open circuit respiration chambers, Climatic and metabolic rooms
Species : Rodents, small ruminants

8. **Haifa University, Department of Biology, Oranim**
Phone Number :
Investigators : A. HAIM, I. IZHAKI, A. RUBAL
Main research topics :
Techniques : Open circuit respiration chambers
Species : Rodents, poultry

9. **Technion, Faculty of Biology, Haifa**
Phone Number
Investigators : Z. ARAD
Main research topics :
Techniques : Open circuit respiration chambers; doubly labelled water
Species : Avian and mammalian species

10. **Agricultural Research Organization, Ministry of Agriculture, Institute of Animal Science, Newe Yaar Research Center, Department of Beef cattle**
Phone number : 972.4.9539523 Fax number : 972.4.836.936
E. Mail : UFNBROSH@volcani.Agri.Gov.il
Investigator : ARIEH BROSH, YOAV AHARONI
Main research topics : Cattle body composition; energy utilization for feedlot and grazing cattle ; energy expenditure of free living cattle
Techniques :
Species : Cattle

ITALY

1. **Istituto di Zootecnia Generale, Facolta di Agraria Universita degli Studi di Milano, via Celoria 2, 20133 Milano Italy**
Phone Number : 392.70.60.01.59 Fax number : 392.7063.80.83
E. Mail : matteo/crovetto@inimi.it
Investigators : G. SUCCI, G.M. CROVETTO, L. RAPETTI, A. TAMBURINI, M. RIGONI, C. CASTROVILLI
Main research topics :
Techniques : Open-circuit respiration chambers
Species : Pigs, ruminants, chickens and rabbits

2. **Dipartemento di Scienze Zootecniche, Universita degli Studi, Via Gradenigo 6, 36131 Padova**
Phone Number : 39.49.8070324
Investigators : G. BITTANTE, M. BONSEMBIANTE, S. DAOLIO, M. RAMANZIN, I. ANDRIGHETTO, R. PARIGI BINI, G. XICCATO
Main research topics :
Techniques : Mass spectrometry and spirometry, Comparative slaughter
Species : Sheep, rabbits

JAPAN

1. **National Institute of Animal Industry, Department of Animal Nutrition, Energy Metabolism Laboratory, Tsukuba Norindanchi, P.O. Box 5, Ibaraki 305**
Phone Number : 81-298-38-8655 Fax number : 81-298-38-8553
E. Mail : sakura04@niai.affrc.go.jp
enishu@ss.niai.affrc.go.jp
kurihara@niai.affrc.go.jp.
Investigators : F. TERADA, M. KURIHARA, O. ENISHI
Main research topics : Efficiency of the energy utilization of various nutrient components under different feeding conditions; Dietary factors affecting methane production in ruminants

Techniques : Open-circuit respiration chambers
Species : Cattle, sheep and goats

2. **Kyushu National Agricultural Experimental Station, Department of Animal Science, Nishigoshi, Kumamoto 861-11**
Phone Number : 81.96.242.3603 Fax number :81.96.249.10.02
E. Mail : shioya@knaes.affrc.go.jp
mtmt@knaes.affrc.go.jp
yukotwm@knaes.affrc.go.jp
Investigators : S. SHIOYA, M. TANAKA, Y. IWAMA
Main research topics : Energy balance studies of cows under heat stress conditions
Techniques : Open-circuit respiration chambers
Species : Cattle

3. **National Grassland Research Institute, Dairy Cattle Feeding Laboratory, Nichinasuno, Tochigi 329-27**
Phone Number : 81-287-37-7806 Fax number : 81-287-36-6629
E. Mail : moto@ngri.affrc.go.jp
ansada@ngri.affrc.go.jp
nishtake@ngri.affrc.go.jp
Investigators : M. ISHIDA, S. ANDO, T. NISHIDA
Main research topics : Factors affecting energy partitions of pregnant cattle and goats
Techniques : Open-circuit respiration chambers, Gradient layer calorimeter
Species : Cattle and sheep

4. **Hokkaido University, Laboratory of Animal Nutrition, Sapporo 060**
Phone Number :
Investigators : M. OHKUBO
Main research topics :
Techniques : Open-circuit respiration chambers
Species : Cattle

5. **Department of Animal Sciences, Obihiro University of Agriculture and Veterinary Medicine, Nishi 2 Sen, Inadacho, Obihiro, Hokkaido O80**
Phone Number : 0155.49.5421 Fax number : 0155.49.5462 (Dept)
E. Mail : junichi@obihiro.ac.jp
0155.495421 (Office)
Investigators : J. TAKAHASHI
Main research topics : The relationship between nitrate poisoning and energy metabolism; the regulation of ruminal methanogenesis by L-cysteine
Techniques : Indirect calorimetry (hood technique)
Species : Cow and sheep

6. **Utsunomiya University, Department of Animal Science, Laboratory of Nutritional Biochemistry, 350 Mine-machi, Utsunomiya-shi, Tochigiken 321**
Phone Number : 0286-36-1515 Fax number : 0286-35-4662
Investigators : K. SUGAHARA
Main research topics : Regulatory mechanism of energy metabolism in domestic fowl
Techniques : Immunohistochemistry; *In situ* hybridization
Species : chicken

7. **Nagoya University, Laboratory of Animal Nutrition, Nagoya, 464-01**
Phone Number : 052.781.5111 Fax number : 052.781.4447
Investigators : J. OKUMURA, T.MURAMATSU, K. KITA
Main research topics :
Techniques : Slaughter method, $^{14}CO_2$ measurement
Species : Poultry, experimental animal

REPUBLIC OF KOREA

1. **Department of Animal Science, College of Agriculture, Seoul National University, Suweon 170**
Phone Number :
Investigators : IN K. HAN, J.K. HA
Main research topics :
Techniques : Slaughter method
Species : Pigs, poultry

2. **National Livestock Experimental Station, Division of Animal Nutrition and Physiology, Suweon, 170-00**
Phone Number :
Investigators : SANG C. LEE, T.Y. THAK
Main research topics :
Techniques : Open-circuit respiration chambers, slaughter method
Species : Cattle, sheep, pigs and poultry

3. **Kangwon National University, Department of Animal Science College of Agriculture, Chunchon, 200-00**
Phone Number :
Investigators : Y.C. RHEE
Main research topics :
Techniques : Open-circuit respiration chambers
Species : Poultry

MALAYSIA

1. **Department of Animal Science, Universiti Putra Malaysia, 43400 upm Serdang, Malaysia**
Phone number : 6039486101 Fax number : 603 9432954
E. mail : jbliang@vet.upm.edu.my
Investigators : J.B. LIANG
Main research topics : Energy metabolism of farm animals under humid tropical conditions
Techniques : Open circuit respiration chamber; Head hood; TOH dilution
Species : *Bos indicus* cattle, water buffalo, mouse-deer

MOROCCO

1. **Laboratory of Animal Physiology, Institut Agronomique et Vétérinaire Hassan II, PO Box 6202 RABAT**
   Phone Number : 00.212.777.17.59 Fax number : 00.212.777.71.19
   Investigators : A. GUEROUALI, S. NAIMA
   Main research topics : Determination of the nutrients requirements of the camel for the production cycle; energy requirements of working equines; energy balance studies in small ruminants
   Techniques : Open circuit-indirect calorimetry with chambers and hood
   Species : Camel, sheep, goats and equines

2. **Department of Animal Production, Institut Agronomique et Vétérinaire Hassan II, P.O. Box 6202 RABAT-INSTITUTS 10100**
   Phone number : 212.7.680606 and 212.7.77.17.58 or 59 Fax number : 212.7.680606 and 212.7.77.58.38
   E. mail : araba@syfed-ma.refer.org
   Investigators : A. ARABA
   Main research topics : Growth of cattle, camels, sheep of different biological types; effect of diet on energy utilization
   Techniques : Deuterium oxide dilution procedure
   Species : Mainly cattle

THE NETHERLANDS

Phone and Fax numbers : a, before and b, from October 1995

1. **Agricultural University Wageningen, Department of Animal Science, Zodiac Building, Marijkeweg 40, 6700 AH WAGENINGEN**
   Phone Number : 31317483120 Fax number : 31317485006
   E. Mail : Joham.Schmama@GENR.VH.WAU.WL
   John.SCHRAMA@GENR.VH.WAU.NL
   Investigators : J.W. SCHRAMA, M.W. BOSCH, M.W.A. VERSLEGEN, H. van den BRAND, M. J.W. HEETKAMP
   Main research topics :
   Techniques : Open-circuit respiration large and small chambers, slaughter method
   Species : Pigs, calves, poultry, goats

2. **Wageningen Agricultural University, Department of Human and Animal Physiology, Haarweg 10, 6709 PJ WAGENINGEN**
   Phone Number : 31317483009/84136 Fax number : 31317404077
   E. Mail :Harry.Boekholt@ALG.FMD.WAU.NL
   Investigators : H.A. BOEKHOLT
   Main research topics : Protein-energy relationship in growth
   Techniques : Indirect calorimetry, stable isotopes
   Species : Cattle, pig, sheep, poultry, human

3. **Wageningen Agricultural University, Department of Human Nutrition Bomenweg 2, 6703 HD Wageningen**
   Phone Number : 31317483634 Fax number : 31317483342
   E. Mail : joop.vanraaij@etO.voed.wau.nl

Investigators : J. Van RAAIJ, P. DEURENBERG, C.P.G.M. de GROOT
Main research topics : Energy inbalance in pregnancy and lactation; Energy inbalance in obesity; Energy metabolism in the elderly; Seasonal variation in energy metabolism
Techniques : Ventilated hood, douglas bag
Species : Humans

4. **Agricultural University Wageningen, Department of Fish Culture and Inland Fisheries, Marijkeweg 40, 6700 Wageningen**
Phone Number :
Investigators : L.T.N. HEINSBROEK, J.A.J. VERRETH
Main research topics :
Techniques : Open-circuit respiration chambers, Slaughter method
Species : Fish

5. **Institute for Animal Science and Health (ID-DLO) , Edelhertweg 15, P.O.Box 65 8200 AB, LELYSTAD**
Phone Number : +31 320 238 238 / +31 320 237 314 Fax number : +31 320 237 320
E Mail : y.vanderhoning@id.dlo.nl
Investigators : Y. VAN DER HONING, G.C.M. BAKKER, J. VAN DER MEULEN, C.W. SCHEELE, A. VAN VUUREN,
Main research topics : Net energy value of fat, non-starch polysaccharides and starch for growing pigs; Evaluation and development of the net energy system for pig diets, Nutrient utilisation by poultry
Techniques : Open-circuit respiration chambers, comparative slaughter technique, blood flow measurements (PDV, liver, udder)
Species : Growing pigs, sows, poultry

6. **University Utrecht, Sect. Veterinary Physiology, Yalelaan 2, 3584 CM UTRECHT**
Phone Number : 31302535308 Fax number : 31302531407
Investigators : M. van KAMPEN, F.M.G. BOSHOUWERS, M.J.M. van GRIENSVEN
Main research topics : The influence of genotype, nutrition, health and environment on energy metabolism and thermoregulation of birds (in all development stages)
Techniques : Open-circuit respiration chambers; Automatic feed and water intake registration; MRI
Species : Quail, guinea fowl, turkey, duck and domestic chick

NEW ZEALAND

1. **Massey University, Department of Animal Science, PALMESTON NORTH**
Phone Number :
Investigators : C.W. HOLMES
Main research topics :
Techniques : Open-circuit respiration chambers
Species : Goats, pigs, cattle

2. **Ministry of Agriculture and Fisheries, Ruakuro Animal Research Station, Private Bag, HAMILTON**
Phone Number :
Investigators : A.M. BRYANT, C.R. PARR
Main research topics :
Techniques : Open-circuit respiration chambers
Species :

3. **DSIR Biotechnology, Private Bag, PALMERSTON NORTH**
   Phone Number :
   Investigators : P.M. HARRIS
   Main research topics :
   Techniques : Open-circuit respiration chambers, isotope dilution; Arterio-venous differences
   Species : Sheep

NORWAY

1. **Agricultural University of Norway, Department of Animal Science, P.O. Box 25, 1432 AAS-NLH**
   Phone Number : 479948000 Fax number : 479947960
   Investigators : F. SUNDSTOL
   Main research topics :
   Techniques : Indirect calorimetry: open-circuit
   Species : Sheep, pig, calf, poultry

2. **AKVAFORSK, Institute of Aquaculture Research, N-6600, SUNNDALSORA**
   Phone Number : 47 71699812 Fax number: 47 71699801
   E. Mail : barb.grisdale@akvaforsk.nlh.no
   stale.helland@akvaforsk.nih.no
   Investigators : B. GRISDALE-HELLAND, S.J. HELLAND
   Main research topics : Nutrient utilization by fish; net energy system
   Techniques : Semi-open/semi-closed respiration chambers; slaughter method
   Species : Salmon, trout, halibut

POLAND

1. **Polish Academy of Sciences, The Kielanowski Institute of Animal Physiology and Nutrition, 05-110 Jabonna near Warsaw**
   Phone Number : 4822.7824422 Fax number : 48227742038
   E. Mail : infiz@atos.warman.compl
   Investigators : H. FANDREJEWSKI
   Main research topics : Body composition of growing pigs and sows; estimation of cost of protein deposition and maintenance requirements; influence of thyroid hormones on energy metabolism
   Techniques : Slaughter method
   Species : Pig

2. **Warsaw Agricultural University, Institute of Human Nutrition, ul. Nowoursynowska 166, O2-766 Warsaw**
   Phone Number : 48.22.43.18.78 Fax number : 48.22.43.18.
   Investigators : J.S. KELLER, J. BUJKO
   Main research topics : Resting metabolic rate, dietary induced thermogenesis and energy cost of physical activity as influenced by diet, training and body composition in man and laboratory rat
   Techniques : Open circuit respiration chambers, ^{15}N measurements in urine
   Species : Human beings and laboratory rats

PORTUGAL

1. **Instituto Nacional De Investigacao Agraria, Estaçao Zootecnica Nacional Fonte Boa. Nutrition and Feeding Department Vale de Santarem 2000 SANTAREM**
   Phone Number : 35.14.3760.202 Fax number : 35.14.37.60.540
   Investigators : J.M.C. RAMALHO RIBEIRO, M.T.V.C.P. DENTINHO, M.M.S. POITIER
   Main research topics : Feed evaluation (chemical analyses *in vitro, in vivo, in situ*); Digestion (kinetics); Metabolic (efficiency of production of meat and milk); Product quality
   Techniques : Slaughter technique
   Species : Chicks, sheep, calves, pigs

ROMANIA

1. **Laboratory of Energy Metabolism in Farm Animals, Institute of Biology and Animal Nutrition, Sos. Bucuresti-Ploiesti km. 18, 8113 BALOTESTI**
   Phone Number : 40.1.312.84.10 Fax number : 40.1.312.84.10
   Investigators : G. BURLACU, D. GROSSU, M. ILIESCU, I. VOICU, M. PIRVU, R. BURLACU
   Main research topics : Energy and protein metabolism; Mathematical modelling of energy metabolism; Feed evaluation
   Techniques : Respiration chambers, Slaughter method
   Species : Cattle, buffaloes, goats, sheep, pig, poultry, fur animals, fish

SOUTH AFRICA

1. **Animal and Dairy Science Research Institute, Private Bag X2, IRENE, 1975**
   Phone Number :
   Investigators : F.K. SIEBRITS, C.Z. ROUX, J. VILJOEN, D. SWART, P.CRONJE
   Main research topics :
   Techniques : Open-circuit respiration chambers, Comparative slaughter
   Species : Cattle, pig and ostrich

2. **University of Stellenbosch, Department of Animal Science, STELLENBOSCH**
   Phone Number :
   Investigators : L.P. VOSLOO
   Main research topics :
   Techniques : Slaughter method
   Species :

SPAIN

1. **Estacion Experimental del Zaidin (C.S.I.C.), Departamento de Nutricion Animal, Profesor Albareda, 1. 18008 Granada**
   Phone Number : 58.121011 Fax number : 58.129600
   E. Mail : aguilera@eez.csig.es
   Investigators : J.F. AGUILERA, C. PRIETO, R. SANZ, J. BOZA, J. FONOLLA, E. MOLINA, F. FERNANDEZ

Main research topics : Monogastric and ruminant nutrition: Energy and protein metabolism
Techniques : Indirect calorimetry : open-circuit and confinement respiration chambers; Comparative slaughter technique; Isotope techniques: ^{13}C-bicarbonate method (CERT) and protein synthesis measurements (L-(2,6^{3}H) phe)
Species : Goat, sheep, poultry, rat

2. **Departamento de Produccion Animal, Escuela Tecnica Superior de Ingenioros Agronomos, Universidad Politecnica de Madrid, Ciudad Universitaria, 28040 Madrid**
Phone number : 1.5492219
Investigators : J.F. GALVEZ MORROS, C. de BLAS BEORLEGUI
Main research topics : Monogastric nutrition: Protein and energy metabolism
Techniques : Indirect calorimetry: closed-circuit respiration chambers; Comparative slaughter technique
Species : Rabbits

SWEDEN

1. **Swedish University of Agricultural Sciences, Department of Animal Nutrition and Management, P.O. Box 7024, S-750 07 UPPSALA**
Phone Number : 46 18 67 21 02 Fax number : 46 18 67 29 95
E. Mail : Jan-Eric-Lindberg@huv.slu.se
Investigator : J.E. LINDBERG
Main research topic : Monogastric nutrition; Partition of digestion; Energy and protein metabolism.
Techniques : Balance; Digestibility; Ileal cannulation; A-V differences
Species : Pigs, horses

SWITZERLAND

1. **Institute of Animal Sciences, Group of Nutrition-Products-Environment, ETH-Zentrum, CH-8092 Zurich**
Phone Number : 41.1.632.32.69 Fax number : 41.1.632.11.28
E. Mail : caspar.wenk@inw.agrl.ethz.ch
michael.kreuzer@inw.agrl.ethz.ch
Investigators : C. WENK, M. KREUZER
Main research topics : Utilization of feedstuffs; Effect of feed additives on energy and nutrient utilization; Regulation of feed intake; Methane formation
Techniques : Open-circuit respiration, Body analysis, Rumen simulation technique
Species : Cattle, pigs, sheep, layers, chickens, man

2. **Université de Lausanne, Institut de Physiologie, Rue Bugnon 7, CH-1022 Lausanne**
Phone Number : 41.21.692.55.00 Fax number : 41.21.692.55.05
Investigators : E. JEQUIER, Y. SCHUTZ, L. TAPPIE
Main research topics :
Techniques : Open-circuit respiration
Species : Humans

UNITED KINGDOM

1. **B.B.S.R.C. Roslin Institute, Roslin, Midlothian EH25 9PS**
   Phone Number : 0131-527-4213 Fax number : 0131-440-0434
   E. Mail : murdo.macleod@bbsrc.ac.uk
   Investigators : M.G. MacLEOD, J. McNAB
   Main research topics : Feed utilisation; Feed evaluation (net energy, metabolisability, digestibility); Nutritional physiology, particularly nutrition/environment interactions; Modelling of energy metabolism
   Techniques : Open-circuit indirect calorimetry; Doubly labelled water technique; ME and digestibility assays
   Species : Poultry and other avian species

2. **The Hannah Research Institute, Ayr, KAT 5HL**
   Phone Number :
   Investigators : D. CHAMBERLAIN, P. MARTIN
   Main research topics :
   Techniques : Close-circuit respiration
   Species : Sheep

3. **Rowett Research Institute, Bucksburn, Aberdeen, AB29SB**
   Phone Number : 44.224.71.27.51 Fax number : 44.224.71.53.49
   E. Mail : id@uk.ac.sari.rri
   gel@uk.ac.sari.rri
   Investigators : J.C. MacRAE, G.E. LOBLEY, E.R. ORSKOV
   Main research topics : Energy metabolism during undernutrition; Tissue/organ metabolism; Intermediary metabolism; Lactation; Growth; Pregnancy
   Techniques : Open circuit respiration chambers, closed circuit respiration chambers, direct calorimeter, doubly labelled water technique, stable isotopes; trans-organ studies and metabolism
   Species : Cattle, sheep, pigs, goats, llamas, alpacas

4. **The Queen's University of Belfast, Agricultural and Environmental Science Division, Newforge Lane, Belfast BT9 5PX**
   Phone Number : 0232.661.166 Fax number : 0232.669.551 E. Mail :
   Investigators : K.J. McCRACKEN, E.F. UNSWORTH
   Main research topics : Feed evaluation; Nutritional physiology; GIT metabolism
   Techniques : Indirect calorimetry and slaughter technique, stable isotopes, AV differences across GIT
   Species : Rats, poultry, pigs, humans

5. **Agricultural Research Institute of Northern Ireland, Large Park, Hillsborough, Co. Down, N. Ireland, BT26 6DR**
   Phone Number : 44 1846 682484 Fax number : 44 1846 689594
   E. Mail : agnewr@dani.gov.uk
   Investigators : R.E. AGNEW, F.J. GORDON, R.W.J. STEEN, C.S. MAYNE, A.R.G. WYLIE
   Main research topics : Feed evaluation, nutritional physiology, GIT metabolism
   Techniques : Indirect calorimetry and slaughter technique, stable isotopes, AV differences across GIT
   Species : Lactating dairy cows, growing cattle, sheep

6. **University of Newcastle upon Tyne, The Dept of Biological and Nutritional Sciences, Newcastle Upon Tyne, NE1 7RU**
Phone Number : Fax number : 44.734.352421
Investigators : D.G.ARMSTRONG, D.S. PARKER, J.S. MALHERS
Main research topics :
Techniques : Metabolism in gastro intestinal tract, Evaluation of bovine feed
Species : Sheep, cattle

7. **University of Reading, Centre for Dairy Research, Department of Agriculture, Earley Gate, Reading, RG6 6AT**
Phone Number : 0118 9318474 Fax number : 0118 9352421
E. Mail : D.E.Beever@Reading.ac.uk
J.D.Sutton@Reading.ac.uk
C.K.Reynolds@Reading.ac.uk
sascromp@Reading.ac.uk
Investigators : D.E. BEEVER, J.D. SUTTON, C.K. REYNOLDS, S.B. CAMMELL, L.A. CROMPTON
Main research topics : Indirect calorimetry, trans-organ studies and organ metabolism, digestive processes, feed utilization
Species : Lactating dairy cows, growing cattle

8. **Institute of Grassland & Environmental Research, Ruminant Nutrition Department, Plas Gogerddan, Aberystwyth, Dyfed SY 23 3EB, WALES**
Phone number : 44.970.82.82.55 Fax number : 44.970.8283.57
E. Mail : scollan@uk.ac.afrc
Investigators : N.D. SCOLLAN
Main research topics : Rumen function, gastro-intestinal trace metabolism. Regulation of fat and lean deposition forages for meat and milk production.
Techniques : Nuclear magnetic resonance; Spectroscopy 13C-MRS, 31P-MRS; A-V difference techniques across gastro-intestinal tract
Species : Sheep, cattle

UNITED STATES OF AMERICA

1. **Nutrient Conservation and Metabolism Laboratory, Livestock and Poultry Sciences Institute, Powder Mill Road, B-200, Room 124, BARC-East, Beltsville, MD 20705**
Phone Number : 301.504.91.00 Fax number : 301.504.81.62
E. Mail : bglenn@asrr.ARSUSDA.gov
rbaldwin@asrr.ARSUSDA.gov
kmcleod@asrr.ARSUSDAs.gov
Investigators : B.P. GLENN, J.B. REEVES, III, R.L. BALDWIN, VI, K.R. McCLEOD
Main Research topics : Ruminant nutrition: protein and energy metabolism, nutritional physiology
Techniques : Indirect respiration calorimetry ; Net nutrient flux; Cell culture. Pyrolysis-gas-chromatography-mass spectrometry; Infrared spectrometry
Species : Beef and dairy cattle

2. **U.S. Department of Agriculture, Roman L. Hruska Meat Animal Research Center, P.O. BOX 166, Clay Center, Nebraska 68933**
Phone Number : 402.762.4274 Fax number : 402.762.4273

E. Mail : Ferrell@MARCVM.MARC.USDA.GOV
Nienaber@MARCVM.MARC.USDA.GOV
Jenkins@MARCVM.MARC.USDA.GOV
FreetlyMARCVM.MARC.USDA.GOV
Investigators : C.L. FERRELL, T.G. JENKINS, J.A. NIENABER, H.C. FREETLY
Main research topics : Nutrition; Livestock stress; Thermal environment
Techniques : Open-circuit respiration, Slaughter method, Mathematical modelling
Species : Sheep, cattle, swine

3. **University of Arkansas, Department of Poultry Science, Poultry Science Center, Fayetteville, Arkansas 7270**
Phone Number : 501.575.4134 Fax number : 501.575.8775
E. Mail : ccoon@comp.uark.edu
Investigator : C.N. COON
Main research topics : Partitioning energy utilization between growth and egg production
Species : Broiler breeders, layers, broilers

4. **University of Alaska, Institute of Arctic Biology, Fairbanks, Alaska 99701**
Phone Number :
Investigators : R.G. WHITE, J. BLIGH
Main research topics :
Techniques : Open-circuit respiration, Confinement, Hood, Dual isotope (THO/DOH)
Species : Reindeer, sheep, wild species (e.g. musk ox, caribou, moose)

5. **University of California, College of Agriculture and Environmental Sciences, Department of Animal Science, University of California, Davis, California 95616**
Phone Number : 1-916-752-0526 Fax number : 1-916-752-0175
E. Mail : rdsainz@ucdavis.edu
Investigators : R.L. BALDWIN, R.D. SAINZ, J.W. OLTJEN
Main research topics : Energy expenditure in growth and lactation; Systems analysis
Techniques : Mechanistic, dynamic modelling
Species : Cattle, sheep, pigs

6. **Colorado State University, Metabolic Laboratory, Department of Animal Sciences, Fort Collins, Colorado 80523**
Phone Number : 970.491.7833 Fax number : 303.491.5826
E. Mail : djohnson@ceres.agsci.colostate.edu
Investigators : D.E. JOHNSON
Main research topics : Maintenance energy requirements; Methane production; Associative effects of feeds
Techniques : Chambers indirect respiration calorimetry; Slaughter balance
Species : Cattle, sheep, cats, rats

7. **Cornell University, Department of Animal Science, Frank Morrisson Hall, Ithaca, New York 14855**
Phone Number :
Investigators : A. BELL, M. THONNEY
Main research topics :
Techniques : Tissue and organ metabolism (A-V differential)
Species : Cattle, sheep

8. **Cornell University, Department of Natural Resources, Fernow Hall, Ithaca, New York**
Phone Number :
Investigators : A.N. MOEN
Main research topics :
Techniques : Slaughter balance
Species : Deer

9. **University of Georgia, Department of Poultry Science, Athens, Georgia 30602**
Phone Number :
Investigators : H.L. FULLER
Main research topics :
Techniques : Open-circuit respiration
Species : Chickens

10. **Iowa State University, 337 Kildee Hall, Ames, Iowa 50011**
Phone Number :
Investigators : R.C. EWAN
Main research topics :
Techniques : Slaughter method
Species : Pigs

11. **University of Kentucky, 805 W.P. Garrigus Building, Lexington, Kentucky KY 40536-0215**
Phone Number : 606.257.2686 Fax number : 606.258-1027
Investigators : D.L. HARMON
Main research topics :
Techniques : Tissue or organ metabolism (A-V differential)
Species : Cattle, sheep

12. **Michigan State University, Food Science and Human Nutrition, East Lansing, Michigan 48824**
Phone Number : 517.353.9511 Fax number : 517.358.8963
E. Mail : DROMSOS@MSU.edu
Investigators : D.R. ROMSOS
Main research topics : Food intake regulation, insulin secretion, lipid and carbohydrate metabolism
Techniques : Open-circuit respiration, Slaughter balance
Species : Rats, dogs, mice

13. **University of Minnesota, St-Paul, Minnesota 55108**
Phone Number : 612 625 3185 Fax number : 612 625.1283
E. Mail : crook00@maroon.te.umn.edu
Investigators : B. CROOKER
Main research topics :
Techniques : Modelling, D_2O dilution techniques, Slaughter, Nutritional physiology
Species : Cattle, pigs

14. **University of Missouri, Department of Dairy Science, Animal Science Research Center, Columbia, Missouri 65203**
Phone Number :
Investigators : J. WILLIAMS

Main research topics :
Techniques : Open-circuit calorimetry
Species : Cattle

15. **South Dakota State University, Department of Animal and Range Sciences, Brookings, South Dakota 57007-0392**
Phone Number : 605/688-5460 Fax number : 605/688-6170
E. Mail : AS22@SDSUMUS.SDSTATE.EDU
Investigators : C.P. BIRKELO
Main research topics : Effects of environment, plane of nutrition, lactation; Feedstuff evaluation
Techniques : Open-circuit respiration calorimetry
Species : Cattle (beef, dairy)

16. **Texas A & M University, Department of Animal Science, College Station, Texas 77843**
Phone Number : 409.845.5065 Fax number : 409.845.5292
E. Mail : g-CARSTENS@tamu.edu
Investigators : F.M. BYERS, G.E. CARSTENS
Main research topics : Beef cattle energy metabolism ; Brown fat thermogenesis
Techniques : Slaughter method, Open-circuit respiration calorimetry
Species : Cattle, sheep

17. **Washington State University, Department of Animal Science, 231 Clark Hall, Pullman, Washington, DC 99164-6351**
Phone Number : 509.335.41.31 Fax number : 509.335.1082
509 335.4113
E. Mail : johnsoka@wsu.edu
mcnamara@wsu.edu
Investigators : K. A. JOHNSON, J.P. McNAMARA
Main research topics : Lipid metabolism; Beef cattle energy metabolism; Methane production
Techniques : Open-circuit respiration, Enzyme kinetics, Modelling
Species : Dairy cattle, sheep, swine, beef cattle

18. **University of Nebraska**
Phone Number :
Investigators : M. NIELSEN
Main research topics :
Techniques : Open circuit indirect calorimetry, Genetic variations in heat production
Species : Mouse

19. **A.R.S., U.S.D.A. Children Nutrition Research Center, Baylor College of Medicine, 1100 Bates Street, Houston, Texas 77030 2600**
Phone Number : 713.798.71.48 Fax number : 713.798.71.71
E. Mail : preeds@BCM.TMC.edu
dburrin@BCM.TMC.edu
Investigators : P.J. REEDS, M.L. FIOROTTO, D.G. BURRIN
Main research topics : Stable isotopics
Techniques : Open circuit respiration chambers, Comparative slaughter
Species : Pigs, humans

20. A.R.S. USDA Livestock and Poultry Sciences Institute, Growth Biology Laboratory, Beltsville Agricultural Research Center, Building 200 RM 201, Beltsville, MD 20705
Phone Number : 301.504.82.22 Fax number : 301.504.86.23
E. Mail : Caperna@GGPL.ARSUSDA.GOV
Investigators : T.J. CAPERNA, A.D. MITCHELL, N.C. STEELE and R.W. ROSEBROUGH
Main research topics : Endocrine control of growth; Body composition methodology
Techniques :
Species : Swine and poultry

21. U.S. Department of Agriculture, ARS, Beltsville Human Nutrition Center, Energy and Protein Nutrition Laboratory, B-308 Barc-East Beltsville, Maryland 20205
Phone Number :
Investigators : W.V. RUMPLER, J.M. CONWAY, J.C. HOWE, J.L. SEALE
Main research topics :
Techniques : Direct and indirect calorimetry, Doubly labelled water
Species : Humans

22. The Ohio State University, Ohio Agricultural Research Development Center, Dept. of Animal Sciences, Wooster, OH 44691
Phone Number : 330 263 3795 Fax number : 330 263 394
E. Mail : palmquist.1@osu.edu
weiss.6@osu.edu
Investigators : D.L. PALMQUIST, W.P. WEISS
Main research topics : Lipid metabolism; Milk synthesis; Feed evaluation
Techniques : Digestion facilities, animal facilities and laboratory equipment for radio and stable isotope studies in whole animals
Species : Cattle and sheep

U.S.S.R

1. All-Union Scientific Research Institute of Animal Husbandry, Dubrovitsy, Raion Podollak, Oblast Moskau
Phone Number :
Investigators : N.I. DENISSOV, T.S. MIELNIKOW, M.V. MIESIUTKINA
Main research topics :
Techniques : Open-circuit respiration chambers
Species :

AUTHOR INDEX

SUBJECT INDEX

Note: page numbers in *italics* refer to figures and tables.

Printed and bound by CPI Group (UK) Ltd, Croydon, CR0 4YY
16/09/2026
14963433-0007